Coal Mechanics

Yuanping Cheng · Qingquan Liu · Ting Ren

Coal Mechanics

Science Press
Beijing

Springer

Yuanping Cheng
School of Safety Engineering
China University of Mining and Technology
Xuzhou, Jiangsu, China

Qingquan Liu
School of Safety Engineering
China University of Mining and Technology
Xuzhou, Jiangsu, China

Ting Ren
School of Civil, Mining and Environmental
University of Wollongong
Wollongong, NSW, Australia

ISBN 978-981-16-3894-7 ISBN 978-981-16-3895-4 (eBook)
https://doi.org/10.1007/978-981-16-3895-4

Jointly published with Science Press, Beijing, China
The print edition is not for sale in China (Mainland). Customers from China (Mainland) please order the print book from: Science Press, Beijing, China.

This Springer imprint is published by the registered company Springer Nature Singapore Pte Ltd.
The registered company address is: 152 Beach Road, #21-01/04 Gateway East, Singapore 189721, Singapore

Preface

Coal is a type of solid-state combustible organic rock, which is transformed from plant remains by complex biochemistry, physical chemistry and geochemistry effects also known as the coal-forming process. Through deposition, coal eventually forms a seam within coal measure strata. Compared to the ordinary sedimentary rocks, coal has the following features: the spacing of coal's structural surface (cleat) is much smaller, when viewed under a microscope; the density of coal cleat can reach several to hundreds of cleats/m; coal can be classified as a discontinuous body consisting of fractures (cleats) and matrix; the average density of coal is about 50% of the rock; the coal strength is much lower than the strength of a typical sedimentary rock which is reflected in lower compressive strength, tensile strength and elastic modulus as well as larger Poisson's ratio; coal is usually porous with well-developed pore structure where the specific surface area of coal micropores and minipores can exceed 80% of the total surface area, providing abundant space for gas adsorption; the maximum adsorption capacity of coal can reach dozens of cubic metres per tonne; after gas adsorption, the sorption-induced swelling effect will lead to weaker coal and reduced permeability. Coal with large gas content is generally weaker with larger Poisson's ratio, leading to greater deformation and increase in strain energy storage under the same stress state. During mining activities in a coal mine, the stress changes occur in coal ahead of the working face, making it easier for the coal body to destabilize causing rib failure, spalling, coal extrusion or bumps, and under the extreme high gas pressure conditions, coal and gas outbursts may occur.

Coal mechanics is a special research field of rock mechanics for underground mining focusing on coal properties such as dual porosity media, change of effective stress caused by pore pressure and the non-mechanical action caused by gas adsorption. Similar to the study of rock mechanics, coal mechanics adopts the research methods of laboratory experiments, field observations and numerical calculations. Coal mechanics field is an essential component of an effective engineering design leading to safe mining practices and improved mine production rates.

The Definition of Coal Mechanics, Coal mechanics is the theoretical and applied science describing the mechanical behaviour of coal containing gas under pressure typically found underground. It provides the theoretical and practical basis for safe production of coal as well as utilization of coal seam gas resources. The study of coal mechanics mainly focuses on the mechanical properties of coal, its permeability and the gas flow in coal seams.

Since the 1950s, scholars began to study the properties of coal mainly focusing on obtaining its mechanical parameters. With the continuous improvements of the testing methods and relevant theories, scholars studied the influence of gas adsorption such as methane and its effects on the coal mechanical properties. As more technological tools became available such as the computed tomography scan (CT scan), mechanical permeability coupling test system and mechanical simulations, the researches gradually entered the stage of microscopic, multifactor coupling and numerical simulation.

Since the 1960s, the scholars began to carry out theoretical and experimental research in the permeability of coal. In the initial models, they did not consider the effect of coal adsorption expansion and geostress and assumed that coal was uniformly continuous single porosity media, but further studies developed the gas–solid coupling model that considers the effective stress and adsorption expansion effect of coal, which was a significant achievement. In recent years, a dual porosity media model has been widely accepted and used by scholars. Based on the coal-simplified models, such as the matchstick model and cubic geometry model, many scholars have established various permeability models concerning the effective stress and adsorption expansion effect based on different boundary conditions. Aiming at the seepage properties after coal body's plastic damage, researchers also have carried out related research work and established the permeability models for different application backgrounds. Based on the studies of the coal permeability, various models for coal seam gas flow have been established from the linear flow model to the gas–solid coupling flow model.

The research team of **National Engineering Research Center for Coal and Gas Control, China University of Mining and Technology** has been engaged in the research of basic theory and application technology of coal mine gas control, carrying out fruitful research achievements in the fields of coal's pore structure, gas adsorption/desorption properties, mechanical properties of coal containing gas, permeability properties of coal and gas flow theory in coal seams. These achievements also have been successfully applied in coal mine gas extraction and disaster control. On the basis of fully referencing the research results, our team have systematically condensed and summarized the research achievements, as well as adopting the cooperation achievements of our team and the School of Mining, University of Wollongong, Australia, to complete this book. The book of *Coal Mechanics* is divided into 11 chapters, which consist of Introduction, Formation and Metamorphism of Coal, Basic Physical Properties and Pore/Fracture Characteristics of Coal, Gas Adsorption and Desorption Properties of Coal, Foundation of Coal Mechanics, Strength and Deformation of Coal Containing Gas, Seepage Characteristics and Permeability Evolution Model of Coal, Geostress and Coal Seam Gas Occurrence, Gas Flow Theory of Coal Seams, Application of Coal Mechanics in Pressure-Relief Gas Extraction and Coal and Gas Outburst Control. This book can be used by teachers and students of safety engineering, mining engineering and coalbed methane engineering, while it can also be used for reference by researchers and technicians of relevant enterprises and research institutes.

This book was written together by Prof. Yuanping Cheng (China University of Mining and Technology), Dr. Qingquan Liu (China University of Mining and Technology) and Prof. Ting Ren (University of Wollongong). Many postdoctorate (Ph.D.) students and Ph.D. candidates of our research team also helped to prepare this book. Ph.D. candidate Wei Zhao participated in writing Chap. 2; Dr. Haina Jiang participated in writing Chap. 3; Ph.D. candidate Haijun Guo participated in writing Chap. 4; Dr. Pinkun Guo participated in writing Chap. 8; Dr. Haidong Chen, Ph.D. candidate Hao Zhang and Ph.D. candidate Zhengdong Liu participated in writing Chap. 10; Dr. Fenghua An and Ph.D. candidate Qingyi Tu participated in writing Chap. 11.

The authors of this book would like to express sincere appreciation to the Ministry of Science and Technology of the People's Republic of China, the Ministry of Education of the People's Republic of China, the National Natural Science Foundation of China and the Science and Technology Department of Jiangsu Province for their funding and encouragement to the scientific researches.

We earnestly ask our readers for comments to correct any shortcomings and errors that may have occurred in this book.

Xuzhou, China — Yuanping Cheng
Xuzhou, China — Qingquan Liu
Wollongong, Australia — Ting Ren
December 2020

Contents

Chapter 1
Introduction

Learning Outcomes

- Coal formation, macroscopic composition of coal/rock, structure of coal, coal seam gas and coal measure strata.
- Pore and fracture structures in coal, simplified physical models of coal and their significance in the study of coal mechanics.
- Comparison of coal and rock structures, strength, adsorption properties and disasters.

1.1 Coal and Coal Measure Strata

1.1.1 Coal and Coal Seam Gas

1. Coal formation

Coal is one type of solid-state combustible organic rocks, which is transformed from plant remains by complex biochemistry, physical chemistry and geochemistry effects (also known as the coal-forming process). Coal is mainly composed of organic matter and a small amount of inorganic minerals. The organic matter in coal is mainly composed of five elements, namely carbon, hydrogen, oxygen, nitrogen and organic sulfur, of which the carbon, hydrogen and oxygen contents constitute more than 95% of the total organic matter. Generally, it is believed that coal is made up of large aromatic rings and fused rings with fat side chains, and the skeletons of these rings are made of carbon elements. With the increasing coal rank, the content of carbon increases while the content of hydrogen and oxygen decreases, and the content of nitrogen slightly reduces. In China, the carbon content is 55–62% for peat (on dry and ash-free basis), 60–76.5% for lignite, 77–92.7% for bituminous coal, and for anthracite coal, carbon is usually above 89%. The inorganic matter in coal is mainly composed of water and minerals [1].

Y. Cheng et al., *Coal Mechanics*,
https://doi.org/10.1007/978-981-16-3895-4_1

The coal-forming process can be divided into two stages. The first stage is peatification or saprofication, which mainly occurs in the peat swamps, lakes and shallow seas that are located on the earth's surface. In this stage, driven by the biogeochemical effects and under the action of various microorganisms, the remains of plants continuously decompose, combine and accumulate, resulting in the lower plants turning to sapropel while the higher plants form peat. The second stage of the coal-forming process is the stage of coalification. In this stage, the peat or sapropel formed in the previous stage is buried deep underground due to the subsidence of the earth's crust. The increase of ground temperature as well as pressure convert the coal-forming process from peatification/saprofication into physi-chemical reactions (namely the coalification). Moreover, coalification can be further divided into diagenesis and metamorphism, of which the transformation process from peat to young lignite is called diagenesis and the transformation process from young lignite to old lignite, bituminite coal even anthracite is called metamorphism.

In China, from the early Palaeozoic stone coal (belongs to the sapropelite) to the Quaternary peat, there exist 14 coal forming periods, of which the most important coal forming periods are: ① early Carboniferous in South China; ② Carboniferous-Permian in North China; ③ Permian in South China; ④ late Triassic in South China; ⑤ early and middle Jurassic in Northwest China; ⑥ late Jurassic to early Cretaceous in Northeast China; ⑦ Tertiary period in Northeast/Southwest/Coastal China. Statistics indicate that the reserves of the early and middle Jurassic coal forming periods contribute for 60% of China's total coal resources, and the reserves of the Carboniferous-Permian coal forming period in North China contribute for 26% of China's total coal resources.

2. Macroscopic composition of coal

The macroscopic composition of coal is the basic unit of coal that can be distinguished by naked eyes, including vitrain, bright coal, dull coal and fusain, of which vitrain and fusain are the simple composition of coal while the bright coal and dull coal are the complex composition of coal [1].

Vitrain is transformed from plants' lignocellulosic tissue through the effect of gelatinization. The microstructure of vitrain is relatively simple, and it is one of the simple macroscopic compositions of coal. The color of vitrain is dark and shiny. Vitrain is also the darkest and the most lustrous element in coal. It has a pure texture, a uniform structure, a shell-like fracture and internal fissure. The vitrain is brittle in texture and easy to break into angular pieces. In coal seams, vitrain always demonstrate as convex lens or ribbon shape, with a thickness ranging from several to 20 mm, sometimes it will also exist in bright coal and dull coal in a lineation form.

Fusain is transformed from plants' woody fibrous tissue which slowly oxidized in a water shortage oxygen-enriched environment or formed due to forest fires. Fusain is also one of the simple macroscopic compositions of coal. Due to the large porosity and strong ability of oxygen adsorption, the coal seams that are rich in fusain would easily be prone to spontaneous combustion. The appearance of fusain is similar to charcoal, with gray color and obvious fibrous structure and silk lustre. With a looser structure and porous properties, fusain is crisp and can be easily crushed as well as dyeing the fingers black. Sometimes, when fusain's cells are filled with minerals, it is called mineralized fusain. Mineralized fusain is hard in texture and dense. Within the coal seam, fusain is usually distributed along the bedding plane in flat lens shape with the thickness of several millimeters; sometimes it can form discontinuous thin layers; in some places of the coal seam, the fusain thickness can reach hundreds of millimetres.

The formation of bright coal is complicated. It is formed through the gelatinization of plants' lignin fibers under the reduction conditions of overlying water, which also needs to be mixed with some other components and mineral impurities introduced by water or wind. The lustre of the bright coal is just next to vitrain, usually showing the color of black. The texture of bright coal is crisp and can be easily crushed, once broken, its cross section is relatively flat. The density of bright coal is also low. The degree of uniformity of the bright coal is less than vitrain and microstratification is visible on bright coal's surface. Sometimes, there are also endogenetic fractures existing in bright coal, however, less developed than vitrain. In coal seams, bright coal is the most common macroscopic coal composition, which usually forms thick layers and sometimes even the whole coal seam.

The composition of dull coal is also complicated. It is formed by transformation of the liptinite, inertinite, or the minerals under the condition of active water aerobics. The dull coal that contains a lot of inertinite or minerals is poor in coal quality; in contrast, with a high content of liptinite, dull coal is usually of good quality and lower density. The dull coal is dim, showing the color of ash black. The texture of dull coal is compacted and hard, with high density and toughness hard to break. Once crushed, the section is rough and rarely has developed endogenous fissures. In coal seams, dull coal is a common macroscopic composition of coal, which often exhibits layers with uneven thickness. Sometimes dull coal can also form a whole coal seam.

3. Structure of coal

The structure of coal can be defined as the characteristics of the coal composition morphology, size, thickness, plant tissue remnants and the relationships among these parameters, which reflect the composition and nature of the original substance as well as the variations in the coal forming process. In low rank coal, the structure of the coal is easy to identify, however, with the increase of coalification, the properties of different kinds of coal composition become more and more similar and of uniform structure.

The coal structure can be divided into two categories, the primary and the secondary structure.

The primary structure of the coal seam retains the primary sedimentary structure and tectonic characteristics. The coal composition, structure, construction and endogenous fissure of the primary coal structure can be clearly distinguished.

The secondary structure refers to the various secondary macrostructures that rose from tectonic stress effects after the formation of the coal seam, which is also known as tectonic coal. Under the effect of tectonic stress, the coal seam undergoes changes in composition, structure and construction resulting in deformation induced by damage, pulverization, thickening, or thinning as well as the metamorphism induced by coal degradation and condensation. The macrostructure of the tectonic coal can be classified as fragmentation structure, fractionation structure, powder particle structure, and mylonitic structure, thus the corresponding tectonic coals can be named as fragmentation coal, fractionation coal, powder particle coal and mylonitic coal respectively.

(1) Fragmentation structure: The coal is cut into pieces by dense secondary fissures, but there is little displacement between the pieces. Moreover, bedding can be found in the coal seam.
(2) Fractionation structure. The coal is broken into granules, and particles where diameter of these granules is mostly bigger than 1 mm. Most of these coal particles lose their edges due to the collision and friction with each other, while the bedding plane of coal is destroyed as well.
(3) Mylonitic structure. The coal is broken into fine powder, and the main particle size is less than 1 mm. Sometimes, coal powder can be re-compressed into one piece, however, there is no bedding or joints, and the coal easy converts back into powder.

Tectonic coal is the combination of the coal's secondary structures, which usually appear in the coalfields under complex tectonic stress conditions, especially near fault zones and fold belts. Figure 1.1 shows the comparison between intergrowth intact coal and tectonic coal (the vitrinite reflectance, also known as the R_o of the intact coal, is 2.77 while that for tectonic coal is 2.83) [2]. Table 1.1 demonstrates the classification method of coal destruction types which is cited from the "Specification for identification of coal and gas outburst mine" (AQ1024-2006) [3]. Referring to the specification, the destruction types of coal can be classified into five categories, namely the un-destructed coal (Type I), destructed coal (Type II), strongly destructed coal (Type III), pulverized coal (Type IV), powdered coal (Type V), of which, Type III to Type V coals may have the coal and gas outburst risk.

4. Coal seam gas

Fig. 1.1 Structure comparison of the No. 3 coal seam in Daning Coal Mine, Shanxi Province, China

Table 1.1 Classification of coal destruction types

Destruction types	Luster	Structure and structural characteristics	Joint properties	Joint plane properties	Fracture properties	Strength
Type I (un-destructed coal)	Bright and semi-bright	Lamellar structure, block structure, clear and obvious ribbon shape	1 group or 2–3 groups of developed joint system, in good order	With filler (calcite), few secondary plane, smooth joint/cleavage planes	Jagged and scalariform, conchoidal, wavy	Hard, difficult to crush with hands
Type II (destructed coal)	Bright and semi-bright	1. With layer features, much more order 2. Obvious strips, sometimes twisted, with staggered movements 3. Irregular blocky, multi-angular 4. Extruded feature	Exhibiting more irregularity joint planes, reticular joints with primary joints	Scratches in joint plane, smooth joints, easy to break	Jagged and polygonal	Easily make small pieces using hands, medium hardness
Type III (strongly destructed coal)	Semi-bright and semi-dull	1. Bending, showing lenticular structure 2. Lamellar structure 3. Tiny fragments, inordinate and unordered beddings	Unclear and undeveloped joints system, denser secondary joints	Lots of scratches	Jagged and granular	Can be twisted into powder, low hardness

(continued)

Table 1.1 (continued)

Destruction types	Luster	Structure and structural characteristics	Joint properties	Joint plane properties	Fracture properties	Strength
Type IV (pulverized coal)	Dim	Grain or small particles cemented, resembling the natural coal pellets	No joint, glued block shaped		Granular	Can be twisted into powder, occasionally hard
Type V (powdered coal)	Dim	1. Soil-like structure, similar to soil as well as coal 2. Like fault gouge			Earthy	Can be twisted into powder, loose

Coal seam gas is the generic term of methane-dominated gas mixture, which is also known as Coalbed Methane (CBM). Coal seam gas was formed from humic type organic matter during the coal-forming process. After the sedimentary accumulation of coal's original parent material, which is also known as humus, the gas-forming process can be divided into two phases: from the plant remains to peat, it belongs to the biochemical gas-forming phase; and from lignite to anthracite under the high pressure/temperature of the strata, it is the gas-forming phase of coal metamorphism. In the biochemical gas-forming phase, the buried depth of the peat layer that was formed by coal-forming materials, was shallow. The cementing degree of upper covering layer was not good, thus the generated gas could easily escape into the ancient atmosphere through seepage/diffusion effects and is hardly preserved in the existing coal seams. From lignite to anthracite, some 300—400 m^3 of gas per tonne of coal was generated, however, under the present coal mining depth, the maximum gas content is only 20–30 m^3 per tonne. When the coal seam has an outcrop or the coal basin is lain under the alluvial layer, there exist two different directions for gas migration from the coal seam. These are the upward migration of the gas from deep coal seam, and the downward migration (seepage, diffusion) of the ground air as well as the gas generated by biochemical/chemical reaction in the surface soil into the deep part of the coal seam, resulting in the vertical zoning characteristics of the gas in coal measure strata. Along the vertical direction (from surface down), the zone of coal seam gas can be generally divided into the gas weathering zone and methane zone. Within the methane zone, the concentration of methane and heavy hydrocarbon is greater than 80%.

The drainage and utilization of coal seam gas have triple benefits: energy, safety and environment [4].

(1) Coal seam gas is one type of clean energy. The calorific value of 1 m^3 pure methane is 35.9 MJ, which is equivalent to the calorific value of 1.2 kg of standard coal. According to the latest evaluation report of China's coalbed methane (coal seam gas) resource [5], the coalbed methane resource of the country (buried depth shallower than 2000 m) is 36.8 Tm^3 (1 $Tm^3 = 10^{12}$ m^3). This is equivalent to the total amount of conventional natural gas resources in China (30 Tm^3), of which, the recoverable resource amount of coalbed methane (buried depth shallower than 1500 m) reaches 10.9 Tm^3.

(2) Coal seam gas is one of the major hazards of coal mining. Gas explosions and coal and gas outbursts are the main gas disasters in coal mines, additionally, gas burning and suffocation can also cause casualties. Without pre-drainage of coal seam gas, during coal mining a lot of gas will flow into the coal mine ventilation system. When methane concentrations reach 5.0–15.0%, the mixture of the methane-dominated combustible gas can be ignited and set off a rapid oxidation reaction, namely the gas explosion. Once the explosion occurs, it not only causes heavy casuanamelylties, but can also destroy the coal mine facilities and ventilation system. Additionally it can cause secondary disasters such as a coal dust explosions, fires, collapses and roof cavings. When the concentration of methane is greater than 15.0%, if there exist sources of ignition, gas combustion

occurs. Coal and gas outbursts involve the release of gas energy and strain energy stored in coal seam, which exhibits as a large ejection of coal, rock and gas into the production space in a very short time. The amount of thrown coal/rock ranges from several tonnes to tens of thousands of tonnes with gas volumes ranging from hundreds of cubic meters to over million cubic meters, which may further induce gas explosion disaster.

(3) Coal seam gas is one kind of strong greenhouse gas. The greenhouse effect of a unit mass of methane is 25 times larger than that of carbon dioxide, thus burning 1 m^3 of methane instead its release to the atmosphere is equivalent to reducing emissions by 15.93 kg of CO_2 gas. According to the coal mine gas grading report of China in 2007 and 2008 [6, 7], the methane emission from coal mining in China amounted to 19.3 Gm^3 and 20.2 Gm^3 respectively (1 Gm^3 = $10^9 m^3$), which can be convert into CO_2 equivalent weight of 344 Tg and 361 Tg respectively (1Tg = 10^6 t), and equivalent to 5% of the CO_2 emissions from burning fossil fuels during the same period.

1.1.2 *Coal Measure Strata*

1. The evolution of coal measure strata

Coal measure strata, which is also known as coal-bearing strata, is a set of sedimentary rocks that have intergrowth relationships and contain coal seams (or coal lines). The sedimentary paleogeographic environment is the direct factor controlling the formation of coal series, when the tectonics, palaeoclimate and palaeophyte conditions are suitable for the development of coal-bearing basins, the sedimentary paleogeographic features of coal basins become the most important condition in determining the coal-forming characteristics. During the coal-formation process, the sedimentary environment before/meanwhile/after peat accumulation can all directly affect the thickness/morphology, lateral distribution of the coal seams, coal/rock composition and coal quality characteristics. The sedimentary environment before the formation of peat bog influences the topography and geomorphological conditions of the coal seam accumulation, thus affecting the thickness variation and distribution of coal seams. The sedimentary environment configuration with the peat bog over the same period not only directly affect the morphology and distribution of coal seams, but also controls the composition of the coal (namely variation of coal and their quality changes) together with the peat bog internal environment. The sedimentary environment after coal-formation will affect the preservation condition for the peat layer and its later denudation, which can indirectly affect coal quality.

The top and bottom interfaces of coal measure strata are not always isochronous interfaces, but can also be diachronous interfaces. This is because coal measure strata are mostly made up of terrigenous clastic sediments, and the formation of sediments is mainly related to the filling occurring near the sedimentary basin edge, namely carrying the sediments from basin edge to intra-basin by water flow. If such progradation effects continuously proceed in stable condition for a long geological time, then the formed coal-bearing series would often present the characteristics of cross-temporal evolution.

Taking the Huaibei Coalfield as an example, brief explanation of the coal measure strata evolution of the coalfield is as follows:

After the early tectonic movements, coal measure strata of Huaibei Coalfield formed the old metamorphic basement of the current coalfield, which provided the foundation for the settlement of coal measures strata and formed a large number of fractures and folds at that time.

At the end of early Proterozoic, strata uplift due to the Lvliang movement occurred and formed an erosion zone, then in the middle period of late Proterozoic, this zone subsided into the sea area. In Xuhuai area which suffered a much deeper subsidence, about 3000 m thick clastic rocks and salt were deposited, forming the sedimentary cover of the Sinian period. During the late period of middle Ordovician (500 Ma, 1 Ma: 1 million anniversary), the Caledonian movement affected the whole north China region and the whole crust of the region rose to land. From the formation of sedimentary cover to the early Carboniferous, no deposits were received by the strata. Until the early period of late Carboniferous (290 Ma), strata began to subside slowly and led to widespread transgression of sea, which formed a 1700 m thick Carboniferous-Permian deposit of coal-bearing clastic sedimentary rocks and carbonate [5].

Figure 1.2 shows a schematic diagram of the burial history and stratigraphic

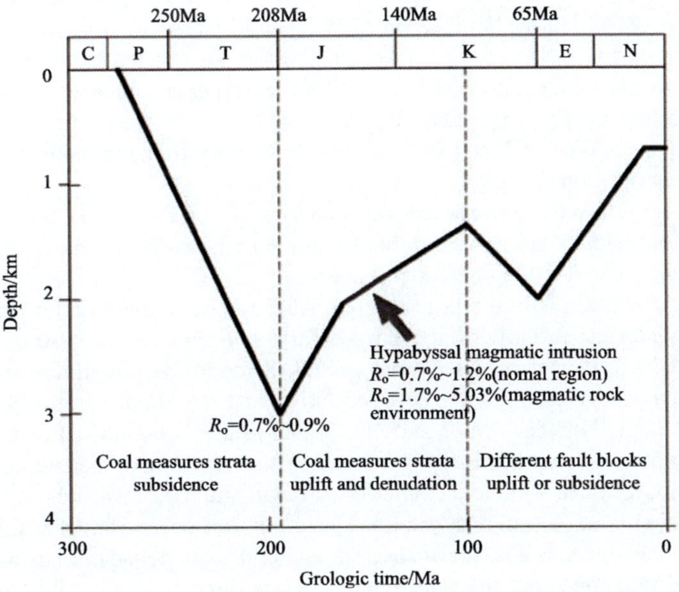

Fig. 1.2 Schematic diagram showing the burial history and stratigraphic evolution of the Carboniferous-Permian coal measure strata in Huaibei Coalfield

evolution of the Carboniferous-Permian coal measure strata in Huaibei Coalfield [8].

In the early Indo-Chinese epoch (250 Ma), with the aggregation of the North China and Yangtze Plates, the coal basin accepted the sedimentation and developed latticed fracture planes in the basement. In the late Indo-Chinese epoch, the maximum burial depth of coal-bearing strata experiencing a maximum earth temperature of 140-180°C reached 3000 m, and the plutonic metamorphism caused the degree of coal metamorphism to reach gas coal and fat coal. During the Yanshanian period (208 Ma), the impact of tectonic activity was strong and the Tanlu Fracture caused the cover of coal measure strata to form NNE direction folds and regional fractures while the coal-bearing strata began to uplift. In the early and middle Yanshanian period, hypabyssal magmatic rocks invaded the region, resulting in the second metamorphism of the coal measure strata. In the later Cretaceous (100 Ma) of the Yanshanian period, the strata were lifted up to the shallowest position and in some areas strata even cropped out the ground surface and suffered heavy denudation where the total erosion thickness of strata reached 2000–2500 m. The Triassic formation of the Huaibei Coalfield was eroded and exhausted during this period, and even parts of the Permian strata were denuded in some places. The partial or total denudation of exposed coal seams may be the main reason for the formation of the coal failure zone in Huaibei Coalfield. In the Himalayan period (65 Ma), the coalfield strata were under a tensile stress field, causing uplift or settlement in different blocks, and finally a structural framework with uplift and subsidence segmented by faults was formed. At this time, the western deep buried area of Huaibei Coalfield subsided rapidly and formed the Zhoukou sag area, but the denudation zone began to rise gradually after the second subsidence. In the Quaternary (2.6 Ma), the whole area of Huaibei Coalfield was still in a subsidence area, however, except for the deeper subsidence area located in the west of Fuyang-Xiayi Fracture, the subsidence degree of the denudation-shallow buried area is smaller, and the location of coal measure strata gradually approached their current position.

Figure 1.3 shows the generalized stratigraphy of the 86 District, Haizi Coal Mine, Huaibei Coalfield. It can be found that the main commercial seams of Haizi Coal Mine are the Nos. 7, 8, 9 and 10 coal seams.

No. 10 coal seam is in the deepest area, whose floor is composed of mudstone, siltstone, sandstone and limestone from the Taiyuan Formation; rock strata between No. 9 and No. 10 coal seams is composited of mudstone, aluminum mudstone, siltstone and sandstone; stratum between No. 8 and No. 9 coal seams consists of mudstone; strata between No. 7 and No. 8 coal seams is composed of mudstone, siltstone and sandstone; the roof of No. 7 coal seam is made up of mudstone, siltstone and sandstone; above which is a super-thick magmatic rock layer whose thickness reaches 120–150 m (140 m on average). The absence of strata from partial Permian, Triassic to Paleogene is a result of stratigraphic uplift and denudation as mentioned above, and thus the super-thick magmatic rock is directly covered by Quaternary unconsolidated formation.

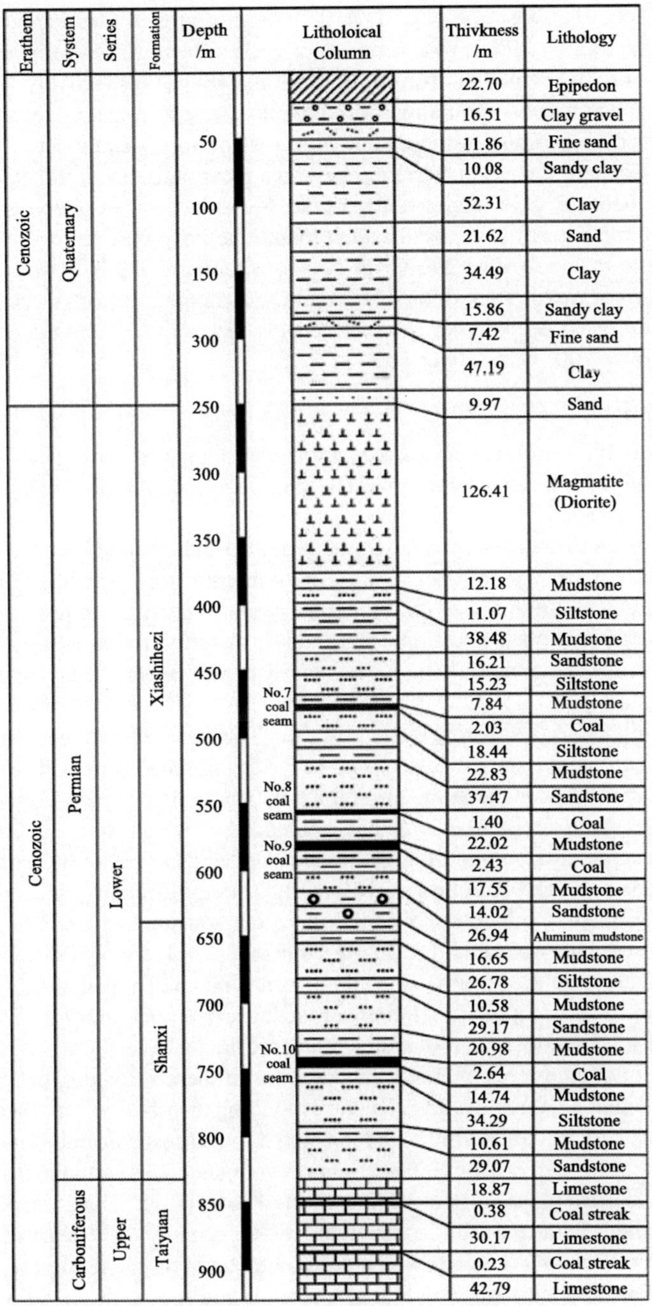

Fig. 1.3 Generalized stratigraphy of the Haizi Coal Mine, Huaibei Coalfield

The distances between the Nos. 7, 8, 9, 10 coal seams of Haizi Coal Mine and magmatic rock are 40–96 m (44 m on average), 56–76 m (64 m on average), 60–80 m (70 m on average), 141–165 m (154 m on average) respectively. The results from thermal conduction simulation of magmatic heat effect in coal measure strata, indicate that the maximum temperature from the thermodynamic activity at No. 7 coal seam reached 350 °C and the temperature remained above 200 °C for about 1500 years; because the distance between No. 8 and No. 9 coal seams is small, the maximum temperature from the thermodynamic activity reached 300 °C and the temperature remained above 200°C for about 1300 years; the No. 10 coal seam is furthest from magmatic rock, thus the maximum temperature from thermodynamic activity could reach is the lowest, and could only reach 235 °C and the temperature remained above 200 °C for about 1000 years [9].

2. Composition of coal measure strata

According to the petrogenetic classification based on the geological viewpoint, rocks are divided into three types, namely magmatic rocks, sedimentary rocks and metamorphic rocks.

Coal measure strata are deposited under humid climate conditions, thus the color of coal measure strata's sedimentary rocks is mainly grey, ash black, black and greyish-green. The lithology of coal measure strata is mainly composed of various kinds of terrigenous clastic rock and clay rocks, mixed with limestone, flint layer, etc.; some coal measure strata can also be mainly composed of limestone; moreover, aluminum bauxite, refractory clay, oil shale, siderite, pyrite and other types of rock are commonly preserved in coal measure strata. The mineral composition of clastic rocks in coal measures strata is determined by the lithological composition of the land source area as well as the tectonic environment. The most common clastic rocks in coal measure strata are quartz sandstone, feldspar quartz sandstone, feldspar sandstone, litharenite, siltstone and conglomerate. Under different sedimentary conditions, compositions and structures vary greatly among different clastic rocks. The coal series formed in inland conditions have more transitional sandstone, such as feldspar quartz sandstone and lithic quartz sandstone; while the coal series formed in near erosion area usually have more conglomerate and coarse sandstone. Clay rocks exhibit a significant proportion in coal series, but often contain a lot of siltstones; limestone is a main constituent of some Paleozoic coal series.

In the formation process of coal measure strata, if there were magmatic activities or volcanic activities, there would exist corresponding distributions of volcanic rocks and pyroclastic rocks. For example, in many of China's Mesozoic and Cenozoic coal measure strata, there are various distributions of volcanic rocks and pyroclastic rocks; some coal measure strata of late Paleozoic often contain pyroclastic rocks as well. The most salient feature of the sedimentary rocks forming coal measure strata are mostly characterized by the various non-horizontal bedding planes and sedimentary structures. Coal series are often rich in plant fossils and sometimes in animal fossils. Moreoer, coal series also contain various carbonate concretions (especially sideritic) and pelitic, silty and sideritic inclusions. During the later stages of the coal-forming phase, due to the tectonic movement of the earth, such as the Yanshan movement of the Jurassic and Cretaceous periods and the Himalayan movement that occurred in the Paleogene and Neogene periods, magma invaded the coal measure strata through tectonic fractures and faults, forming magmatic rocks when it cooled down. These formations exist in the form of rock bed, rock wall, rock stock and rock vein.

In the coal measure strata the roofs and floors of coal seams are mainly composed of mudstone, shale, sandstone, siltstone, limestone and other sedimentary rocks as well as hypabyssal rocks formed by magma intrusion.

3. Mechanical parameters of coal/rock in coal measure strata

The mechanical parameters of coal/rock of coal measure strata are shown in Table 1.2. The data ranging from mudstone to coarse sandstone are the test results of the geological exploration drilling's rock samples in Zhaoji Coal Mine, Huaibei coalfield and data of magmatic rocks and limestone are cited from literature [10]. Intact coal and tectonic coal samples were collected from the underground working face in Daning Coal Mine, Shanxi Province [11]. By using intact coal blocks standard intact coal samples were made. In the laboratory, tectonic coal samples were made by compressing tectonic coal powders gotten based on special methods, and 0.074-0.25 mm sized coal power sieved from pulverized intact coal are converted into coal briquette samples by compacting. As shown in Table 1.2, the mechanical parameter differences of coal and rocks in coal measure strata can be distinguished in the following aspects:

(1) The densities of different rocks range from 2.40 to 3.01 m^3/t, and 2.69 m^3/t on average; however, the densities of intact coal and tectonic coal are 1.49 m^3/t and 1.32 m^3/t respectively, thus the average density of coal is 1.41 m^3/t.
(2) Magmatic rock has the highest uniaxial compressive strength while mudstone is the weakest, the strength of sandstone is in the middle range and limestone's strength varies greatly. The uniaxial compressive strength of intact coal is 17.54 MPa which is similar to the softest mudstone; the uniaxial compressive strength of tectonic coal is 1.15 MPa, and 0.37 MPa for coal briquette samples, both much weaker than that of rocks and intact coal.
(3) Rock has a certain tensile strength, magmatic rock has the largest tensile strength while mudstone has the smallest tensile resistance, the tensile strengths of sandstone and limestone range in the middle. Compared to rocks, the tensile strength of intact coal is much smaller while the tectonic coal has almost no tensile strength.
(4) Within different rock types, magmatic rock has the largest cohesion while mudstone's cohesion is the smallest, the cohesion of sandstone is in the middle and that for limestones vary greatly. The cohesion of intact coal is 3.79 MPa, but for tectonic coal the measured cohesion was only 0.44 MPa.
(5) Compared to rocks, coal has a much smaller elastic modulus. The elastic modulus of intact coal is 4050 MPa and that for tectonic coal is only 290 MPa.
(6) Compared to rocks, coal has a larger Poisson's ratio. The Poisson's ratio of intact coal is 0.26 and that for tectonic coal is 0.34, indicating that coal samples will generate larger axial deformation under the same axial stress condition.

Table 1.2 Mechanical parameters of coal and rocks in coal measure strata

Lithology	Density ρ/(t/m^3)	compressive strength σ_c /MPa	Tensile strength σ_t/MPa	Friction angle φ/(°)	Cohesion c/MPa	Elasticity modulus E /(10^4 MPa)	Poisson's ratio μ
Mudstone[a]	2.68–2.69	15.82–27.77	1.80–2.24	31–33	1.69–2.21	1.46–2.06	0.19–0.24
Siltstone[a]	2.68–2.69	19.69–55.35	1.18–3.66	34–37	1.25–2.83	2.47–3.78	0.15–0.21
Fine sandstone[a]	2.68–2.69	53.14–66.00	4.37–4.70	36–40	3.67–4.05	2.87–3.76	0.14–0.17
Medium sandstone[a]	2.70–2.73	70.93–98.46	3.67–4.26	38–40	4.14–5.31	3.98–4.14	0.10–0.17
Coarse stone[a]	2.69	91.85	4.34	40	4.05	4.56	0.12
Magmatite[b]	2.41–3.01	102.30–161.91	6.78–16.94	37.73–48.29	8.47–17.46	1.69–3.96	0.16–0.19
Limestone[b]	2.40–2.69	29–245	2.90–4.90	35–50	9.8–49	2.40–3.80	0.19–0.35
Intact coal[c]	1.49	17.54		47	3.79	0.405	0.26
Deformed coal[c]	1.32	1.15		34	0.44	0.029	0.34
Coal briquette[c]	1.13	0.37		34	0.30	0.010	0.37

Note [a]Cited from Exploration Report for Coal Resource of Zhaoji Coal Mine, Huaibei Coal Basin (completed by No. 3 Exploration Team of Anhui Coal Geology Bureau)
[b]Cited from ref [10]
[c]Cited from ref [11]

1.2 Structure of Coal and Its Simplified Physical Model

1.2.1 Pores and Fractures in Coal

Gas adsorption capacity, permeability and strength of coal is highly dependent on the pore properties, pore size, connectivity and development degree of the coal.

Studies indicate that coal has a porous system with different orders of magnitude that range from 0.5 nm to several millimeters.

According to coal's composition and structure properties, the pores within coal can be divided into macroscopic pores, microscopic pores and molecular pores. Macroscopic pores were formed by bedding, cleats and secondary fissures which can be distinguished by the naked eye. The maximum resolution of naked eyes is approximately 0.1 mm, therefore the dimension of macroscopic pores is generally classified in millimeters. Figure 1.4 shows a bright coal sample collected from underground, polished in the laboratory and photographed with the scaleplate, from which we can clearly see the fracture distribution on the coal sample's surface. Microscopic pores are the pores that can be distinguished by optical microscope and scanning electron microscope. After a magnification of 300–10,000 times the microscopic pores can be clearly observed and measured, thus the microscopic pores are micrometer level pores. Figure 1.5 shows the SEM photos of four different coal samples. The figure clearly shows the different genetic pores in coal. Of which, Fig. 1.5a, b are #2 long frame coal and Jimusar lignite from Inner Mongolia respectively [12]; Fig. 1.5c, d are anthracite from Bailongshan Coal Mine, Diandong Coalfield, Yunnan Province

Fig. 1.4 Photograph showing the fractures distribution on coal surface (Dot spacing: 5 mm) The photograph was sourced from the Outburst Risk Determination and Associated Factors (Sigra Job Reference Number: 398, 2015)

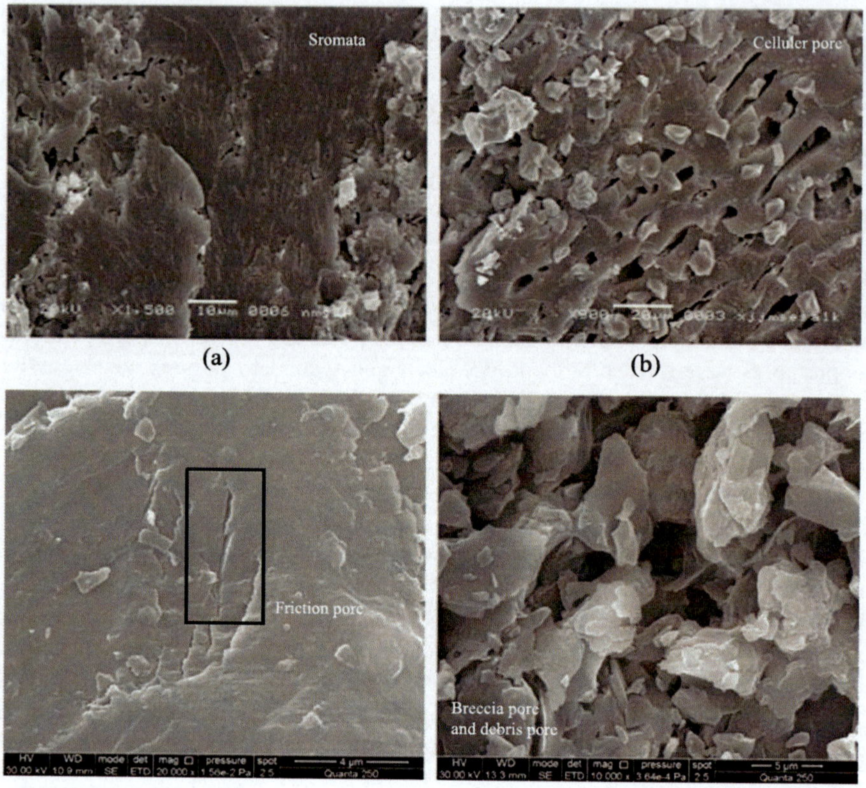

Fig. 1.5 Scanning electron microscope images of coal. **a** Stomata. **b** Cellular pore. **c** Friction pore. **d** Breccia pore and debris pore

[13]. Molecular pores are the ultra-micro pores made up by the molecular structure of coal, and the size is generally of nanometer level.

To better study the gas flow condition in coal, researchers regard the coal as a dual porosity medium. They divide the pores of coal into cleat pore and matrix pore. The pores in which gas migration regularity conforms to the Darcy's seepage law are regarded as cleat pores (short for cleat, including macroscopic fractures and micro-cracks), and cleat can be treated as the boundary cutting the matrix apart. The pores in which gas migration regularity conforms to the Fick's law are regarded as matrix pores, which occur in the coal matrix. In this way, coal can be simplified as dual porosity media consisting of matrix and cleats.

The pore volume of coal is the volume sum of matrix pore and cleat pore. Thus, defining the porosity as the ratio of the pore volume to coal volume, defining the matrix porosity as the ratio of the matrix pore volume to coal volume and the cleat porosity as the ratio of the cleat pore volume to coal volume, then the porosity of coal can be expressed as the sum of matrix porosity and cleat porosity. The results of the coal samples from Daning Coal Mine indicate that [11], the porosity of intact coal is 3.9%, of which the matrix porosity is 3.5%, and the cleat's porosity is 0.4%; the porosity of tectonic coal is 13.6%, of which the matrix porosity is 4.2%, and the cleat's porosity is 9.4%. Seidle [14] obtained the cleat porosity of coal based on coal's permeability calculation model and the permeability data as well as matrix scale. Li et al. [15], Chatterjee and Paul [16], Zhou et al. [17] studied the cleat's porosity of coal using dual lateral logging data and coal's volume model. Keshavarz et al. [18] applied a certain pressure difference between both ends of the water-saturated coal sample until the flow stabilized, then by replacing the water in coal with Helium gas under the same conditions he pointed out that the displacement volume of water was the initial cleat porosity of the coal. The cleat porosity data of coal is shown in Table 1.3, from which it can be found that the cleat porosity of coal is usually 0.04–1.50%.

1. Pores in coal

The most widely used classification method of coal pore structure in China is the decimal classification method proposed by Hodot (former Soviet researcher) in 1961. Coal pore structure is divided into four categories, namely micropores (pore size $\leq$ 10 nm), minipores (10–100 nm), mesopores (100–1000 nm) and macropores (pore size $\geq$ 1000 nm). According to this classification method, some of the larger pores will have the properties of cleat's porosity [19].

Due to the differences in testing methods and instruments, the characterization methods of coal's pore are mainly categorized into the radiation method and fluids seepage method, of which the most commonly used methods are the mercury intrusion porosimetry and physical adsorption method such as N_2 (77 K) adsorption method, CO_2 adsorption method. And different test methods provide different ranges of pore size determination, for example, mercury intrusion porosimetry is applicable for the pore size ranging from several nanometers to 100,000 nm; N_2 (77 K)

Table 1.3 Cleat porosities of coal

Literature resources	cleat porosity/%	Matrix scale /mm	Cleat aperture /μm	Sampling location
Seidle [14]	1.2	3–5	24	Arkoma Basin
Li [15]	0.8–1.5	0.15–1.72	1.16–11.156	Southeast Qinshui Basin
Chatterjee and Paul [16]	0.52–0.67	20	62.72–80.1	Rangamati area, Raniganj coalfield
Zhou and Yao [17]	0.04–0.74	0.39–0.67	0.229–0.398	Southeast Qinshui Basin
Keshavarz et al. [18]	0.48–0.72	–	–	Dawson Mine Central, Queensland Affinity coal mine, West Virginia

adsorption method provides the best resolution in the pore size of 0.35-400 nm; CO_2 adsorption method suits the pore size of 0.3–1.5 nm. Therefore, it is recommended to select the corresponding test methods based on the test requirement of pore size, or comprehensively study the pore characteristics of coal using the combination of different test methods.

To characterize the pore structures of coal, two parameters, namely the specific surface area and pore volume, are widely used. The specific surface area of coal is defined as the total surface area of the coal pores per unit of mass, with unit of m^2/g and the pore volume of coal is defined as the total volume of the coal pores per unit of mass, with unit of mL/g. Meanwhile, to study the pore characteristics of a certain pore size interval, the specific surface area ratio and pore volume ratio are defined as well: the specific surface area ratio is the percentage of the specific surface area of a certain pore interval to coal's total specific surface area; the pore volume ratio is the percentage of the pore volume of a certain pore interval to coal's total pore volume. Figure 1.6 shows the pore volume and specific surface area distributions tested by low pressure nitrogen adsorption method of an anthracite ($R_o = 2.94$) sampled from Jiulishan Coal Mine, Jiaozuo Coalfield [20]. From the pore size distribution in Fig. 1.6a, it can be found that the pore volume ratio of minipores is 37.4%, pore volume ratio of mesopores is 31.6%, pore volume ratio of macroores is 23.4%, and the sum of the pore volume ratio of mesopores/macropores reach 55.0%, which indicates that pore volume of coal is mainly concentrated in the mesopores and macropores, while the pore volume of micropore is small. From specific surface area distribution in Fig. 1.6b, it can be found that the specific surface area ratio of micropores is 56.1%, specific surface area ratio of minipores is 39.8%, and the sum of specific surface area ratio of micropores/minipores reaches 95.9%. This indicates that although the pore volume of coal is mainly concentrated in the mesopores and macropores, the specific surface area of coal is mainly concentrated in micropores and minipores. This also certifies the viewpoint that coal seam gas is mainly adsorbed on the surface of micropores and minipores. Pore structure characterization results from other ranks of coal also indicate that, although corresponding results from different rank coals vary greatly, the distribution regularities of pore volume and specific surface area are basically the same.

2. Cleats in coal

The fractures of coal refer to the fractures which were caused by various stress effects from nature since the coal-formation process [21], coal's fractures can be divided into endogenous fractures and exogenous fractures based on their origins. The endogenous fractures are caused by coal dehydration, devolatilization and shrinkage during the coalification process; as well as the fractures induced by water /gas that is produced by coalification or abnormal high pressure that is caused by temperature rising/pore swelling. The exogenous fractures are caused by stress effects during the period of tectonic deformation after the formation of the coal seam.

In the late nineteenth century, the British mining industry first named the natural fractures in coal as “cleat”, and this term has been widely used in countries such as the UK, USA and Australia. The cleat of coal is also known as endogenous

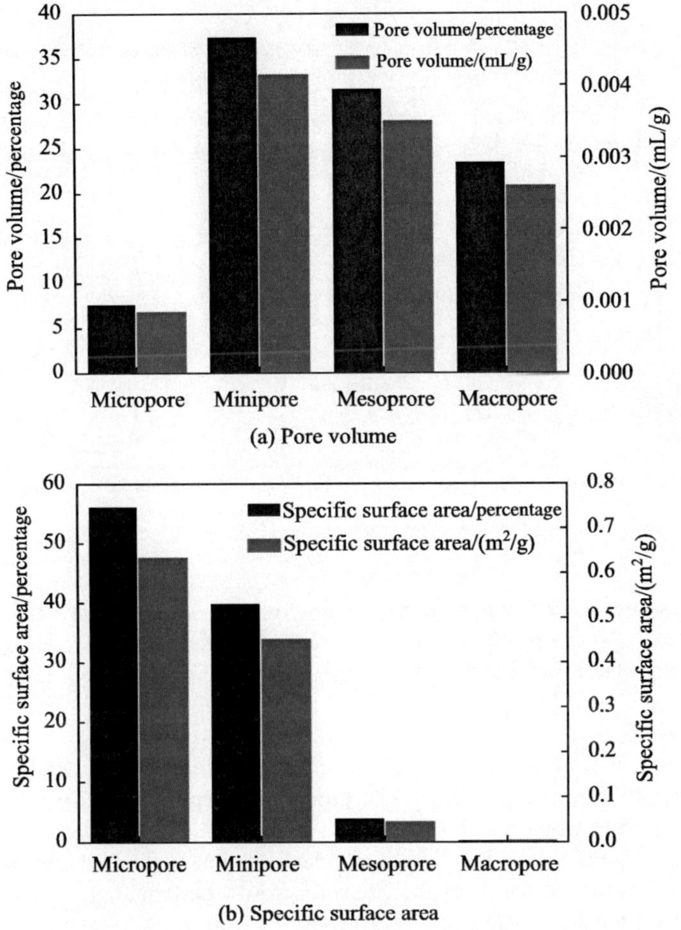

Fig. 1.6 Pore volume/specific surface area distribution of the anthracite sampled from Jiulishan Coal Mine, Jiaozuo Coalfield

fracture, which is generally exhibited as perpendicular or roughly perpendicular to the bedding plane, and only develops in the ribbon or layer of the vitrain or bright coal. The cleat plane is flat and smooth, sometimes showing the trace of the augen tension bedding plane that formed by very fine ring-like grains. Usually, there are two groups of cleats that are roughly parallel to each other (see Fig. 1.7) [22], of which, the well-developed group is called face cleat and the other is called butt cleat. The face cleat is approximately parallel to the bedding plane, generally extends as flat-shaped with good continuity, which is the main endogenous fracture in the coal seam. The butt cleat is only developed between two adjacent face cleats, which is approximately perpendicular to the bedding plane, and whose continuity is poor and irregular, which is the secondary endogenous fracture in the coal seam. The face cleat

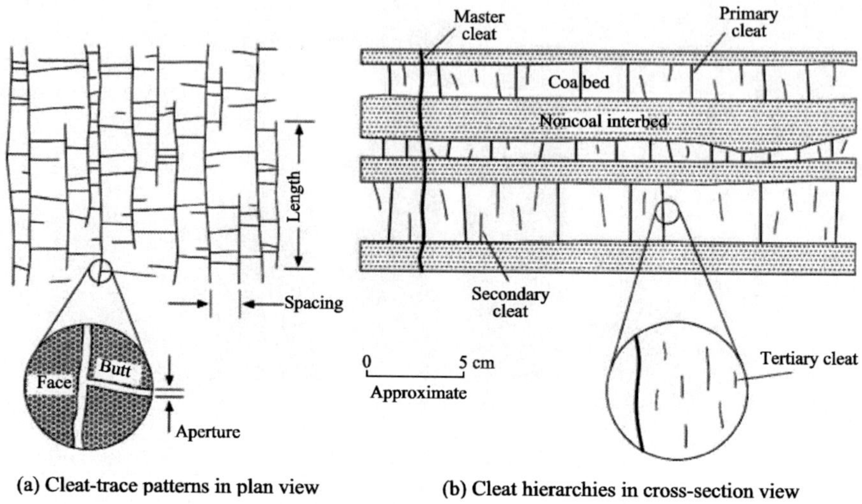

Fig. 1.7 Schematic illustration of coal cleat geometries [22]

is obvious to recognize; but butt cleat is not obvious, and intersects with the face cleat in an angle of approximately 90°. Because of the development of face/butt cleats in coal, although their individual scale is very small, the cleats still can intersect with each other in a three- dimensional network in space, cutting the coal into a series of rhombus matrix.

Generally, the parameters obtained by optical microscope observations and statistics are used to describe the cleats [23]. Procedural steps are described here: First, polish the surface of the block of coal into a polished section, place the polished section under the objective lens and adjust the direction of the fractures perpendicular to the direction of the ruler, then starting from a certain angle of the polished section, move the scaleplate from left to right (or from right to left) and count the number of fractures getting through the cross point of the eyepiece; pass through the polished section with the row spacing of 5 mm, record the number of rows, the length of each row and the total length.

To describe the cleat, widely used parameters include the cleat's aperture, length, height, spacing, density and so on.

(1) The cleat aperture means the width of the cleat. Under the ground surface conditions, cleat aperture is generally less than 0.1 mm; and the estimated range of cleat aperture is 0.001–20 mm [22]. However, at present, information about the cleat aperture is mostly from the study of outcrop coal and microscopic observations of the coal samples under stress free condition, reliable information about the aperture of coal cleat under the geostress field is seriously lacking. Referring to the inversion method of permeability models or logging curve, the range of cleat's aperture is 1.16–80.1 μm (Table 1.3).

(2) The length and height of the cleat. Treating the cleat as a square, then the length of the cleat means the side along the extend direction (parallel to the bedding plane of the coal), and the height is the side vertical to the length side. At the present, there is very little data on the length and height of the cleat because the sample collection behavior itself destroys these data, and the field observation is very difficult as well. Many of the cleat height and length data are given in centimetres level. In the San Juan Basin of New Mexico, USA, the length and height of the cleats of the fruitland formation coal seam range from several micrometers to several meters [22].

(3) The spacing of the cleats means the distance between two adjacent cleats. The level of cleat spacing was given in centimeters. The cleat spacing varies with the change of coal rank, from lignite to medium volatile bituminous coal the cleat shows the trend of spacing reduction, and from medium volatile bituminous coal to anthracite the cleat spacing shows an upward trend, thus forming the bell-shaped distribution of the cleat spacing. Based on the data from the outcrop and core of North America coals, the face cleat spacing is found to increase from about 22 cm in lignite (R_o: 0.25–0.38%) to 0.2 cm in anthracite (R_o > 2.6%) [22]. Laboratory test results of the coal samples from Southwestern Indiana coal mine, USA indicate that the face cleat spacing is 2.15–5.95 cm and the butt cleat spacing is 2.75–8.38 cm [24].

(4) The density of cleat means the number of cleats intersecting a line of a certain length that perpendicular to the direction of the cleat extension. The length of the line depends on the cleat density and usually 50 mm is appropriate. The cleat density of coal increases with the increase of coal rank, and it reaches the maximum value in coal of the middle rank, then with the further increase of coal rank it shows either the decreasing trend or remain the same value [25]. The face cleat spacing of the coal from Southwestern Indiana coal mine, USA is 8.4–23.5 cm and the butt cleat spacing is 6.0–18.2 cm [24]. The average distribution density of the face cleats in anthracite from Qinshui Basin, China is 6.5–71/5 cm [26]. The average face cleat density of the coking coal from Liulin Basin, China is 8–18/5 cm and average butt cleat density is 4–14/5 cm [27]. The average face cleat density of the Permain coal samples from Queensland, Australia is 3 to 5 cm ($R_o = 0.67\%$), 4.4 to 5 cm ($R_o = 0.74\%$), 33.4 to 5 cm ($R_o = 1.24\%$) and 7.7 to 5 cm ($R_o = 1.85\%$) respectively [28].

1.2.2 Simplified Physical Model of Coal

1. Gas migration in coal

From the above description of coal pore and cleat system, it can be seen that coal is a typical porous medium in the natural world and the migration pattern of gas in coal seam (diffusion or seepage) is closely related to the pore size and the pore size distribution of the coal seam. When the coal seam is not affected by mining and extraction activities, gas in the fractures and matrix are in a dynamic equilibrium state, although there exists mass exchange between the fractures and matrix, but from the macroscopic point of view there is no transport of mass. However, once the coal seam is affected by mining or extraction, the adsorbed gas and free gas of coal seam will migrate towards the coal face in the form of Fick diffusion and Darcy seepage respectively under the effect of concentration or pressure differences between the coal seam and the coal face. Meanwhile, because the gas seepage velocity in the cleat system is much faster than the gas diffusion rate in the pore system, the dynamic adsorption equilibrium state of the coal seam will be destroyed resulting in the matrix gas pressure being greater than the cleat gas pressure, thus the gas in the pore system will diffuse into the surface of matrix and then further into the cleats, which means that mass exchange will occur between the pore system and the cleat system. There will be outflow for the pore system and inflow into cleat system, thus the pore system is the uniformly distributed internal mass source of the cleat system.

It can be found that the gas migration in actual coal seams is a combined parallel and series connection process of diffusion/seepage, however, since the amount of gas that diffuses into the borehole or roadway is small, this book simplifies the gas migration in the coal seam into a continuous two-step process, namely the diffusion and seepage are a series of processes: ① Gas diffuses from the coal matrix to the fractures through the form of Fick diffusion; ② Gas flows from fractures into roadways or boreholes through the form of Darcy seepage. As shown in Fig. 1.8, coal is cut into different sized matrix by face cleat and butt cleat, matrix contains a large number

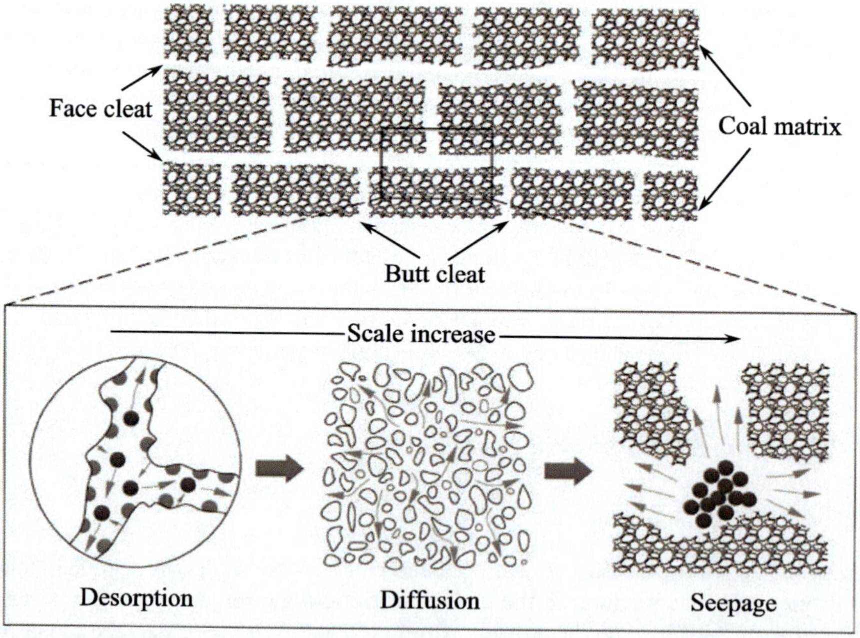

Fig. 1.8 Schematic diagram showing the migration modes of gas in coal

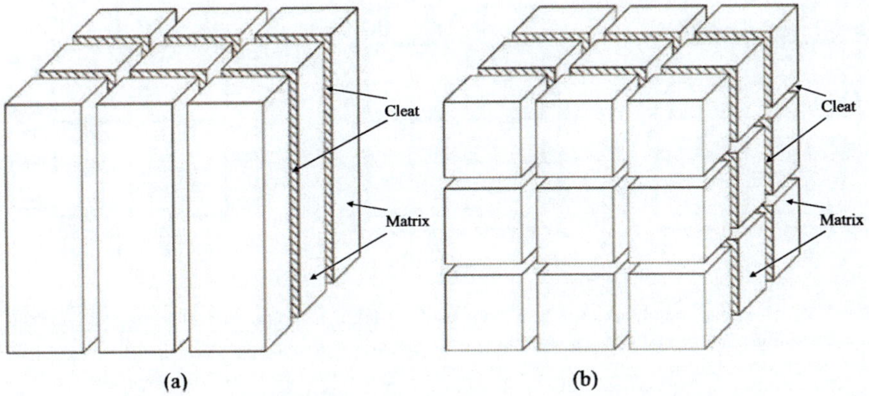

Fig. 1.9 Physical model of coal. **a** Matchstick model. **b** Cubic geometry model

of pores mainly comprised of micropores and minipores, and the coal seam gas is mainly stored on the surface of matrix pores in an adsorption state. When the gas adsorption equilibrium state changes, the gas molecules first desorb from the matrix pore surface and then diffuse into the fractures and finally flow into the exposed coal surface (borehole or coal wall) by seepage.

2. Simplified physical model of coal

In order to quantitatively analyze the effect of effective stress and adsorption expansion deformation in permeability evolution, when establishing the permeability model, it is necessary to simplify the physical model of coal on the basis of Fig. 1.8. Two of the most commonly used simplified models are "matchstick model" and "cube geometry model" [29, 30] as shown in Fig. 1.9. The matchstick model considers that the coal is divided by two orthogonal fractures, thus the geometry of coal matrix is separated into cuboids which are similar to tied together matchsticks; cube geometry model considers that the coal is divided by three orthogonal fractures, and thus the matrix is a cube. In addition to the hypothesis of coal structure, referring to the actual strata conditions the boundary conditions of isovolumetric, uniaxial strain and triaxial stress are also proposed, as shown in Fig. 1.10. In the process of coalbed methane surface exploitation the overburden load is basically kept constant, the coal seam is constrained by horizontal stress and the horizontal displacement is basically unchanged, then in this condition, the coal seam strain only occurs in the vertical direction, as a result it can be simplified as a uniaxial strain condition. Based on the theoretical analysis and field survey, literature [30] proves that the volume of coal seam almost keeps the same size in the process of the coalbed methane surface exploitation, therefore the coalbed methane extraction can also be simplified as an isovolumetric condition. Due to the limitation of experimental conditions, the confining pressure applied on the coal sample must be able to seal the internal gas, as a result the coal sample is in a boundary condition of triaxial stress.

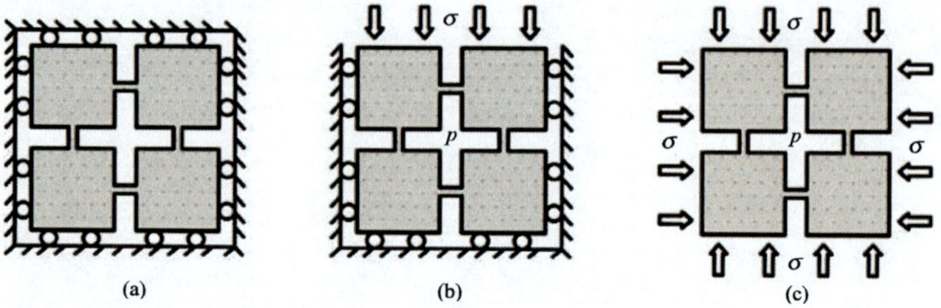

Fig. 1.10 Boundary conditions for coal's physical model. **a** Isovolumetric condition. **b** Uniaxial strain condition. **c** Triaxial stress condition

1.3 Comparison of Coal and Rock

1.3.1 Comparison of Structure and Strength

Rock is a natural object formed by minerals or rock debris in a certain gathering regularity under geological effects. Rock body is a geological mass that is composed of rock structure body and structural surface, including rocks and different geological structural features such as the structural surfaces of joints, fractures and folds [10]. The structural surface spacing of the rock body is usually very large, and rock samples used in laboratory are usually complete rock blocks without fractures which are called rock or rock materials. The structural types of rock body can be divided into four categories: whole block structure (Type I), stratified structure (Type II), cataclastic structure (Type III) and granular rock structure (Type IV). Except for the granular rock structure distributed in fault fracture zones and other places, even the Type III level cataclastic rock body which has complex lithologies, the upper limit of structural surface space is 50 cm. For our purpose, complete rock block (rock) can be regarded as a isotropic continuum with sparse or no structural surface [31]. In contrast, the spacing of coal's structural surface (cleat) is much smaller. Under a microscope the average density of coal cleat can reach several to hundreds of cleats/5 cm, while the scale of coal matrix is typically measured in millimeters. Therefore the cleats of coal samples (50 mm in diameter and 100 mm in height) used in the laboratory cannot be excluded, hence coal is a discontinuous body consisting of fractures (cleats) and matrix. Rocks are generally much stronger than coal with larger compressive strength, tensile strength and elastic modulus as well as smaller Poisson's ratio (Table 1.2).

1.3.2 Comparison of Adsorption Properties

1. Pore and adsorption capacity of coal

Using the combined method of mercury intrusion porosimetry and N_2 (77 K) adsorption the pore volumes and pore size distributions of long flame coal ($R_o = 0.65\%$), gas coal ($R_o = 0.98\%$), fat coal ($R_o = 1.12\%$), cooking coal ($R_o = 1.67\%$), lean coal ($R_o = 1.86\%$) and anthracite ($R_o = 2.94\%$) were determined. Results indicate that pore volumes range from 0.011 to 0.038 mL/g and the specific surface area ranges from 0.44 to 13.44 m^2/g. Because the gas is adsorbed on the coal's pore surface as monolayer adsorption, the specific surface area of coal can well reflect the adsorption ability of gas molecules on coal's pore surface. The specific surface area of coal is mainly concentrated in the micropores (pore size $\leq$ 10 nm) and minipores (pore size: 10-100 nm) where the specific surface area ratio reaches 80.41–96.57% [20].

The gas adsorption process on coal pore surface can be expressed by adsorption isotherms, that conforms to the Langmuir equation, of which the parameters describing the adsorption capacity are the Langmuir volume (also called ultimate gas adsorption volume, m^3/t) and the Langmuir pressure (MPa). The Langmuir volumes of the above-mentioned six coal samples range from 14.6 to 75.56 m^3/t, the Langmuir pressures range from 0.51 to 6.6 MPa. In fact, when using the indirect method to determine the gas content the measurements should include the adsorbed gas content and the free gas content. Generally, the percentage of adsorbed gas content is above 90%. Under the current depth of cover in Chinese coal mines, the measured maximum gas pressure of the coal seam is 8–10 MPa and the maximum gas content reaches 20–30 m^3/t [4].

2. Adsorption expansion of coal and its weakening effect on strength

After gas adsorption, the volume of coal will expand. The relationship between coal volume expansion and gas pressure is similar to the function relationship of Langmuir form. The maximum volumetric strain of the intact coal samples from Daning Coal Mine, Shanxi Province is 0.001, while the maximum volumetric strain of the tectonic coal samples from the coal mine is 0.03 [11]. The maximum volumetric strain of coal samples cited from literature [32, 33–35] is 0.016–0.058. The volume expansion of coal after gas adsorption not only weakens the strength of coal, but can also reduces the permeability of coal.

The influence of gas on coal strength mainly consists of two aspects: on one hand, it is the mechanical effect induced by free gas; on the other hand, it is the non-mechanical effect induced by adsorbed gas. The mechanical effect of free gas can be regarded as internal force according to the principle of effective stress; after the gas adsorption, the volume of coal expands, which will reduce the cohesion within the coal matrix leading to decrease of coal strength; meanwhile, the gas adsorbed on the surface of fractures between coal matrix will form adsorption membrane that has the effect of lubrication, hence reducing the coal surface friction leading to coal strength reduction.

Using coals from Daning Coal Mine, Shanxi Province, triaxial compressive strength tests of the intact and tectonic coal samples containing gas and without gas were conducted in the laboratory. The results indicated that under 3 MPa confining pressure and 1 MPa pressure of adsorbed gas, the intact coal sample strength reduced by 8.44 MPa or 21.32% when compared to the strength of sample with no gas adsorption, of which the free gas and the adsorbed gas contributed 14.32% and 7.0% respectively to the strength reduction. The same test on the tectonic coal samples with adsorbed gas indicated strength drop of 2.76 MPa or 20.9% when compared to the strength of sample with no gas adsorption, of which the free gas contributed 20.68% to the strength weakening while the contribution of adsorbed gas was 0.22% [11].

3. Adsorption capacity and gas storage property of rock

Generally, rock has no gas adsorption capacity except for the shale gas reservoirs, the natural gas reserves in the sandstone stratum with high porosity and large amount of gas are stored in the limestone of the coal measure strata in some mining areas of China.

The Marine shale gas reservoirs are generally dominated by nanometer pores, with pore size concentrated in 20-80 nm of certain adsorption capacities. The identified average gas content of the southern Paleozoic shale in China is 2.1–2.4 m^3/t. The average gas content of American shale is 4–8 m^3/t [36].

For the Bashijiqike sandstone formation in China (upper portion of Lower Cretaceous), in the shallow tectonic belt (burial depth < 5500 m), the porosity of the sandstone reservoir is greater than 12% and the permeability is greater than 20.93mD, which is a conventional gas pool; however, in the deep tectonic belt (burial depth: 5500–8000 m), the porosity of the sandstone reservoir is 3.5%-9.0% and the permeability is 0.1–3.86 mD, which is a tight gas pool [36].

The K_3 and K_4 limestone in the coal measure strata of Yangquan Coal Basin, China contain a lot of gas, which becomes a major source of gas emission of adjacent layers of the coal mine. According to the field measurements results, the maximum gas pressure in the limestone stratum is 0.5 MPa, and the maximum value of borehole gas emission is 21 m^3/min [37].

1.3.3 Comparison of Gas Outburst and Rock Burst Disasters

In deep Chinese mines gas outburst and rock burst disasters are sometimes break out. During the coal mining process, coal and gas outburst and rock burst are two common dynamic disasters. In the coal measure strata coal and rock are in the same stress environment, due to the lower strength and larger Poisson's ratio coal can produce greater deformation and store more strain energy. If the deformed coal exists in the coal seam, due to its much lower strength and stiffness, it can cause areas of uneven stress distributions and partial coal failure areas in front and adjacent to the working face in the process of excavation. This action can destabilize coal, causing rib spalling, extrusions or bumps, and can potentially induce coal and gas outbursts in high gas pressure areas [4]. There are also some rock and gas outbrusts during coal mining in China.

Rock burst is a sudden dynamic failure of the surrounding rock that can occur during excavation underground. It is an instability phenomenon of the rock fracture process with the sudden release of strain energy stored in the rock mass. Small rock bursts only peel off rock slices, without the rock ejection; however, serious rock bursts can cause earthquakes of up to 3.8 magnitude as well as loud noise. Serious rock bursts can destroy the underground production facilities, cause casualties and damage the ground building. Conditions in which rockbursts suddenly occur require high geostress and brittle rock with high elasticity. Under such conditions, once the original equilibrium state of rock mass is destroyed by underground engineering activities, the energy accumulating in the rock mass will be released and cause rock failures and ejections of broken rocks [38].

In most cases a typical rock failure is of a progressive nature forming small fractured zones. However, when mining coal seam with roof consisted massive hard strata, rock burst can occur. Under certain conditions a large brittle rock mass may be broken with rapid release of stored potential energy, disintegration and ejection of rock fragments follows causing potential damage. This process can also initiate combined disasters of coal burst and gas outburst.

1.4 Research Contents of Coal Mechanics

1.4.1 Definition of Coal Mechanics

The internationally recognized definition of rock mechanics was first proposed by the American institute of rock mechanics in 1964, then in 1974, it was amended to: rock mechanics is a theoretical and applied science of the mechanical behavior of rock and rock masses; it is that branch of mechanics concerned with the response of rock and rock masses to the force fields of their physical environment.

Rock mechanics has evolved for over one hundred years since the end of the nineteenth century. Its development has experienced different stages including initial development, empirical theory, classical theory and the modern development stage. Based on the introduction of elastic mechanics, plastic mechanics, fracture mechanics, non-continuum mechanics, fluid dynamics, modern mathematical science and numerical calculation methods, and combining with the rock engineering practice, rock mechanics has developed into a complete and independent subject. Aimed at the rock engineering field with different application backgrounds, rock mechanics has developed various science branches that serves petroleum engineering, underground mining engineering, hydropower engineering and other special fields [39–41].

The research target of rock mechanics for underground mining is to ensure the overall stability of the excavation space, provide reasonable support for all kinds of roadways and protect the recoverability of the unmined resources. The needs of expanding the scale and increasing the productivity of underground mining, meeting the effective mining structure design and exploiting the mineral resources in adverse environments, provide great power to the development of rock mechanics for underground mining. In recent years, world has been paying more attention to the protection of mineral resources and production safety, which exhibits the improving of the recovery rate of mineral resources, in-depth study of disaster developing mechanism in the process of mining, development of construction method and engineering technology to protect the safety of workplaces, and protection of workers' safety [40].

Coal mechanics is a special research field of the rock mechanics for underground mining. It takes the coal containing gas as the research object, studying the effect of pore structure of coal on gas adsorption and desorption properties, the damage mechanism of coal containing gas and its interaction with the permeability of coal seams, the flow characteristics of coal seam gas and so on. And finally, the above theoretical results are applied to the exploitation and safe production of coal and coal seam gas resources.

Coal mechanics is defined as a theoretical and applied science of the mechanical behavior of coal containing gas under the stress field environment of underground mining from the perspective of control of coal mine gas disaster and gas resource utilization. It provides the theoretical basis for the exploitation and safe production of coal as well as coal seam gas resources.

Similar to the study of rock mechanics, coal mechanics adopts the combined research methods consisting of laboratory experiment, field observation and numerical calculation. However, in the study of coal mechanics, it pays more attention to the dual porosity media properties of coal, the change of the effective stress caused by pore pressure and the non-mechanical action caused by gas adsorption. Based on the numerical calculations and other available methods of rock mechanics, effective engineering methods can be devised to prepare for gas disasters and gas resource exploitation.

1.4.2 Development of Coal Mechanics

1. Mechanical properties of coal

Since the 1950s, scholars such as Evans and Pomeroy [42], Murrell [43], Hobbs [44, 45] have carried out a large number of uniaxial and triaxial compression tests of coal samples. Considering the scale effect of the tested samples Bieieniawski [46, 47], Medhurst and Brown [48], Van der Merwe [49] et al., carried out mechanical tests of large scale coal samples both in the field and in the laboratory. For the requirements of pillar design, Mark and Barton et al. [50] conducted over 4000 groups of uniaxial compression tests for different coal samples, and obtained the uniaxial compressive strength of over 60 coal seams. Medhurst and Brown et al. [51] carried out triaxial compressive strength tests of different scale coal samples under different confining pressure conditions. Chinese scholar Minggao Qian [52] introduced the properties and mechanical effects of coal and gas outburst from the perspective of the mechanics process. However, the early researches who tested the mechanical properties of coal were limited by the practical factors such as the service objectives and test methods, which mainly focused on obtaining the mechanical parameters of the coal.

The continuous enrichment of the testing methods and the development of relevant theories paved the foundation for further study of the gas effects (adsorbing gas) [53–65], mechanical path [66–73], confining pressure [74, 75], temperature [76–78], anisotropy [79, 80], loading/ unloading rate [65, 81] and moisture content [59, 82, 83] on the mechanical properties of coal. For example, publishing of the suggested method for laboratory test of rock mechanics by the International Society for Rock Mechanics provides a basis for the unification of mechanical test conditions of coal [84]. Gas adsorbtion and the mechanical path are the two most important influencing factors of coal behavior. The relevant research results include:

(1) The influence of gas and other adsorbing gases on the mechanical properties of coal

By carrying out the triaxial compression tests of coal under different confining pressure and different pore gas pressure, Ates and Barron [65], Yuping Yao and Shining Zhou [61], Jiang Xu and Xuefu Xian [60], Bing Liang [85], have concluded that the deformation and destruction of coal containing gas are affected by both free gas and adsorbed gas. These conditions not only change the mechanical parameters of the coal, but also affect the constitutive relation of the coal. Larsen and Ranjith and their research team [53–59] conducted a large number of experimental studies investigating the effect of CO_2 on the mechanical properties of coal. Their results indicate that after the adsorption of CO_2, both the strength and elastic modulus of the coal decrease greatly, and the reduction degree is closely related to the pressure and state of the CO_2.

(2) The influence of mechanical path on the mechanical properties of coal

Considering the differences between mechanical paths of loading and unloading, Chen et al. [86] studied the spatial and temporal evolution of triaxial stress in the protected coal seam during the protective seam mining using numerical simulation. He transformed the experimental regularity into the mechanical path of laboratory test to further study the damage and permeability development in unloading coal during the process of protective seam mining. Guangzhi Yin et al. [68] carried out experimental tests of the mechanical properties of coal containing gas under different loading/unloading conditions. His work indicated that the mechanical characteristics of coal containing gas in different loading/unloading conditions varied greatly.

In recent years, with the wide application of the techniques like the industrial CT, mechanic-seepage coupling characteristic test system and mechanical simulation, research focusing on the mechanical properties of coal has gradually entered the stage of the microscopic, multi-factor coupling and numerical simulation. The mesoscopic evolution mechanism of coal fractures [87–90], the co-evolution mechanism of mechanical and permeability of coal [69, 91, 92] and the ability to numerically simulate mechanical behavior [93, 94] have attracted more and more attention of the scholars, which are expected to provide powerful support to this study of mechanical behavior of coal subjected to deep ground geostress conditions.

2. Seepage properties of coal

The permeability models of coal can be mainly divided into two categories as the model of elastic segment and model of plastic segment. Referring to the differences in the simplified physical model of the coal, the elastic segment model can further be divided into single porosity media model and dual porosity media model.

(1) Single porosity media permeability model

At the beginning of the permeability study, the empirical formula and the elastic theory of single porosity media were first used to establish the permeability model. Based on radial gas flow and Darcy's law, Zhou and Sun [95] proposed a model for the calculation of gas permeability coefficient, which describes the gas flow pattern in the coal seam and has been widely used. Lin and Zhou [96], Yaoqing Hu and Yangsheng Zhao [97], Sun and Ling [98], Li et al. [99, 100] established empirical formulas of permeability according to the evolution regularities of permeability obtained from laboratory tests. Ping Lu et al. [101] deduced the permeability model that considers the body strain and initial porosity based on the basic definition of porosity. Xiangchun Li et al. [102], Wu [103] constructed the permeability model that considers the coal matrix deformation caused by adsorption. Yunqi Tao et al. [104] took the coal's adsorption deformation and thermal expansion effect of temperature into consideration to establish the permeability evolution model of coal before the dilatancy stage of coal mass. Zhou [105] constructed the permeability model considering the effective stress and matrix shrinkage. Zhang et al. [106] established the permeability model of coal based on the theory of pore elasticity, which was further developed and perfected by Liu et al. [107].

(2) Dual porosity media permeability model

Based on the simplified model of coal such as the matchstick model and cubic geometry model, many scholars have established various permeability models considering different boundary conditions. Based on the matchstick model, Ma et al. [108] presented a simple permeability model under the assumption of fixed volume. Gray [109] established the theoretical model of permeability considering the influence of adsorption deformation. On the above assumption, Palmer and Mansoori [110], Shi and Durucan [111], Cui and Bustin [112] established their own permeability models used widely to predict the actual permeability of the site. Under the triaxial strain assumption, Warren and Root [29] established the permeability model based on the cubic geometry model, Robertson and Christiansen [113], Connell et al. [91], Liu et al. [114], Guo et al. [115], Wang et al. [116] perfected this type of model. Yao et al. [117], Gu and Chalaturnyk [118], Pan and Connell [119] et al. established the permeability model that considered the anisotropy of coal body. Pan and Connell [120] conducted a detailed summary analysis of the permeability models in the form of a review.

(3) Permeability model of plastic segment

Yang et al. [121], Xu et al. [122], Lianchong et al. [123], Zhu et al. [124] introduced the damage factor to quantitatively describe the permeability behavior of coal after damage, and established the permeability models of coal body under different damage mechanisms. Using the rock mechanics of damage and fracture, Xie et al. [125] considered the comprehensive effects of bearing stress, gas pressure and adsorption deformation and built the calculation formula of the permeability for four different types of coal. Yin et al. [126] constructed the permeability evolution model of coal samples under loading/unloading conditions from the perspective of effective stress. An et al. [127] established the permeability model of coal-body's plastic damage based on the equivalent plastic strain. Yuanping Cheng et al. [128] established the permeability evolution model for coal's unloading damage which considered the effects of the effective stress and coal's adsorption deformation. Lei Meng [129] established a statistical damage permeability model of the coal containing gas. Xingguang Liu [130] proposed a permeability evolution model of deformation and destruction for coal containing gas based on the theory of flat-bed fracture seepage.

3. Gas flow in coal seam

(1) Flow model of single porosity media

By assuming the coal as a uniformly continuous single porosity media, Zhou and Sun [95] established the linear flow theory of gas in coal seams without considering the adsorption expansion of coal and the effect of geostress. On this basis, Tan and Yuan [131] replaced the ideal gas equation with the real gas state equation and established the real gas seepage equation of coal seam, which considers the influence of gas and porosity on gas seepage. Peide Sun [132] further revised and improved the mathematical model of gas flow in coal seam. Guangyang Zhang and Xueshu Tan [133] studied the influence of geostress on coal seam gas migration, and built a gas seepage equation of coal seams with geostress applying the basic principle of seepage theory. Yangsheng Zhao et al. [134, 135] considered the influence of coal body deformation, effective stress and pore gas pressure on the permeability coefficient of coal seam, and established the gas–solid coupling mathematical model of gas flow in coal seam. Zhu et al. [136] established the gas–solid coupling model of coal and gas considering the effective stress and adsorption expansion effect of coal.

(2) Flow model of dual porosity media

Recent work on the dual porosity media model of cleat/matrix pores has been widely recognized and used by scholars. The physical significance of the dual porosity media model is: the pores in which the gas migration regularity obeys the Darcy's seepage law can be treated as the cleat pores, which is also considered as the boundaries of the matrix segmentation; while the pores in which the gas migration regularity obeys the Fick's diffusion law can be treated as the matrix pores, which is considered to exist in the coal matrix [137, 29, 138].

Based on coal's dual porosity media model, Warren and Root [29] illustrated the characteristics of the dual porosity media model, and established the control equation for the series process of gas diffusion and seepage. Shining Zhou [138] established a gas flow equation without considering the adsorption expansion of coal and geostress. Valliappan and Zhang [139] established a mathematical model of coupled gas diffusion, seepage and deformation considering the effective stress. Gilman and Beckie [140] established the gas–solid coupling model considering the effective stress of coal and the effect of gas adsorption expansion. Based on coal's dual porosity media model, coalbed methane reservoir simulation softwares like COMET, COALGAS, GRUSSP and so on have been widely applied in the exploration and development of coalbed methane [141].

In the aspect of nonlinear gas flow, Klinkenberg [142] studied the effect of molecular slip flow on the gas flow properties in low permeability coal/rock mass, which is known as the Klinkenberg effect. Afterwards, a number of scholars introduced the Klinkenberg effect [136, 143–145] in their gas flow mathematical models.

1.4.3 Structure of This Book

The principle of this book is not simply cite scientific research reports and references, but from an angle of helping learning, emphatically explain the basic concept and basic principle (geology, rock mechanics, fluid mechanics, surface chemistry and other basic theory). On this basis, the content of this book is properly extended and combined with the engineering practice (engineering and technology of mining and coal mine gas control) to be more systematic and scientific.

In this book, Chaps. 2 to 4 mainly introduce the generation and metamorphism of coal, the basic physical properties of coal, the characteristics of pore/cleat, the gas adsorption and desorption properties of coal. Besides the introduction of coal forming effect, microstructure, metamorphic type and gas generation in coal, Chapter 2 also introduces the influence of magma intrusion on coal metamorphism and its control effect on coal seam gas occurrence. In addition to the explanation of basic physical properties of coal, the pore and fracture characteristics of coal, Chap. 3 also introduces the coal matrix and its scale characteristics. Chap. 4 mainly introduces the gas adsorption/desorption properties of coal, the gas diffusion characteristics of coal particles as well as gas diffusion model and its application.

Chapters 5 to 7 of this book mainly introduce the coal mechanics basis, the strength and deformation of coal containing gas, the permeability characteristics of coal and the evolution model of permeability, of which, Chap. 5 introduces the basic concept of stress and strain, the common strength criterions of rock and the effective stress law of porous media. In addition to introduction of the mechanical property tests of coal containing gas, the strength and deformation characteristics of coal containing gas, the constitutive equation of coal containing gas and the macroscopic failure characteristics of coal containing gas. Chapter 6 introduces the progressive failure mechanism of damaged and dilated coal. Chapter 7 introduces the basic concept of permeability and the theoretical basis and evaluates classical permeability evolution models, then further introduces the test instruments, the principle and the method of coal's permeability characteristic, and the influence of effective stress/ adsorption deformation on coal permeability, the influence of stress disturbance on coal permeability and permeability evolution model of coal in gas migration process within the stress disturbance zone.

Chapters 8 to 11 of this book mainly introduce the relationship between geostress and coal seam gas occurrence, the gas flow theory of coal seam and the applications of coal mechanics in the pressure-relief gas extraction and coal & gas outburst control, of which, Chap. 8 mainly introduces the basic concepts and measurement methods of geostress, the evolution law of geostress with buried depth and the changing regularity of coal measure strata tectonic stress and gas parameters with buried depth. In addition to the introduction of the basic rules of coal seam gas migration, gas diffusion theory and seepage theory, Chapter 9 builds the gas–solid coupling model for coal and gas, and analyzes the evolution regularities of coal sample pore pressure and permeability in adsorption equilibrium process as well as the gas flow field characteristics during the extraction process. Chapter 10 mainly introduces the basic concept of pressure-relief gas extraction, space–time evolution characteristics of stress and strain in pressure relief coal seam, the space and time evolution of permeability in the protected seam, the pressure-relief gas extraction and flow field characteristics of the protected seam, the extraction method for pressure-relief gas and its application. In additional to the introduction of general rule and representative hypothesis of coal & gas outburst, Chap. 11 further introduces the effects of gas, geostress and physical mechanics of coal on outburst tendency, the development stage of outburst and the effect of rapid desorbed gas of pulverized coal on the transport of outburst coal.

References

1. Shao, Z. (1993). *Coal geology*. China Coal Industry Publishing House. (in Chinese).
2. Lu, S., Cheng, Y., Li, W., et al. (2015). Pore structure and its impact on CH_4 adsorption capability and diffusion characteristics of normal and deformed coals from Qinshui Basin. *International Journal of Oil Gas & Coal Technology, 10*(1), 76–78.
3. State Administration of Work Safety. (2006). *AQ1024–2006 Specification for identification of coal and gas outburst mine*. (in Chinese).
4. Qixiang, Yu., & Yuanping, C. (2012). *Coal mine gas control*. China University of Mining and Technology Press. (in Chinese).
5. Xianbo, Su., & Xiaoying, L. (2009). *Coalbed Gas Geology [M]*. China Coal Industry Publishing House. (in Chinese).
6. State Administration of Coal Mine Safety. (2008). *Summary report on 2007 coal mine gas grade accreditation*. (in Chinese).
7. State Administration of Coal Mine Safety. (2009). *Summary report on 2008 coal mine gas grade accreditation*. (in Chinese).
8. Wang, L., Cheng, Y., An, F., et al. (2014). Characteristics of gas disaster in the Huaibei coalfield and its control and development technologies. *Natural Hazards, 71*(1), 85–107.
9. Wang, L., Cheng, L., Cheng, Y., et al. (2014). Thermal effects of magmatic sills on coal seam metamorphism and gas occurrence. *Bulletin of Volcanology, 76*(4), 1–16.
10. Chuanxin, R., & Donglin, W. (2014). *Rock mechanics*. Wuhan University Press. (in Chinese).
11. Shouqing, Lu. (2016). *Mechanical failure and permeability evolution mechanism of coal based on equivalent matrix scale and its application*. China University of Mining and Technology. (in Chinese).
12. Qiao, J. (2009). *Research on the low rank coal of pore characteristics and desorption laws*. Xi'an University of Science and Technology. (in Chinese).
13. China University of Mining and Technology. (2016). *Effect of pore features of anthracite in Bailongshan coal mine on gas desorption characteristics*. China University of Mining and Technology. (in Chinese).
14. Seidle, J. (2011). *Fundamentals of coalbed methane reservoir engineering*. PennWell Books.
15. Li, J., Liu, D., Yao, Y., et al. (2011). Evaluation of the reservoir permeability of anthracite coals by geophysical logging data. *International Journal of Coal Geology, 87*(2), 121–127.
16. Chatterjee, R., & Paul, S. (2013). Classification of coal seams for coal bed methane exploitation in central part of Jharia coalfield, India—A statistical approach. *Fuel, 111*, 20–29.
17. Zhou, F., & Yao, G. (2014). Sensitivity analysis in permeability estimation using logging and injection-falloff test data for an anthracite coalbed methane reservoir in South-east Qinshui Basin, China. *International Journal of Coal Geology, 131*, 41–51.
18. Keshavarz, A., Yang, Y., Badalyan, A., et al. (2014). Laboratory-based mathematical modelling of graded proppant injection in CBM reservoirs. *International Journal of Coal Geology, 136*, 1–16.
19. Hodot, B. B., & Song, S. (Trans). (1966). *Outburst of coal and coalbed gas*. China Industry Press. (in Chinese).
20. Fei, W. (2016). *Kinetics characteristic of methane adsorption/desorption in coal and its application in the rapid determination of methane parameters*. China University of Mining and Technology. (in Chinese).
21. National Development and Reform Commission. (2006). MT/T968-2005 Description methodology fracture in coal. (in Chinese).
22. Laubach, S. E., Marrett, R. A., Olson, J. E., et al. (1998). Characteristics and origins of coal cleat: A review. *International Journal of Coal Geology, 35*(1), 175–207.
23. General Administration of Quality Supervision, Inspection and Quarantine of the People's Republic of China, Standardization Administration of the People's Republic of China. (2008). *GB/T 16773-2008 Method of preparing coal samples for the coal petrographic analysis*. (in Chinese).

24. Solano-Acosta, W., Mastalerz, M., & Schimmelmann, A. (2007). Cleats and their relation to geologic lineaments and coalbed methane potential in Pennsylvanian coals in Indiana. *International Journal of Coal Geology, 72*(3), 187–208.
25. Bi, J., Su, X., Han, D., et al. (2001). The relation between cleat frequency and coal rank. *Journal of China Coal Society, 26*(4), 346–349. (in Chinese).
26. Liu, H., Li, G., Wang, F., et al. (2008). Characteristics and genetic mechanisms of coal cleat systems in QinShui Basin. *Natural Gas Industry, 28*(3), 36–39. (in Chinese).
27. Liu, H., Wang, H., Zhang, J., et al. (2000). Evaluation method of cleats in coal reservoir bed. *Natural Gas Industry, 20*(4), 27–29. (in Chinese).
28. Dawson, G., & Esterle, J. S. (2010). Controls on coal cleat spacing. *International Journal of Coal Geology, 82*(3), 213–218.
29. Warren, J. E., Root, P. J. (1963). The behavior of naturally fractured reservoirs. *Society of Petroleum Engineers Journal, 3*(03), 245-255.
30. Wang, J., & Park, H. D. (2002). Fluid permeability of sedimentary rocks in a complete stress-strain process. *Engineering geology, 63*(3), 291–300.
31. Xu, Z. (2007). *Rock Mechanics* (3rd ed.). China Water & Power Press. (in Chinese).
32. Zhao, X. (1991). *An experimental study of methane diffusion in coal using transient approach.* The University of Arizona.
33. Liu, J., Chen, Z., Elsworth, D., et al. (2010). Evaluation of stress-controlled coal swelling processes. *International Journal of Coal Geology, 83*(4), 446–455.
34. Liu, H., & Rutqvist, J. (2009). A new coal-permeability model: Internal swelling stress and fracture-matrix interaction. *Transport in Porous Media, 82*(1), 157–171.
35. Harpalani, S., & Schraufnagel, R. A. (1990). Shrinkage of coal matrix with release of gas and its impact on permeability of coal. *Fuel, 69*(5), 551–556.
36. Xie, K., Qiu, Z., Jin, Q., et al. (2014). *Strategic research on unconventional natural gas development & utilization in China.* Science Press. (in Chinese).
37. Bao, J., Su, S., & Li, G. (1996). *Gas control technology in Yangquan coal mine.* China Coal Industry Publishing House. (in Chinese).
38. Qi, Q., & Dou, L. (2008). *Theory and technology of rock burst.* China University of Mining and Technology press. (in Chinese).
39. Fu, J. (2012). *Petroleum engineering rock mechanics.* Petroleum industry press. (in Chinese).
40. Brady, B. H., Brown, E. T. (2013). *Rock mechanics: For underground mining.* Springer Science & Business Media.
41. Dong, X. (2004). *Rock mechanics in hydraulic engineering.* Beijing: China Water & Power Press. (in Chinese).
42. Evans, I., & Pomeroy, C. D. (1958). The strength of cubes of coal in uniaxial compression. *Mechanical Properties of Non-Metallic and Brittle Materials*, 5–28.
43. Murrell, S. (1958). The strength of coal under triaxial compression. *Mechanical Properties Of Non-metallic Brittle Materials*, 123–153.
44. Hobbs, D. W. (1960). The strength and stress-strain characteristics of Oakdale coal under triaxial compression. *Geological Magazine, 97*(05), 422–435.
45. Hobbs, D. W. (1964). The strength and the stress-strain characteristics of coal in triaxial compression. *The Journal of Geology*, 214–231.
46. Bieniawski, Z. T. (1968). The effect of specimen size on compressive strength of coal. *International Journal of Rock Mechanics & Mining Science & Geomechanics Abstracts, 5*(4), 325–335.
47. Bieniawski, Z. T. (1968). In situ strength and deformation characteristics of coal. *Engineering Geology, 2*(5), 325–340.
48. Medhurst, T. P., & Brown, E. T. (1996) Large scale laboratory testing of coal. In *7th Australia New Zealand Conference on Geomechanics: Geomechanics in a Changing World: Conference Proceedings.* Institution of Engineers, Australia.
49. Van der Merwe, J. N. (2003). A laboratory investigation into the effect of specimen size on the strength of coal samples from different areas. *Journal-South African Institute of Mining and Metallurgy, 103*(5), 273–280.

50. Mark, C., & Barton, T. M. (1997). Pillar design and coal strength. *Proceedings New Technology for Ground Control in Retreat Mining*, 49–59.
51. Medhurst, T. P., & Brown, E. T. (1998). A study of the mechanical behaviour of coal for pillar design. *International Journal of Rock Mechanics and Mining Sciences, 35*(8), 1087–1105.
52. Qian, M. (1955). Introduction to the modern theory of the nature and mechanics of coal and gas outbursts. *Journal of China University of Mining & Technology, 3*, 92–100. (in Chinese).
53. Larsen, J. W. (2004). The effects of dissolved CO_2 on coal structure and properties. *International Journal of Coal Geology, 57*(1), 63–70.
54. Viete, D. R., & Ranjith, P. G. (2006). The effect of CO_2 on the geomechanical and permeability behaviour of brown coal: Implications for coal seam CO_2 sequestration. *International Journal of Coal Geology, 66*(3), 204–216.
55. Viete, D. R., & Ranjith, P. G. (2007). The mechanical behaviour of coal with respect to CO_2 sequestration in deep coal seams. *Fuel, 86*(17), 2667–2671.
56. Perera, M., & Ranjith, P. G. (2012). Carbon dioxide sequestration effects on coal's hydro-mechanical properties: A review. *International Journal of Energy Research, 36*(10), 1015–1031.
57. Ranjith, P. G., & Perera, M. (2012). Effects of cleat performance on strength reduction of coal in CO_2 sequestration. *Energy, 45*(1), 1069–1075.
58. Perera, M., Ranjith, P. G., & Viete, D. R. (2013). Effects of gaseous and super-critical carbon dioxide saturation on the mechanical properties of bituminous coal from the Southern Sydney Basin. *Applied energy, 110*, 73–81.
59. Poulsen, B. A., Shen, B., Williams, D. J., et al. (2014). Strength reduction on saturation of coal and coal measures rocks with implications for coal pillar strength. *International Journal of Rock Mechanics and Mining Sciences, 71*, 41–52.
60. Xu, J., Xian, X., Du, Y., et al. (19936). An Experimental Study on the Mechanical Property of the Gas-filled Coal. *Journal of Chongqing University (Natural Science Edition), 16*(5), 42–47. (in Chinese).
61. Yao, Y., & Zhou, S. (1988). The mechanical property of coal containing gas. *Journal of China Institute of Mining & Technology, 1*, 1–7. (in Chinese).
62. Yin, G., Wang, Z., & Zhang, D. (2010). Experiment of the gas effect on coal mechanical properties under zero effective confirming pressure. *Journal of Chongqing University (Natural Science Edition), 33*(11), 129–133. (in Chinese).
63. Yin, G., Wang, D., Zhang, D., et al. (2009). Test analysis of deformation characteristics and compressive strengths of two types of coal specimens containing gas. *Chinese Journal of Rock Mechanics and Engineering, 28*(2), 410–417. (in Chinese).
64. Maphala, T. (2012). *Effects of carbon dioxide storage in coal on the physical and chemical properties of coal*. Faculty of Engineering and the Built Environment, University of the Witwatersrand.
65. Ates, Y., & Barron, K. (1988). The effect of gas sorption on the strength of coal. *Mining Science and Technology, 6*(3), 291–300.
66. Su, C., Gao, B., Nan, H., et al. (2009). Experimental study on acoustic emission characteristics during deformation and failure processes of coal samples under different stress paths. *Chinese Journal of Rock Mechanics and Engineering, 28*(4), 757–766. (in Chinese).
67. Peng, R., Ju, Y., Gao, F., et al. (2014). Energy analysis on damage of coal under cyclical triaxial loading and unloading conditions. *Journal of China Coal Society, 39*(02), 245–252. (in Chinese).
68. Yin, G., Li, W., Li, M., et al. (2013). Experimental study of mechanical properties of coal containing methane under different loading-unloading conditions. *Chinese Journal of Rock Mechanics and Engineering, 32*(5), 891–901. (in Chinese).
69. Xu, J., Li, B., Zhou, T., et al. (2012). Experimental study of coal deformation and permeability characteristics under loading-unloading conditions. *Journal of China Coal Society, 37*(09), 1493–1498. (in Chinese).
70. Chen, H., Cheng, Y., Pu, Y., et al. (2013). Study on reconstituted coal damage features during unloading process. *Coal Science and Technology, 41*(03), 84–87. (in Chinese).

71. Zhao, H., & Wang, J. (2011). Experimental study of evolution law of mechanical properties of coal containing gas under unloading confining pressure. *Rock and Soil Mechanics, S1*, 270–274. (in Chinese).
72. Lu, Y., & Qin, H. (2012). Investigation into mechanical responses and energy dissipation properties of coal containing methane to confinement unloading. *Journal of China Coal Society, 37*(09), 1505–1510. (in Chinese).
73. Cao, S., Liu, Y., Li, Y., et al. (2009). Experimental study on acoustic emission characteristics of coal rock at different confining pressure. *Journal of Chongqing University (Natural Science Edition), 32*(11), 1321–1327. (in Chinese).
74. Zhao, H., Yin, G., & Zhang, Z. (2009). Study of creep characteristics and constitutive relation of briquette under confining pressure. *Rock and Soil Mechanics, 30*(8), 2305–2308. (in Chinese).
75. Liu, Q., Liu, K., Zhu, J., et al. (2014). Study of mechanical properties of raw coal under high stress with triaxial compression. *Chinese Journal of Rock Mechanics and Engineering, 33*(1), 24–34. (in Chinese).
76. Ma, Z., Mao, X., Li, Y., et al. (2005). Experimental study on the effect of temperature on coal mechanical properties. *Ground Pressure and Strata Control, 22*(3), 46–48. (in Chinese).
77. Xu, J., Zhang, D., Peng, S., et al. (2011) Experimental research on influence of temperature on mechanical properties of coal containing methane. *Chinese Journal of Rock Mechanics and Engineering, 30*(S1), 2730–2735. (in Chinese).
78. Guo, W., & Li, X. (2007). Research status of deep coal and rock mass mechanic properties under high temperature and high confining pressure. *Journal of Henan Polytechnic University (Natural Science), 26*(1), 16–20 (in Chinese).
79. Pan, J., Meng, Z., Hou, Q., et al. (2013). Coal strength and Young's modulus related to coal rank, compressional velocity and maceral composition. *Journal of Structural Geology, 54*, 129–135.
80. Li, S., Tang, D., Pan, Z., et al. (2014). Influence and control of coal facies on physical properties of the coal reservoirs in Western Guizhou and Eastern Yunnan, China. *International Journal of Oil, Gas and Coal Technology, 8*(2), 221–234.
81. Yin, G., Jiang, C., Wang, W., et al. (2011). Experimental study of influence of confining pressure unloading speed on mechanical properties and gas permeability of containing-gas coal rock. *Chinese Journal of Rock Mechanics and Engineering, 30*(1), 68–77. (in Chinese).
82. Van Eeckhout, E. M. (1976). The mechanisms of strength reduction due to moisture in coal mine shale. *International Journal of Rock Mechanics and Mining Sciences & Geomechanics Abstracts.*
83. Perera, M., Ranjith, P. G., & Peter, M. (2011). Effects of saturation medium and pressure on strength parameters of Latrobe Valley brown coal: Carbon dioxide, water and nitrogen saturations. *Energy, 36*(12), 6941–6947.
84. Brown, E. T. (1981). *Rock characterization, testing & monitoring*. International Society for Rock Mechanics by Pergamon Press: Published for the Commission on Testing Methods.
85. Liang, B., Zhang, M., Pan, Y., et al. (1995). The experimental research on the effect of gas on mechanical properties and mechanical response of coal. *Chinese Journal of Geotechnical Engineering, 17*(5), 12–18. (in Chinese).
86. Chen, H., Cheng, Y., Zhou, H., et al. (2013). Damage and permeability development in coal during unloading. *Rock Mechanics and Rock Engineering, 46*(6), 1377–1390.
87. Mao, L., An, L., Wang, Z., et al. (2010). Experimental study on relationship between coal mechanical characteristics and interior crack evolution using CT technique. *Journal of Liaoning Technical University (Natural Science), 29*(3), 408–411. (in Chinese).
88. Ge, X., Ren, J., Pu, Y., et al. (1999). A real-in-time Ct triaxial testing study of meso-damage evolution law of coal. *Chinese Journal of Rock Mechanics and Engineering, 18*(5), 497–502. (in Chinese).
89. Zhao, H., & Gong, W. (2009). Characterization on anisotropic fractures of coal and rocks by computed X-ray tomography based on image segmentation. *Coal Geology & Exploration, 37*(6), 14–18. (in Chinese).

90. Wolf, K. A., van Bergen, F., Ephraim, R., et al. (2008). Determination of the cleat angle distribution of the RECOPOL coal seams, using CT-scans and image analysis on drilling cuttings and coal blocks. *International Journal of Coal Geology, 73*(3), 259–272.
91. Connell, L. D., Lu, M., & Pan, Z. (2010). An analytical coal permeability model for tri-axial strain and stress conditions. *International Journal of Coal Geology, 84*(2), 103–114.
92. Huang, Q., Yin, G., & Jiang, Y. (2010). Experimental statistics. *Chinese Journal of Rock Mechanics and Engineering, 29*(8), 1639–1848. (in Chinese).
93. Gao, F., Stead, D., & Kang, H. (2014). Numerical investigation of the scale effect and anisotropy in the strength and deformability of coal. *International Journal of Coal Geology, 136*, 25–37.
94. Deisman, N., Ivars, D. M., Darcel, C., et al. (2010). Empirical and numerical approaches for geomechanical characterization of coal seam reservoirs. *International Journal of Coal Geology, 82*(3), 204–212.
95. Zhou, S., & Sun, J. (1965). Coal seam gas flow theory and its application. *Journal of China Coal Society, 1965*(1). (in Chinese).
96. Lin, B., & Zhou, S. (1987). Experimental investigation on the of the coal samples containing permeability methane. *Journal of China Institute of Mining & Technology, 16*(1), 21–28. (in Chinese).
97. Hu, Y., Zhao, Y., Wei, J., et al. (1996). Experimental study of permeating law of coal mass gas under action of 3-dimension stress. *Journal of Xi'an University of Science & Technology, 16*(4), 308–311. (in Chinese).
98. Sun, P., & Ling, Z. (2000). Experimental study of the law for permeability of coal under action of 3-triaxial compression. *Journal of Chongqing University (Natural Science Edition), 23*(S1), 28–31. (in Chinese).
99. Li, S., & Xu, J. (2001). Experimental study on permeability of soft coal sample with electro-hydraulic servo-controlled testing system. *Chinese Journal Geotechnical Engineering, 23*(1), 68–70. (in Chinese).
100. Li, S., Qian, M., & Shi, P. (2001). Permeability-strain equation relation to complete stress-strain path of coal sample. *Coal Geology & Exploration, 29*(1), 22–24. (in Chinese).
101. Lu, P., Shen, Z., Zhu, G., et al. (2002). Characterization of and experimental study on the permeability of rock-samples during complete stress-strain course. *Journal of University of Science and Technology of China, 32*(6), 678–684. (in Chinese).
102. Li, X., Guo, Y., & Wu, S. (2005). Analysis of the relation of porosity, permeability and swelling deformation of coal. *Journal of Taiyuan University of Technology, 36*(3), 264–266. (in Chinese).
103. Wu, S. (2006). *Research of methane-coalbed coupling movement theory and its application.* Northeastern University. (in Chinese).
104. Tao, Y., Xu, J., Peng, S., et al. (2010). Experimental study of of influencing factor of porosity and effective stress of gas-filled coal. *Rock and Soil Mechanics, 31*(11), 3417–3422. (in Chinese).
105. Zhou, J., Xian, X., Jiang, Y., et al. (2009). A permeability model considering the effective stress and coal matrix shrinking effect. *Journal of Southwest Petroleum University (Science & Technology Edition), 31*(1), 4–8. (in Chinese).
106. Zhang, H., Liu, J., & Elsworth, D. (2008). How sorption-induced matrix deformation affects gas flow in coal seams: A new FE model. *International Journal of Rock Mechanics & Mining Sciences, 45*(8), 1226–1236.
107. Liu, J., Chen, Z., Elsworth, D., et al. (2011). Evolution of coal permeability from stress-controlled to displacement-controlled swelling conditions. *Fuel, 90*(10), 2987–2997.
108. Ma, Q., Harpalani, S., & Liu, S. (2011). A simplified permeability model for coalbed methane reservoirs based on matchstick strain and constant volume theory. *International Journal of Coal Geology, 85*(1), 43–48.
109. Gray, I. (1987). Reservoir engineering in coal seams: Part 1-The physical process of gas storage and movement in coal seams. *SPE Reservoir Engineering, 2*(01), 28–34.

110. Palmer, I., & Mansoori, J. (1996). How permeability depends on stress and pore pressure in coalbeds: A new model. In *SPE Annual Technical Conference and Exhibition*. Society of Petroleum Engineers.
111. Shi, J. Q., Durucan, S. (2004). Drawdown induced changes in permeability of coalbeds: A new interpretation of the reservoir response to primary recovery. *Transport in Porous Media, 56*(1), 1–16.
112. Cui, X., & Bustin, R. M. (2005). Volumetric strain associated with methane desorption and its impact on coalbed gas production from deep coal seams. *Aapg Bulletin, 89*(9), 1181–1202.
113. Robertson, E. P., & Christiansen, R. L. (2008). A permeability model for coal and other fractured, sorptive-elastic media. *SPE Journal, 13*(13), 314–324.
114. Liu, J., Chen, Z., Elsworth, D., et al. (2011). Interactions of multiple processes during CBM extraction: A critical review. *International Journal of Coal Geology, 87*(3), 175–189.
115. Guo, P., Cheng, Y., Jin, K., et al. (2014). Impact of effective stress and matrix deformation on the coal fracture permeability. *Transport in Porous Media, 103*(1), 99–115.
116. Wang, K., Zang, J., Wang, G., et al. (2014). Anisotropic permeability evolution of coal with effective stress variation and gas sorption: Model development and analysis. *International Journal of Coal Geology, 130*(4), 53–65.
117. Yao, Y., Liu, D., Tang, D., et al. (2009). Fractal characterization of seepage-pores of coals from China: An investigation on permeability of coals. *Computers & Geosciences, 35*(6), 1159–1166.
118. Gu, F., & Chalaturnyk, R. (2010). Permeability and porosity models considering anisotropy and discontinuity of coalbeds and application in coupled simulation. *Journal of Petroleum Science & Engineering, 74*(3–4), 113–131.
119. Pan, Z., & Connell, L. D. (2011). Modelling of anisotropic coal swelling and its impact on permeability behaviour for primary and enhanced coalbed methane recovery. *International Journal of Coal Geology, 85*(3), 257–267.
120. Pan, Z., & Connell, L. D. (2012). Modelling permeability for coal reservoirs: A review of analytical models and testing data. *International Journal of Coal Geology, 92*, 1–44.
121. Yang, T., Xu, T., Liu, J., et al. (2005). Coupling model of stress-damage-flow and its application to the investigation of instantaneous seepage mechanism for gas during unloading in coal seam with depth. *Chinese Journal of Rock Mechanics and Engineering, 24*(16), 2900–2905. (in Chinese).
122. Xu, T., Tang, C., Song, L., et al. (2005). Numerical simulation of coupled gas flow in failure process of gassy coal-rock. *Chinese Journal of Rock Mechanics and Engineering, 24*(10), 1667–1673. (in Chinese).
123. Lianchong, L., Tianhong, Y., Zhengzhao, L., et al. (2011). Numerical investigation of groundwater outbursts near faults in underground coal mines. *International Journal of Coal Geology, 85*(3), 276–288.
124. Zhu, W. C., Wei, C. H., Liu, J., et al. (2013). Impact of gas adsorption induced coal matrix damage on the evolution of coal permeability. *Rock Mechanics and Rock Engineering, 46*(6), 1353–1366.
125. Xie, H., Gao, F., Zhou, H., et al. (2013). On theoretical and modeling approach to mining-enhanced permeability for simultaneous exploitation of coal and gas. *Journal of China Coal Society, 38*(7), 1101–1108. (in Chinese).
126. Yin, G., Li, W., Li, M., et al. (2014). permeability properties and effective stress of raw coal under loading-unloading conditions. *Journal of China Coal Society, 39*(8), 1497–1503. (in Chinese).
127. An, F., Cheng, Y., Wang, L., et al. (2013). A numerical model for outburst including the effect of adsorbed gas on coal deformation and mechanical properties. *Computers and Geotechnics, 54*, 222–231.
128. Cheng, Y., Liu, H., Guo, P., et al. (2014). A theoretical model and evolution characteristic of mining-enhanced permeability in deeper gassy coal seam. *Journal of China Coal Society, 39*(8), 1650–1658. (in Chinese).

129. Meng, L. (2013). *Research on coal damage characteristics and gas migration law of coal containing gas*. China University of Mining and Technology, Beijing. (in Chinese).
130. Liu, X. (2013). *Study on the deformation and failure characteristics and the permeation behavior of gas-saturated coal*. Xu Zhou: China University of Mining and Technology. (in Chinese).
131. Tan, X., & Yuan, J. (1986). Investigation of the seepage equations of the real gas in coal seam in mines. *Journal of Chongqing Institute of Architecture and Engineering, 8*(1), 25–28. (in Chinese).
132. Sun, P. (1989). The basic model of coal gas dynamics. *Journal of Xi'an University of Science & Technology, 9*(2), 7–13. (in Chinese).
133. Zhang, G., Tan, X., Xian, X, et al. (1994). Mathematical model of methane transport in coal seam. *Journal of Chongqing University (Natural Science Edition), 17*(4), 53–57. (in Chinese).
134. Zhao, Y., Qin, H., & Bai, Q. (1994). A mathematical model for solid-gas coupled problems on the methane flowing in coal seam. *Chineses Journal of Solid Mechanics, 15*(1), 49–57. (in Chinese).
135. Zhao, Y. (1994). Coupled mathematical model on coal mass-methane and its numerical method. *Chinese Journal of Rock Mechanics and Engineering, 13*(3), 220–239. (in Chinese).
136. Zhu, W. C., Liu, J., Sheng, J. C., et al. (2007). Analysis of coupled gas flow and deformation process with desorption and Klinkenberg effects in coal seams. *International Journal of Rock Mechanics and Mining Sciences, 44*(7), 971–980.
137. Reiss, L. H. (1980). *The reservoir engineering aspects of fractured formations*. Editions Technip.
138. Zhou, S. (1990). Mechanism of gas flow in coal seams. *Journal of China Coal Society, 15*(1), 15–24. (in Chinese).
139. Valliappan, S., & Zhang, W. (1996). Numerical modelling of methane gas migration in dry coal seams. *International Journal for Numerical and Analytical Methods in Geomechanics, 20*(8), 571–593.
140. Gilman, A., & Beckie, R. (2000). Flow of coal-bed methane to a gallery. *Transport in Porous Media, 41*(1), 1–16
141. Sun, K. (2004). *Fluid-solid coupling theory of exploiting coal methane an improving coal methane production by gas injection in low permeability reservoir and its application*. LiaoNing Technical University. (in Chinese).
142. Klinkenberg, L. J. (1941). The permeability of porous media to liquids and gases. *Socar Proceedings, 2*(2), 200–213.
143. Liu, Q., Cheng, Y., Zhou, H., et al. (2015). A mathematical model of coupled gas flow and coal deformation with gas diffusion and klinkenberg effects. *Rock Mechanics and Rock Engineering, 48*(3), 1163–1180.
144. Wu, Y., & Pruess, K. (1998). Gas flow in porous media with Klinkenberg effects. *Transport in Porous Media, 32*(1), 117–137.
145. Hu, G., Wang, H., Fan, X., et al. (2009). Mathematical model of coalbed gas flow with Klinkenberg effects in multi-physical fields and its analytic solution. *Transport in Porous Media, 76*(3), 407–420.

Chapter 2
Coal Formation and Metamorphism

Learning Outcomes

- Coal formation, including coalification effect, peatization effect, diagenesis effect and metamorphism effect;
- Metamorphism types and maceral composition of coal;
- Generation of coal seam gas during coal metamorphism, basic composition of coal seam gas, and the existence forms of adsorbed gas and free gas.
- Controlling effect of magma intrusion on the coal metamorphism and coal seam gas occurrence.

2.1 Coal Forming Process

Coal forming is the process of natural plant remains change from deposit to coal which includes the peatification (saprofication) and the coalification [1]. Coal produced from higher plants raw material is called humic coal. Coal made primarily from lower plants raw material as well as plankton is called sapropelic coal. Coal formed from a combination of higher and lower plants is called humic-sapropelic coal or sapropelic-humic coal depending on the proportion of these two plants. The coal forming process can be divided into five stages as shown in Fig. 2.1.

Coal transformation from the plant remains deposited in the swamp, is controlled by multiple factors such as the ancient plant, ancient climate, ancient geology and ancient structure. The required coal formation conditions are: large number of growing plants to provide material for the coal to form; warm and humid climate to give chance for plants to grow and peat swamps to develop; large coal basin swamps to provide space for the storage and peatification of plant remnants; occurrence of the crustal subsidence for the plant remains to be successfully deposited underground; and the subsidence speed of burial that affects the thickness of coal seams as well as the roof and floor characteristics of the seams.

Y. Cheng et al., *Coal Mechanics*,
https://doi.org/10.1007/978-981-16-3895-4_2

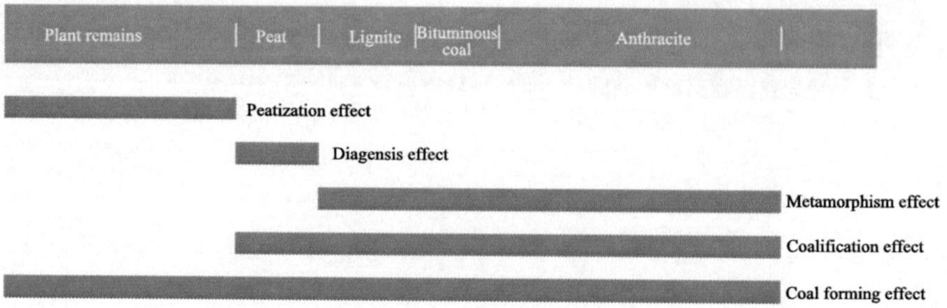

Fig. 2.1 Classification of coal forming process

For the entire geologic age, there are three periods which meet the above mentioned requirements for coal forming: ① Paleozoic Carboniferous and Permian (Coal forming plants are mainly spore bearing plants, main types of coal are bituminous coal and anthracite); ② Mesozoic Jurassic and Cretaceous (Coal forming plants are mainly gymnosperms, main types of coal are lignite and bituminous coal); ③ Cenozoic Tertiary (Coal forming plants are mainly angiosperms, main types of coal are lignite, then peat as well as some young bituminous coal (as listed in Table 2.1).

Table 2.1 Geological age, coal forming plants and main coal types [2]

Era (Erathem)	Period (System)	Before present /Ma	Main coal forming period around the world ●	Main coal forming period in China ▲	Coal forming plants	Coal types
Cenozoic	Quaternary	1.6			Angiosperms	Peat
	Neogene	23	●			Mainly lignite and some bituminous coal
	Paleogene	65				
Mesozoic	Cretaceous	135	●		Gymnosperms	Mainly lignite, bituminous coal and some anthracite
	Jurassic	205		▲		
	Triassic	250				
Paleozoic	Permian	290	●	▲	Ferns	Bituminous coal and anthracite
	Carboniferous	355		▲		
	Devonian	410			Psilophytes	

Around the world, there are three main coal forming periods: Paleozoic Carboniferous and Permian, Mesozoic Jurassic and Cretaceous and Cenozoic Tertiary. For China, the main coal forming periods are located in Carboniferous, Permian and Jurassic. The coal forming period of Carboniferous is mainly located in the late Carboniferous Epoch, forming the coal measure strata in the North China, East China and central-south China. During the early Permian Epoch, the coal measure strata of Shanxi formation was formed spanning large area of North China, while during the late Permian Epoch, the coal measure strata of Longtan formation was formed within the Guizhou Province. In the Jurassic, due to the countrywide spreading of Yanshan geological movement, the maximum number of coalfields were formed and mainly concentrated in North China and Northwestern China. Additionally, the massive volcanic activities induced by Yanshan movement also resulted in special features for some coalfields like Huaibei Coalfield, such as the complex structures, changeable coal types and difficult coal mining conditions.

2.2 Coalification

After the peatification process when higher plant remains accumulated in the swamps into peat, the further process which transform the peat into lignite, bituminous coal and anthracite, or transform the sapropelite into saprodite, sapanthracon and sapanthracite is called coalification. The coalification occurs in the second stage of the coal forming process, where inner mechanisms are mainly attributed to the physic-chemistry effects, instead of the biochemical effects. Coalification is not a process of evenly linear variation, but a process with jumping transition. The main factors that influence the coalification process are temperatures, pressures and time of duration. The coalification can be further divided into the diagenesis and metamorphism of coal.

2.2.1 Diagenesis of Coal

The process that makes peat or sapropelite turned into lignite or saprodite is called the diagenesis of coal. Diagenesis occurs at the initial stage of coalification. When the peat or sapropelite is covered by the upper deposits and buried into the ground due to the basin subsidence, the effects of compression, dehydration and solidification gradually converts the humic acid of peat/sapropelite to humus. Meanwhile, the carbon content of peat/sapropelite increases and the elements like oxygen, nitrogen, hydrogen decrease, the colloid ages and the peat/sapropelite color becomes darker. Finally, the peat or sapropelite is converted to the lignite or saprodite.

2.2.2 Metamorphism of Coal

The geochemical process that converts lignite to bituminous coal or anthracite under the effect of geo-temperature, geo-stress and time is called the metamorphism. This process includes two different stages, the chemical coalification stage and the physical coalification stage. In the chemical coalification stage, humic materials experience a further polymerization and lose a large number of oxygen-containing functional groups. The humic acid is further reduced which makes the acid humus to become neutral and more humus complex is produced. Whilst in the physical coalification stage, the diagenetic gelation is finalized and forms the gelatinous component, so that the plant remains disappear. The stable components experience asphaltization, making the epidermis wax of leaves and the outer layers of sporo-pollenin lose methoxy group, which forms the asphaltene with the properties of easy softening, strong plasticity and caking, as well as weak luster. Under the continuous influence of temperature and stress, the humus complexes continue to polymerize, which increase the fragrant ring system and reduce the side chain. With the increase of aromatization degree and the regularity of the molecular arrangement, the metamorphism is continuously improved, finally the lignite is converted to bituminous coal or even anthracite.

According to the metamorphism factors and modes of actions, the coal metamorphism can be divided into plutonic metamorphism, magmatic metamorphism and dynamic metamorphism. The plutonic metamorphism means that under the normal conditions of low temperature, the metamorphism of coal will increase with the increasing of coal seam subsidence, ground temperature and lasting time. The magmatic metamorphism includes the regional magma thermal metamorphism and contact metamorphism. Regional magma thermal metamorphism means the accelerated metamorphism effect caused by the raised ground temperature, which is attributed to the intruded magma and the gas/liquid it carries when igneous activities exist in the coal basin. The contact metamorphism means that when the coal seam is directly intruded by the magma, the high temperature gas, liquid and high pressure induced by the magma intrusion will promote the metamorphism of coal. The dynamic metamorphism means the metamorphism associated with the dynamic pressure and structural changes during the tectonic movement of folds and fractures.

Because the coal is much more sensitive to the changes of temperature and pressure when compared to the surrounding rock, when coal converts from lignite to bituminous coal or even anthracite, the surrounding rock usually will not suffer the effect of metamorphism. Therefore, in fact, the process from lignite to bituminous coal or even anthracite is roughly equivalent to the diagenesis of sedimentary rocks; while the effect that converting the anthracite to graphite or cokeite is equivalent to the metamorphism of sedimentary rocks. The effect that makes coal become graphite is called graphitization. Because the graphite does no longer belong to coal, the metamorphism of coal does not include the graphitization stage [1].

2.3 The Maceral Composition and Metamorphic Type of Coal

2.3.1 The Maceral Compositon of Coal

The Maceral composition of coal is the coal's basic constituent element that can be distinguished under the ordinary microscopy [3]. Classified by the optical characteristics of coal under microscopy, the smallest unit of distinguishable organic component is called maceral composition. Similarly, the mineral unit is called the mineral composition. The term of maceral composition was firstly proposed by M. Stopes in 1935. In China, based on the classification method of International Committee for Coal Petrology (ICCP), scholars divided the maceral composition of coal into the vitrinite, inertinite and liptinite according to the origins of coal as well as their different optical characteristics under microscopy, as listed in Table 2.2 (GB/T15588-2013). The mineral categories and components of mineral composition are abundant, generally include clay, carbonates and sulfides.

The vitrinite is the maceral component converted from the plants' lignocellulosic tissue through the effect of gelatinization in the reduction environment of overlying water. Vitrinite is the main component of coal, which usually has larger particle size and uniform surface. The reflectance of vitrinite increases rapidly with the degree of metamorphism, which can be easily determined. Thus, the reflectance of vitrinite is usually used as an ideal index to characterize the metamorphism degree of coal, higher vitrinite reflectance means higher metamorphism degree of coal. The vitrinite can be further divided into telinite, collinite and vitrodetrinite.

The inertinite is the maceral component converted from the plants' remains under the effect of fusinization. Because of the pre-oxidation influence, the inertinite changes little during the coalification process. Inertinite includes the components of fusinite, semifusinite, funginite, secretinite, macrinite, micrinite and inertodetrinite.

The liptinite is the maceral component converted from higher plants' reproductive organs, bark, secretions as well as the algae, with the lowest reflectance value. With the proceeding of coalification, the contour, convex and structure of liptinite become gradually unclear. The liptinite usually has high hydrogen content and volatile matter, therefore heating can produce a lot of tar and gas. Also, the caking property of liptinite is poor or may not even have caking ability. Commonly, liptinite is classified as low rank coal, and becomes rarely seen in higher coalification stage. The liptinite also includes the subcomponents as: sporinite, cutinite, resinite, suberinite, barkinite, biturninite, exsudatinite, fluorinite, alginite and exodetrinite [1].

2.3.2 Metamorphism Types of Coal

The metamorphism degree of coal means the degree of change in coal's physi-chemical properties under the interaction of temperature, pressure and duration.

Table 2.2 Classification of maceral composition of bituminous coal

Names of maceral composition (abbreviation)	Components of maceral composition (abbreviation)	Subcomponents of maceral composition (abbreviation)
Vitrinite (V)	Telinite (T)	Telinite 1 (T1) Telinite 2 (T2)
	Collinite (C)	Telocollinite (TC) Degradinite (DC) Corpocollinite (CC) Gelocollinite (GC)
	Vitrodetrinite (VD)	–
Inertinite (I)	Fusinite (F)	Pyrofusinite Oxyfusinite (OF)
	Semifusinite (Sf)	–
	Funginite (Fu)	–
	Secretinite (Se)	–
	Macrinite (Ma)	–
	Micrinite (Mi)	
	Inertodetrinite (ID)	–
Liptinite (L)	Sporinite (Sp)	Macrosporinite (Sp1) Microsporinite (Sp2)
	Cutinite (Cu)	–
	Resinite (Re)	
	Suberinite (Sub)	
	Barkinite (Ba)	
	Bituminite (Bt)	
	Exsudatinite (Ex)	
	Fluorinite (Fl)	
	Alginite (Alg)	Telalginite (Alg1) Lamalginite (Alg2)
	Exodetrinite (Ed)	–

The indexes classifying coal's metamorphism types include volatile matter, vitrinite reflectance, carbon content, hydrogen content, moisture content, heat productivity, etc.

The commonly used indexes of vitrinite reflectance are the average maximum vitrinite reflectance (R_o) and the average random vitrinite reflectance (R_r), which represent the reflectance value of the point with the largest reflectance and the random selected point of the coal thin slice under the oil immersion condition, respectively. International classification methods of coal types mostly refer to the average random reflectance, while in China the average maximum vitrinite reflectance (hereinafter referred to as vitrinite reflectance) is chosen as the main classification index. There is a certain statistical relationship between the vitrinite reflectance and the average random vitrinite reflectance. According to the Chinese national standards (GB/T 6948-2008. Method of determining microscopically the reflectance of vitrinite in coal), when $R_o \leq 2.5$, $R_o = 1.0645R_r$; while $2.5 \leq R_o \leq 6.5$, $R_o = 1.2858R_r - 0.3963$.

In 2005, the formal international standards ISO11760 (*Classification of coals*) was established and published with the participation of China, USA, UK, Germany and some other countries. In this standard, coal is classified to different types based on the degree of metamorphism (vitrinite reflectance), petrographic constitute (vitrinite content) and mineral content of coal (ash yield rate on dry basis). According to the value of average random reflectance, coal can be divided into three categories and ten subcategories, which are low rank coal ($R_r \leq 0.5$), middle rank coal ($0.5 \leq R_r \leq 2.0$) and high rank coal ($2.0 \leq R_r \leq 6.0$). China adopted a different standard published in 2009 (GB/T 5751-2009,*Chinese classification of coals*), which takes the coalification degree indexes (including volatile and hydrogen content on an dry ash-free basis) and the technological property indexes (including caking index, maximum thickness of the glial layer, Arnu-Audibert's dilatometer and the higher heating value on a moist ash-free basis) as the classification indexes. In China, the coals are classified into three categories as lignite, bituminous coal and anthracite, of which the bituminous coal can be further classified into twelve subcategories (including gas coal, fat coal, coking coal, lean coal, etc.), as listed in Table 2.3. Li [4] summarized the corresponding relationship between the vitrinite reflectance and the coal types in China, and gave a general range of virtinite reflectance for every kind of coal, which is conducive to the comparison with international common standard (see Table 2.4).

2.4 Gas Generation in Coal

Coal seam gas is generated from the humic organic matter during coalification. After the deposition of original parent material humus, the gas generation process can be divided into two main stages: one is the biochemistry-related gas generation period which turns the plant remains into peats; the other is the coal metamorphism-related gas generation period under the effect of high pressure and high temperature of the strata, which accompanied the transformation from lignite to bituminous coal or anthracite. The amount of generated gas depends on the composition of the original parent material and the stage of coalification [5–7].

2.4.1 *Gas Generation During the Biochemical Coalification Period*

The biochemical coalification period begins when the primary organic matter of coal deposits is in the swamp and delta facies environment. Under a temperature

Table 2.3 Coal classification of China

Category	Number	Classification indexes					
		Volatile matter on dry ash-free basis V_{daf}/%	Caking index of bituminous coal G_{RL}	Maximum thickness of bituminous coal's glial layer Y/mm	Arnu-Audibert's dilatometer of bituminous coal B/%	Transmittance of bituminous coal P_m/%	Higher heating value on a moist ash-free basis $Q_{gr,maf}$ /MJ/kg
Anthracite	01,02,03	≤ 10					
Meagre coal	11	> 10.0–20.0	≤ 5				
Meager-lean coal	12	> 10.0–20.0	5–20				
Lean coal	13,14	> 10.0–20.0	> 20–65				
Coking coal	24 15, 25	> 20.0–28.0 > 10.0–28.0	> 50–65 > 65①	≤ 25.0	(≤ 150)		
Fat coal	16,26,36	> 10.0–37.0	(>85)①		①		
1/3 Coking coal	35	> 28.0–37.0	> 65①	≤ 25.0	(≤ 220)		
Gas-fat coal	46	> 37.0	(>85)①	> 25.0	> 220		
Gas coal	34 43,44,45	> 28.0–37.0 > 37.0	> 50–65 > 35	≤ 25.0	(≤ 220)		
1/2 Mid-caking coal	23,33	> 20.0–37.0	> 30–50				
Slight caking coal	22,32	> 20.0–37.0	> 5–30				

(continued)

Table 2.3 (continued)

Category	Number	Classification indexes					
		Volatile matter on dry ash-free basis V_{daf}/%	Caking index of bituminous coal G_{RL}	Maximum thickness of bituminous coal's glial layer Y/mm	Arnu-Audibert's dilatometer of bituminous coal B/%	Transmittance of bituminous coal P_m/%	Higher heating value on a moist ash-free basis $Q_{gr,maf}$ /MJ/kg
None caking coal	21,31	> 20.0–37.0	≤ 5				
Long flame coal	41,42	> 37.0	≤ 35			> 50	
Lignite coal	51 52	> 37.0 > 37.0				≤ 30 > 30–50	≤ 24

Note 1. If $G_{RL} > 85$, Y or B can be used to distinguish fat coal, gas-fat coal and other coal types. If $Y > 25.0$mm, the sample should be classified as fat coal or gas-fat coal; otherwise, it should be further classified by V_{daf} value

When classifying coal types with the B value, if $V_{daf} > 28\%$ and $B > 150\%$, then the sample should be fat coal; if $V_{daf} > 28\%$ and $B > 220\%$, it should be fat coal or gas-fat coal. If contradiction appears with the Y and B value, take the Y value as the final decision

2. For coal with $V_{daf} > 37\%$ and $G_{RL} < 5$, use P_m to decide which one is long flame coal or lignite coal

3. For coal with $V_{daf} > 37\%$ and $P_m > 30\%$ -50%, the value of $Q_{gr,maf}$ should be measured. If its value is larger than 24 MJ/kg, the coal should be long flame coal

4. The measurement of the volatile content on ash-free dry basis, caking index, maximum thickness of glial layer, Arnu-Audibert's dilatometer, transmittance, heating value should follow the standards of GB212-977, GB5447-985, GB479-964, GB5450-985, GB2566-981, GB213-979, respectively

Table 2.4 Relationship between the vitrinite reflectance and the coal type in China

R_o/%	Chinese coal category	International coal category	R_o/%	Chinese coal category	International coal category
< 0.5	Lignite coal	Mid-low rank coal	1.3–1.4	Coking coal Fat coal	Middle rank coal
0.5–0.6	Long flame coal None caking coal Gas coal	Middle rank coal	1.4–1.5	Coking coal	Middle rank coal
0.6–0.7	Gas coal Long flame coal None caking coal Gas-fat coal	Middle rank coal	1.5–1.6	Coking coal Lean coal Meager-lean coal	Middle rank coal
0.7–0.8	Gas coal Gas-fat coal Slight caking coal None caking coal 1/2 Mid-caking coal	Middle rank coal	1.6–1.7	Lean coal Coking coal Meager-lean coal	Middle rank coal
0.8–0.9	1/3 Coking coal Gas coal Fat coal Gas-fat coal Slight caking coal None caking coal	Middle rank coal	1.7–1.8	Lean coal Meager-lean coal Coking coal Meager coal	Middle rank coal
0.9–1.0	1/3 Coking coal Gas coal Fat coal Gas-fat coal 1/2 Mid-caking coal	Middle rank coal	1.8–1.9	Meager-lean coal Lean coal Meager coal	Mid-high rank coal
1.0–1.1	Fat coal 1/3 Coking coal	Middle rank coal	1.9–2.0	Meager-lean coal Meager coal Lean coal	High rank coal
1.1–1.2	Fat coal 1/3 Coking coal Coking coal	Middle rank coal	2.0–2.5	Meager coal	High rank coal

(continued)

Table 2.4 (continued)

R_o/%	Chinese coal category	International coal category	R_o/%	Chinese coal category	International coal category
1.2–1.3	Coking coal Fat coal 1/3 Coking coal	Middle rank coal	> 2.5	Anthracite coal	High rank coal

of less than 65 °C, the primary organic matter is gradually decomposed into gas by the anaerobic microorganisms. This process can be summarized by the chemical equation of cellulose as follows:

$$4C_6H_{10}O_5 \rightarrow 7CH_4\uparrow + 8CO_2\uparrow + 3H_2O + C_9H_6O$$

or

$$2C_6H_{10}O_5 \rightarrow CH_4\uparrow + 2CO_2\uparrow + 5H_2O + C_9H_6O$$

During this stage, the peat layer formed by the organic matter is buried at a shallow depth. The cementation and solidification of its overlying cap layer is still weak and the generated gas can easily escape to the ancient atmosphere by diffusion and seepage. Thus, the gas generated during the biochemical coalification period is usually not preserved in the current coal seams. Then, with the subsidence of peat layer, the overlying cap layer becomes thicker and the organic matter is progressively exposed to higher temperature and pressure. The biochemical process gradually weakens and finally terminate, and the peat begins conversion into lignite under high pressure and temperature, after which, the coal metamorphism period begins.

2.4.2 Gas Generation During the Coal Metamorphism Period

With the further subsidence of lignite coal seam, the effects of temperature and pressure intensify dramatically, and the gas generation period is converted into the coal metamorphism-related gas generation period. Under the influence of high ground temperature of about 100 °C and corresponding strata pressure, thermo-metamorphism gas generation will take place in the coal. In the initial metamorphism period, the basic organic elements of coal are condensed, fused with aromatic hydrocarbons including hydroxyl (–OH), methyl (–CH_3), carboxyl (–COOH), ether (–O–) and other side chain or functional groups. Coal carbon elements are mainly concentrated in the fused ring. Normally, the fused ring chemical bonds have strong and stable binding forces, while the side chains or functional groups have weak and unstable binding. Therefore, with the subsidence of strata (as well as the increase of

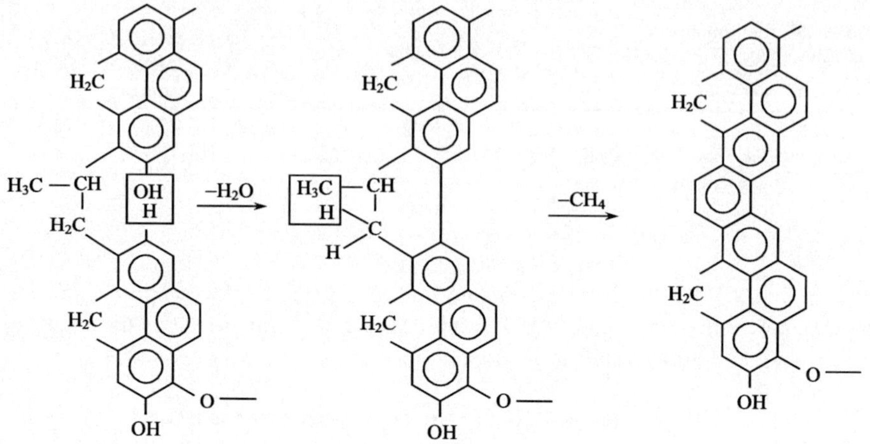

Fig. 2.2 Reaction diagram of gas generation during the coal metamorphism period (*Carbon content* 83–92%)

ground temperature and pressure), the side chains and functional groups will continuously break and fall off, generating volatile gases such as CO_2, CH_4, H_2O among others, as shown in Fig. 2.2.

The basic reaction concerning the gas generation during coalification is the process when the organic matter decomposes the methyl side chain and produce CO_2, CH_4 and H_2O. This process can be described by the following chemical formula as:

$$\underset{\text{Peat}}{4C_{16}H_{18}O_5} \rightarrow \underset{\text{Lignite}}{C_{57}H_{56}O_{10}} + 3CH_4 + 4CO_2 + 2H_2O$$

$$\underset{\text{Lignite}}{C_{57}H_{56}O_{10}} \rightarrow \underset{\text{Bituminous coal}}{C_{54}H_{42}O_5} + 2CH_4 + CO_2 + 3H_2O$$

$$\underset{\text{Bituminous coal}}{C_{15}H_{14}O} \rightarrow \underset{\text{Anthracite}}{C_{13}H_4} + 2CH_4 + H_2O$$

From the formula above and Fig. 2.2, it can be concluded that the main gas category generated from coal during coalification is methane. Meanwhile, with the generation of gas, the aromatic nucleus is further condensed and the carbon is further concentrated in the carbon network. With the deepening of coal metamorphism, the number of condensed aromatic nuclei of the basic structural units is ever-increasing. When coal becomes anthracite, the structure is mainly made of aromatic nuclei. From lignite to anthracite, the higher the coal rank is, the larger amount of gas will be generated. However, for different coalification stage, the components of the generated gas differ very much both in category and quantity.

Fig. 2.3 General model of gas generated in humic coal during metamorphic stage

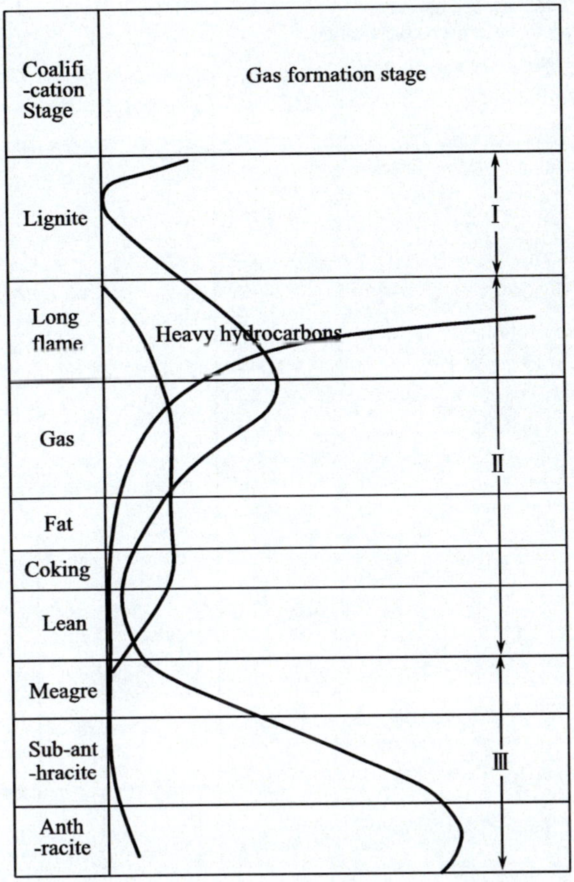

Figure 2.3 gives a general model for the gas generation in humic coal during the metamorphic stage, from which it can be noticed that the generation of methane is continuous in the whole metamorphic stage, while the generation of heavy hydrocarbons are not. Previous experiments show that the result trend of the gas generation model based on the artificial thermal evolution is consistent with the measured results. The ability order of the organic microstructures to produce hydrocarbons is ranked as: liptinite > vitrinite > inertinite.

Uspenski (former Soviet scholar) calculated the amount of methane produced by coal in each coalification stage based on the principle of balance between reactants and products in geochemical and coalification processes, as shown in Fig. 2.4.

Because the gas generated in the period from peat to lignite can escape easily, the calculation of methane generation amount usually takes the lignite as a starting point. However, the actual coalification process in nature is far more complicated than the theoretical calculations with many hypotheses, therefore, the above-mentioned data can only be approximate result, and for reference only.

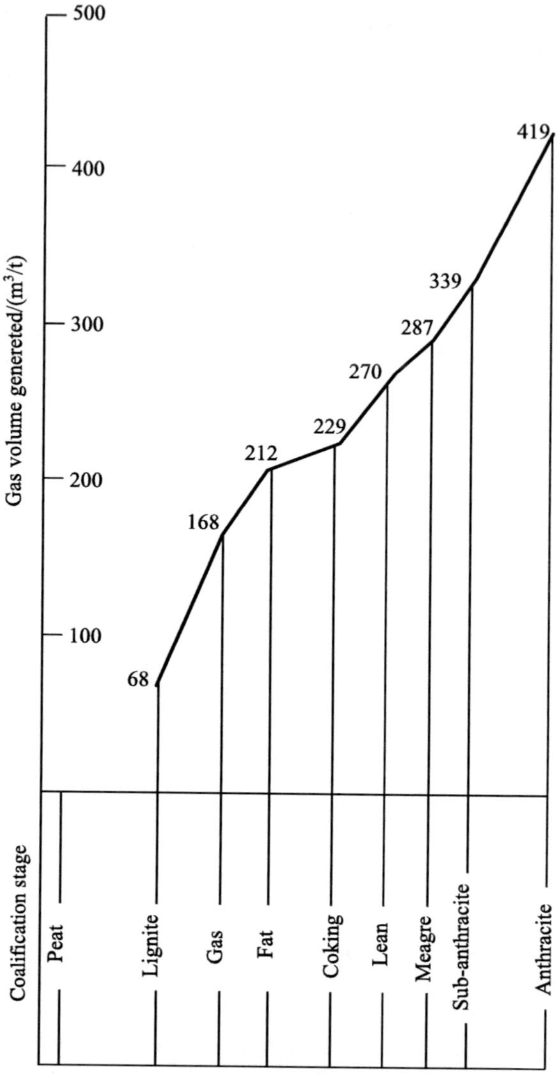

Fig. 2.4 The amount of methane in each coalification stage

Additionally, in some mining areas where coal and petroleum are symbiotic, sometimes the origin of coal seam gas is related to the gas intrusion from oil and gas fields. For examples, gas of the No. 10 coal seam, Zhongliangshan Coal Mine, Sichuan Province is believed to have some relationships with the floor gas intrusion from the limestone caves and the gas intrusion into the Jiaoping Coal Mine, Tongchuan Coalfield, Shaanxi Province is thought to originate from the sandstone oil layers found in the roof and floor strata.

Generally speaking, the gas category in coal mines worldwide is mainly methane CH_4. But some coal seams also contain heavy hydrocarbon gases like C_2H_6, C_3H_8 and other gases such as CO_2. For example, the No. 2 coal seam in Yaojie Coal Mine, Gansu Province contains large amount of CO_2. The geochemical characteristics of the CO_2 gas indicate that this CO_2 is formed by the crust thermodynamics metamorphism, and the periodic activity characteristics of the F19 fault zone in the eastern coal field controls the occurrence of CO_2.

2.4.3 *Coal Seam Gas Composition*

A large number of experiments show that the coal seam gas can be composed of about 20 kinds of different gas components, including methane and its homologous hydrocarbon gases (ethane, propane, butane, pentane, hexane, etc.), carbon dioxide, nitrogen, sulfur dioxide, hydrogen sulfide, carbon monoxide and rare gases like helium, neon, argon, krypton. Among them, methane and its homologues as well as carbon dioxide are the primary products of coalification process. When the coal seam is deeper than the depth of gas weathering zone, the main component of coal seam gas is methane (> 80%) [7].

Figure 2.5 and Table 2.5 give the qualitative identification of hydrocarbon gas components in coal seams by Coal Science Research Institute Fushun Branch. Tables 2.6 and 2.7 list the gas components analysis results from several coal seams in China. Tables 2.8 and 2.9 list the results of coal seam gas components analysis in some western countries.

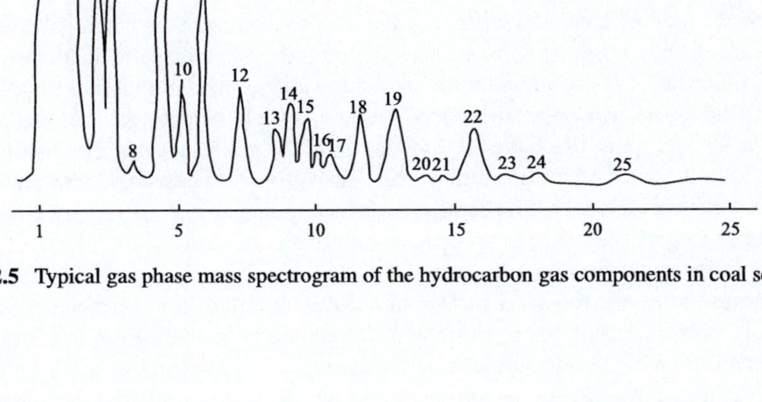

Fig. 2.5 Typical gas phase mass spectrogram of the hydrocarbon gas components in coal seams

Table 2.5 Qualitative mass spectral identification of hydrocarbon gas components in coal seams

Peak No.	Component	Peak No.	Component
1	Methane	14	Cyclohexane
2	Ethane	15	2-methylhexane
3	Propane	16	3-methylhexane
4	Isobutane	17	2-methylcyclopentane
5	N-butane	18	N-heptane
6	Isopentane	19	Methylcyclohexane
7	N-pentane	20	1-invert 2-cis-4-trimethylcyclopentane
8	2,2-dimethyl butane	21	1-invert-2-cis–3-trimethylcyclopentane
9	Dimethylpentane	22	Toluene
10	Trimethylpentane	23	3,2-dimethylhexane
11	N-hexane	24	1-cis-3-dimethylhexane
12	Methyl cyclopentane	25	N-octane
13	Benzene		

2.4.4 Occurrence Status of Coal Seam Gas

There are two kinds of occurrence status of gas in coal: namely the adsorbed gas and the free gas. Generally, the adsorbed gas may account for nearly 80–90% of the total gas amount, while the free gas only accounts for 10–20%. Most of the adsorbed gas is adsorbed on the coal surface. When the external conditions are constant, the adsorbed gas and free gas stay in a dynamic equilibrium state in coal, which means that the adsorbed gas molecules and free gas molecules keep exchanging all the time. When the external conditions change, such as the variation of temperature and pressure or impact and oscillation, the energy of molecules will change and the equilibrium state will be broken, which will eventually result in a new equilibrium state. Therefore, when it comes to the study of gas transport through coal, it cannot be simply asserted that the adsorbed gas is hard to transport in coal whilst the free gas is not. But in the short period of coal and gas outburst, the free gas may diffuse first, then Replenished by the rapid desorption of adsorbed gas.

Previous studies have shown that, when the gas pressure in not high, the gas molecules can enter the pores that have similar or larger sizes than the mean free path of gas molecules, but cannot enter the interspaces between coal material molecules or aromatic layers. However, if the gas pressure is high enough, gas can wedge or enter the smaller pores that have the similar size of the gas molecules, and finally are absorbed in the form of solid solution which are difficult to desorb. Under high gas pressure, methane in coal has no capillary condensation, but stay on the surface of the coal pore in a quasi-liquid homogeneous monomolecular film. Different gases, may possess different adsorption characteristics, which are related to the forces between the gas molecules and the solid surface molecules as well as the thermal activity of gas molecules. In recent years, with the development of technology, more accurate experiments like X-ray diffraction analysis have been used to test the characteristics of coal. Studies have shown that the gas occurrence in coal not only has the adsorbed

Table 2.6 Analysis of gas components from several coal seams in China

Sampling location	Coal seam	Coal type	Gas composition (Volume/%)					
			N_2	CO_2	CH_4	C_2H_4	C_3H_8	i-C_4H_{10}
Beipiao, Taiji mine, −550 m level	4	Gas-fat	7.10	0.39	92.03	0.0908	0.0079	0.0187
Beipiao, Taiji mine, −550 m level, Dongsan cross cut	5A	Gas-fat	1.42	1.60	73.07	16.18	5.49	0.7130
Beipiao,Guanshan mine, −58 m level	5C	Gas-fat	7.28	0.93	91.57	0.0704	0.0018	0.0052
Tiefa, Dalong mine western region Naner district	7	Gas	12.27	1.08	84.92	1.6868	0.0060	0.0003
Jixi, DiDaomine, No. 2 vertical shaft	18	Coking	12.85	1.07	85.87	0.0453	0.0042	0.0004
Zhongliangshan, northern mine, 2443 mining face	K_4	Coking	4.61	3.33	91.35	0.6708	0.0135	0.0020
Tianfu southern mine, No. 6 cross cut	K_9	Coking	5.05	2.95	91.92	0.0344	0.0260	0.0026
Tianfu southern mine, + 110 m	K_2	Coking	2.98	2.64	93.78	0.5477	0.0043	0.0006
Nantong Zhishu No. 2 mine, 2504 mining face	5	Lean	2.96	1.97	87.44	6.2711	1.3121	0.0050
Shenyang, Hongyang No. 3 mine 860 drilling	7	Lean	5.45	5.50	87.58	1.3143	0.0888	0.0031
Shenyang, Hongyang No. 3 mine 895 drilling	13	Lean	3.73	2.02	92.79	1.3777	0.0607	0.0025
Yangquan, No.1 mine	3	Anthracite	0.93	2.29	96.72	0.0500	0.0036	0.002
Sonezao mine, + 430 m level, 1356 mining face	K_3	Anthracite	14.17	0.32	84.84	0.5485	0.0060	
Baisha, Hongweitanjiachong mine	6	Anthracite	9.07	12.14	73.72	4.12	0.0348	0.0027
Jiaozuo, Lifengda mine	2	Anthracite	9.15	9.14	77.82	2.97	0.0205	0.0001

state and the free state, but also contains the liquid and solid solution state of gas. However, the amount of adsorbed and free state gas, generally accounts for 85% or more of the total gas content, thus under normal conditions, the overall gas behavior of coal still demonstrates the characteristics of adsorbed and free state gases, which shows no conflicts with the traditional viewpoint.

Table 2.7 Analysis of gas components from several coal seams in China

Sampling location	Coal seam	Coal type	Gas composition (Volume/%)					
			n-C_4H_{10}	i-C_5H_{12}	i-C_5H_{12}	C_6H_{14}	C_7H_{16}	C_8H_{18}
Beipiao, Taiji mine, −550 m level	4	Gas-fat	0.0228	0.0118	0.0169	0.0236	0.0271	
Beipiao, Taiji mine, −550 m level, Dongsan cross cut	5A	Gas-fat	0.6520	0.1730	0.1580	0.0990	0.1540	0.064
Beipiao, Guanshan mine, −58 m level	5C	Gas-fat	0.0112	0.0060	0.0112	0.0223	0.0648	0.0194
Tiefa, Dalong mine western region Naner district	7	Gas	0.0005	0.0019		0.0108		
Jixi, DiDaomine, No. 2 vertical shaft	18	Coking	0.0120				0.1590	
Zhongliangshan, northern mine, 2443 mining face	K_4	Coking	0.0057	0.0006	0.0003		0.0052	
Tianfu southern mine, No. 6 cross cut	K_9	Coking	0.0069	0.0011	0.0018	0.0023	0.0065	
Tianfu southern mine, + 110 m	K_2	Coking	0.0015	0.0005	0.0006			
Nantong Zhishu No. 2 mine, 2504 mining face	5	Lean	0.0127	0.0011	0.0021	0.0011	0.0281	0.0166
Shenyang, Hongyang No. 3 mine 860 drilling	7	Lean	0.0076	0.0012	0.0036	0.0053	0.0517	
Shenyang, Hongyang No. 3 mine 895 drilling	13	Lean	0.0090	0.0003	0.0007		0.0062	
Yangquan, No. 1 mine	3	Anthracite	0.0020					
Sonezao mine, + 430 m level, 1356 mining face	K_3	Anthracite	0.0013	0.0004			0.0949	
Baisha, Hongweitanjiachong mine	6	Anthracite	0.0101	0.0019	0.0049	0.0063	0.0143	
Jiaozuo, Lifengda mine	2	Anthracite		0.0004	0.0014	0.0023	0.0332	

Note 1. In accordance to the rate of Oxygen to Ammonia in the air, the content of mixed air is equal to the Oxygen content of coal sample
2. Coal samples except for that from exploration drilling in Hongyang No. 3 mine, are from new-exposed mining face

Table 2.8 Coal seam gas components of the former Soviet Union, Belgium and UK

Country	Mine	Gas composition (Volume/%)					
		CH_4	N_2	CO_2	H_2	O_2	Other
Former Soviet Union	Donbass	91.56–98.4	0.71–7.97	0–0.47			
Belgium	Statistic from 60 mines	92.9–99.6	0–6.27	0.03–3.38	0–0.24		0.02–2.79
UK	Baulz mine	95.8	3.07	0.68	0.20	0.20	

Table 2.9 Coal seam gas components of the USA (Volume/%)

Gas composition	Maximum	Minimum	Average	Gas composition	Maximum	Minimum	Average
CH_4	99.22	63.1	91.89	CO_2	14.75	0.06	3.58
C_2H_6	1.59	0	0.48	N_2	35.96	0.05	4.017
C_3H_8	0.0109	0	0.00149	O_2	0.5	0	0.148
C_4H_{10}	0.0022	0	0.00033	H_2	0.02	0	0.002
C_6H_{12}	0	0	0	He	0.27	0	0.03

Note Statistic from the GC results of 84 gas samples of 6 coal seams

When the coal seam has an outcrop or the coal basin is lain under the alluvial layer, there exist two different directions for gas migration in the coal seam. These are the upward migration of the gas from deep coal seam, and the downward migration (penetration, diffusion) of the ground air. There is also some gas generated by biochemical/chemical reactions in the surface soil that may enter into the deeper parts of the strata, which result in the vertical zoning characteristics of the gas in coal measure strata. The zonal distribution of coal seam gas is the basis for the prediction of coal gas content, gas emission, as well as the key point of gas management. Along the vertical direction (from surface to the deep), the zoning of coal seam gas can be generally divided into the gas weathering zone and methane zone [5, 6].

As shown in Fig. 2.6, the gas weathering zone is the general designation of the CO_2–N_2 zone (I), N_2 zone (II) and N_2–CH_4 zone (III). Different zones, possess different gas components and gas contents. The depth of gas weathering zone is the result of the long-term geological evolution in the coal field, which is comprehensively caused by the geological factors such as denudation weathering, properties of surrounding rock, coal seam outcrop, fault development, dip angle of coal seam, and groundwater activity. The denudation process can lead to a shallower depth of the gas weathering zone. Long term weathering gives more time for gas to escape from coal seam and thus results in a greater depth of gas weathering zone. In addition, greater permeability of the surrounding rock, development of the open faults and intense groundwater activity, will all result in greater depth of the gas weathering zone. The depth of gas weathering zone may also change with the specific conditions of the coal measure strata. For different mining areas or coalfields, even for different seams

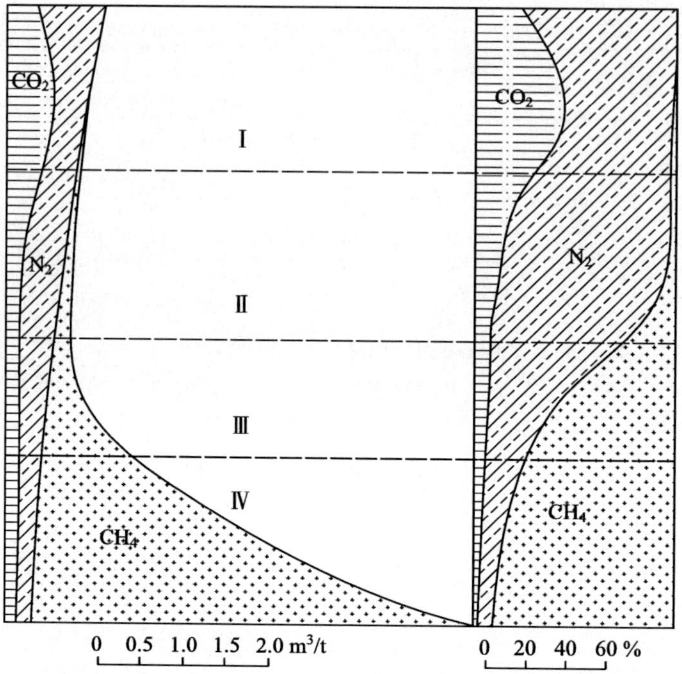

Fig. 2.6 Vertical gas distribution in coal seams

in the same coal mine, this value may vary a lot. For example, the lower boundary elevation of gas weathering zone at No. 7 coal seam is -400 m, whilst the elevation at the No. 10 coal seam is -420 m in the Taoyuan Coal Mine, Huaibei Coalfield.

The lower boundary of gas weathering zone can be determined by the following rules:

I. The volume concentration of methane and other heavy hydrocarbons should be equal to 80%;
II. Gas pressure should lie within the range of 0.1–0.15 MPa;
III. Relative gas emission quantity: $q_{CH_4} = 2\text{–}3\ m^3/t$;
IV. Gas content of coal seams:

$$X = 1.0 - 1.5\,\mathrm{m^3/t \cdot r}\ \text{(Long flame coal)}$$

$$X = 1.5 - 2.0\,\mathrm{m^3/t \cdot r}\ \text{(Gas coal)}$$

$$X = 2.0 - 2.5\,\mathrm{m^3/t \cdot r}\ \text{(Fat or Coking coal)}$$

$$X = 2.5 - 3.0\,\mathrm{m}^3/\mathrm{t}\cdot\mathrm{r}\,(\text{Lean coal})$$

$$X = 3.0 - 4.0\,\mathrm{m}^3/\mathrm{t}\cdot\mathrm{r}\,(\text{Meagre coal})$$

$$X = 5.0 - 7.0\,\mathrm{m}^3/\mathrm{t}\cdot\mathrm{r}\,(\text{Anthracite})$$

where X represents the gas volume per tonne of coal on a dry and ash-free basis.

The determination of the gas weathering zone depth is helpful to predict the gas emission quantity, obtain the gas occurrence and transport properties, and choose an effective gas management strategy. The coal mine or district located within the gas weathering zone is classified as a low gas coal mine or district, however if met with the situation like poor ventilation or ventilation failure, suffocation (CO_2, N_2) and gas explosion risks still exist.

The methane zone (IV, CH_4) is located below the lower boundary of gas weathering zone. In the methane zone, the gas pressure and content in coal seams typically increase with depth, as well as the relative gas emission quantity. The rate of gas content increase differs in different geological situations and coal types. Apart from the general gas emission, at a certain depth within the methane zone special hazardous gas emissions like the gas blowout or coal and gas outburst can also occur in some mines. Thus in the methane zone, dangers (like suffocation, gas explosion and gas combustion) appear not only when the ventilation is poor or stopped, but under normal ventilation conditions, when a special gas emission phenomenon occurs. Accidents like suffocation, gas explosion, workers buried by ejected coal can happen but with good management these occurrences can be minimized. In conclusion, only when the gas occurrence and gas transport regularities are mastered, as well as the corresponding technical measures are taken, then the general and special hazardous gas emissions can be reduced to a bare minimum.

2.5 Effect of Magma Intrusion on Coal Metamorphism and Gas Occurrence

2.5.1 Ways of Magma Intrusion into Coal Seams

Magma intrusion has a profound influence on the coal metamorphism and gas occurrence. According to the different magmatic activities, the magmatic rocks formed from the cooled metamorphic magma can be divided into intrusive rocks and ejected rocks. During the sedimentary process of coal seam formation magma can invade the coal seam along the geological structures such as faults. Hot magma intrusion will cause the thermal metamorphism of the coal seam and affect the coal quality and gas occurrence. According to the shape and distribution, the intrusion rocks can be further divided into sill, batholith, laccolith, dyke and so on, as shown in Fig. 2.7 [8].

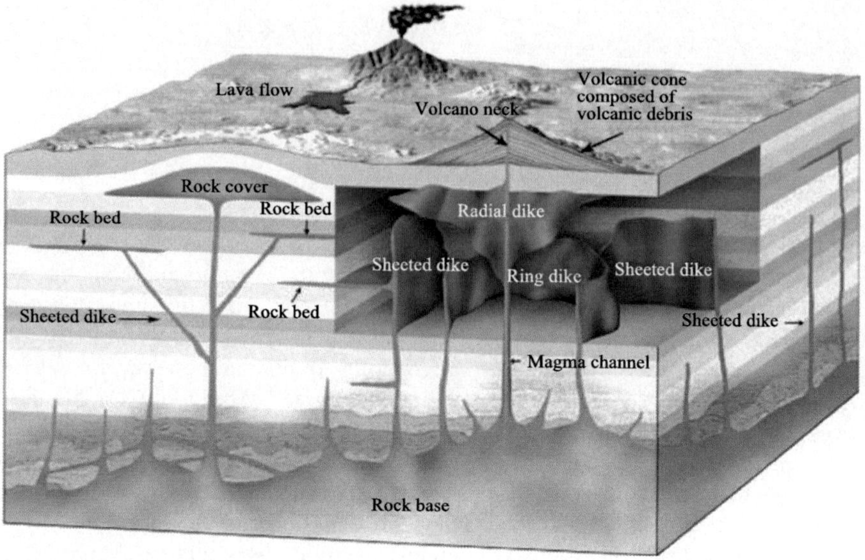

Fig. 2.7 Magmatism and the occurrence of magmatic rocks

1. Sill
   It is also known as intrusive sheet. The invaded magma spreads along the level, forming a plate-like intrusive body which has conformable contact with the floor of coalbed. Sill generally has a thin thickness but a larger area.
2. Batholith
   It is generally a large intrusive rock mass with an exposed area of above 100 km^2. It is usually composed of granite.
3. Laccolith
   It is an interlayer intrusion body which has conformable contacts with the strata layers. Laccolith usually has flat bottom, arched top, with thick center and thin edge, showing circular shape in the plane. For the acid rock mass, it can be called as the laccolith; but for the mafic rock of similar form, it is called rock plate.
4. Dike/Dyke
   It is a lode intrusion body which is heterotropic with the bedding or schistosity of the surrounding rocks, whose length and depth are much greater than the thickness. The plate-like dike which is almost vertical standing is called dyke.

2.5.2 Thermal Temperature Field Analysis of Magma

After magma invasion of coal seams, there will be a large temperature difference between the intrusion and the surrounding rock. The heat stored in magma is transferred into the surrounding rocks and coal seams. The magma intrusion brings a brief high temperature into the strata, which is a typical unsteady process. The heat transfer pattern of magmatic rocks is dominated by heat conduction. It will also be influenced by factors such as the heat transfer direction, heat conduction mode, temperature level, temperature gradient and magmatic scale, affecting the characteristic variation of coal metamorphism in the thermal evolution area.

The main heat source in the crust is the deep heat flow, which is the sum of radioactive elements decay heat flows in the crust, mantle and sediments. The heat transfer in the crust includes three kinds of modes, namely the heat conduction mode which plays a dominant role in the total flow, then the heat convection and thermal radiation. The magmatic intrusion conveys its heat to the nearby coal/rock and gradually cools itself down. Many studies have shown that the heat conduction transfer is still dominant in this process. Although heat convection and heat radiation also exist under certain conditions, these heat exchanges are much weaker than heat conduction. Current researches widely believe that the cooling process of magma is conducted mainly by heat conduction [9, 10]. The magmatic cooling process can be divided into two stages. The first stage is the process of continuous cooling of the magma and the continuous heating process of the surrounding rock. At this stage, the magmatic temperature is significantly higher than that of the surrounding rock, so there is a rapid heat exchange and the magma's cooling rate is fast. As a result, the surrounding rock continues to heat up, and the thermal evolution range begins to gradually expand. However, duration of this stage is relatively short. The second stage is the slowly cooling process of magma and surrounding rock. During this stage, the magma temperature and the surrounding rock temperature are similar. The heat exchange of magma and surrounding rock is weak, with the property of low cooling efficiency, but the duration is longer than cooling during the first stage.

Taking the Haizi Coal Mine as an example,which has a wide distribution of magma rocks, the temperature evolution of magma rocks and coal seam is analyzed. The Haizi Coal Mine is located in the Huaibei Coalfield. The coal-bearing strata of Haizi Mine is Permian, including the Shanxi Formation, the lower Shihezi Formation and the Upper Shihezi Formation. In the Shanxi Formation (P_{1s}), No. 10 coal seam is a well-developed seam, which is also one of the primary mineable coal seams in the mine. The lower Shihezi Formation (P_{1xs}) and the Upper Shihezi Formation (P_{2ss}) are the main coal-bearing groups with six seams. Among them, Nos. 7, 8, 9 coal seams are minable, while Nos. 4, 5, 6 seams are too thin to mine. No. 5 coal seam in the western part of the mine was intruded by the magmatic rocks.

The Haizi Coal Mine is located in the south of the Subei Fault which connects to the Daliujia Fault (the edge of the coalfield). During the Yanshanian era, magma was injected along the Subei Fault from shallow crustal magma chambers and intruded into the No. 5 coal seam in the Haizi Coalfield along the Daliujia Fault. The intruded magmatic rock showed the occurrence pattern of sill and brought great influences on the underlying coal seams (Nos. 7, 8, 9 and 10), as shown in Fig. 2.8.

Figures 2.9 and 2.10 gives the temperature variations in the cooling process 50 and 2000 years (50a and 2000a) after magma intrusion. In Fig. 2.9, the rectangle represents an intruded magmatic rock, the black lines represent Nos. 7, 8, 9 and 10 coal seams, and the background colors represent isothermals. From Fig. 2.10, it can be seen that 50 years (50a) after the magma intrusion, the maximum temperature of internal magma decreased rapidly to 786 °C and the temperature of the surrounding rocks began to increase. But the affected range was limited to the roof and floor of magmatic rocks. The coal seams underlying Nos. 7, 8, 9 seams were previously influenced by thermal evolution, while the No. 10 coal seam located further barely

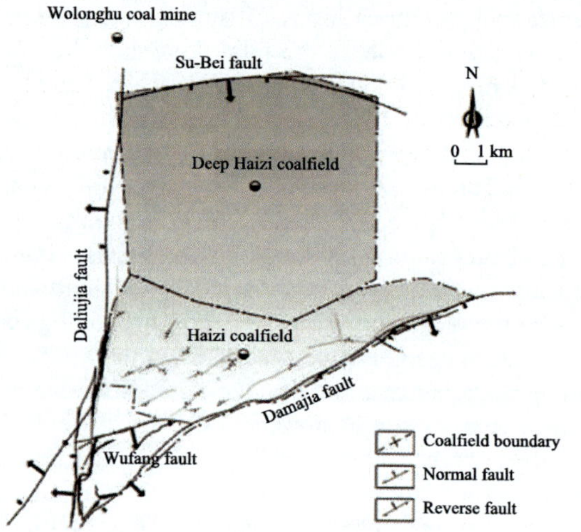

(a) Planar map of magma intrusion

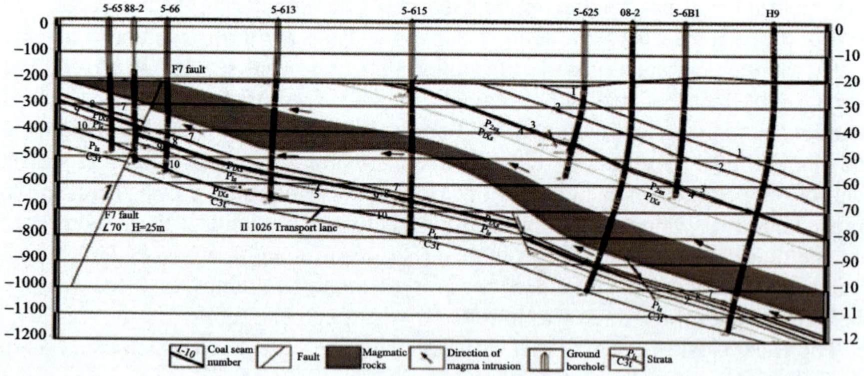

(b) Profile map of magma intrusion

Fig. 2.8 Schematic diagram showing magmatic activity in Haizi Coal Mine

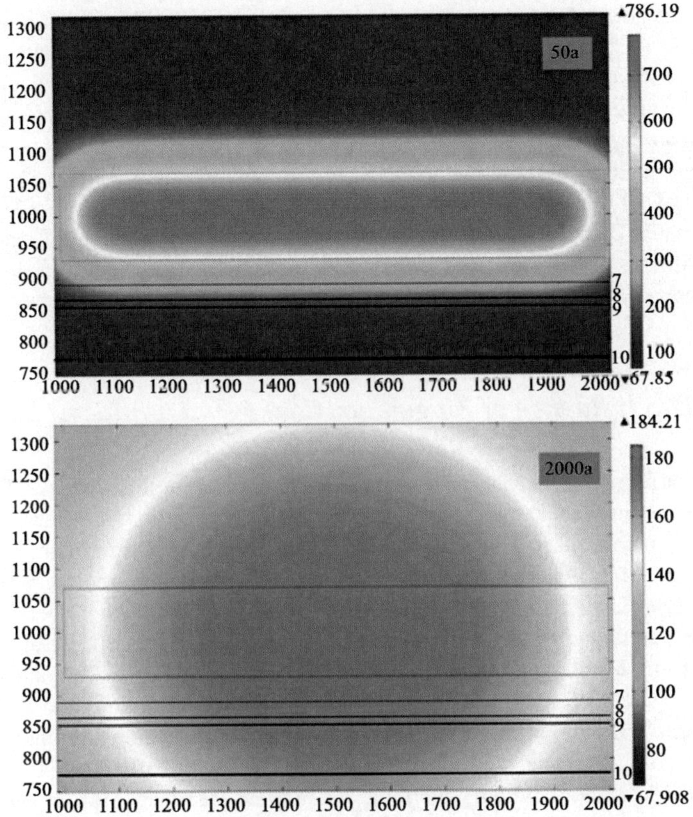

Fig. 2.9 Comparison of temperature distribution in 50a and 2000a during magma cooling

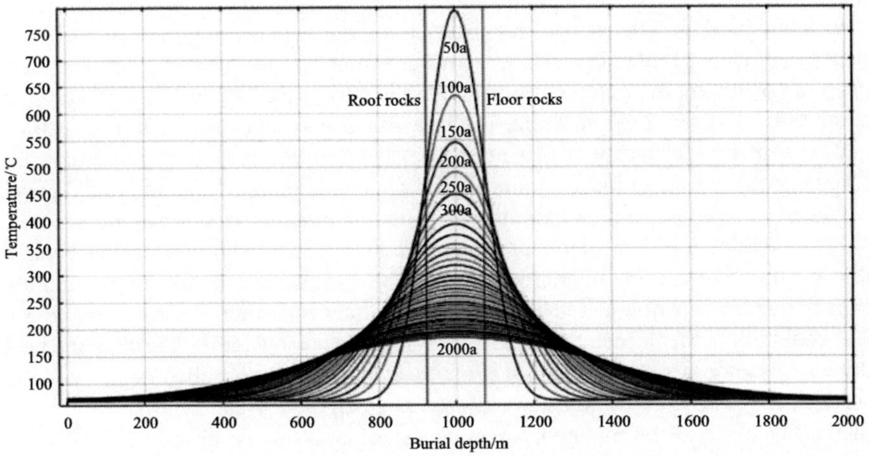

Fig. 2.10 Temperature distribution during magma cooling

suffered that effect. With the magmatic rocks cooling down, the temperature of the surrounding rocks continuously increased and the affected range of strata gradually expanded. 400 years after intrusion, the temperature of magmatic rocks decreased to 374 °C. The temperature of the middle coal group (No. 7, 8, 9 seams) increased to the maximum value, while No.10 coal seam began to be affected by the heat conduction. Then, the temperature of the magma and the surrounding rocks decreased further to 250 °C until 1000 years after intrusion and the temperature of Nos. 7, 8, 9 and 10 coal seams also decreased. 2000 years after intrusion, the temperature of magma and the surrounding rocks decreased to below 200 °C, while the Nos. 7, 8, 9, 10 coal seams gradually cooled down as well [11].

2.5.3 The Influence of Magma on Coal Metamorphism

The heat warming process plays a leading role in the metamorphism of coal. Based on the differences of heat sources, heating ways and the metamorphism characteristics, the metamorphism of coal can be divided into plutonic metamorphism, magmatic metamorphism (regional magmatic thermal metamorphism and contacting metamorphism) and dynamic metamorphism. After the burial of peats, the coal-bearing strata were subjected to diagenesis and plutonic metamorphism in the process of sedimentation, forming lignite and bituminous coal. Some parts of China were subjected to regional magmatic intrusion thermal influence and formed middle and high grade metamorphic coal. According to the nature, infiltration depth, position and scale of magma invading the coal-bearing strata, the regional magmatic thermal metamorphism can be further divided into three categories: the deep-, middle-deep- and shallow-magmatic thermal metamorphism [12–14]. The deep magmatic thermal metamorphism is mainly from a large concealed intrusion body, mostly acidic magmatic rock with a strong gas–liquid interaction, with intrusion depth normally more than 3000 m. This effect may produce a large area of high-quality coal belt, which mainly distributed in southeastern Shanxi and central Henan region. The mid-deep-magmatic thermal metamorphism often occurs in the Ordovician or Cambrian strata below the coal-bearing strata. The invasion depth ranges from 1500 to 3000 m, and the intrusion forms are primarily middle-acid and alkaline stock, sills or bosses, forming medium-sized metamorphic zones such as Hanxing Coalfield and Xishan Coalfield. The shallow-magmatic thermal metamorphism effect often occurs in a shallow coal seam with the depth of about 0–1500 m. The overlying deposited layer is thin and the magma is mainly intruded into the coal strata directly, forming carbonite. Magmatic rocks are dominated by small medium acid and middle base magma rocks. The coal rank in the influenced area is generally in a narrow range. Through the heat driven metamorphism effect, the magma makes the metamorphism degree of coal show the regularity of zoning, for example as a ring belt. Near the magmatic rock, increasing trend in the metamorphism degree of coal can be observed. Under the effect of plutonic metamorphism, an increase in coal rank can requires an increase of approximately 1000 m at depth, while in the magma intrusion area, different ranks of coal may appear in the range of less than 100m.

According to the vitrinite reflectance of coal (R_o), the maximum value of paleo temperature (T_{peak}) can be estimated, which can also reflect the intensity of primitive magmatic activity. Using the relationship established by Barker and Morrow, the R_o of coal which represent the coal metamorphism is decided by the highest paleo temperature T_{peak} that coal seam was subjected to, as shown in Eq. (2.1) [8, 15]:

$$T_{peak} = [\ln(R_o) + 1.19]/0.00782 \quad (2.1)$$

Taking Haizi Coal Mine as an example, the paleo temperature estimated from the coal samples collected from different coal seams and different distances between magmatic rocks is shown in Table 2.10. Referring to Eq. (2.1), the maximum temperature during magma intrusion in Haizi coal mine can be obtained, which are: 283 °C (No. 7 coal seam, 44 m to magma rocks), 258 °C (No. 8 coal seam, 64 m to magma rocks), 253 °C (No. 9 coal seam, 70 m to magma rocks), 229 °C (No. 10 coal seam, 154 m to magma rocks), 228 °C (No. 10 coal seam, 156 m to magma rocks), respectively.

In addition, the thickness of sill, which represents the intensity of primitive magmatic activity, can also illustrate the influence of magma on the coal metamorphism. In the No. 7 coal seam, the coal metamorphism differs with the thickness of

Table 2.10 Petrographic analyses and paleo temperature estimation results

No	Coal seam	Distance to sill/m	Elevation/m	Organic microscopic components/%		R_o/%	T_{peak}/°C
				Vitrinite	Inertinite		
HZ 1#	7	44	−658	95.40	2.20	2.78	283
HZ 2#	7	46	−660	94.80	2.30	2.74	281
HZ 3#	8	64	−680	94.30	3.20	2.30	258
HZ 4#	8	66	−681	94.32	3.10	2.27	257
HZ 5#	9	70	−683	94.48	2.60	2.20	253
HZ 6#	9	73	−690	94.13	2.40	2.15	250
HZ 7#	10	154	−638	93.69	3.80	1.82	229
HZ 8#	10	156	−620	91.37	5.80	1.81	228
HZ 9#	10	–	−515	83.31	8.91	1.50	204

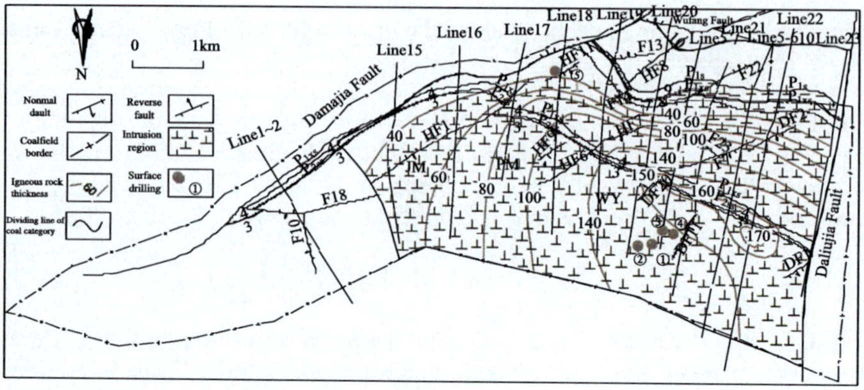

Fig. 2.11 Distribution of magma rock and coal category in Haizi Coal Mine. WY—Anthracite; PM—Lean coal; JM—Coking coal

sill, as shown in Fig. 2.11. In the west of No. 19 Line, the magmatic rock has a large thickness which is basically above 140 m, and the rank of underlying No. 7 coal seam is higher (anthracite, WY). While in the east of No. 19 Line, the coal ranks decrease with the decreasing magma thickness (lean coal, PM). In the magmatic rock thinning area and the magmatic rock margin, the degree of metamorphism becomes coking coal (JM). The degree of metamorphism in Nos. 8, 9, 10 coal seam also experiences the influence from the up magma sill. Therefore, we may conclude that the magma intrusion controls the coal category distribution in Haizi Coal Mine [11, 16].

2.5.4 *The Influence of Magma on Gas Occurrence*

After the magma intrusion, influenced by the size, occurrence, location of the magmatic rock as well as spacing to the coal seam, the range of thermal evolution and the intensity of thermal action both differ greatly. As a result, coal in the thermal evolution zone possesses different degrees of metamorphism, pore structures and ad/desorption characteristics. Meanwhile, the occurrence of rock mass also affects the possibility of escaping and storing of gas in coal [17]. The control effect of magmatic rocks on gas occurrence includes the heat evolution effect, heat metamorphism effect, pushing effect and trapping effect, as shown in Fig. 2.12.

1. Thermal evolution and heat metamorphism

The heat brought by the magma intrusion increases the degree of coal metamorphism and develops the pore volume and the specific surface area of coal, increasing the gas adsorption capacity [18–22]. During magma intrusion, the thermal evolution of magmatic rocks and the plutonic metamorphism are superimposed on each other,

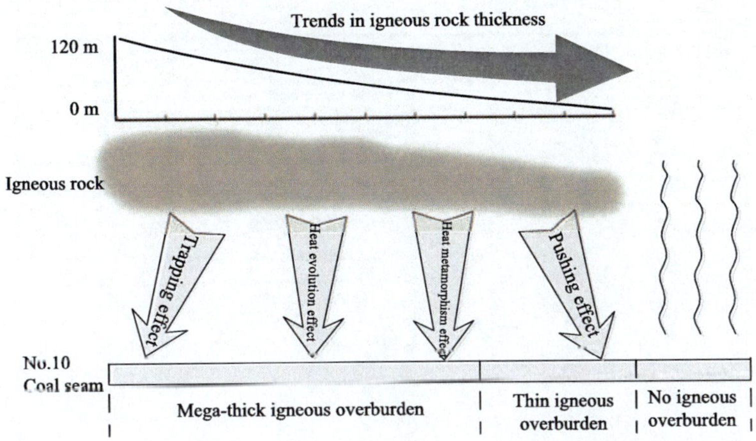

Fig. 2.12 The control effect of thick sills on the gas occurrence of underlying coal seam

making the low-rank and shallow buried coal seam produce additional amount of gas. The coal seam under the superimposed effect is characterized by high gas production and rate, and the effect have an important influence on gas occurrence.

Figure 2.13 schematically gives the evolution and hydrocarbon generation process of the coal seams under magma sill in the Haizi Coal Mine. From the Permo-Carboniferous period to the late Indosinian, the coal measure strata kept subsiding. When the coal seam subsided to the deepest position (about 3000 m deep), (with the ground temperatures of approximately 140 °C), the degree of the plutonic metamorphism reached to gas or fat coal. At the stage of plutonic metamorphism, approximately 220 m^3/t of methane was generated. In the Yanshanian, the basin was tectonically uplifted and because of the decreasing ground temperature, the coalification temporarily stopped. However, during the middle Yanshanian, magma intrusion produced variable heating of the coal seams. The magma invaded the No. 5 coal seam and its high-temperature thermal effects metamorphised Nos. 7, 8, 9, and 10 coal seams located under the magmatic rock, causing the coal seams in the Haizi Coal Mine to reach their highest temperatures in coalification. In particular, Nos. 7, 8, 9 and 10 coal seams reached temperature of approximately 350 °C, 310 °C, 300 °C and 235 °C, respectively, which improved the coal rank in the influenced area. In the region without magmatic intrusion, the heat flow induced by the tectonic movement of Yanshanian also improved the metamorphism of coal, but the level was far lower than the coal in the thermal evolution area.

After magma intrusion, the metamorphism of coal has further increased and secondary hydrocarbon generation also occurred. The total amount of magma-induced hydrocarbon gas generation was approximately 340 m^3/t. As the sill finally cooled to ambient temperature and the strata were uplifted, coalification finally terminated, and the metamorphism of coal remained basically the same until now.

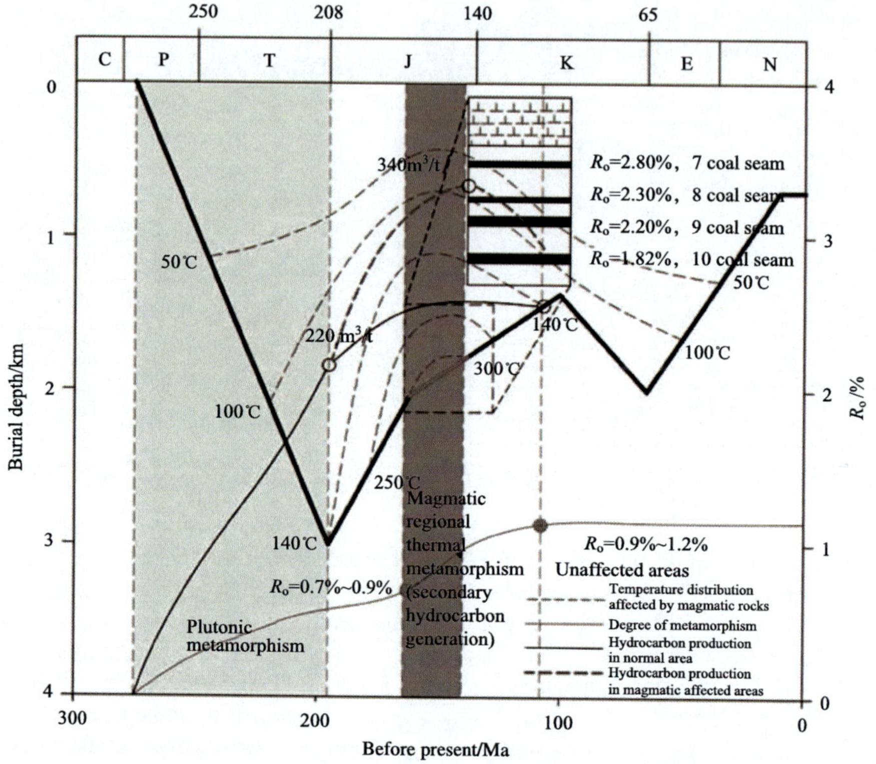

Fig. 2.13 Evolution and hydrocarbon generation characteristics of the Nos.7, 8, 9, and 10 coal seams in the Haizi Coal Mine

From Indosinian to present time the coal measure strata has been uplifted about 1200 m, which lead to largest erosion of overlying strata and escape of coal seam gas. However, due to the effective trapping and sealing effect of the magmatic rock in the magma intrusion area, the escape rate of gas was reduced resulting in higher gas pressure/content of the coal seam. This increased the coal and gas outburst risk for the underlying coal seams [11, 23–26].

2. Stress effect

Geostress is the internal stress in the crust, mainly composed of geostatic stress, tectonic stress, pore stress, thermal stress and residual stress. Magma activity, as one of the tectonic activities, creates a complicated tectonic stress field in the surrounding rocks during its intrusion, condensed contraction and diagenesis processes. The magma in the molten state is in a high hydrostatic pressure state, applying a uniform pressure to the surroundings. But with the gradually condensed contraction of intruded magma, from the contact interface to the internal position, different thermal expansion coefficients and thermodynamic processes, complex changes of stress in

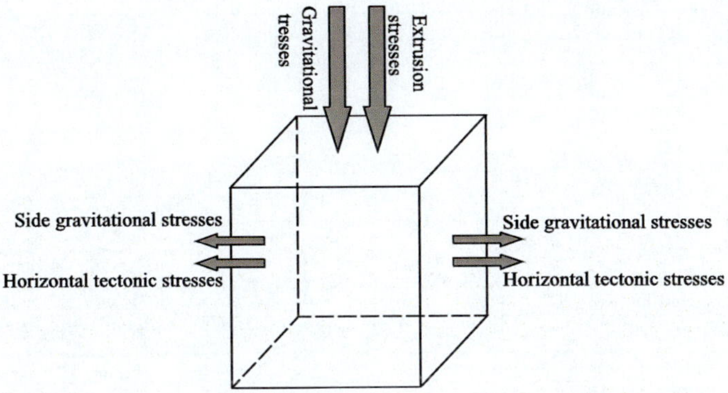

Fig. 2.14 Force analysis of magmatic rock mass

the magma itself as well as in the surrounding rocks occur. The rock mass is not only affected by the background gravitational and tectonic stresses, but also by stresses induced by high pressure magma expansion that upon cooling may become locked in place. The thicker the sill is, the greater expansion stress the underlying rocks are subjected to, as shown in Fig. 2.14. The dynamic expansion force can contribute to increase in a number and scale of pore, fracture structures, which can change the adsorption capacity and the gas storage ability of the coal seam. Moreover, tectonic stresses also have a certain effect on the coalification, resulting in stress degradation and stress polycondensation effect. The stress degradation due to the tectonic stress will act on the coal organic macromolecules in the form of mechanical or kinetic energy to break down the lower chemical bonds of the side chains and functional groups of the coal aromatic ring structure, and degrade them into free radicals with smaller molecular weight, forming an out flow of organic matter. The stress polycondensation effect is that, under the effect of the anisotropic tectonic stress, the coal aromatic rings tend to have a parallel arrangement with continuous rotation and displacement. Then, the basic structure unit will produce an oriented growth and a preferential superposition, increasing the aromatic fused ring system [25, 27].

For example, in the No. 10 coal seam of Haizi Coal Mine, the volatiles and moisture content of coal decrease monotonously with the increase of magma sill's thickness, which is reduced from 24.19% to 10.04% and from 0.63% to 0.27%, respectively. While the vitrinite reflectance, ash and fixed carbon content have an increased trend with the sill thickness, which is from 0.6901% to 2.1262%, 11.16% to 18.73%, 64.02% to 70.97%, respectively (Fig. 2.15).

Also, the magma activity has a significant influence on the pore structure. From the area without sill to the area covered by sill, the Brunauer–Emmett–Teller (BET) specific surface area increases from 1.5221 to 20.04 m^2/g, and the D-R micropore size increases from 5.31×10^{-4} to $7.13 \times 10^{-3} cm^3/g$ (see Fig. 2.16).

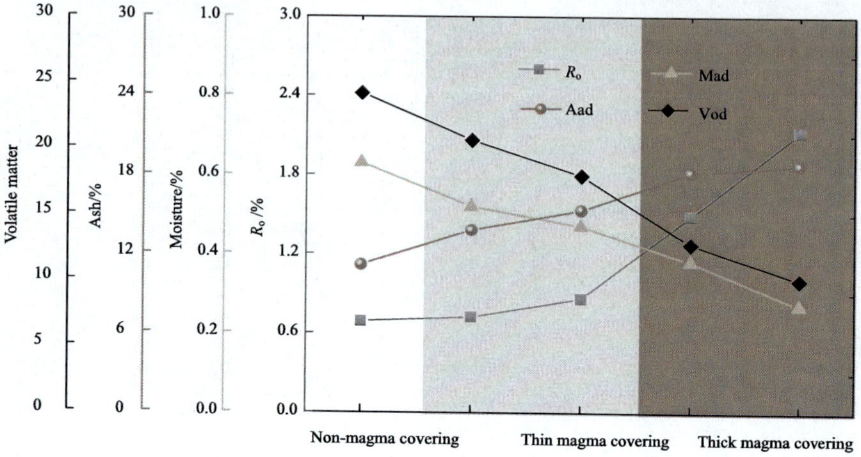

Fig. 2.15 Variation in vitrinite reflectance and proximate analysis results

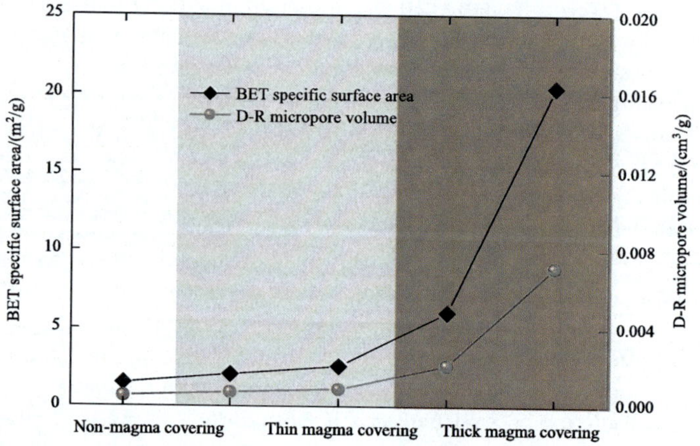

Fig. 2.16 Variation in BET specific surface area and micropore volume

3. Trapping effect

Magma rock is one kind of dense rock mass with low permeability coefficient, whose permeability is much lower than that of coal seam. Therefore, the magmatic rock itself can be regarded as a sealing body. When the magmatic rock acts as the roof rock overlying the coal seam, the gas in underlying coal seam may be trapped possibly forming a huge gas reservoir in the seam and may increase the risk of coal and gas outburst. Figures 2.8 and 2.11 illustrates the trapping effect of thick sill cover in Haizi Coal Mine, Huaibei Coalfield. Figure 2.17 shows the ring-like trapping effect of sill in Wolonghu Coal Mine of Wanbei Coalfield. The magma intrusion in these

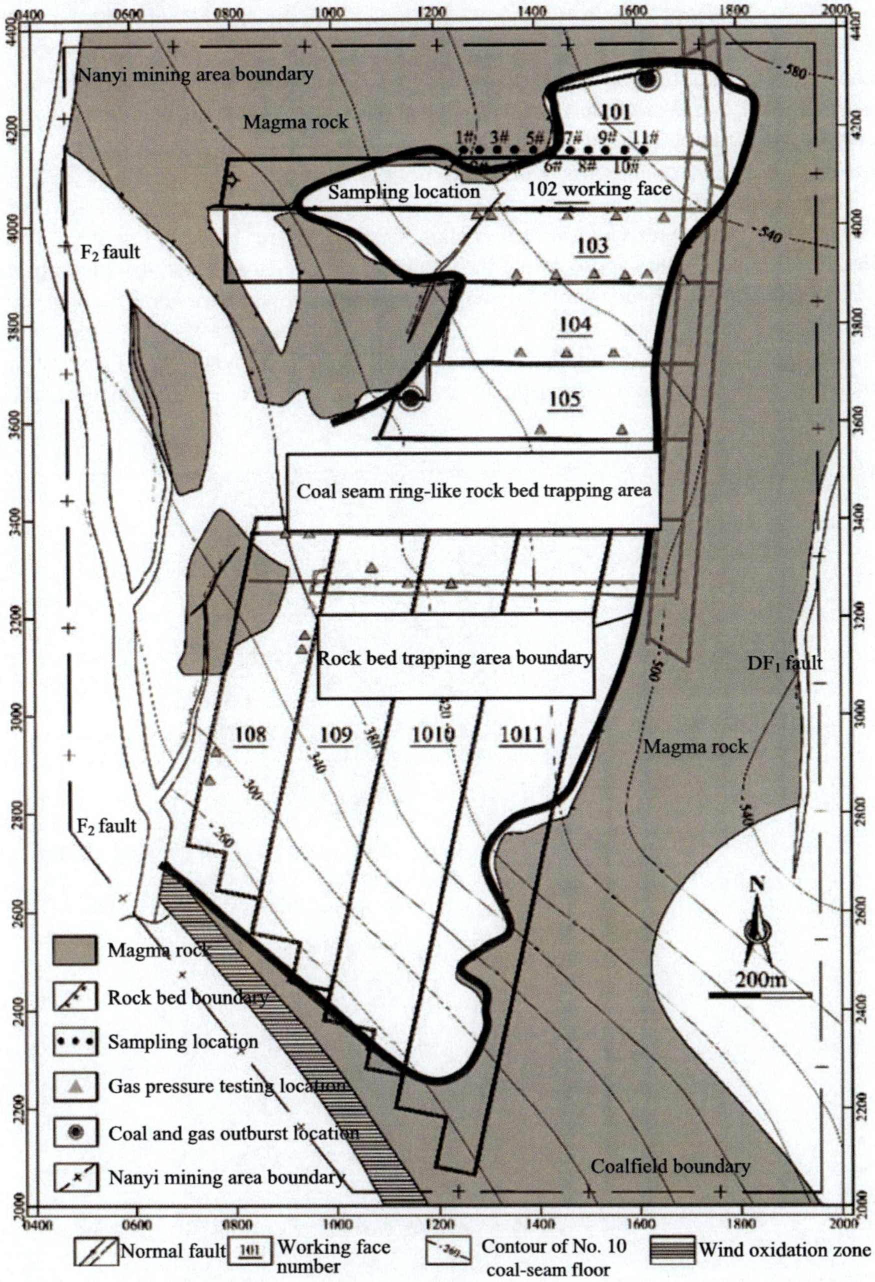

Fig. 2.17 The ring-like sill in Wolonghu Coal Mine

two examples are all in the form of sills: the thick sill in the Haizi Coal Mine entered along the roof of No. 5 coal seam forming a layer above the Nos. 7, 8, 9, 10 coal seams; and sill in Wolonghu Coal Mine invaded along the No. 10 coal seam. The difference is that the sills in Wolonghu Coal Mine look like a horizontal ring surrounding the No. 10 coal seam, but the sills in the Haizi Coal Mine look like a giant cover above the Nos. 7, 8, 9, 10 coal seams sealing the coal group in the vertical direction [8, 15].

The measured adsorption capacities of the collected samples from the affected coal seam in Wolonghu Coal Mine and Haizi Coal Mine are shown in Fig. 2.18. The data indicate that when approaching the sill in Wolonghu Coal Mine, the adsorption constant (*a*) initially increases and then decreases rapidly; while for Haizi Coal Mine, a variable increase trend was noticed. In the Wolonghu Coal Mine, the contact-metamorphism effect of magmatic sill reduced the methane adsorption capacity of the coal in the contact-metamorphism area (0–5 m from the sill). However, the

Fig. 2.18 Adsorption parameter (*a*) versus the distance to magma rocks

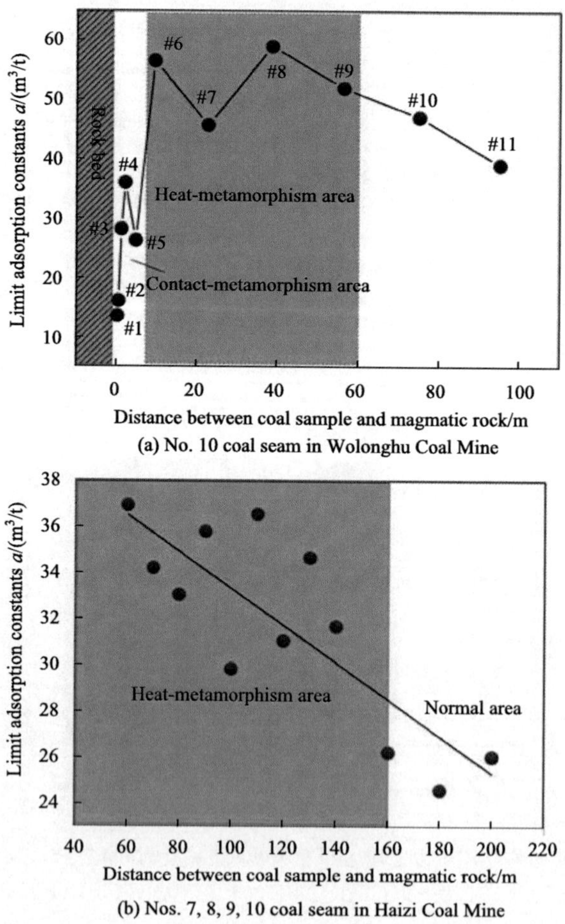

thermal evolution effect increased the methane adsorption capacity of the coal in the thermal evolution area (5–60 m from the sill). For the Haizi Coal Mine, the thermal evolution effect significantly increased the gas adsorption capacity of coal in the thermal evolution area within 60–160 m from the sill.

References

1. Li, Z. X. (2009). *Coal geology*. China Coal Industry Publishing House. (in Chinese).
2. Zhang, S. Q. (2004). *Coal chemistry*. China University of Mining and Technology Press. (in Chinese).
3. Han, D. X. (1996). *Coal and rock science in China*. China University of Mining and Technology Press. (in Chinese).
4. Li, W. H. (2006). Correspondence between mean maximum reflectance of vitrinite and classification of bituminous coals. *Journal of China Coal Society, 03*, 342–345. (in Chinese).
5. Yu, Q. X., & Cheng, Y. P. (2011). *Coal mine gas control*. China University of Mining and Technology Press. (in Chinese).
6. Cheng, Y. P. (2010). *Theory of coal mine gas control and its engineering application*. China University of Mining and Technology Press. (in Chinese).
7. Yu, B. F., & Wang, Y. A. (2000). *Technical manual for mine gas disaster prevention and utilization*. China Coal Industry Publishing House. (in Chinese).
8. Jiang, J. Y. (2012). *Control effect of igneous intrusion on methane reserve and outburst disaster prevention technology based on Huaibei Coalfield*. China University of Mining and Technology. (in Chinese).
9. Webber, K. L., Falster, A. U., Simmons, W. B., et al. (1997). The role of diffusion-controlled oscillatory nucleation in the formation of line rock in pegmatite-plite dikes. *Journal of Petrology, 38*(12), 1777–1791.
10. Barker, C. E., Bone, Y., & Lewan, M. D. (1998). Fluid inclusion and vitrinite-reflectance geothermometry compared to heat-flow models of maximum paleo temperature next to dikes, western onshore Gippsland Basin, Australia. *International Journal of Coal Geology, 37*(1), 73–111.
11. Zhang, X. L. (2015). *Research on coal seam occurrence and its dynamic disasters prevention and control technologies under an extremely thick magmatic rock*. China University of Mining and Technology (in Chinese).
12. Yang, Q. (1999). Superimposed metamorphism of Chinese Coal. *Earth Science Frontiers* (in Chinese).
13. Yang, Q. (1987). Hot metamorphism of regional magma and its effect on coal quality in China. *Modern Geology* (1) (in Chinese).
14. Yang, Q., & Ren, D. Y. (1981). Discussion on the coal deterioration in China. *Coal Geology and Exploration* (1) (in Chinese).
15. Jiang, J. Y., Cheng, Y. P., Wang, L., et al. (2011). Petrographic and geochemical effects of sill intrusions on coal and their implications for gas outbursts in the Wolonghu Mine, Huaibei Coalfield, China. *International Journal of Coal Geology, 88*(1), 55–66.
16. Zhang, X., Cheng, Y., Wang, L., et al. (2015). Research on the controlling effects of a layered sill with different thicknesses on the underlying coal seam gas occurrence. *Journal of Natural Gas Science and Engineering, 22*, 406–414.
17. Wang, W. (2014). Controlling effect of extremely thick igneous rock on gas occurrence of underlying coal seam. *Journal of Mining and Safety Engineering, 01*(1), 154–160. (in Chinese).
18. Liu, D., Yao, Y., Tang, D., et al. (2009). Coal reservoir characteristics and coalbed methane resource assessment in Huainan and Huaibei coalfields, Southern North China. *International Journal of Coal Geology, 79*(3), 97–112.

19. Lu, P., Bao, J., & Shen, Z. W. (2001). Pore structure characteristics of coal seam in magmatic rock erosive area and its effect on gas storage. *China Safety Science Journal, 11*(6), 41–44. (in Chinese).
20. Yao, Y., & Liu, D. (2012). Effects of igneous intrusions on coal petrology, pore-fracture and coalbed methane characteristics in Hongyang, Handan and Huaibei coalfields, North China. *International Journal of Coal Geology, 97*(3), 72–81.
21. Yao, Y., Liu, D., & Huang, W. (2011). Influences of igneous intrusions on coal rank, coal quality and adsorption capacity in Hongyang, Handan and Huaibei coalfields, North China. *International Journal of Coal Geology, 88*(2), 135–146.
22. Chen, J. (2014). Mineralogical and geochemical responses of coal to igneous intrusion in the Pansan Coal Mine of the Huainan coalfield, Anhui, China. *International Journal of Coal Geology, 124*(4), 11–35.
23. Wang, L. (2009). *Fracture evolution and seepage characteristics of distant pressure relief coal-rock masses and application in gas extraction under an extremely thick igneous rock*. China University of Mining and Technology (in Chinese).
24. Wang, L. (2011). Effects of an extremely thick igneous rock on gas occurrence and outburst disasters in a coal seam. *Journal of china university of mining and technology, 1*, 29–34. (in Chinese).
25. Xu, C., Cheng, Y., Ren, T., et al. (2014). Gas ejection accident analysis in bed splitting under igneous sills and the associated control technologies: A case study in the Yangliu Mine, Huaibei Coalfield, China. *Natural Hazards, 71*(1), 109–134.
26. Wang, L. (2013). The controlling effect of thick-hard igneous rock on pressure relief gas drainage and dynamic disasters in outburst coal seams. *Natural Hazards, 66*(2), 1221–1241.
27. Yang, Q., Li, S. T., & Chen, Z. H. (1979). *Coal field geology*. Geological Publishing House. (in Chinese).

Chapter 3
Basic Physical Properties and Characteristics of Coal Pores and Fractures

Learning Outcomes

- Moisture, ash content, volatiles, density and hardness of coal
- Pore, fracture structures of coal and characterization methods
- Coal matrix and its scale features.

3.1 The Basic Physical Properties of Coal

The basic physical properties of coal described in this chapter include the indexes of proximate analysis, density and hardness. Proximate analysis refers to the moisture, ash content and volatiles of coal. Density of coal includes the true density and apparent density. Hardness of coal is usually characterized using f value.

The proximate analysis of coal is determined using a 5E-MAG6600 automatic proximate analyser (Changsha Kaiyuan Instruments, China) in the National Engineering Research Center (NERC) for coal and gas control, China University of Mining and Technology (CUMT). The instrument contains two parts: Analyzer I for volatiles measurement and Analyzer II for moisture and ash content measurement. The instrument design and the whole process of measurements are performed according to the China National Standard GB/T 212–2001.

Moisture determination: place 0.5–0.6 g of air-dry coal samples in a shallow wall crucible and put the crucible on the ceramic rotary table of the heating furnace for moisture and ash measurement (Analyzer II). Heat the sample in the dry nitrogen flow of 107 °C until the mass is constant, then the percentage of moisture is calculated according to the mass loss of the coal sample.

Ash determination: heat the sample residue of moisture determination in the oxygen flow to 500 ± 10 °C according to the specified rate of heating, and keep the temperature constant for 30 min. After that, increase heat of the heating furnace to 815 ± 10 °C, and burn the sample to a constant mass under this temperature, then the percentage of ash can be obtained according to the mass of burned residue.

Y. Cheng et al., *Coal Mechanics*,
https://doi.org/10.1007/978-981-16-3895-4_3

Volatiles determination: place a number of 0.5–0.6 g air-dry coal samples in a deep wall crucible and put the crucible on the ceramic rotary table of the constant temperature furnace for volatiles measurement (Analyzer I). Heat the high temperature furnace to 900 ± 10 °C, and keep the temperature constant for 2 min. Place the No. 0 crucible (empty, as a reference basis) into the high temperature furnace for 7 min, then put the No. 0 crucible back to the constant temperature furnace again. After that, for every 7 min send another sample to the high temperature furnace for burning, and then put back into the constant temperature furnance. When all samples are analyzed, heat the constant temperature furnace and maintain at 107 °C. Dry all samples for 30 min then cool down the furnace. The volatiles of coal can be obtained by the difference value between the percentage of mass loss and the moisture content (on air dry basis) of each coal sample.

In this chapter, the physical properties of coal are described together with their corresponding coal ranks (vitrinite reflectance of coal, R_o). In the process of metamorphic evolution, physical and chemical properties of coal have experienced substantial changes. In this case, based on the variation range of vitrinite reflectance, coal can be divided into lignite/non metamorphic coal ($R_o < 0.5\%$), low metamorphic bituminous coal ($0.5\% < R_o < 0.9\%$), medium metamorphic bituminous coal ($0.9\% < R_o < 1.6\%$), high metamorphic bituminous coal ($1.6\% < R_o < 2.5\%$) and anthracite ($R_o > 2.5\%$) and so on.

The data of moisture, ash and volatiles in this chapter are derived from 3 parts:

(1) The experimental data from NERC in CUMT;
(2) The data published in Chinese literature;
(3) The data published in English literature.

Due to the number of literature involved [1], they are not listed in detail in this book, if necessary, readers can find them in the reference.

3.1.1 *Moisture of Coal*

Generally, the coal moisture can be divided into inherent moisture and external moisture [2]. The inherent moisture is formed from the moisture in the plants when the plants were converted into coal, and exists in the less-developed pores. External moisture is formed in the process of mining and transport and usually attached to the surface of coal particles or well-developed capillary pores and fractures. Generally, the coal samples adopted for proximate analysis are air-dried coal and the corresponding moisture content is called moisture on air dry basis, and expressed as M_{ad}, the value of which is consistent with the numerical value of inherent moisture. The relationship between air-dried moisture (M_{ad}) and the vitrinite reflectance (R_o) is shown in Fig. 3.1.

As shown in Fig. 3.1, the moisture content of coal is high when the coal rank is in low metamorphic phase. However, with the increase of metamorphism degree, the relationship between the coal moisture (M_{ad}) and vitrinite reflectance (R_o) can be

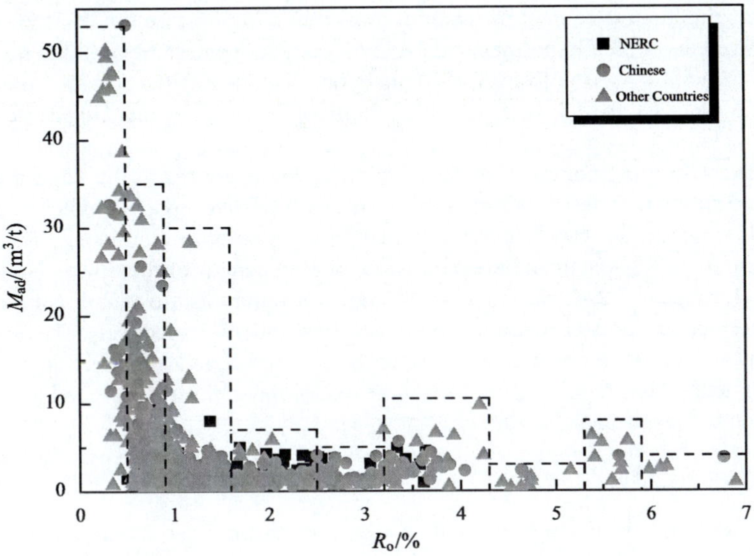

Fig. 3.1 Relationship between moisture of coal and coal rank (*Data size* NERC-189, Chinese-869, Other Countries-658)

presented by the exponential form as shown in Eq. (3.1):

$$M_{\text{ad}} = k_{mad} \exp(R_o/c_{mad}) + b_{mad} \tag{3.1}$$

where M_{ad} is coal moisture on air dry basis, %; k_{mad}, c_{mad}, b_{mad} are coefficients ($c_{mad} < 0$, $k_{mad} > 0$).

Taking the first and second derivative of Eqs. (3.1), (3.2) and (3.3) are obtained.

$$M'_{\text{ad}} = \frac{k_{mad}}{c_{mad}} \exp(R_{\text{o}}/c_{mad}) \tag{3.2}$$

$$M''_{\text{ad}} = \frac{k_{mad}}{c^2_{mad}} \exp(R_{\text{o}}/c_{mad}) \tag{3.3}$$

As $c_{mad} < 0$ and $k_{mad} > 0$, the first derivative [Eq. (3.2)] is always less than zero and the second derivative [Eq. (3.3)] is always greater than zero. Thus, it can be noticed that Eq. 3.1 is a monotone decreasing exponential function. That is, the inherent moisture of coal increases with the decreasing metamorphism degree and the decrease rate gradually. In the low metamorphic phase the inherent moisture of coal decreases markedly, that is, the degree of metamorphism has an inverse effect on the moisture of the coal. In Fig 3.1 it can be seen that the maximum value of moisture from non-metamorphic coal to high-metamorphic bituminous coal are generally less than the values of $M_{\text{ad,max}} = 2.217 + 83.84 \exp(-R_o/0.64)$. The non-metamorphic coal has the highest content of inherent moisture, with a maximum of 55%; while the low-metamorphic bituminous coal shows a slight decrease trend in the inherent moisture, with a maximum of 35%. This is because the capillary pores from non-metamorphic coal to high-metamorphic bituminous coal are inclined to adsorb more water. Inherent moisture of mid-metamorphic bituminous coal shows an ordinal decrease with a maximum of 30%, which may be attributed to the large number of fused ring structure containing hydrogen that have hydrophobic effect in coal's organic structure. The inherent moisture of high-metamorphic bituminous coal shows a maximum of 8% and the inherent moisture of anthracite is further lower. However, in the R_0 range of 3.2–4.2% and 5.2–5.9%, the inherent moisture of anthracite shows obvious increase, and the maximum can reach 10% and 9% respectively. The possible reason for this phenomenon maybe due to the carbon network of anthracite that has already tended to graphitization, thus there exists aphanitic structure and the space between the layers can absorb more water [3].

Reasons for the changing of inherent moisture can be comprehensively summarized as shown below:

(1) The inherent moisture of coal is adsorbed on the internal surface of coal's pore, the higher the internal surface area is, the higher capability of moisture absorption will be, as a result the moisture content of coal will be higher.
(2) The more polar oxygen-containing functional groups are in the structure of the coal, the ability of coal for moisture adsorption will be higher.
(3) The low metamorphic bituminous coal usually has developed internal surface area, as well as large numbers of polar oxygen-containing functional groups in the molecular structure, thus the inherent moisture of coal is extremely high. With the increase of coal metamorphism, both the internal surface areas and the oxygen-containing functional groups show decreasing trend, and the inherent moisture of coal decreases as a result. When the metamorphism of coal comes to the anthracite, because of the increase of coal's internal surface area, the inherent moisture of coal increases slightly.

3.1.2 Ash in Coal

Ash is the remaining residue after complete combustion of coal, and almost all of it comes from minerals, usually, the ash content is proportional to the content mineral of coal. Ash is not an inherent component of coal but converted from the mineral of coal, however there also exist a big difference between ash and mineral. Firstly, the ash content is lower than corresponding mineral content in coal, and secondly the composition changes when the mineral is converted to ash. Minerals turn into ash after chemical reactions such as decomposition, oxidation and combination under high temperature. Because the moisture in air dried coal changes with air humidity, the measured data of ash content will change as well. But in terms of the absolutely dried coal, the ash content keeps constant. Therefore, the ash content on dry basis (A_{ad}) is commonly used in the practice. Figure 3.2 shows the relationship between

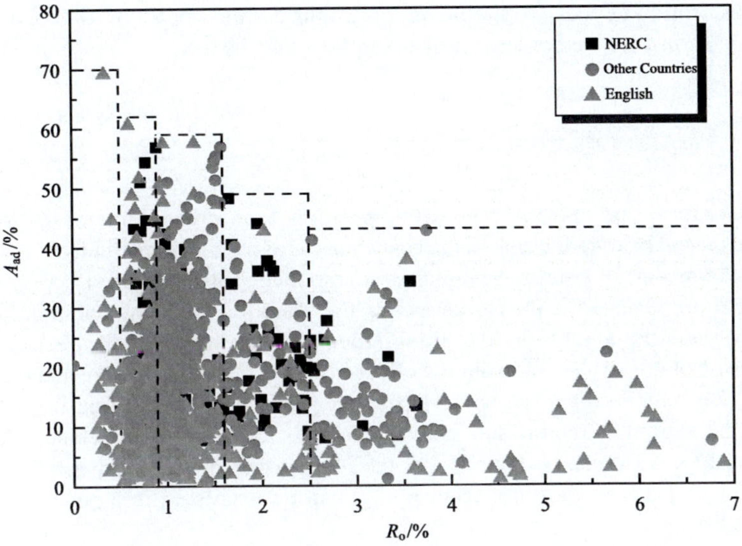

Fig. 3.2 Relationship between ash content and vitrinite reflectance (*Data size* NERC-189, Chinese-868, Other Countries-697)

A_{ad} and the vitrinite reflectance (R_o).

Data in Fig. 3.2, show that because the ash content of coal is affected greatly by the geological evolution conditions, the data points are much more dispersed than the moisture trend shown in Fig. 3.1. Generally, the ash content of coal shows a decreasing trend with the increased metamorphism where the maximum value does not exceed the curve described by this formula: $A_{d,max} = -8.657R_o + 68.435$. Usually, the ash content in subbituminous coal is below 70%, and below 62% in high volatile bituminous coal, below 59% in mid volatile bituminous coal, below 49% in low volatile bituminous coal and below 43% in anthracite.

Existence of ash in coal has adverse effects on coal's adsorption capacity. The influencing mechanism of ash on coal's adsorption can be mainly attributed to:

(1) According to the rule of "like dissolves like", the properties of methane are similar to the organic matter with weak polarity and better adsorption character, but ash is mostly inorganic with weak adsorption capacities, the presence of ash reduces the adsorption content of methane in coal;
(2) The presence of silica, iron, calcium, magnesium and other components, especially the presence of metal elements in coal will result in the basal cementation tendency, and the increasing cements will block the coal pores and fissures, reducing the porosity of coal as well as the methane adsorption sites, resulting in decreased adsorption capacity of methane in coal.

The combined effects of the two aspects reduce the ability of organic matter to adsorb methane in coal. In view of this, in the evaluation of ash effect on the adsorption of coal, we should not only consider the impact of ash content, but also consider the possession of adsorption sites by ash.

3.1.3 Volatiles of Coal

Volatiles are the products of heat decomposition from organic matter and part of minerals when heating the coal in the condition of high temperature and air isolation. Volatiles are not all from the permanent components of coal but also part of the pyrolysis products, so it is also known as the volatile yield. The commonly used index for volatiles are: air-dried basis volatile (V_{ad}), dry basis Volatile (V_d), dry ash-free basis volatile (V_{daf}). To avoid the effect from the inorganic matter like moisture or ash, the volatile matter is usually characterized on a dry ash-free basis (V_{daf}). Figure 3.3 shows the relationship between volatile and vitrinite reflectance (R_o).

From Fig. 3.3 it can be noticed that the relationship between dry ash-free basis volatile (V_{daf}) and vitrinite reflectance (R_o) can be expressed in a power function form as Eq. (3.4).

$$V_{daf} = k_{vdaf} R_o^{b_{vdaf}} \tag{3.4}$$

where V_{daf} is the volatile matter on a dry ash-free basis, %; k_{vdaf}, b_{vdaf} are the coefficients in Eq. (3.4), where $k_{vdaf} > 0$, $b_{vdaf} < 0$.

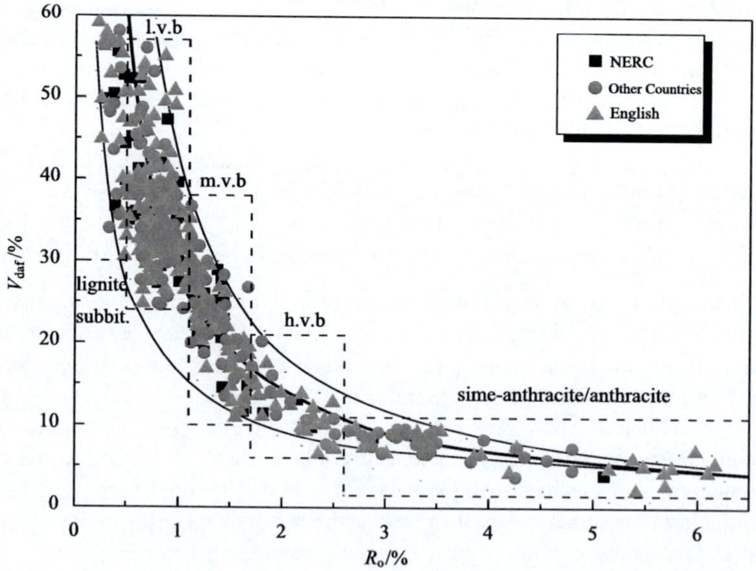

Fig. 3.3 Relationship between volatile content and vitrinite reflectance (*Data size* NERC-189, Chinese-195, Other Countries-371)

The first and second order derivatives of Eq. (3.4) are shown as Eqs. (3.5) and (3.6) respectively:

$$V'_{\text{daf}} = k_{v\text{daf}} b_{v\text{daf}} R_{\text{o}}^{b_{v\text{daf}}-1} \tag{3.5}$$

$$V''_{\text{daf}} = k_{v\text{daf}} b_{v\text{daf}} (b_{v\text{daf}} - 1) R_{\text{o}}^{b_{v\text{daf}}-2} \tag{3.6}$$

As $k_{v\text{daf}} > 0$ and $b_{v\text{daf}} < 0$ in Eq. (3.4) its first derivative [Eq. (3.5)] is always less than zero, while its second derivative [Eq. (3.6)] is always greater than zero. According to the properties of function, it can be concluded that Eq. (3.4) exhibits as Monotone decreasing power function. Therefore, with the increase of coal rank the volatile matter of coal decreases, and compared with the higher coal rank, the volatile at the low metamorphic grade shows range of a rapid decrease. The reason for this is because the volatile matter is mainly composed of the small molecular compounds broken from unsaturated group in the fat side-chain and the oxygen-containing functional groups in coal molecules, as well as the hydrogen gas produced by the condensation of organic polymers. With the increase of coal metamorphism, the content of fat side chain and oxygen-containing functional groups in coal both decrease. From Fig. 3.3, it is also noticed that the changing trend of V_{daf} and R_{o} has upper/lower limits, where the upper limit can be represented using the equations of $V_{\text{daf,max}} = 42.39R_{\text{o}}^{-1.183}$ while the lower limit can be expressed using $V_{\text{daf,min}} = 16.39R_{\text{o}}^{-0.756}$, and the corresponding average curve is $V_{\text{daf},avr} = 29.63R_{\text{o}}^{-1.05}$. Likewise, it is clear that the dry ash-free basis volatile V_{daf} for the non-metamorphic coal is generally larger than 30%; for low metamorphic bituminous coal, V_{daf} ranges between 25 and 58% which shows disperse distribution; for medium metamorphic bituminous coal, V_{daf} ranges between 10 and 38% which indicates that due to the complex forming environment and coal/rock composition, the differences in this kind of coal composition are very significant; for high metamorphic bituminous coal, V_{daf} ranges between 5 and 20% which shows narrow distribution; and for anthracite, V_{daf} is usually smaller than 10%, which is the smallest value among all kinds of coal rank.

3.1.4 Density of Coal

The coal density means the mass of coal per unit volume, and the units are defined as g/cm^3 or t/m^3, which is a key parameter to reflect coal's physical property and structure. The interactions between molecules are a function of the distance between the molecules and the density depends on the degree of compaction of molecular structure and molecular arrangement [4]. The density of coal can be described by the following two methods.

True density is the mass of coal per unit volume under the absolutely compacted condition, the volume of which refers to the skeleton volume by subtracting the intragranular pore volume and interparticle void volume from the volume of coal.

Apparent density is the mass of coal per unit apparent volume where the apparent volume is described as the sum of skeleton volume, intragranular pore volume and interparticle void volume.

The UltraPYC 1200e automatic true density analyzer produced by Quantachrome, USA has been currently used in the National Engineering Research Centre (NERC), CUMT. This instrument uses helium gas instead infiltration fluid, because helium is a stable inert gas with small molecule diameter (0.26 nm) and characteristics of easy diffusion and good permeability. Thus the helium molecule can quickly penetrate into the internal pores in coal and overcome the disadvantages of conventional fluid methods that were time consuming with less reliable penetration of the internal pores and irregular surfaces. In this case, the test results provide values that are closer to the actual volume of the coal sample to determine the true coal density value. Defining the skeleton volume of coal sample as V_p and the mass of sample as m, the true density of coal can be calculated as:

$$\rho_t = \frac{m}{V_p} \tag{3.7}$$

where ρ_t means the true density of coal, g/cm^3.

True density of coal depends on the coal-forming material, degree of metamorphism, petrographic components, coal properties and mineral contents. The density of humic coal is higher than that of sapropelic coal and the density variation is not clear when the metamorphism changes from lignite to bituminous coal. The true density is usually 1.3–1.5 g/cm^3 for lignite, 1.3–1.4 g/cm^3 for bituminous coal and 1.35–1.9 g/cm^3 for anthracite.

To obtain the apparent density of coal, the measurement of apparent volume (namely the total volume of coal) is needed. For out-of-shape coal, wax-sealing method is generally adopted to seal pores. Then, fluid-discharge method is used to measure the apparent volume. The apparent density of coal is calculated as follows:

$$\rho_a = \frac{m}{V_t} \tag{3.8}$$

where ρ_a is the apparent density of coal, g/cm^3; V_t is the apparent volume of coal, cm^3.

According to the definition of true density and apparent density, Eqs. (3.7) and (3.8), the porosity of coal (ϕ_t) can be obtained as follows:

$$\phi_t = \frac{V_t - V_p}{V_t} \cdot 100\% = \frac{\rho_t - \rho_a}{\rho_t} \cdot 100\% \tag{3.9}$$

3.1.5 Hardness of Coal

The coal hardness is generally characterized by the hardness coefficient of coal, which is also known as the Protodyakonov coefficient and is usually expressed with a dimensionless f value. The hardness coefficient of coal is defined as the relative value to characterize the resistance of solid coal to crushing, which is equivalent to 1/10 of the uniaxial compressive strength of coal. The hardness coefficient was first proposed by Russian scholar in 1926. Because the test method is simple and practical, it is widely used in the Chinese coal mining industry. The hardness coefficient is mainly applied to petrographic classification and the risk analysis of coal and gas outburst.

The hardness coefficient reflects the mechanics and strength properties of solid coal, which is based on the strength theory of brittle materials, indicating that the work consumed by crushing is related to the increase in surface area of the crushed material. The hardness coefficient measurements follow the standard: "*Methods for determining the physical and mechanical properties of coal and rock—Part 12: Methods for determining coal hardness coefficient*" (GB/T 23,561.12-2010). The measurement procedures are as follows:

(1) Crush the coal samples into smaller particle sizes 20–30 mm using small hammer, and then sieve them to the needed quantities and particle sizes using the 20–30 mm sieves. For the soft coal, the 1–3 mm sieves are recommended instead of the 20–30 mm sieves. Divide the sieved samples into five 50 g portions.

(2) Place the pounding barrel on the concrete floor or 2 cm thick iron plate, and put one portion of coal sample into the barrel. Lift the 2.4 kg weighted heave hammer to the height of 600 mm, then release the heave hammer and let it strike each portion of coal sample three times. After that, put the pulverized samples (5 portions) together into the same container.

(3) Sieve out the < 0.5 mm sized coal powder in the pulverized samples (5 portions), until no particles leak out from sieves.

(4) Put the sieved coal (< 0.5 mm) into metering cylinder and finger-tap to make the coal powder compact tight. Carefully insert the cylindrical calibrated scale into the metering cylinder until lightly in contact with the coal powder surface. Read the height of powder in the metering cylinder from the scale and record as h.

If $h \geq 30$mm, the number of strikes for each sample portion can be recorded as 3 times, and other groups of samples can be tested according to the above steps.

Else if $h < 30$mm, discard the first samples and increase the number of the heavy hammer striking to 5 times.

(5) Calculate the hardness coefficient using the following equation:

$$f = 20n/h \tag{3.10}$$

where f is the hardness coefficient, mm^{-1}; n is the striking number for each sample portion; h is the measured height of pulverized samples after sieving, mm.

Parallel tests are required to obtain the most probable value of hardness coefficient. One parallel test needs three groups of parallel samples (5 portions for each group) and the arithmetic mean value is regarded as the value of hardness coefficient.

(6) The determination of hardness coefficient

If the particle size of sieved coal samples does not reach the required size (20–30 mm), the 1–3 mm coal samples can also be used to conduct the test according to the above-mentioned procedure and requirements. However, the results must be converted using the following equations:

When $f_{1-3} > 0.25$,

$$f = 1.57 f_{1-3} - 0.14 \tag{3.11}$$

When $f_{1-3} \leq 0.25$,

$$f = f_{1-3} \tag{3.12}$$

where f_{1-3} refers to value of hardness coefficient of 1–3 mm sized coal samples.

Figure 3.4 exhibits data from NERC, CUMT [5] showing the relationship between hardness coefficient and dry ash-free basis volatiles (V_{daf}). As the data suggest, the

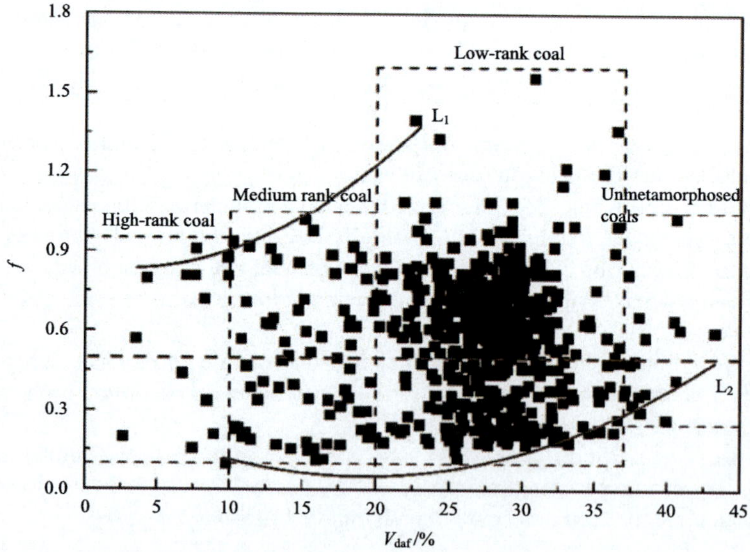

Fig. 3.4 Relationship between hardness coefficient and coal rank

hardness coefficient is mainly controlled by the geological structure, which has no link with coal rank. Usually, the firmness coefficient of intact coal is greater (generally >0.7) and tectonically deformed coal is smaller (minimum value can reach 0.1, which is controlled by the tectonic deformation degree). According to the "*Identification Standards of Coal and Gas Outburst*" of China the coal seam can be identified as an outburst prone coal seam if: the destruction type of coal belongs to the type III to type V (as shown in Table 1.1); the coal seam gas pressure is > 0.74 MPa and the initial speed of methane diffusion $\Delta P \geq 10$ mm Hg; and the hardness coefficient $f \leq 0.5$.

3.2 Coal Pore Features

3.2.1 Classification of Coal Pore

Coal is a porous medium with exceptionally well developed pores, distributed on the surface or in the body mainly composed of organic matter and minerals. The pore size in coal is not uniform and scholars in China and abroad divided coal pores according to their size. At present, the decimal classification criterion proposed by B. B. Hodot is commonly applied, i.e., micropores (<10 nm), minipores (10–100 nm), mesopores (100–1000 nm) and macropores (>1000 nm) [6].

There exist abundant pores in the coal body forming a complex porous medium. In the coal reservoir, pores act as the primary site for gas adsorption, storage, diffusion and migration. Hence, the features of coal pore are one of the main factors that affect the storage and migration of coal seam gas. According to the connectivity of pores, the pore types in coal can be classified as interconnected pores, passing pores, dead end pores and closed pores as shown in Fig. 3.5a. And according to the shapes, pores in coal can also be classified as cylindrical pores, conical pores, slits pores, spherical (ink bottle) pores, and interstices pores, as shown in Fig. 3.5b.

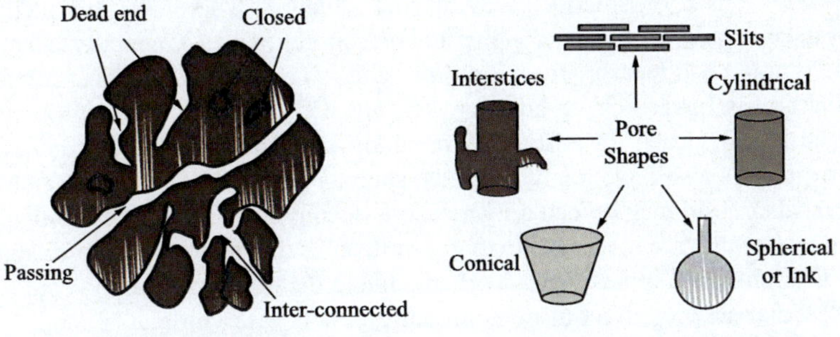

(a) Types of by pore connectivity (b) Types of pore by pore shapes

Fig. 3.5 Different types of coal pore

Six different samples (whose metamorphic degree ranges from long-flame coal to anthracite) were collected from different coal mines and transported to the laboratory. In the laboratory, the vitrinite reflectance, proximate analyses, true density, apparent density and porosity of the coal samples were tested and the results are shown in Table 3.1. In addition, the pore volume and specific surface area (SSA) of the samples were also investigated using the mercury intrusion porosimetry (MIP) and low-pressure nitrogen gas adsorption (LP-N_2GA) with corresponding results shown in Tables 3.2 and 3.3 [7].

As shown in Tables 3.2 and 3.3, in general, with the increase of coal rank the pore volumes of micropores and minipores firstly decreased and then increased, but the pore volume variation of mesopores shows no obvious change. For SSA, macropores and mesopores firstly showed an increased trend and then decrease, but no obvious changes were observed in micropores and minipores. The pore volume of coal is mainly concentrated in the macropores and mesopores therefore the ratio of pore volumes of macropore and mesopore can account for at least 55.00% (Sample 6#) of the total pore volume (TPV) and the maximum value can reach 90.24% (Sample 4#). On the contrary, the SSA of coal is usually concentrated in the minipores and micropores, and the SSA for micropores and minipores accounts for at least 80.41% (Sample 4#), and the maximum value can reach 96.57% (Sample 1#). Therefore, it can be concluded that coal seam gas is mainly adsorbed on the surface of coal's micropores and minipores.

3.2.2 Coal Pore Characterization Methods

The characterization methods for coal pore include MIP, adsorption method, Small Angle X-ray diffraction (SAXS), microscopic method, nuclear magnetic resonance (NMR) and CT scanning method (CT) as summarized in Fig. 3.6 [8]. Here is a brief introduction to these methods:

The minimum pore diameter that MIP method can measure is 7 nm, but the higher mercury intruding pressure (> 10 MPa) may cause potential destruction or compression effect on smaller pores of coal structure. Therefore, this method is mainly used to measure the macropores and mesopores of coal. Generally, adsorption method is applied to measure the micropores and minipores, and the most commonly used method is LP-N_2GA. In addition, H_2, He, CO_2 and CO can also be applied as the molecular probe for adsorption method. After the test, the interpretation of isotherm curves also have many methods, such as BET, BJH, t-plot, DFT and HF, etc. Gas adsorption method can only measure open pores, and for the smaller pores (closed pores) that are inaccessible to gas molecules this method is not suitable. By using different kinds of gases as adsorbent, due to the differences in molecules size, the molecular sieving effect of pore structure in coal reservoir can be studied. The advantage of SAXS method is the non-destructiveness to the sample and its ability to measure open as well as closed pores, but it is not easy to accurately measure the particle size because of the errors caused by multiple scattering are difficult to

Table 3.1 Basic physical properties of coal

Sample number	Sources	Coal rank	R_o/%	M_{ad}/%	A_{ad}/%	V_{daf}/%	True density/(g/mL)	Apparent density/(g/mL)	Porosity/%
1#	Yuanzigou coal mine	Lone-flame coal	0.65	6.22	13.28	43.10	1.4281	1.3142	7.98
2#	Renlou coal mine	Gas coal	0.98	1.60	9.89	38.10	1.3824	1.3368	3.30
3#	Pingmeiwu coal mine	Fat coal	1.12	0.93	27.60	28.86	1.5708	1.5328	2.42
4#	Zhongxing coal mine	Coking coal	1.67	1.10	7.82	18.37	1.4411	1.3900	3.55
5#	Yangzhuang coal mine	Meagre coal	1.86	1.59	17.85	9.10	1.6158	1.5246	5.64
6#	Jiulishan coal mine	Anthracite	2.94	3.59	19.96	7.98	1.5412	1.4618	5.15

Table 3.2 Analysis results of the pore volume distribution of coal

Coal sample	Macropore		Mesopore		Minipore		Micropore		Total	
	mL/g	%	mL/g	%	mL/g	%	mL/g	%	mL/g	%
1#	0.0220	25.25	0.0331	37.99	0.0204	23.43	0.0116	13.32	0.0871	100
2#	0.0122	49.64	0.0064	26.04	0.0058	23.78	0.0001	0.54	0.0246	100
3#	0.0107	55.68	0.0059	30.70	0.0023	12.21	0.0003	1.41	0.0192	100
4#	0.0187	62.73	0.0082	27.51	0.0029	9.65	0.0004	0.12	0.0298	100
5#	0.0135	35.84	0.0150	39.82	0.0080	21.42	0.0011	2.92	0.0377	100
6#	0.0026	23.44	0.0035	31.56	0.0041	37.38	0.0008	7.62	0.0111	100

Table 3.3 Analysis results of the SSA of coal

Coal sample	Macropore		Mesopore		Minipore		Micropore		Total	
	m^2/g	%	m^2/g	%	m^2/g	%	m^2/g	%	m^2/g	%
1#	0.0324	0.24	0.4248	3.18	2.9376	22.03	9.9413	74.54	13.3360	100
2#	0.0150	2.25	0.0757	11.31	0.48492	72.44	0.0937	14.00	0.6694	100
3#	0.0142	2.83	0.0622	12.44	0.2017	40.32	0.2222	44.41	0.5003	100
4#	0.0212	4.80	0.0652	14.79	0.29259	66.32	0.0622	14.09	0.4412	100
5#	0.0201	0.99	0.1690	8.37	0.9293	46.03	0.9007	44.61	2.0191	100
6#	0.0028	0.25	0.0435	3.85	0.4503	39.83	0.6339	56.07	1.1305	100

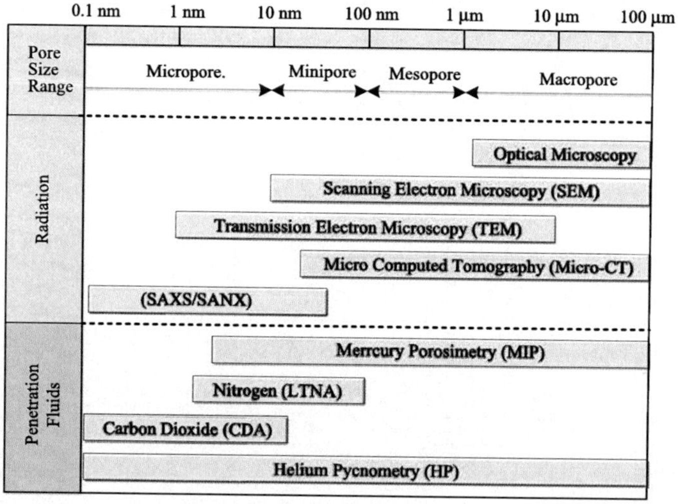

Fig. 3.6 Characterization technologies for coal pore

eliminate. The use of microscopy is intuitive. Micropores and mesopores can be observed through scanning tunneling microscope (STM) and transmission electron microscopy (TEM), and macropores can be observed through scanning electron microscope (SEM) as well as optical microscope (OM). By using the microscopy methods, the size, distribution, cross sectional area of pore structure, and smoothness of pore wall can be obtained, but the further interpretation of data needs plenty of statistical analysis and there are also some requirements for the image recognition technology. In addition, NMR and CT scan is also widely applied in the studies of pore permeability of coal in the reservoir of coalbed methane (CBM). The range of pore characterization technologies is shown in Fig. 3.6.

At present, the commonly used methods for pore structure analysis are MIP and LP-N_2GA. Their experimental principles and applications are introduced as follows:

1. Mercury intrusion porosimetry (MIP)

MIP is a common method for the characterization of pore size distribution (PSD). Generally, mercury cannot intrude into pore structure in the absence of external pressure. If there is external pressure, mercury may intrude into pore structure by overcoming the resistance from surface tension. Assuming that coal pore is cylinder-shaped, the pore radius and length are defined as r and l, respectively, the work generating wetted area (S) can be denoted as:

$$W_1 = -S\sigma \cos\theta = -2\pi rl\sigma \cos\theta \tag{3.13}$$

where S means wetted area of mercury or the specific surface area of cylinder-shaped pore, m^2; σ means surface tension of mercury, N/m; θ means wetting contact angle of mercury, °; r means pore throat radius, m; l means length of cylindrical pore, m.

The reason that introducing $\cos\theta$ into Eq. (3.13) is to interpret the fact that the force preventing mercury from entering pore shows an effect on the contact angle θ.

In addition, the work forcing mercury to intrude into cylinder-shaped pore is equal to the product of the value of pressure, sectional area of pore and length of pore:

$$W_2 = p\pi r^2 l \tag{3.14}$$

Because W_1 must be equal to W_2, then by combining the Eqs. (3.13) and (3.14) the following expression can be obtained:

$$p = \frac{-2\sigma \cos\theta}{r} \tag{3.15}$$

where p is capillary pressure, N/m^2.

The Eq. (3.15) is also known as the Washburn's equation, and the radius of pore throats can be calculated by the capillary pressure of the intruded mercury. Furthermore, if assuming that θ and r are constant, then the larger the radius of pore throats is, the smaller the capillary resistance and the injected pressure are. Therefore, with the increase of injected pressure, mercury intrude gradually from macropores into micropores. Under the condition of equilibrium pressure, the injected mercury volume is equal to the pore volume under this pressure.

The MIP tests using a PoreMaster 33-GT-12 automated mercury intrusion porosimeter (Quantachrome Instruments, United States) were conducted in the NERC, CUMT laboratory. The measuring range of low pressure station for this instrument is 1.5–350 kPa, and for high pressure station this range is 0.14–227 MPa. Thus the measurable range of pore diameter is 7 nm–1000 μm. Dried and non-corrosive N_2 is normally adopted as the compressed gas for test, with a pressure of 420 kPa. The liquid nitrogen is used as a coolant in the cold trap. Before the test, the samples were degassed in the vacuum pump at 50 °C for 6 h, and weighed. The surface tension of mercury was set to 0.48 N/m and contact angle of mercury for coal samples was 140°.

The morphology of MIP curves can reflect the pore development and pore connectivity of coal, which may provide theoretical foundation for the gas adsorption/desorption properties of coal. From the MIP results (shown in Fig. 3.7 [9]) of Machang samples (V_{daf} = 21.42%, particle size: 0.08–0.2 mm), Yangquanwu samples (V_{daf} = 10.68%, particle size: 0.2–0.25 mm) and Bailongshan samples (V_{daf} = 8.84%, particle size: 0.2–0.25 mm), it can be noticed that different.

MIP curves show similar trends. That is, the intrusion curves increase faster at the low pressure stage, slowing down as the pressure grows. At the high pressure stage the intrusion and extrusion curves seem to be close and overlapped, indicating that three samples show the characteristics of the open pore at both ends for this certain particle size. Figure 3.8 summarizes the MIP results of Machang samples with various particles sizes of 1.0–3.0 mm and 3.0–4.0 mm. These results are markedly

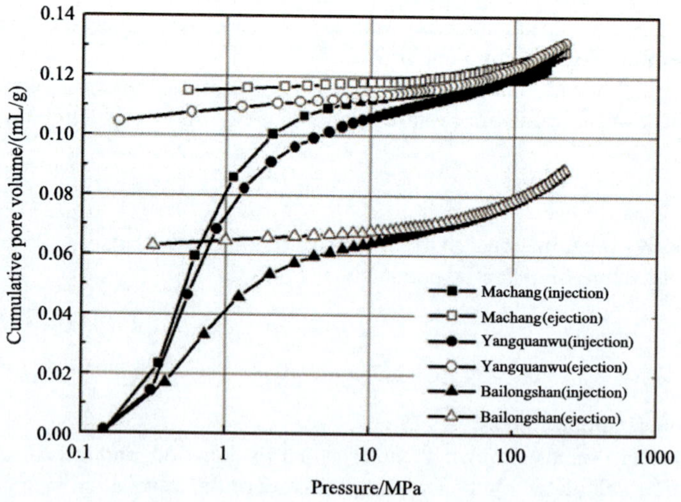

Fig. 3.7 MIP curves of coal samples collected from Machang, Yangquanwu and Bailongshan Coal Mines

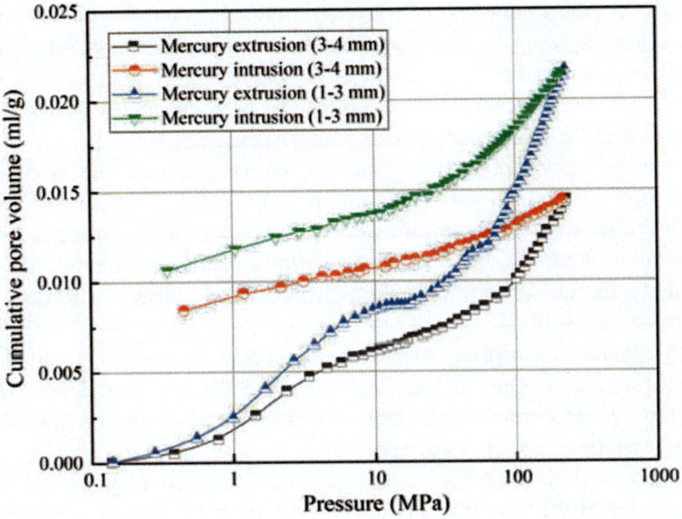

Fig. 3.8 MIP curves of large particle sized coal sample from Machang Coal Mine

different from the results of Fig. 3.7. The intrusion curves increase slowly at the low pressure stage and increase faster with the growing pressure. Then, the intrusion and extrusion curves separate at the high pressure stage, indicating that the samples exhibit the characteristics of open pore at one side only.

2. Low-pressure CO_2 adsorption (LPGA-CO_2);
   Low-pressure N_2 adsorption (LPGA-N_2)

The principle of LP-N_2GA for pore characterization is the method of equivalent volume substitution, namely regarding the filling volume of liquid nitrogen as the pore volume. It can be seen from capillary condensation theory that the pore size range of capillary condensation increases with the relative pressure ratio p/p_0. For a given p/p_0, there exists a critical value of pore radius r_k, and when pore size is less than r_k then the capillary condensation is triggered; else if the pore size is larger than r_k, capillary condensation does not occur. This critical radius of r_k can be calculated by Kelvin Eq. (3.16):

$$r_k = \frac{-0.414}{\ln(p/p_0)} \tag{3.16}$$

where r_k is critical radius triggering the capillary condensation under the given relative pressure p/p_0, which is also known as the Kelvin radius. It can also be understood that when the relative pressure ratio is lower than p/p_0, then the condensed liquid nitrogen in the pores whose radius is above the corresponding r_k will gasify and desorb. However, before the real condensation, the internal surface of pore actually has adsorbed a certain thickness of nitrogen molecules, and thickness of adsorbed layer changes with p/p_0. As a consequence, the calculation of pore size distribution needs an appropriate correction.

For the pore range that cannot be determined by MIP, especially for the measuring of nanoscale pore, the most widely used characterization method is LP-N_2GA. Using the nitrogen as the adsorbent gas, the relative pressure is gradually elevated to measure adsorption capacity of coal under the constant temperature. Graphing the correlation between the relative pressure and the adsorption capacity, the nitrogen adsorption isotherm can be obtained. After that, by generally decreasing the relative pressure and measuring the gas desorption volume, the corresponding desorption isotherm can be obtained as well.

For the physical adsorption experiments in NERC laboratory, an Autosorb-IQ_2-MP Full-automatic Specific Surface Area and Micropore Diameter Analyzer (produced by Quantachrome Instrument, USA) is used. Nitrogen and other non-perishable gases (such as CO_2, Ar, H_2) are adopted as adsorbate gas for the instrument. Before the test, the experimental samples are first degassed in the vacuum pump (150–200 °C for 10 h). The instrument is equipped with a double analysis station and can conduct BET analysis, pore structure analysis, ultralow specific surface area (using Kr) and ultramicropore analysis (using N_2, Ar, CO_2). The measurement range of pore size is 0.35–400 nm, and the specific surface area this instrument can determine is more than 0.0005 m^2/g (using Kr) and 0.01 m^2/g (using N_2). The measurement environment is liquid nitrogen (77 K), liquid argon (87 K), ice-water bath (273 K) and other cold bath.

As shown in Fig. 3.9 (particle size of coal is the same as that in Fig. 3.7) [9], the adsorption branches of LP-N_2GA isotherms of Machang, Yangquanwu, Bailongshan

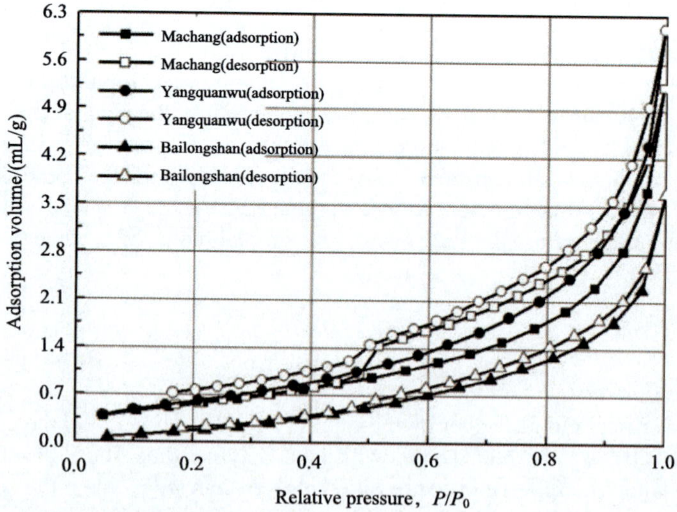

Fig. 3.9 LP-N_2GA isotherms of Machang, Yangquanwu, and Bailongshan coal samples

coal samples increase slowly in low pressure stage. Then with the increase in relative pressure, the adsorption isotherms increase sharply, indicating that larger pores participate in the capillary condensation. For Machang, Bailongshan coal samples, their adsorption/desorption branches are closed in the low pressure stage. However, for the Yangquanwu coal samples, the branches do not close in the low pressure stage.

3. Full-scale analysis of coal pore

Because the measuring ranges of MIP and LP-N_2GA are different in pore structure characterization and neither of them can cover the full-scale of pore system, therefore choosing the optimum analytic ranges of each method and conduct the analysis may provide much more accurate results. That is, for the pores whose diameter < 20 nm, the pore structure is characterized by LP-N_2GA; and for the pores whose diameter > 20 nm; the pore structure is measured by MIP. Using Eqs. 3.15 and 3.16 to transform the data of x-coordinate in Figs. 3.7 and 3.9 into pore radius r, and then the cumulative pore volume in Figs. 3.7 and 3.9 are adopted for taking the derivative on pore radius r to get the pore volume per pore radius $D_v(r)$, namely pore size distribution density function. The full-scale pore analysis of $D_v(r)$ related with r for Machang, Yangquanwu, Bailongshan coal samples are presented in Fig. 3.10 [9].

From Fig. 3.10, the pore size distribution density function curves of three coals basically correspond to the logarithmic normal distribution. When the pore radius range is between 0.5 and 5000 nm, with the increase of pore radius the distribution density of pore volume firstly increases and then decreases, and shows a peak value at the pore radius of 1.6 nm. This phenomenon demonstrates that at the pore radius of 0.8–1.6 nm, the increasing trend for cumulative pore volume becomes faster and

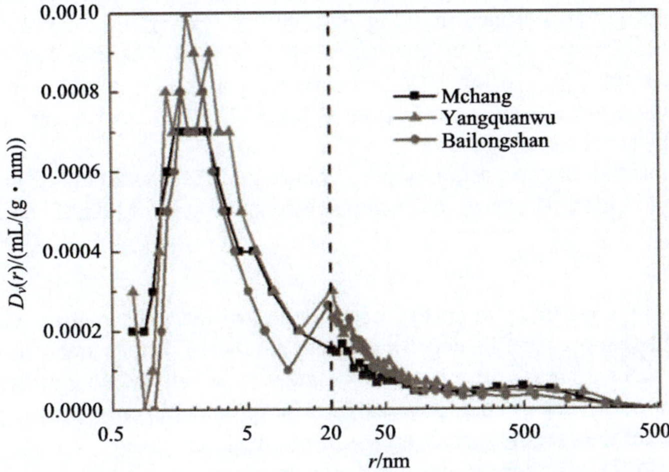

Fig. 3.10 Full-scale pore analysis of $D_v(r)$ related with r for Machang, Yangquanwu, Bailongshan coal samples

reaches a peak value at the radius of 1.6 nm; then at the pore radius range of 1.6–5000 nm, the increasing speed of cumulative pore volume gradually slows. This pore volume distribution pattern suggests that more pores exist at a radius of 1.6 nm and and thus can lead to more specific surface area as well as stronger capacity for gas adsorption. However, with the increase or decrease of pore radius, the number of pores decreases, and as a result the specific surface area and capacity for gas adsorption both decrease as well. From these, it can be speculated that the number of micropores and minipores of coal is larger than that of mesopores and macropores.

3.2.3 *Pore Structure Characterization*

The commonly used parameters for the characterization of pore structure contain specific surface area, pore volume, fractal dimension, pore length, pore size distribution, pore connectivity, etc. Micropores and minipores are the main sites for gas adsorption and occurrence. The macropores and fractures are the main percolation and migration channels for the methane molecules diffusing from coal matrix into fracture system during the process of coal-bed methane exploitation and coal mine gas drainage. Furthermore, during this process, the porosity of coal reservoir is dynamically changing. With decrease in coal reservoir pressure, the compressed effect of coal body induced by the effective stress and the shrinkage effect from methane molecules desorption both lead to the porosity change in coal reservoir.

1. Pore volume

Pore volume (mL/g) refers to the total pore volume per coal mass, which is an important parameter of pore structure to reflect gas adsorption capacity. Integrating the pore volume density function in Fig. 3.10 with the pore radius, the pore volume of corresponding pore diameter interval can be obtained. The calculation results, including the total pore volume (V_t), micropore volume (V_w), minipore volume (V_g), mesopore volume (V_z), macropore volume (V_d) and the relevant pore size distributions, are shown in Table 3.4.

From Table 3.4, the proportion of pore volume and their distributions are mainly concentrated in mesopores and macropores, which account for 80.846–86.292%.

2. Specific surface area

Specific surface area (m^2/g) refers to the surface area of coal per coal mass. Generally, coal is a kind of porous media, which contains a number of surface areas (also known as internal surface areas). Specific surface area, as an important indicator of the physical characteristics of coal, governs gas adsorption and penetration capacity in coal. The larger the specific surface area is, the stronger gas adsorption capacity coal will have, and the micropores and minipores are the main contributors of specific surface area.

Table 3.4 Pore volume and the distributions (Machang, Yangquanwu, Bailongshan coals)

Coal samples	V_t/(mL/g)	The proportion of pore volume and the distributions							
		V_w	%	V_g	%	V_z	%	V_d	%
Machang	0.094	0.004	4.606	0.009	9.103	0.044	47.044	0.037	39.248
Yangquanwu	0.129	0.013	10.337	0.011	8.818	0.047	36.061	0.058	44.785
Bailongshan	0.076	0.004	4.714	0.010	12.879	0.027	36.180	0.035	46.227

Based on the density function of pore volume distributions ($D_v(r)$) and assuming the shape of coal pore to be cylindrical, the density function of specific surface area distributions ($S(r)$) can be calculated as follows:

$$S(r) = \frac{D_v(r)}{r} \tag{3.17}$$

Integrating the density function of specific surface area distributions with the pore radius, the specific surface area of corresponding pore diameter interval can be calculated. The results, including the total specific surface area (S_t), micropore specific surface area (S_w), minipore specific surface area (S_x), mesopore specific surface area (S_z), macropore specific surface area (S_d) and the relevant specific surface area distributions, are shown in Table 3.5.

From Table 3.5, the proportion of specific surface area mainly concentrates in the micropores and minipores, accounting for 91.345–93.042% of the total specific surface area, which further illustrates that the methane molecules are mainly adsorbed by the surface of micropores and minipores.

3. Fractal dimension

Coal particle has very complex surface morphology and porous structure. The internal pore surface area of coal particle usually accounts for about 95% of its total surface area. In this case, the traditional Euclidean geometry model is difficult to accurately describe coal's irregular structure. The fractal theory is suitable to describe the physical structure that has no characteristic scale but the pore self-similarity can quantitatively and accurately describe the geometric structure of irregular and rough surfaces with nonlinear characteristics. Therefore, fractal theory can better describe the characteristics of internal pore shape in coal.

Fractal dimension (d_f) is a parameter to characterize the roughness of solid surface. Extremely flat surface area is a function of the feature dimension, which can be obtained by calculation. For a nonporous sphere of radius R, the corresponding surface area is $4\pi R^2$, and by increasing the surface roughness or adding cavities into the sphere, the surface area of sphere will increase. Successively adding internal pore into the sphere, sphere will gradually become a porous medium, and the volume of which is the sum of matrix volume and internal pore volumes. At this moment, the surface area of porous sphere is proportional to the volume. According to the fractal theory, the actual surface area of sphere is proportional to R^{d_f}, and fractal dimension (d_f) is between 2 and 3. When the fractal dimension is 2, it represents the plane; when the fractal dimension is close to 3, it represents the extremely high surface roughness or the cube with gradually reducing anisotropy.

The equation for fractal dimension measurement by MIP method is shown as Eq. (3.18) [10]:

$$-\frac{\mathrm{d}V}{\mathrm{d}r} = k_1 r^{(2-d_f)} \tag{3.18}$$

Table 3.5 Specific surface area and the distributions (Machang, Yangquanwu, Bailongshan coals)

Coal samples	S_t/(m^2/g)	The proportion of specific surface area and the distributions							
		S_w/(m^2/g)	%	S_x/(m^2/g)	%	S_z/(m^2/g)	%	S_d/(m^2/g)	%
Machang	3.178	2.558	80.492	0.345	10.853	0.240	7.561	0.035	1.095
Yangquanwu	3.898	3.054	78.356	0.558	14.320	0.223	5.716	0.063	1.609
Bailongshan	2.691	2.061	76.608	0.442	16.434	0.145	5.374	0.043	1.583

where k_1 is a constant; r is pore radius, nm; d_f is fractal dimension of porous media.

The expression of coal pore radius is derived from Eq. (3.15) as:

$$r = -\frac{2\sigma \mathrm{con}\theta}{P} \tag{3.19}$$

The derivative of Eq. (3.19) is as follows:

$$\mathrm{d}r = \frac{2\sigma \mathrm{con}\theta}{P^2}\mathrm{d}P \tag{3.20}$$

Substituting Eqs. (3.19) and (3.20) into (3.18), the following expression is obtained:

$$\log\frac{\mathrm{d}V}{\mathrm{d}P} = k_2 + (d_f - 4)\log(P) \tag{3.21}$$

where k_2 is a constant.

The equation for fractal dimension measurement by LP-N_2GA method is shown in Eq. (3.22) [10]:

$$\ln\left(\frac{V}{V_0}\right) = k_3 + (d_f - 3)\left[\ln\left(\ln\left(\frac{P_0}{P}\right)\right)\right] \tag{3.22}$$

where V is gas adsorption volume under the equilibrium pressure, m^3; V_0 is saturated adsorption volume, m^3; k_3 is a constant; P_0 is saturated vapor pressure, MPa.

Thus, straight line can be obtained from the relationship between $\log(\mathrm{d}V/\mathrm{d}P)$ and $\log(P)$ of Eq. (3.21) as well as the $\ln(\mathrm{d}V/\mathrm{d}V_0)$ versus $\ln\left(\ln\left(P_0/P\right)\right)$ of Eq. (3.20), with the slope of straight line representing the fractal dimension (d_f).

Table 3.6 gives the fractal dimensions of internal pore surface for various particle sizes [9] of Machang, Yangquanwu, Bailongshan coals. From the table, it is noticed

Table 3.6 Fractal dimension of internal pore surface of Machang, Yangquanwu, Bailongshan coals

Partical size D/mm	Machang coals from Guizhou		Yangquanwu coals from Shanxi		Bailongshan coals from Yunnan	
	< 20 nm	> 20 nm	< 20 nm	> 20 nm	< 20 nm	> 20 nm
< 0.01	2.518	1.487	2.353	1.665	2.605	1.176
0.01–0.08	2.292	1.863	2.483	2.306	2.058	2.021
0.08–0.2	2.318	1.949	2.378	2.283	1.727	2.544
0.2–0.25	2.373	2.053	2.318	2.378	1.626	2.442
0.25–0.5	2.203	2.308	2.248	2.479	1.770	2.507
0.5–1	2.259	2.445	2.093	2.560	1.801	2.743
1–3	2.310	2.650	2.246	2.700	1.946	2.852
3–4	2.203	2.659	2.267	2.898	1.935	2.843

that when pore radius is less than 20 nm, the fractal dimensions of Machang, Yangquanwu, Bailongshan coal samples are 2.203–2.518, 2.093–2.483 and 1.626–2.605, respectively; when pore radius is larger than 20 nm, the corresponding fractal dimensions of the three samples are 1.487–2.659, 1.487–2.659 and 1.176–2.843, respectively. In general, the fractal dimension of less than 2, indicates that the two-dimensional plane has many pores; and if the fractal dimension is between 2 and 3, then there are more pores in the three-dimensional space. Due to the different measurement mechanism of MIP and LP-N_2GA, the obtained fractal dimensions from these two methods have a certain variation. When calculating the fractional fractal diffusion model of particle size of coal samples, the larger values in the fractal dimension of the pore size intervals of < 20 nm and > 20 nm are used for calculation. It can be seen from each particle size that fractal dimension increases with particle size. The fractal dimensions of Machang, Yangquanwu, Bailongshan coals are 2.292–2.659, 2.353–2.898, 2.058–2.852, respectively. The minimum values of fractal dimension appear at the particle size intervals of 0.01–0.08 mm, < 0.01 mm and 0.01–0.08 mm respectively while the maximum values appear at the particle size intervals of 3–4 mm, 3–4 mm and 1–3 mm, respectively.

3.3 Coal Fracture Features

3.3.1 Classification of Coal Fracture

The coal fractures refer to the fractures which were caused by various natural stress effects since the coal-formation process began [11]. According to their origins, they can be divided into two categories of endogenic fractures and exogenous fractures. The endogenic fractures were caused by coal dehydration, devolatilization and shrinkage during the coalification process as well as the action of water/gas produced by coalification or abnormal high pressure caused by temperature rise/pore swelling. The exogenous fractures were caused by tectonic deformation stress effects after the coal seam formation. As shown in Fig. 1.7, the endogenic fractures are also known as cleat, typically represented by two groups of cleats namely face cleat and butt cleat that are roughly perpendicular to each other. The face cleat is the main endogenous fracture in the coal seam. It is approximately perpendicular to the bedding planes, generally flat-shaped and extends through coal mass with good continuity. The butt cleat which is not easily recognized is the secondary endogenous fracture in the coal seam. It is approximately perpendicular to the bedding plane and also to the face cleat and spans between two adjacent face cleats with poor and irregular continuity. The fracture structure of coal reservoir is dominated mainly by cleat system, but tectonic movement can cause coal seam deformation promoting the development of additional fractures, and therefore affect the regular fracture system [12, 13].

Coal fractures are formed by a combination of internal and external forces [14, 15], the morphology and distribution of which are very complicated. Because of the differences in origins and development degrees, the face cleat and butt cleat have multiple composite types in space. According to the cleat distribution patterns in different coalfields, Xianbo Su classified the cleat system into reticular, polygonal, isolated and stochastic types [16–18]. Exogenous fractures are the product of tectonic stress and can intercept the bedding plane in any angle. Therefore, exogenous fractures can be classified into three types: ① horizontal fractures parallel to the plane, including natural and deposited fractures (diagenetic fractures) as well as the interbedded fractures induced by tectonism; ② vertical fractures perpendicular to the Bedding plane; ③ oblique crossing fractures intersecting the Bedding plane at any angle.

3.3.2 Fracture Distribution Characteristics in Coal

Because the distribution of fractures in the coal reservoir is very complicated and random, researches study the origin and size classifications of the fractures, but usually do not describe the distribution regularities. In engineering geology, there are some traditional statistical methods, including extremely dense graph method for joints, rose diagram method, frequency histogram method and equal angle statistical method, which supply a statistical analysis for the orientation, spacing and length of the joints.

The parameters to describe cleat in coal seam contain aperture, length, height, spacing and density, the variation ranges of these parameters can be referred to in the Sect. 1.2.1 of this book.

Taking the rose diagram method for example, coal samples are firstly cored from the surface through the overburden or sampled in the underground coal mine and the coal seam occurrence of sampling sites are recorded. Then, in the laboratory, samples are polished along the bedding plane of coal, and high resolution photographs are taken with labelling rulers. The quantity and direction of fractures are measured with the help from a computer, then a further statistics are adopted to get the fracture quantity and developed direction of each group with accuracy of 5°. Marking the above results in the geographical azimuth circle with a radius length to denote the quantity of fractures of the corresponding group (the proportion of fracture quantity in total), then by connecting the adjacent sub-points, a rose diagram which reflects the distribution properties of coal fractures can be obtained. Through the rose diagram, the development condition and direction of coal fractures can be easily obtained.

In fact, the real fractures in coal are much more complex than the above mentioned situation. Figure 3.11 shows the photographs of the coal samples after polishing/labelling rulers in the laboratory which collected from ground drilling. Figure 3.11a exhibits the top of coal sample and Fig. 3.11b exhibits the bottom of the sample [19]. Figure 3.12 shows the rose diagrams of surface fracture distributions of two samples [19]. As shown in Figs. 3.11 and 3.12, the top of coal samples are a

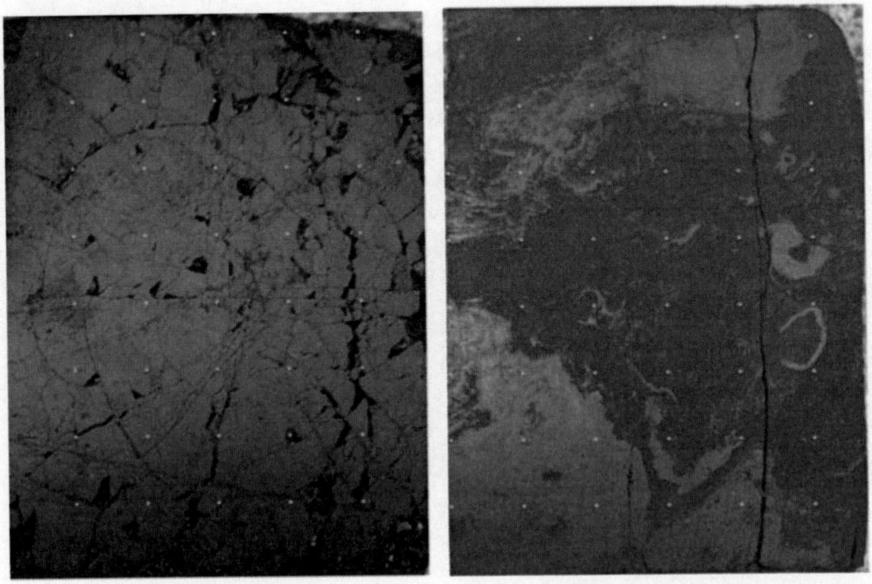

(a) Top of coal sample (b) bottom of coal sample

Fig. 3.11 Fracture distribution of coal surface (The dot interval is 5 mm)

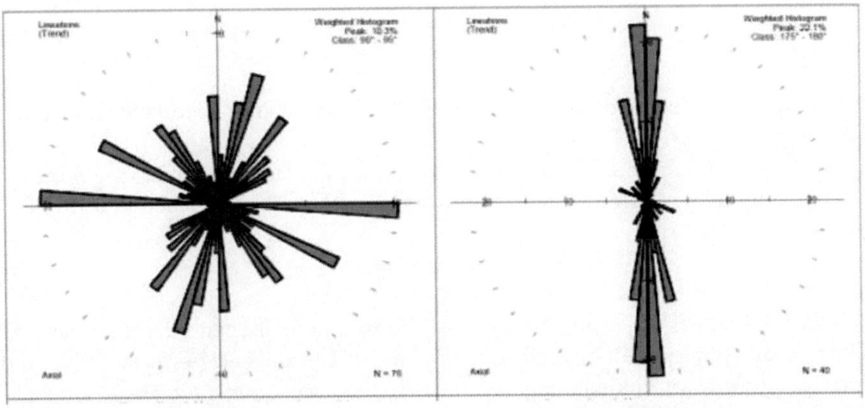

(a) Top of coal sample (b) bottom of coal sample

Fig. 3.12 Rose diagram of fracture distribution on coal surface

type of light coal with reticular fractures. The fractures whose length below 2 mm account for more than 80%. The direction of fractures is randomly distributed, the most likely direction for fracture distribution is located within 90°–96° and the quantity of fracture distributed in this interval accounts for 10.3% of the total number. The bottom of coal sample is a type of dim coal with the poorly developed fractures, only showing a little number of fractures whose length ranges from 7 to 10 mm, and the direction is mainly concentrated in 165°–190° (approximate north–south in these photographs). The amount of fractures distributed in 175°–180° accounted for 22.1% of the total number, while fractures distributed in 180°–185° accounted for 20.4%.

The method of geological statistics can only recognize the macro-fractures with millimeter-level size, but for the smaller scaled fractures there is no valid measurement and analytical method. In the study of coal mechanics, people pay more attention to the fracture porosity of coal, which refers to the percentage of the fracture porosity volume to the volume of coal.

The method of gas flooding water is generally adopted to measure the fracture porosity of coal. Firstly, after subjected to vacuum, coal samples are saturated with water, and then helium under pressure is used to displace water out of the coal sample. In this case, the flooded water refers to flow water, which represents the fracture volume of coal. The fracture porosity is calculated as follows:

$$\phi_f = \frac{V_f}{V_t} \cdot 100\% \tag{3.23}$$

where ϕ_f is the fracture porosity of coal, %; V_f is the fracture volume of coal, mL; V_t is the total volume of coal, mL.

The matrix porosity of coal can be expressed as:

$$\phi_m = \frac{V_m}{V_t} \cdot 100\% \tag{3.24}$$

where ϕ_m is the matrix porosity of coal, %; V_m is the matrix pore volume of coal, mL.

Therefore, if defining the porosity of coal as ϕ_t, the relationship among porosity of coal, fracture porosity and matrix porosity of coal can be described as follows:

$$\phi_t = \phi_f + \phi_m \tag{3.25}$$

Table 3.7 shows the results of porosity, fracture porosity and matrix porosity of Liulin coal samples and Anyang coal samples in China [11]. From the table, the fracture porosity of coking coals are 0.94–1.65%, accounting for 10–18% of the total pore volume. The matrix porosity of coking coals are 5.84–10.51%, accounting for 82.00–89.78% of the total pore volume. The fracture porosity from lean coals to anthracite coals are 0.47–0.79%, accounting for 4.8–11.52% of the total pore volume. The matrix porosity from lean coals to anthracite are 3.93–11.23%, accounting for 88.48–91.00% of the total pore volume. With the increase of coal ranks, the fracture porosity and its proportion in total pore volume show the decreasing trend.

Table 3.7 Porosity, fracture porosity and matrix porosity of Liulin and Anyang coal samples in China

Sample number	Coal rank	$\phi_t/\%$	$\phi_f/\%$	$\phi_f/\phi_t/\%$	$\phi_m/\%$	$\phi_m/\phi_t/\%$
A-II 1-3	Coking coal	7.20	1.36	18.00	5.84	82.00
LQ-8-1	Coking coal	12.00	1.49	12.42	10.51	87.58
LQ-8-5	Coking coal	8.50	1.17	13.76	6.33	86.24
LQ-10-1(1)	Coking coal	9.20	0.94	10.22	8.26	89.78
LQ-10-1(2)	Coking coal	9.10	1.12	12.31	7.98	87.69
LQ-10-2	Coking coal	11.10	1.65	14.86	9.45	85.14
AH-II 1-2	Lean coal	7.20	0.79	10.97	6.41	89.06
AH-II 1-1	Lean coal	4.60	0.53	11.52	4.07	88.48
AH-II 1-2	Meagre coal	8.00	0.73	9.00	7.27	91.00
AH-II 1-1	Meagre coal	11.80	0.57	4.80	11.23	95.20
AL-II 1-3	Anthracite	4.40	0.47	10.68	3.93	89.32

3.4 The Matrix Characteristics of Coal

3.4.1 Definition of Coal Matrix

Figure 1.8 demonstrates the simplified conceptual model of gas migration in coal. The coal is cut into different sized matrix by face cleat and butt cleat. Coal matrix contains a large number of pores mainly composed of micropores and minipores, and the coal seam gas is mainly stored on the surface of matrix pores in an adsorption state. When the gas adsorption equilibrium state changes, the gas molecules first desorb from the matrix pore surface and then diffuse into the fractures, and finally flow into the exposed coal surface (borehole or excavation) by seepage flow.

When the boundary conditions of coal seam gas flow change, gas migration begins in the fractures, and then gas exchange follows between the coal matrix and the fracture system. In the coal composed of coal matrix and fracture network, mass exchange of gas exists called a fracture-pore dual medium. Because the scale of pores in coal matrix is different from the fractures, the gas migration and storage in coal matrix is also different from the gas flow in fractures. In the case of gas migration in coal, the coal matrix with a large amount of pores can be regarded as the gas source, and the large scale fracture system and good connectivity are the main channels for gas seepage. The coal matrix where the fracture system exists, also influences the gas flow in the fractures. This is because the swelling deformation of coal matrix from gas adsorption will compress the fracture system and reduce the fracture apertures. But, when gas desorption occurs within the coal matrix, shrinkage deformation may lead to increased fracture apertures. The variation in coal's fracture aperture will affect the seepage characteristics of coal. Further research studying the matrix adsorption deformation indicates that the physical model which only considers fractures between matrix is different from the real coal body. Due to the fracture surface irregularity, the matrix cannot be completely separated by the fractures and some contacts between the fracture surfaces will still exist. Since the contact segments of matrix are constrained, the swelling deformation will generate expansion force.

3.4.2 Coal Matrix Scale

The diffusion flux formula for mass exchange between coal matrix and fracture system is described here [20]:

$$q = D\sigma_s V_m (c_m - c_f) \tag{3.26}$$

where q refers to diffusion flux, kg/s; D refers to diffusion coefficient, m^2/s; σ_s refers to shape factor, m^{-2}; V_m refers to the volume of matrix block, m^3; c_m refers to gas concentration of matrix block, kg/m^3; c_f refers to gas concentration of fractures, g/mL.

In the above diffusion flux formula, there is an important parameter named shape factor. Shape factor reflects the characteristics of coal matrix that is cut by fractures, which is not only related to the shape of coal matrix, but also the gas diffusion coefficient and the mass concentration in the matrix block. For the matrix shape factor, different expressions can be obtained under different simplifying assumptions, including the expressions based on matrix, fractures, quasi-stationary and unsteady, respectively. Different shape factor expressions used in the model can get different results. For gas migration in coal seam, the diffusion path of gas is controlled by the size and geometry of coal matrix, which sequentially affects the gas desorption and seepage flow.

Coal matrix is defined as the basic gas storage and diffused flow unit under the assumption conditions of flow model, and different assumptions correspond to different shapes/scales of coal matrix. For the convenience of analysis and application, we take the widely used cubic geometry model as the basis to define the equivalent matrix scales of coal as: simplifying the physical structure of coal to the cubic geometry model, and when the external factors like the effective stress and swelling are determined, the side length of cube is defined as the equivalent matrix scales, and the corresponding fracture width is defined as the equivalent fracture width.

In the cubic geometry model, defining the length of the coal matrix as a_c and the fracture width as b_c, then the fracture porosity ϕ_f is denoted as:

$$\phi_f = \frac{V_f}{V} \approx \frac{3b_c}{a_c} \tag{3.27}$$

In the cubic geometry model, the flow in one direction is controlled by two sets of perpendicular fractures. According to Darcy's law, the quantity of flow should be twice of that for a set of parallel fracture, thus the permeability is expressed as

$$k = \frac{b_c^3}{6a_c} \tag{3.28}$$

By simultaneously solving Eqs. (3.27) and (3.28), it can be obtained as:

$$\begin{cases} a_c = \dfrac{9}{\phi_f}\sqrt{\dfrac{2k}{\phi_f}} \\ b_c = 3\sqrt{\dfrac{2k}{\phi_f}} \end{cases} \tag{3.29}$$

As long as the fracture porosity ϕ_f and permeability k can be measured without the effect from external factors (initial moments) such as effective stress and adsorption swelling, the equivalent matrix scales and the fracture width of coal can be solved by Eq. (3.29). The initial fracture porosity of coal ϕ_f can be determined directly, but the initial permeability of coal k cannot be directly determined. To solve k, we use helium (because helium has no adsorption capability on coal) as the seepage gas of coal samples to measure the change in permeability under different effective stress. Then by fitting the formula of permeability and effective stress, the initial permeability can be calculated when the effective stress is set to zero. The formula to solve the initial permeability of coal is [21]:

$$\ln(k) = \ln(k_0) - 3C\Delta\sigma_e \tag{3.30}$$

where k_0 refers to initial permeability of coal; σ_e refers to effective stress; C refers to coefficient.

To study the equivalent coal matrix scales, raw coal blocks and tectonically deformed coal were collected from Daning Coal Mine, Shanxi Province. In the laboratory, the raw and tectonic coal samples were prepared into the standard cylindrical samples with a height of 100 mm and a diameter of 50 mm. Meanwhile, the raw coal blocks were also crushed and sieved into three sizes of coal powder, 0.074–0.25, 0.25–0.5 and 0.5–1 mm. Then the coal powder and a small amount of water were placed into the mold and pressurised to 100 MPa for 30 min in the compression machine. After that, the pressure-hardened moulded coal samples were dried and polished into the standard coal samples. The matrix and fracture porosities of these coal samples are shown in Table 3.8 [22]. The relationship between helium permeability and effective stress is shown in Fig. 3.13 [22].

The results from Fig. 3.13 were used for fitting through Eq. (3.30) to obtain the function relationship between helium permeability and effective stress of coal samples. The initial permeability of coal sample is shown in Table 3.9 [22]. The results of fracture porosity in Table 3.8 and initial permeability in Table 3.9 were

Table 3.8 Matrix and fracture porosities of coal sample

Coal samples	Apparent density/(g/mL)	Pore volume/mL	Fracture volume/mL	Matrix porosity	Fracture porosity
Raw coal	1.4983	0.0233	0.0027	0.0349	0.0040
Tectonic coal	1.3178	0.0319	0.0712	0.0420	0.0938
Moulded coal (0.074–0.25 mm)	1.1262	0.0232	0.1413	0.0261	0.1591
Moulded coal (0.25–0.5 mm)	1.1088	0.0201	0.0788	0.0223	0.0874
Moulded coal (0.5–1 mm)	1.1825	0.0202	0.0612	0.0239	0.0724

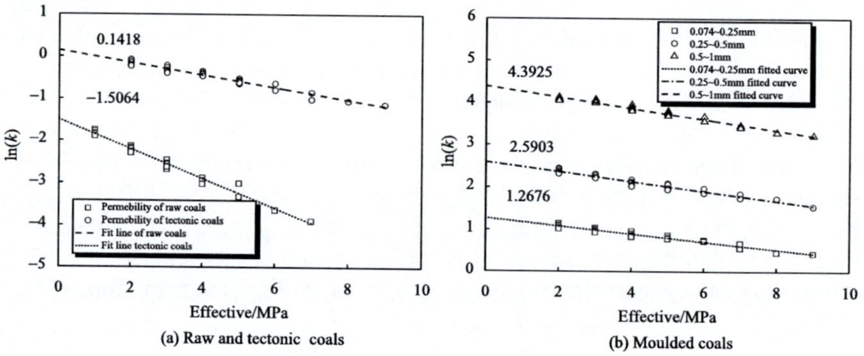

Fig. 3.13 Relationship between helium permeability and effective stress

Table 3.9 Initial permeability of coal sample

Coal samples	Function relationship	Initial permeability/mD
Raw coal	$\ln(k) = -1.5064 - 0.3481\Delta\sigma_e$	0.2217
Tectonic coal	$\ln(k) = 0.1418 - 0.1457\Delta\sigma_e$	1.1524
Moulded coal (0.074–0.25 mm)	$\ln(k) = 1.2676 - 0.0925\Delta\sigma_e$	3.5524
Moulded coal (0.25–0.5 mm)	$\ln(k) = 2.5903 - 0.1139\Delta\sigma_e$	13.3341
Moulded coal (0.5–1 mm)	$\ln(k) = 4.3925 - 0.1308\Delta\sigma_e$	80.84

used in Eq. (3.30) to calculate the equivalent matrix scale and fracture aperture as shown in Table 3.10 [22].

From Table 3.10, the equivalent matrix scale and equivalent fracture aperture of raw coal is 0.7317 mm and 0.9867 μm, respectively. Compared with the structure characterization results of coal sample in this area by Zhou's logging data [23], the matrix scale is basically the same. However, fracture aperture is slightly larger than that in the literature. The equivalent matrix scale of raw coal sample is much larger than that of tectonic coal sample and moulded coal sample.

Table 3.10 Results of equivalent matrix scale and fracture aperture

Coal samples	Initial permeability/mD	Fracture porosity	Equivalent matrix scale/mm	Equivalent fracture aperture/(μm)
Raw coal	0.2217	0.0040	0.7317	0.9867
Tectonic coal	1.1524	0.0938	0.0149	0.4671
Moulded coal (0.074–0.25 mm)	3.5524	0.1591	0.0119	0.6297
Moulded coal (0.25–0.5 mm)	13.3341	0.0874	0.0565	1.6465
Moulded coal (0.5–1 mm)	80.84	0.0724	0.1847	4.4548

References

1. Cheng, Y., Jiang, H., Zhang, X., et al. (2017). Effects of coal rank on physicochemical properties of coal and on methane adsorption. *International Journal of Coal Science & Technology* 1–18.
2. Meng, Z., Zhu, S., Jia, L., et al. (2011). Relationship between approximate analysis of coal and log parameters and its models. *Coal Geology & Exploration, 39*(2), 1–6. (in Chinese).
3. Zhang, S. (2004). Coal chemistry. China University of Mining and Technology Press (in Chinese).
4. Li, P., Xiao, H., Zheng, Y., et al. (2014). Result impact analysis of lump coal and granular coal in apparent relative density measurement. *Coal Geology of China, 07*, 33–35. (in Chinese).
5. Jiang, H., Cheng, Y., & Wang, L. (2014). Effect of coal rank on gas diffusion initial velocity and hardness coefficient. *Safety in Coal Mines, 12*, 11–13. (in Chinese).
6. Cheng, Y. (2010). *Theories and engineering applications on coal mine gas control*. China University of Mining and Technology Press (in Chinese).
7. Wang, F. (2016). Kinetics characteristic of methane adsorption/desorption in coal and its application in the rapid determination of methane parameters. Xuzhou China University of Mining and Technology Press (in Chinese).
8. Zhao, Y., Liu, S., Elsworth, D., et al. (2014). Pore structure characterization of coal by synchrotron small-angle X-ray scattering and transmission electron microscopy. *Energy & Fuels, 28*(6), 3704–3711.
9. Jiang, H. (2015). Pore structure effect on ad-desorption dynamics of methane in pulverized coal. Xuzhou China University of Mining and Technology Press (in Chinese).
10. Friesen, W. I., & Mikula, R. J. (1987). Fractal dimensions of coal particles. *Journal of Colloid and Interface Science, 120*(1), 263–271.
11. Guan, J., & Hou, R. (1993). Coal pore properties of matrix and cleat and their porosity measurement methods. *Journal of North China Institute of Water Conservancy & Hydroelectric Power, 20*(1), 23–27. (in Chinese).
12. Li, H. (2001). Major and minor structural features of a bedding shear zone along a coal seam and related gas outburst, Pingdingshan coalfield, northern China. *International Journal of Coal Geology, 47*(2), 101–113.
13. Gamson, P. D., Beamish, B., & Johnson, D. P. (1993). Coal microstructure and micropermeability and their effects on natural gas recovery. *Fuel, 72*(1), 87–99.
14. Zhang, S., & Li, B. (1996). Mechanism of coalbed cleat and Ins significance in evaluation of coalbed methane exploration and development. *Coal Geology of China, 8*(1), 72–77. (in Chinese).
15. Jian, B., Xian, S., Han, D., et al. (2001). The relation between cleat frequency and coal rank. *Journal of China Coal Society, 26*(4), 346–349. (in Chinese).

16. Su, X., Feng, Y., & Chen, J. (2002). The classification of fractures in coal. *Coal Geology & Exploration, 30*(4), 21–24. (in Chinese).
17. Su, X., Feng, Y., Chen, J., et al. (2001). The characteristics and origins of cleat in coal from Western North China. *International Journal of Coal Geology, 47*(1), 51–62.
18. Li, H., Shimada, S., & Zhang, M. (2004). Anisotropy of gas permeability associated with cleat pattern in a coal seam of the Kushiro coalfield in Japan. *Environmental Geology, 47*(1), 45–50.
19. Gray, I. (2015). Outburst risk determination and associated factors. Sigra Job Reference Number 398.
20. Fan, Z., & Xia, Z. (2009). The theory solution of matrix factor in coalbed reservoir. *Coal Geology & Exploration, 37*(3), 15–18. (in Chinese).
21. Seidle, J., Jeansonne, M., Erickson, D., et al. (1992). Application of matchstick geometry to stress dependent permeability in coals. *SPE Rocky Mountain Regional Meeting* 433–444.
22. Lu, S. (2016). *Mechanical failure and permeability evolution mechanism of coal based on equivalent matrix scale and its application.* China University of Mining and Technology Press. (in Chinese).
23. Zhou, F., & Yao, G. (2014). Sensitivity analysis in permeability estimation using logging and injection-falloff test data for an anthracite coalbed methane reservoir in Southeast Qinshui Basin, China. *International Journal of Coal Geology, 131*, 41–51.
24. National Development and Reform Commission. (2006). MT/T968-2005 description method of coal fracture (in Chinese).
25. Laubach, S. E., Marrett, R. A., Olson, J. E., et al. (1998). Characteristics and origins of coal cleat: A review. *International Journal of Coal Geology, 35*(1), 175–207.

Chapter 4
Gas Adsorption–Desorption Properties of Coal

Learning Outcomes

- Gas adsorption mechanism of coal and its influencing factors;
- Test principles and methods of gas adsorption properties of coal;
- Gas desorption mechanism of coal, its influencing factors and the representative desorption models;
- Test methods of gas desorption properties and its applications in coal mine engineering;
- Gas diffusion properties of coal, diffusion models and its applications.

4.1 Gas Adsorption Properties of Coal

4.1.1 Gas Adsorption Mechanism in Coal

Adsorption is a type of interfacial phenomena, which includes physical adsorption, chemical adsorption and absorption. The gas adsorption in coal belongs to physical adsorption [1–3], which means that the gas molecules concentrate in the surface of coal pores, i.e. the gas molecule concentration at the interface between gas and coal is larger than elsewhere. For the physical adsorption, the adsorption force between the solid surface and the gas is the van der Waals force, and the adsorption volume depends mainly on the pressure, temperature and the size of coal surface area [4]. Similar to the gas liquefaction and water vapor condensation, physical adsorption is also reversible. The heat released by gas adsorption is small, and generally, there is about 10–20 kJ of heat released when 1 mol of gas molecules are adsorbed, which is similar to that of gas liquefaction. For the chemical adsorption, chemical bonds are formed between the solid molecule and the gas molecule on the solid surface, i.e. there exists electron exchange between gas molecules and solid surface molecules. The main difference between chemical adsorption and physical adsorption is that the chemical adsorption is irreversible, and the generated heat of chemical adsorption

Y. Cheng et al., *Coal Mechanics*,
https://doi.org/10.1007/978-981-16-3895-4_4

is similar to the thermal effect of chemical reaction, which is generally ten times to several tens of times larger than the physical adsorption heat. A large number of experiments on coal adsorption show that the gas adsorption/desorption in coal is a reversible process and the volume of adsorbed gas is approximately the same as the desorbed gas. The gas adsorption heat measured by experiments ranges from 12.6 to 20.9 kJ/mol [5], which is approximately equal to the heat released by methane liquefaction.

In addition to adsorbed gas on coal's pore surface, there is also absorbed gas in coal. Gas absorption can be described as gas molecules penetrating into the coal lattice through the coal micropores and form a substance in a solid solution state. The macroscopic difference between absorption and adsorption is that the equilibrium time for the absorbtion is longer and the swelling deformation induced by absorption is larger as well. However, gas adsorption and absorption in coal is not easy to distinguish, therefore, in the research field of coal mine gas, the absorbed gas is generally put into the same category as the adsorbed gas rather than studied separately.

4.1.2 Gas Adsorption Law of Coal

The major ingredient of gas in coal seam is methane (CH_4) and the gas is mainly present in coal pores in the adsorbed state, therefore, choosing the suitable adsorption model for CH_4 isotherms not only can help accurately understand the relationship between coal adsorption capacity and reservoir pressure, but also can indirectly understand the gas adsorption mechanism in coal. A large number of experiments in China and abroad show that the gas adsorption isotherm conforms to the Langmuir equation when the coal adsorbs gas, as shown in Eq. (4.1).

$$X = \frac{abp}{1 + bp} \tag{4.1}$$

where X is the adsorption capacity of the unit coal mass (or volume) on the dry ash free basis at pressure p and a certain temperature, m^3/t; p is the gas pressure at adsorption equilibrium, MPa; a is the gas adsorption constant, which is the ultimate gas adsorption volume in coal, that is, the maximum gas adsorption volume in coal at a temperature when the gas pressure approaches infinity, m^3/t; and b is also the gas adsorption constant, MPa^{-1}.

In some studies, the gas adsorption constant a is also known as the Langmuir volume and the gas adsorption constant b can be characterized as $b = 1/p_L$, where p_L is the Langmuir pressure which refers to the pressure when the adsorption volume is in accordance with half of the limit adsorption capacity.

Six coal samples with the different metamorphic degree (shown in Table 3.1) were used to conduct isothermal adsorption experiment under the temperature of 30 °C and the result is shown in Fig. 4.1 [6]. It is found that the gas adsorption volume

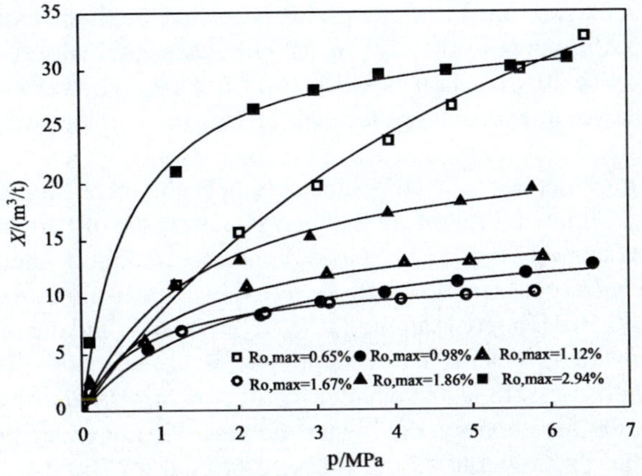

Fig. 4.1 Adsorption isotherms of coal with different metamorphic degrees

of different coal samples all increase with the increasing adsorption equilibrium pressure, but the growth rate decreases gradually. When the gas pressure increases infinitely, the gas adsorption volume of coal tends to a certain limit value and this limit refers to the gas adsorption constant *a* (i.e. the Langmuir volume). Figure 4.2 [6] demonstrates the relationship among the Langmuir volume, Langmuir pressure and vitrinite reflectance of these six coal samples. The comparative analysis of Figs. 4.1 and 4.2 shows that with the increasing metamorphic degree, the Langmuir volume decreases first and then increases gradually, namely the adsorption capacity of coal

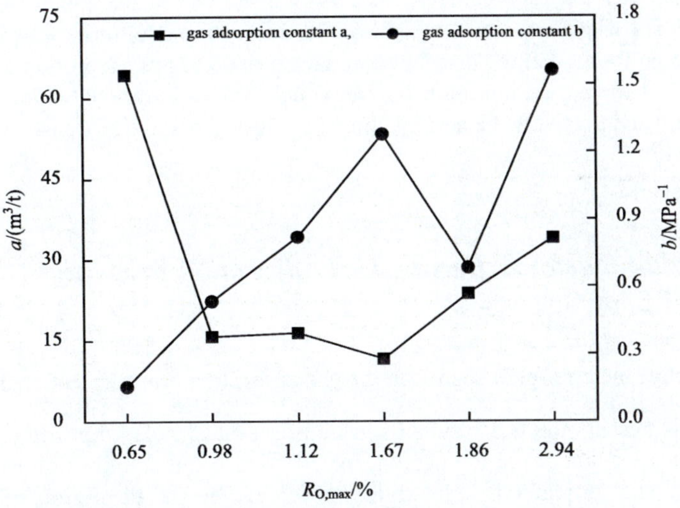

Fig. 4.2 Relationship between gas adsorption constants and metamorphic degree

to gas decreases first and then increases. This is because with the increase of metamorphic degree, the pores in the coal of medium rank gradually decrease, which leads to the decrease of adsorption capacity. But from the high-rank bituminous coal to anthracite, the volume of micropores and the specific surface area both increase gradually, the directional orientation and anisotropy of coal molecules and the carbon atoms significantly increase as well, which results in the increasing gas adsorption capacity of coal. Figure 4.2 reveals that the maximum value of Langmuir volume is 64.32 m^3/t (coal sample 1#, $R_{o,max} = 0.65\%$) and the minimum value of Langmuir volume is 11.57 m^3/t (coal sample 4#, $R_{o,max} = 1.67\%$); the maximum value of Langmuir pressure is 6.50 MPa (coal sample 1#, $R_{o,max} = 0.65\%$) and the minimum value of Langmuir pressure is 0.64 MPa (coal sample 6#, $R_{o,max} = 2.94\%$). The smaller the value of Langmuir pressure is, the smaller the gas adsorption equilibrium pressure of coal will be. On the contrary, the bigger the value of Langmuir pressure is, the greater the gas adsorption equilibrium pressure of coal will be (by comparing the coal sample 1# and sample 3# in Fig. 4.1).

Figure 4.3 reveals the variation of gas adsorption constants a and b in coal with the coal rank [7]. Figure 4.3a reveals that the distribution property of gas adsorption constant a is very disperse, the minimum and maximum values of which are about 8 m^3/t and 60 m^3/t, respectively. Normally, the value of gas adsorption constant a is less than 50 m^3/t. The general change trend of gas adsorption constant a is that: from low rank coal to medium rank coal, the constant a decreases gradually and reaches the lowest value when it is in the metamorphic degree of medium rank; from medium rank coal to high rank coal, the value of constant a increases gradually with the increasing metamorphic degree. The changing trend of the gas adsorption constant a comprehensively reflects the variations of coal's pore volume and specific surface area during the metamorphic process of coal. The bigger the gas adsorption constant b is, the smaller the Langmuir pressure P_L will be and at the smaller Langmuir pressure it is easier for the coal to achieve the adsorption equilibrium. In Fig. 4.3b, it can also be found that the distribution characteristic of gas adsorption constant b is also very disperse, the minimum and maximum values are about 0.15 MPa^{-1} and 2.25 MPa^{-1}, respectively. Generally, the value of gas adsorption constant b is less than 2.0 MPa^{-1}.

4.1.3 Main Factors Affecting Gas Adsorption Properties of Coal

1. Effect of metamorphic degree on the gas adsorption properties of coal

Figure 4.4 presents the test results of gas adsorption of coal with different metamorphic degrees under the temperature of 30 °C and the equilibrium pressure of 2 MPa [8]. The vertical axis in the figure represents the gas adsorption volume per tonne on a combustible basis and the horizontal axis represents the volatile matter content of coal on a dry ash-free basis which reflects the metamorphism degree of

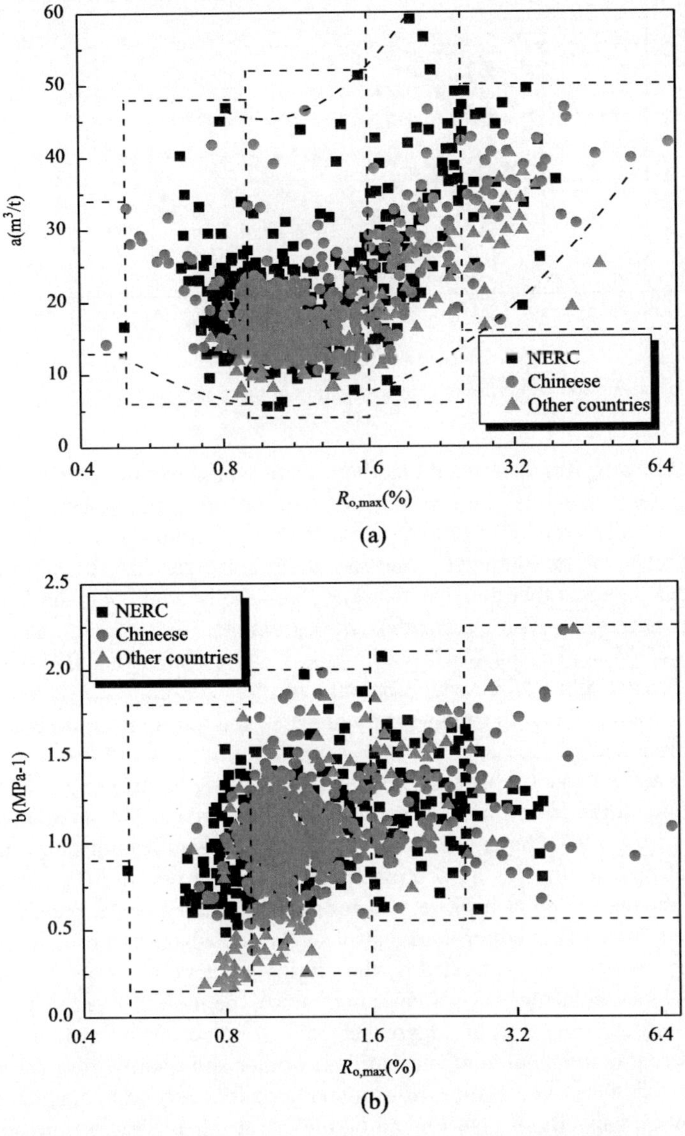

Fig. 4.3 Relationship between gas adsorption constants and metamorphic degree of coal (NERC data from the Chinese National Engineering Research Center for Coal Gas Control)

coal. Figure. 4.4 reveals that under the same gas pressure, the higher the degree of metamorphism of coal, the greater amount of gas that coal can adsorb. When coal's volatile matter content is similar, the gas adsorption volume can also vary significantly and the difference value can even exceed 20 m^3/t, which indicates that it is difficult to determine the coal gas adsorption volume only based on the volatiles of

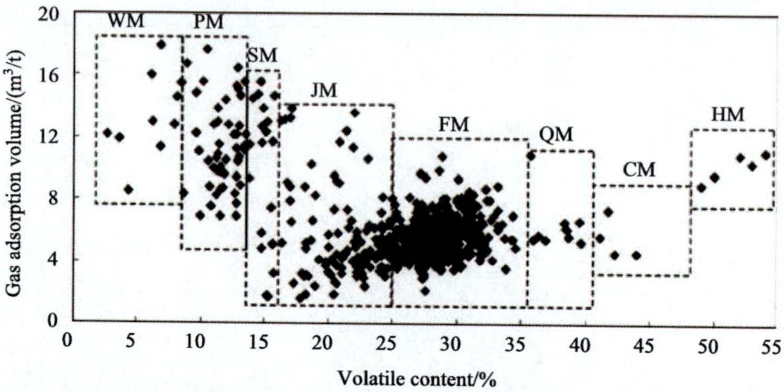

Fig. 4.4 Gas adsorption properties of coal with different metamorphic degree

coal. Additionally, it is also found that there's a big crossover between the measured gas adsorption volume of the coal with or without being the outburst prone coal, therefore it can be concluded that there exists no significant difference between the outburst prone and non-outburst prone coal seam in the gas adsorption properties.

It should be noted that the gas adsorption volume of coal is drastically reduced when the anthracite is further transferred to graphite. For example, the coal from Dongxi Coal Mine, Baisha Coalfield has the volatile matter of 3.0–4.0% and the apparent density of 1.8 t/m^3, which belongs to the super-anthracite. However, gas adsorption volume under the temperature of 30 °C and the equilibrium pressure of 2 MPa is only 8.2 m^3/t.

The isothermal adsorption experiments of coal using the mixed gas of CH_4 and CO_2 with the different compounding concentration were carried out and results are shown in Fig. 4.5 [9]. Under the same gas adsorption equilibrium states, the effect of the coal rank on the gas adsorption volume is slight because the metamorphic degree of the coal samples is close. But for the same coal sample, the CO_2 adsorption capacity for coal is larger than that of CH_4. The adsorption properties of coal for multicomponent gas is affected by the original content of every component, the diameter of gas molecule, the critical temperature, the molecule polarity, the diffusion rate and the metamorphic degree of coal. As a result, adsorptions effect of every different component compete with each other and finally affect the total gas adsorption volume as well as the adsorption volume for every component. Generally, the adsorption capacity of coal for multicomponent gas is between the adsorption volume of strong adsorbate and the adsorption volume of weak adsorbate. The larger the ratio of strong adsorbate that is present in the multicomponent gas, the closer the adsorption curve is to the adsorption curve of strong adsorbate, and vice versa.

2. Effect of organic macerals on the gas adsorption properties of coal

In addition to the metamorphic degree of coal, the gas adsorption capacity of coal is also related to the organic macerals of coal. The organic maceral of coal consists of vitrinite, inertinite and liptinite. Vitrinite is a gummy matter transformed from the

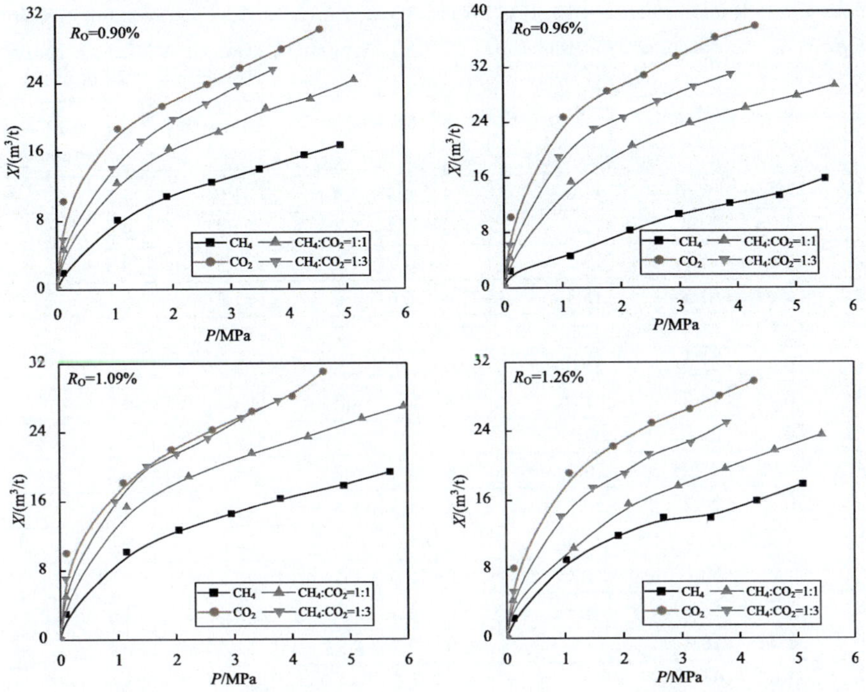

Fig. 4.5 Gsas adsorption isotherms of coal for the mixed gas of CH_4 and CO_2 with different compounding concentration

plants' lignocellulosic tissue through the effect of gelatinization under the reduction conditions. Inertinite is a product of fusainization, in the condition of oxidation, due to the loss of oxidated atomic group, plant remains experienced dehydrogenation and dehydration and as a result, the content of carbon in plant remains at a higher level. Then by a further olefins reduction and condensation, this matter is finally formed. Inertinite can be divided into three categories: Type I inertinite is the structureless fusinite, which includes vitrofusinite, fusinite splitter, matrix fusinite, megaspore, microsome and round fusinite; Type II inertinite is the fusinite with cellular structure but no filling, which includes fusinite and wood vitrinite; Type III inertinite is the cellular structure filled by the organic matter or mineral substance. Liptinite is converted from the part of coal-forming plant with strong chemical stability, such as spore, pollen, phellem layer, resin, cuticle and so on.

Table 4.1 shows the gas adsorption volume of the coal sample with different metamorphic degree and organic maceral, which was measured by the Xi'an Research Institute of China Coal Technology Engineering Group [10]. From the table, it is found that liptinite has the weakest gas adsorption capacity (Group 10); when the content of Type II inertinite is low (Group 1, 3, 7), the higher the content of vitrinite,

Table 4.1 Relationship between the organic maceral content in coal and the gas adsorption capacity

Group	R_O /%	Vitrinite/%	Inertinite/%			Liptinite/%	Gas adsorption constant a/(cm^3/g)	Gas adsorption volume at the equilibrium pressure of 5 MPa/(cm^3/g)
			I	II	III			
1	0.55	77.8	11.6	4.4	0.8	5.4	31.0	24.5
	0.55	14.4	72.1	5.0	0.2	8.3	23.9	19.4
2	0.62	75.9	13.1	9.1	0.7	1.2	25.3	20.8
	0.62	61.3	15.3	21.6	0.6	1.2	31.0	25.6
	0.62	43.4	41.3	12.3	0.3	2.7	22.7	18.9
	0.62	31.0	36.0	30.5	–	2.5	26.7	22.2
3	0.70	94.5	1.0	0.3	–	4.2	26.4	20.5
	0.70	29.0	62.2	1.1	5.5	2.2	20.7	16.5
4	0.95	78.9	10.3	5.6	–	5.2	14.2	11.3
	0.95	34.7	50.0	2.6	1.3	11.4	11.1	9.5
5	1.2	64.9	22.5	4.7	6.6	1.2	19.0	15.0
	1.2	52.1	31.3	1.2	12.9	1.0	16.2	13.2
6	1.52	83.3	9.4	6.0	1.3	–	22.0	18.0
	1.52	78.3	17.6	0.9	2.8	0.85	14.6	11.8
7	1.67	93.8	–	–	6.2	–	27.5	21.6
	1.67	78.6	–	–	21.4	–	22.6	18.5
	1.67	72.2	–	–	27.8	–	21.7	17.5
8	1.85	80.6	17.8	1.4	0.2	–	25.5	21.0
	1.85	64.1	21.9	13.6	0.4	–	27.4	23.5
	1.85	56.5	39.6	2.3	1.6	–	23.5	19.0
9	2.85	66.5	26.3	7.2	–	–	39.5	34.1
	2.85	58.7	31.9	9.4	–	–	36.5	31.5
	2.85	31.1	63.4	5.3	0.2	–	35.7	31.0
10	0.73	15.9	7.0	–	–	77.1	13.9	10.8

the larger the gas adsorption volume. The gas adsorption capacity of Type II inertinite is higher than that of vitrinite.

3. Effect of temperature on the gas adsorption properties of coal

Six coal samples presented in Table 3.1 were used to conduct the isothermal adsorption experiments at the temperature of 30 °C, 45 °C and 60 °C, and the results are shown in Fig. 4.6 [6]. From the figure, it can be noticed that no matter what the degree of metamorphism is, with the variation of temperature the adsorption isotherms of

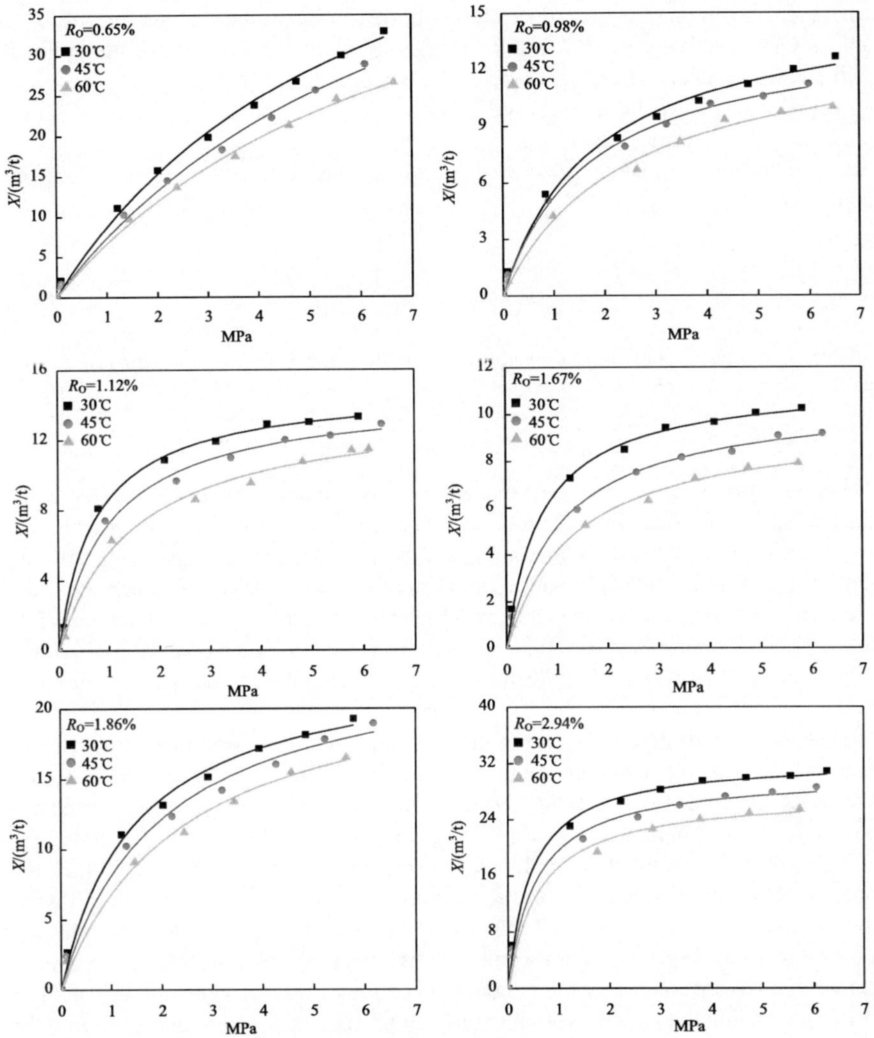

Fig. 4.6 Adsorption isotherms of coal under different temperature

coal show similar change trend. That is, with the temperature increase, the adsorption amount of methane molecules to coal gradually decrease, demonstrating that the temperature increase reduces the adsorption capacity of methane molecules to coal.

To determine the effect of temperature on the gas adsorption volume, the below empirical formula [11] can be used:

$$X_t = X_0 e^{-nT} \tag{4.2}$$

where X_t and X_0 are the gas adsorption volume of coal at the temperature of T °C and 0 °C, respectively, m^3/t; T is the temperature of coal, °C; n is a constant related to the gas pressure, $n = \frac{0.02}{0.993+0.07p}$; p is the gas pressure, MPa.

In order to remove the error caused by the temperature conversion, the temperature used for the isothermal adsorption experiments in the laboratory should be equal to the temperature of coal seam. If not, the adsorption volume should be calculated using Eq. (4.3).

$$X_T = X_{T_1} e^{n(T_1 - T)} \tag{4.3}$$

where X_{T_1} is the gas adsorption volume measured under the temperature of T_1 °C, m^3/t.

4. Effect of moisture on the gas adsorption properties of coal

There are three types of moisture that exist in coal, namely the crystal water, equilibrium water and free water. Crystal water refers to the water combining with coal composition in the form of compound, such as calcium sulfate ($CaSO_4 \cdot 2H_2O$) and kaoline ($Al_2O_3 \cdot 2SiO_2 \cdot 2H_2O$) and so on. Generally, the crystal water cannot be separated out until they are placed in the extra high temperature. Equilibrium water refers to the water reaching the saturation state in coal pores under the condition of normal temperature and dryness. Equilibrium water will volatilize under the condition of heating drying. Free water is the water existing on the surface and in the big fractures of coal, which will gradually evaporate if the coal is stored for a long time. Generally, the moisture in coal can weaken the gas adsorption ability of coal, because the water molecules will compete with the methane molecules and reduce the gas adsorption position, meanwhile, the adsorption potential of coal to water molecules is larger than that to methane molecules [6]. Moreover, the moisture will block the pore throat of coal which can prevent the entrance of methane molecules. Taking the six coal samples detailed in Table 3.1 as the examples again, after different pretreatments of drying, water equilibrium, water injection (3% and 5%, respectively), isothermal adsorption tests were conducted and the results are shown in Fig. 4.7 [6]. It was found that the gas adsorption volume of all six coal samples decrease with the increasing moisture content.

At present, the adsorption isotherm of coal is usually carried out by using dry coal samples. In order to take the effect of moisture on gas adsorption of coal into account, different empirical equations were proposed. In China, the Aiqiegeer equation is commonly used [12]:

$$X_W = \frac{X_d}{1 + 0.31 M_{ad}} \tag{4.4}$$

where X_W is the gas adsorption volume of wet coal, m^3/t; X_d is the gas adsorption volume of the dry coal, m^3/t; M_{ad} is the moisture content of coal, %.

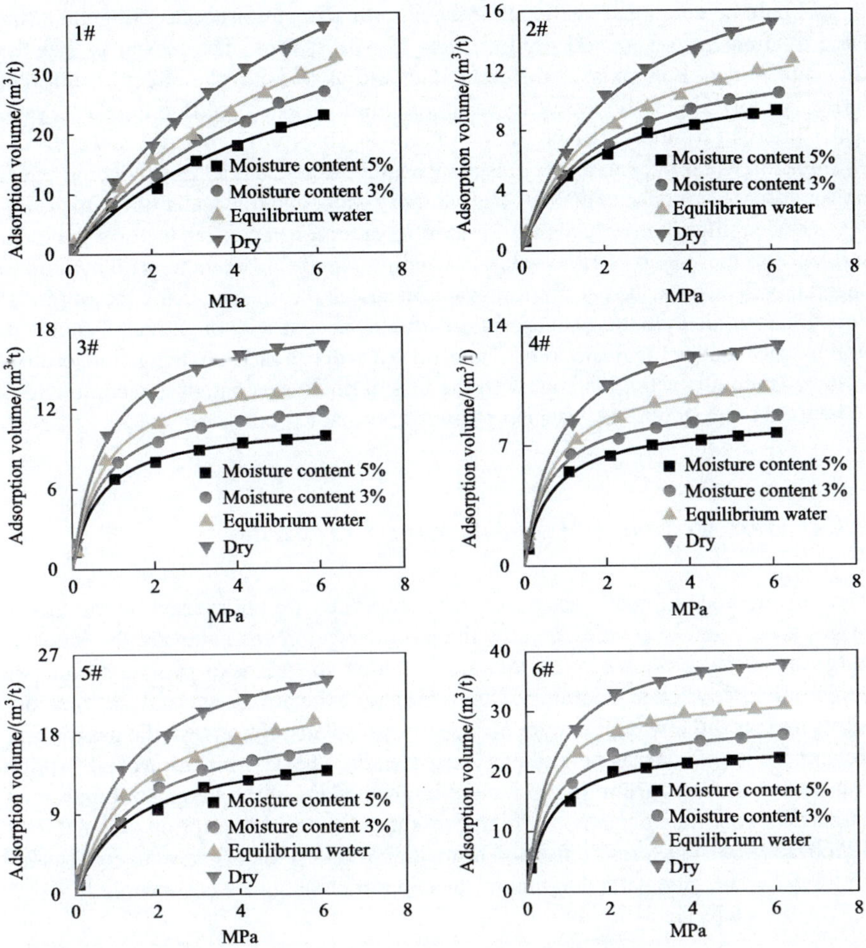

Fig. 4.7 Gas adsorption isotherms of coal with different moisture content

Equation (4.4) does not involve the effect of metamorphic degree on the gas adsorption volume of coal with the different moisture content. The scholars in the Fushun Research Institute of China Coal Technology Engineering Group optimized the above equation by measuring the gas adsorption isotherms of coal with the different moisture content [11]:

$$X_{\mathrm{W}} = \frac{X_{\mathrm{d}}}{1 + (0.10 + 0.0058V_{\mathrm{daf}})M_{\mathrm{ad}}} \tag{4.5}$$

where V_{daf} is the volatile content of coal, %.

It should be noted that nearly all of the empirical equations concerning the influence of moisture on gas adsorption show big deviations. The reason is that the moisture in coal may exist in different state and as a result, the effect of different kinds of moisture on the gas adsorption volume is also very different. The crystal water in coal or mineral has no effect on the gas adsorption volume. However, the equilibrium water in coal micropores in an adsorbed state has a large reducing effect on gas adsorption capacity of coal. The capillary water and free water filling in the big pores and capillary can only dissolve trace concentration of gas, but has no significant influence to the gas adsorption volume. Owing to the fact that there is no method to separately determine the equilibrium water in coal at the present, using the empirical equations to calculate the gas adsorption amount in coal with the effect of moisture will usually induce big deviations. Therefore, in order to measure the gas adsorption volume of coal exactly, the coal with the in-situ moisture content is recommended to be used to carry out the gas adsorption isothermal experiment.

4.1.4 Test Methods of Gas Adsorption Properties

The test methods of coal's gas adsorption properties are built based on the theory of gas adsorption in coal. Generally, the gas adsorption constants are the index to judge the gas adsorption capacity of coal. In China, the gas adsorption constants are commonly measured in laboratory. Combining with the proximate analysis, mercury intrusion test and low temperature nitrogen physical adsorption test, the gas adsorption properties of coal can be studied systematically. The determination of adsorption constants a and b is according to the national standard of the People's Republic of China as: Experimental method of high-pressure isothermal adsorption to coal (GB/T 19560-2008) and Determine method of methane adsorption capacity in coal (MT/T 752-1997). The schematic diagram of the experimental device is shown in Fig. 4.8.

The test steps are as follows:

1. Coal sample preparation instructions

Crush the selected coal samples in a pulverizer and sieve into the desired particle size of 0.2–0.25 mm (about 100 g). Place the specimen in a vacuum drying oven (100 °C) for 1 h and then cool down the sample in a desiccator. After that, the cooled sample will be stored in the ground-glass stoppered flask and sealed.

2. Adsorption experiment

(1) Put approximately 50 g of coal sample into the coal sample container and place in the thermostatic water bath with temperature of (60 ± 1) °C. Use the vacuum pump to degas the coal sample. When the vacuum level in the coal sample container drops below 4 Pa, turn off the valve of the coal sample container and the vacuum pump.

(2) After degassing, adjust the temperature of thermostatic water bath to (30 ± 1) °C.

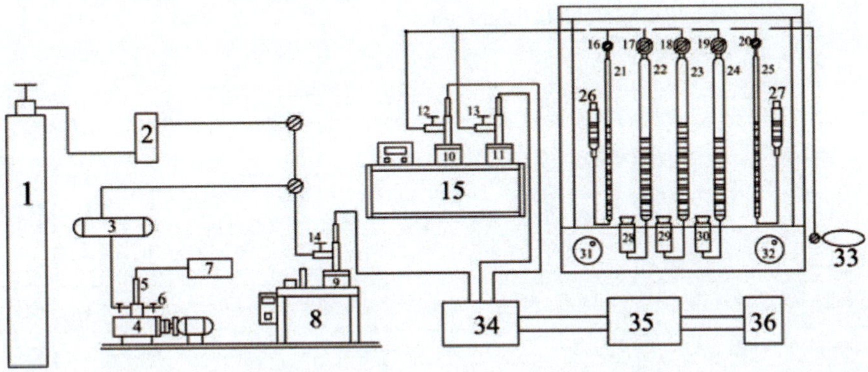

Fig. 4.8 Schematic diagram of the experimental device for gas adsorption by high-pressure capacity method. 1—High pressure CH_4 cylinder; 2—Pressure buffer tank; 3—Vacuum tube system; 4—Vacuum pump; 5—Vacuum gauge tube; 6—Vacuum diaphragm valve; 7—Compound vacuum gauge; 8—Thermostatic water bath (60 °C); 9–11—Coal sample containers; 12–14—Valves; 15—Thermostatic water bath (30 °C); 16–20—Glass valves; 21–25—Graduated cylinder; 26–30—Leveling bottle; 31–32—Lifter knob; 33—Gas sample bag; 34—Data acquisition system; 35—Computer; 36—Output result

(3) Open the inflation valve and the control valve of reference container, let the high pressure gas in the CH_4 cylinder flow into the reference container and connection pipe. Close the inflation valve and record the pressure value p_{1i} in the reference container.

(4) Slowly open the valve of the sample container and let the gas of the reference container flow into the sample container. Close the valve of sample container immediately when the gas pressure in the container increases to the desired value (generally set $n = 7$ pressure interval points within the test pressure range of 0–6 MPa, every pressure interval thereby approximately to be the $1/n$ of the maximum gas pressure). Record the pressure value p_{2i} in the reference container and the room temperature T_1. The gas volume charged into the coal sample container, Q_{ci} can be calculated using the Eq. (4.6).

$$Q_{ci} = \left(\frac{p_{1i}}{Z_{1i}} - \frac{p_{2i}}{Z_{2i}}\right)\frac{273.2V_0}{0.101325(273.2 + T_1)} \tag{4.6}$$

where Q_{ci} is the volume of gas charged into the sample container in the standard condition, cm^3; p_{1i} and p_{2i} are the absolute pressure of the sample container before/after gas filling; Z_{1i} and Z_{2i} are the compressibility coefficients of the gas under the temperature of T_1 and the pressure of p_{1i}, p_{2i} respectively; V_0 is the standard volume of the reference container and communication pipe, cm^3.

(5) Balance the sample for at least 8 h until the gas adsorption equilibrium state is reached, record the adsorption equilibrium pressure p_i and calculate the volume Q_{di} of residual free gas in coal sample container. The gas adsorption volume Q_i of the total coal samples and the gas adsorption volume X_i per

gram of coal sample can be obtained.

$$Q_{di} = \frac{273.2V_d p_i}{0.101325Z_i(273.2 + T_3)} \tag{4.7}$$

$$Q_i = Q_{ci} - Q_{di} \tag{4.8}$$

$$X_i = \frac{Q_i}{m_r} \tag{4.9}$$

where V_d is the dead volume of the sample container, cm^3; Z_i is the compressibility coefficients of the gas when the temperature is T_3 and the pressure is p_i; T_3 is the experimental temperature, °C; M_r is the mass of coal sample, g.

(6) Repeat the steps of (3), (4), (5) and gradually increase the gas pressure, different values of Q_{ci}, Q_{di}, Q_i and X_i can be obtained.
Thus, the total gas adsorption volume Q_c is:

$$Q_c = \sum_{i=1}^{n} Q_{ci} \tag{4.10}$$

(7) Plot p_i and $x_i$for every coal sample group as Langmuir adsorption isotherms, and fit the data of (p_i, X_i) according to the Eq. (4.11) by the least square method, the gas adsorption constants a and b can be calculated:

$$\frac{p}{X} = \frac{p}{a} + \frac{1}{ab} \tag{4.11}$$

Now take the gas adsorption experiment of the coal sample from No. 10 coal seam of Wolonghu Coal Mine as an example to introduce the test method and the application of coal's gas adsorption properties. According to the steps of (2) and (3) mentioned above, the gas adsorption volume of the coal sample under different adsorption equilibrium pressure is shown in Table 4.2.

Table 4.2 Gas adsorption volume of the coal sample from No. 10 coal seam in Wolonghu Coal Mine under different adsorption equilibrium pressure

Equilibrium pressure p_i/MPa	0.47	1.35	2.29	3.17	3.92	4.65	5.51	6.48
Adsorption volume X_i/(m^3/t)	18.91	31.39	38.41	40.87	42.21	43.15	44.97	46.66
p_i/X_i	0.0249	0.0430	0.0596	0.0776	0.0929	0.1078	0.1225	0.1389

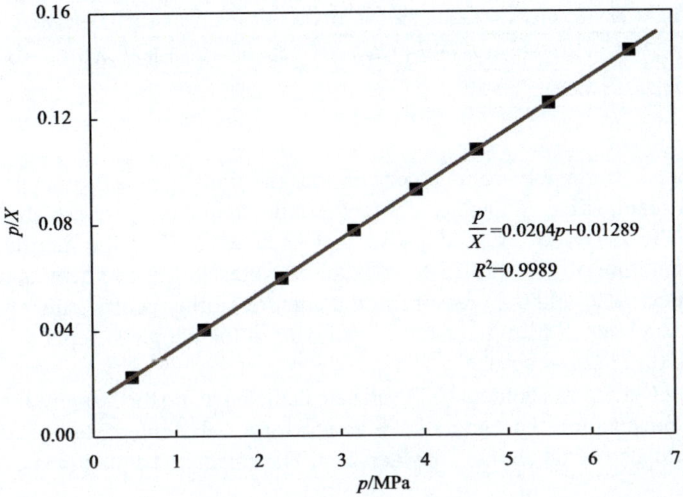

Fig. 4.9 Fitting result of the gas adsorption data for the coal sample from No. 10 coal seam, Wolonghu Coal Mine

Plot the data points on the figure with the relation of *p*/*X* versus *p* and fit the data as shown in Fig. 4.9. Then, the Eq. (4–12) can be obtained for the coal sample from No. 10 coal seam of Wolonghu Coal Mine:

$$\begin{cases} \frac{1}{a} = 0.0204 \\ \frac{1}{ab} = 0.01289 \end{cases} \tag{4.12}$$

Solving the Eq. (4.12), the adsorption constants of the coal sample from No. 10 coal seam, Wolonghu Coal Mine are $a = 49.02$ m^3/t and $b = 1.58$ MPa^{-1}.

Using the gas adsorption constants, the gas content of the coal sample from No. 10 coal seam, Wolonghu Coal Mine can be calculated by the indirect method. The steps are as follows.

The gas content calculated by using the indirect method is equal to the sum of free gas content and adsorbed gas content:

$$X = X_x + X_y \tag{4.13}$$

where X_x is the adsorbed gas content, m^3/t; X_y is the free gas content, m^3/t.

(1) The free gas content (X_y) is calculated by the ideal gas state equation:

$$X_y = VpT_0/(Tp_0Z) \tag{4.14}$$

Table 4.3 Basic physical parameters of the No. 10 coal seam in Wolonghu Coal Mine

M_{ad}/ %	A_{ad}/ %	V_{daf}/ %	True density /(g/cm^3)	Apparent density /(g/cm^3)	Porosity/ %
3.92	14.28	10.80	1.37	1.28	6.60

where V is the pore volume per unit mass coal, m^3/t; p is the gas pressure of coal seam, MPa; T_0 and p_0 are the thermodynamic temperature and pressure in STP, $T_0 = 273$ K and $p_0 = 0.101325$ MPa; T is the thermodynamic temperature of gas, 303.15 K; Z is the compressibility coefficients under the temperature t and the pressure p, the compressibility coefficients of methane can be obtained from the national standard of the People's Republic of China (GB/T 17747.3–1999).

(2) The adsorbed gas content (X_x) can be calculated using the Langmuir equation, and meanwhile, the factors such as moisture, ash content, temperature, etc. should also be taken into consideration. The calculated equation is as follows:

$$X_x = \frac{abp}{(1+bp)} e^{n(T_0-T)} \frac{1}{(1+0.31M_{ad})} \times \frac{(100 - A_{ad} - M_{ad})}{100} \tag{4.15}$$

where T_0 is the temperature of coal seam, 30 °C; A_{ad} is the ash content of coal, %.

Table 4.3 shows the determined result of the basic physical parameters of No. 10 coal seam in Wolonghu Coal Mine.

Using the data in Table 4.3 as well as the coal seam gas pressure at different elevation, the gas content of No. 10 coal seam (elevation above −450 m) in Wolonghu Coal Mine is obtained in Table 4.4.

4.2 Gas Desorption Properties of Coal

4.2.1 Gas Desorption Mechanism

The adsorbed gas in coal has formed a stable balance with the compressed free gas in pores after a long geological age. Underground mining or drilling will damage the original stress balance in the strata, resulting in stress change around the working face or drilled borehole, which will eventually damage coal seam, induce fractures in coal and increase the coal seam permeability. That is, when conditions affecting the sorption-desorption equilibrium states are changed, the adsorption equilibrium state of the methane-coal adsorption system will be changed will convert into and escape from the adsorption system, resulting in the decrease of adsorption amount in adsorption/desorption dynamic equilibrium system. In the process of coal mining and coal bed methane (CBM) exploitation, desorption is mainly achieved by reducing

Table 4.4 Calculated gas pressure and gas content of the No. 10 coal seam in Wolonghu Coal Mine (elevation above −450 m)

Elevation/m	Calculated gas pressure p /MPa	Free gas content X_y/(m^3/t)	Adsorbed gas content X_x/(m^3/t)	Calculated gas content X/(m^3/t)
−270	0.01	0.01	0.28	0.29
−285	0.15	0.07	3.47	3.54
−290	0.26	0.12	5.27	5.39
−300	0.38	0.17	6.79	6.97
−310	0.51	0.23	8.08	8.31
−328	0.74	0.34	9.76	10.10
−330	0.76	0.35	9.88	10.23
−340	0.88	0.41	10.53	10.94
−370	1.26	0.59	12.05	12.64
−380	1.38	0.64	12.41	13.05
−390	1.51	0.71	12.76	13.46
−400	1.63	0.76	13.04	13.80
−410	1.76	0.83	13.31	14.14
−420	1.88	0.88	13.54	14.43
−430	2.01	0.95	13.77	14.71
−440	2.13	1.01	13.95	14.96
−450	2.26	1.07	14.14	15.21

the reservoir pressure. Most CBM exist in the matrix pores of the coal in the form of physical adsorption, once the pressure of coal reservoir drops below the critical desorption pressure, CBM begins to desorb and is converted from the adsorption state to the free state.

When the gas contacts coal surface, the methane molecules do not immediately contact with all surfaces of pores and fractures, but form pressure gradient and concentration gradient in the coal. The methane pressure gradient in coal would cause flow which follows Darcy's law, and this process dominates in large pore system; while in small pore and micropores systems, methane molecules diffuse from high concentration to low concentration under concentration gradients. During the penetration and diffusion of methane in coal, adsorption and desorption also occurs on the surface of coal pores. Therefore, it is a whole and comprehensive process of penetration-diffusion and adsorption-desorption [13], as shown in Fig. 4.10:

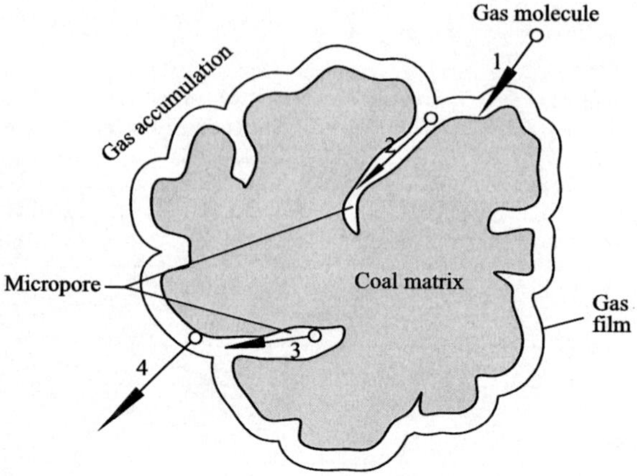

Fig. 4.10 Schematic diagram of methane adsorption process in coal matrix. 1—external diffusion process; 2—internal diffusion process; 3—internal anti-diffusion process; 4—external anti-diffusion process

(1) The penetration process is the first step in the whole process of adsorption. Under the methane pressure gradient, the methane molecules penetrate toward the large pore system of coal and form a methane gas film on the outer surface of coal matrix;
(2) The external diffusion process refers to the process that the methane molecules in the peripheral space of the coal matrix diffuse to the surface of the coal matrix through the gas film, in the direction indicated by the symbol "1";
(3) The internal diffusion process refers to the process that the methane molecules enter micropores of coal matrix and diffuse into the inner surface of the coal matrix, in the direction indicated by the symbol "2";
(4) The adsorption process refers to the process that the methane molecules which reach the inner surface of pores in coal matrix adsorb on the surface of coal pores;
(5) The desorption process refers to the process that the adsorbed methane molecules on the surface of pores leave the inner surface of pores in coal matrix. The desorption process and adsorption process are carried out simultaneously;
(6) The internal anti-diffusion process refers to the process that the desorbed methane molecules diffuse to the outer surface of coal matrix, in the direction indicated by the symbol "3";
(7) The external anti-diffusion process refers to the process that the methane molecules in gas film diffuse into gas phase space of methane, in the direction indicated by the symbol "4".

The adsorption/desorption equilibrium is the dynamic conversion of the above processes, dominated by different steps in different circumstances.

4.2.2 Main Factors Affecting Gas Desorption Properties

1. Effect of gas pressure moisture on the desorption properties of coal

The original coal gas pressure not only characterizes the gas content in the coal, but also provides the driving force for gas desorption of coal. Studies of desorption characteristics of coal under different pressure were performed for the No.7 coal seam, Qinan Coal Mine, Huaibei Coalfield. Figure 4.11 shows the relationship between the gas desorption amount and desorption time under different gas pressure. These results show that the desorption characteristics of coal samples are different under different adsorption equilibrium pressures. The relationship between the gas desorption amount and time shows the positive correlation in a parabolic form. The high pressure curves are located above the low pressure curves within desorption time, and the common characteristic of these curves is that desorption amount gradually increases with desorption time, however the desorption rate gradually decreases. At the initial moment, the gas desorption rate is very high but the decay is also fast. Under different adsorption equilibrium pressures, desorption curves are different. In the same desorption time interval, desorption amount increases with the increasing of adsorption equilibrium pressure.

2. Effect of coal destruction type on the gas desorption properties

The gas desorption process research of the tectonic coal at normal atmospheric pressure shows that the initial gas desorption rate or the accumulative gas desorption amount in the given desorption time are both faster that for the normal coal under the

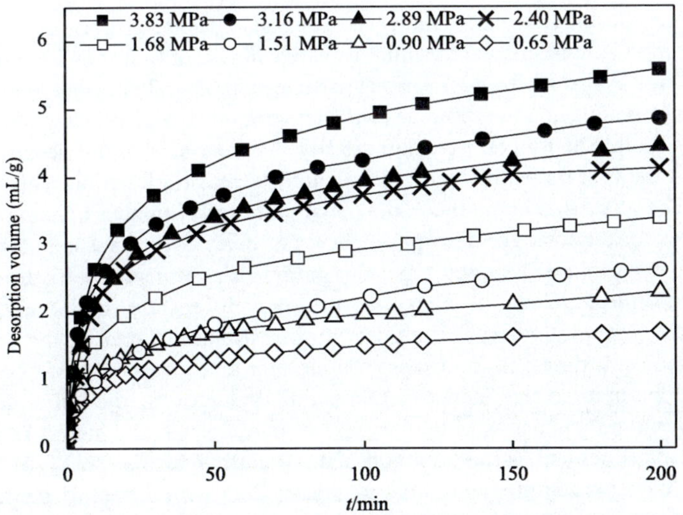

Fig. 4.11 Desorption curves under different gas pressure of the coal sample from the No. 7 coal seam, Qinan Coal Mine

same adsorption equilibrium pressure. In addition, the differences in the coal destruction type have the same effect on the non-isobaric desorption process of coal in the coal slurry medium. The effect of coal destruction type on coal's desorption properties is similar to that on coal adsorption characteristics. The reason is that although the Langmuir volume of the deformed coal is similar to the normal coal with the same metamorphism, the Langmuir pressure of the deformed coal is much smaller than that of normal coal. Therefore the adsorption equilibrium rate for tectonic coal is faster than the normal coal and the gas desorption rate is also faster. Taking the normal and deformed coal from Daning Coal Mine in Shanxi Province as example, in the Daning Coal Mine, Shanxi Province, China, the Langmuir volume and the Langmuir pressure of the normal coal are 51.89 m^3/t and 1.38 MPa, respectively; while the Langmuir volume and the Langmuir pressure of the deformed coal is 50.77 m^3/t and 1.10 MPa, respectively. Studying the gas desorption characteristics of tectonic coal with different destruction types can provide theoretical basis for the prediction of coal and gas outburst, the estimation of coal seam gas pressure and the prediction of gas emission from coal mining.

3. Effect of particle size on the gas desorption properties of coal

The particle size of the coal samples firstly affects the total surface area of coal and then affects the internal pores where the gas molecules enter. Chinese researchers studied the effects of coal particle size with different destruction types on desorption process [14, 15]. They conducted experimental analysis on gas desorption process of coal from No. 6 coal seam in Liwangmiao Coal Mine and No. II_1 coal seam in Baizhuang Coal Mine, using constant desorption temperature (30 °C) and constant adsorption equilibrium pressure (0.5 MPa). These results are shown in Figs. 4.12 and Fig. 4.13.

These results indicate that: ① under the same adsorption equilibrium pressure, the gas desorption amount increases with the decreasing particle size in the same period of time;② for coal samples with various particle size, the relationship between total desorption amount and time coincides with a monotonically increasing function with the upper limit. The function is continuously differentiable in the desorption time [0, *t*], and the first derivative is monotonically decreasing function with the lower limit; ③ There is a maximum horizontal asymptote and a number of extreme values asymptotes for the total gas desorption curves of coal; ④ for coal samples with the same coal rank and destruction type, the particle size reflects the path resistance for gas desorption. When all other desorption conditions are the same, the larger the particle size, the greater is the desorption resistance, therefore the desorption intensity per unit time and the desorption amount at a given time will be smaller. But when the particle size increases to a certain value of d_0, the initial desorption rate V_1 would not change with the increase of particle size any more. As shown in Fig. 4.14, d_0 is defined as the minimum natural particle size of coal [16], which is related to the destruction type of coal structure. Coal with different coal ranks and different destruction types has different minimum natural particle sizes.

4. Effect of moisture on the gas desorption properties of coal

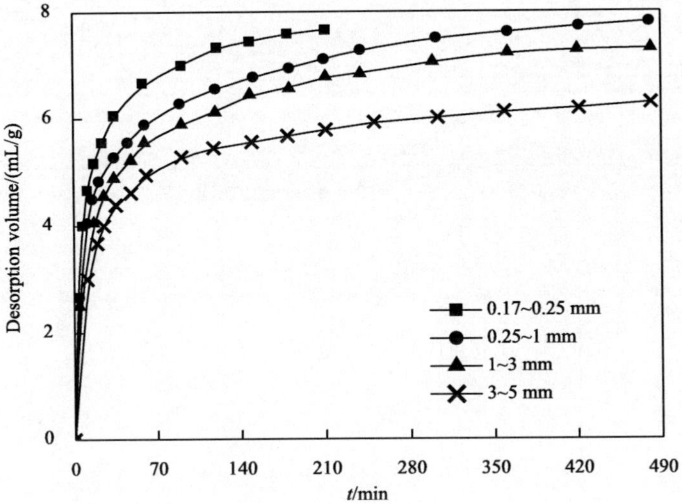

Fig. 4.12 Gas desorption curves of coal samples with different particle sizes from No.6 coal seam, Liwangmiao Coal Mine

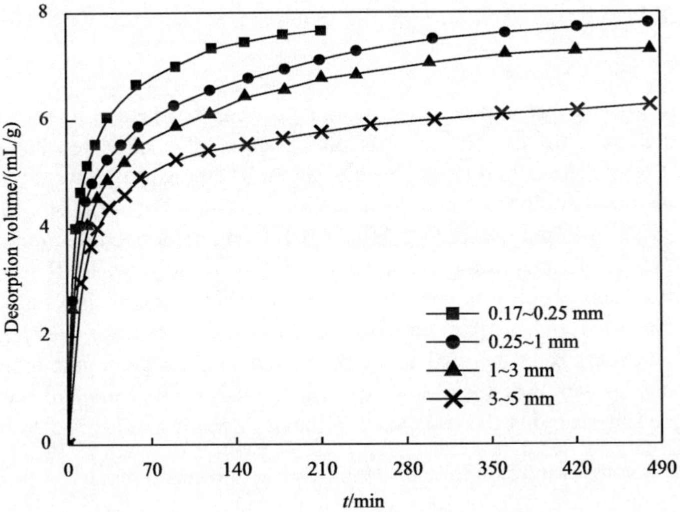

Fig. 4.13 Gas desorption curves of coal samples with different particle sizes from No.II_1 coal seam, Baizhuang Coal Mine

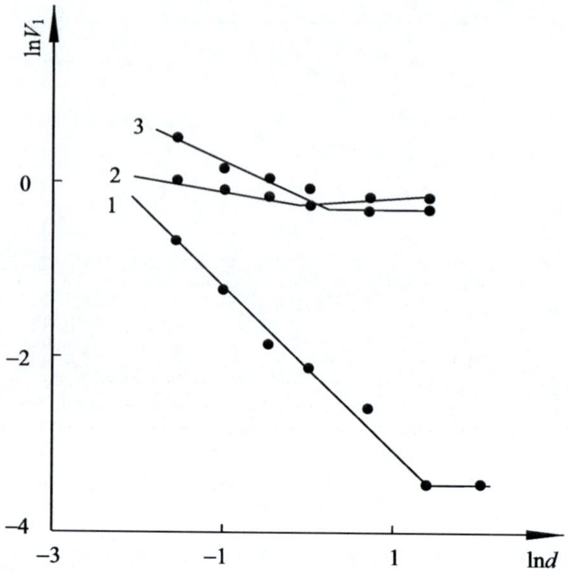

Fig. 4.14 Relationship between gas desorption rate V_1 and particle size d. 1—Soft coal from Baizhuang coal mine; 2—Coal from Liwangmiao coal mine; 3—Hard coal from Baizhuang coal mine.

Research results show that, desorption capacity of coal decreases with the increasing moisture content until the critical moisture content of coal is reached [17–19]. Figure 4.15 gives the experimental results of the relationship between desorption amount and moisture content in coal under the constant temperature (30 °C) and adsorption equilibrium pressure (1.5 MPa) [20]. It can be noticed that moisture acts as one of the factors restraining adsorption and desorption process. The desorption rate and limit desorption amount decrease with the increasing moisture content. Although the moisture has effect on gas adsorption and desorption, the effect can be ignored when measuring the coal seam gas content during geological exploration. Because once the sampling locations are the same, the inherent moisture content are basically the same, so that the resistance to the desorption process is also constant.

5. Effect of temperature on the gas desorption properties of coal

When other conditions are constant, with temperature increase, the desorption rate and desorption amount increase as well. The Tiefa mining area (Liaoning, China) once had such a test phenomenon: two ground drilling boreholes with the distance less than 500 m, using the same equipment, the same personnel and the same method, but the coal seam gas content measured in one borehole constructed in autumn was 11.43 m^3/t (estimated lost content was 4.12 m^3/t) and the coal seam gas content measured in the other borehole constructed in winter at −20 °C was only 7.6 m^3/t (estimated lost content was 0.76 m^3/t). The temperature difference of the slurry in the two boreholes was less than 5 °C due to low temperature effect during the construction

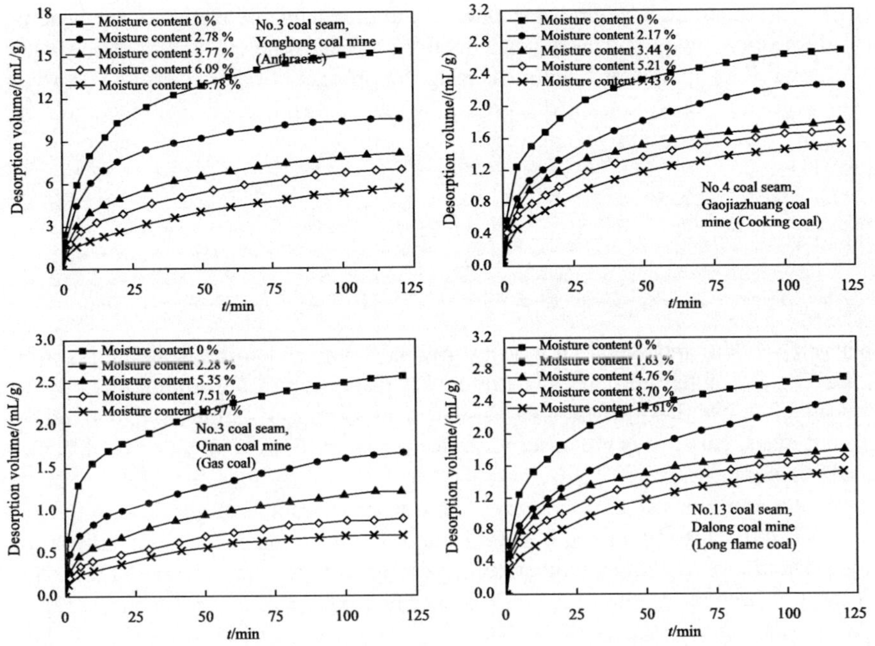

Fig. 4.15 Variation characteristics between desorption amount and moisture content for coal from different coal mines

in autumn and winter, so the lost gas amount of coal samples during sampling process should be similar. The temperature of the desorption measurement was close to the temperature of the slurry in the borehole constructed in autumn, while the coal samples taken from the borehole in winter desorbed at −20 °C. Because of the low temperature in winter, the measured gas desorption amount was very small, and the calculated amount of lost gas was only 1/6 of that in autumn. The estimated difference between the lost gas volume was the fundamental cause of the difference between the gas content of two boreholes. Therefore, in the cold area, when measuring coal seam gas content in winter, insulation measures of the coal sample should be taken to simulate the temperature of the drilling medium, otherwise incorrect results may be obtained.

4.2.3 Gas Desorption Models of Coal

Vast amount of studies in gas desorption of the granular coal in air medium have been conducted by many scholars in China and abroad. Numerous statistical or empirical equations were proposed based on gas desorption law. Because of the differences in perceived views as well as the objects/conditions of these studies, the proposed empirical equations have both reasonable and unreasonable elements at revealing gas desorption law [14]. The typical equations are listed here:

Barrer (UK) concludes from measuring the process of natural zeolite absorbing various gas that adsorption and desorption are reversible process and the accumulative amount of gas adsorption and desorption is in direct proportion to the square root of time [21]:

$$\frac{Q_t}{Q_\infty} = \frac{2s}{V}\sqrt{\frac{Dt}{\pi}} = k\sqrt{t} \tag{4.16}$$

where Q_t is the accumulative gas desorption amount from the beginning to time t, cm^3/g; Q_∞ is the ultimate gas amount of adsorption or desorption, cm^3/g; s is the sample's external surface area of unit mass, cm^2/g; V is the sample's volume of unit mass, cm^3/g; t is the adsorption or desorption time, min; D is the diffusion coefficient, cm^2/min.

Winter and Janas et al. (Germany) found that the amount of gas desorbing from the coal in the adsorption equilibrium state depends on the gas content, adsorption equilibrium pressure, time, temperature, particle size etc. and the correlation of gas desorption amount vs. time can be expressed in the form of power function [22]:

$$Q_t = \left[\frac{u_1}{1-k_t}\right] t^{1-k_t} \tag{4.17}$$

where u_1 is the gas desorption rate when $t = 1$, cm^3/(g·min); k_t is the characteristic index for changing of the gas desorption rate.

In Eq. (4.17), k_t cannot be equal to 1. In the early stage of gas desorption, the calculated value is consistent with the measured value. When time t gets longer, the error between calculated value and measured value shows the increasing tendency.

А.З.Петросян (former Soviet Union) [23] thinks that there will be a big difference between calculated value and measured value if we calculate the gas desorption using Darcy Law. Without intensive theoretical study but based on the statistical analysis of measured values, he proposes an empirical equation for calculation which is in agreement with the measured value.

$$Q_t = u_0\left[\frac{(1+t)^{1-n_1} - 1}{1-n_1}\right] \tag{4.18}$$

where u_0 is the gas desorption velocity at $t = 0$, cm^3/(g·min); n_1 is the coefficient that depends on the coal property and so on.

Chinese scholars, Wenshu Su, Liangchen Yu et al. proved the validity of Eq. (4.18) by testing coal samples from the Yangquan, Jiaozuo, Huainan and other coalfields.

Airey (UK) [24] took a coal body as an aggregate consisting of separate "blocks" with fractures to study the coal seam gas emission. Based on the Darcy Law and the assumption that every single block varies in size, he proposed the following empirical equation for the gas desorption amount vs. time in coal:

$$Q_t = Q_\infty\left[1 - e^{-\left(\frac{t}{t_0}\right)^{n_2}}\right] \tag{4.19}$$

where t_0 is the time constant; n_2 is the constant relative to the extent of fracture development in coal.

Airey's empirical equation emphasizes the coal blocks, especially the blocks with intensive fractures, and believes the fractures between blocks constitute the permeable pore volume in coal. However, viewing from the practice of production, when the damage degree of coal body is not very fierce, it's relatively reasonable to simplify coal seam as the block structure. Whilst, if coal damage is very strong (e.g. pulverized coal or soft stratification) or if it is the small particle sized samples gathered by manpower, there will be a relatively large proportion of micropores and transition pores in coal, consequently decreasing the proportion of permeable pore volume. For such situation, it might be better to use Fick's law of diffusion for the description of coal's gas desorption.

Bolt and Jinnes et al. (Australia) [25] concludes from the experimental tests concerning gas desorption process for coal of different metamorphism degree that the gas desorption process in coal is very similar to the diffusion process where gas passes through zeolite:

$$\frac{Q_t}{Q_\infty} = 1 - Ae^{-\lambda t} \tag{4.20}$$

where A and λ are empirical constants.

Wang and Yang [26] measured coal sample's gas desorption rate by gravimetric method and indicated that the change of The gas desorption amount of coal cuttings with time conforming to the equation type of Langmuir:

$$Q_t = \frac{A_t Bt}{1 + Bt} \tag{4.21}$$

where A_t and B are desorption constants.

Sun [27] studied coal cuttings' gas desorption rule and demonstrated that when the particle size of coal was small, then the gas desorption in coal is mainly a diffusion process and the relation between gas desorption amount and time can been expressed by power function:

$$Q_t = A_i t^i \tag{4.22}$$

where A_i and i are constants related to coal's gas content and structure.

Many scholars in China maintain the viewpoint that the decay process for coal cuttings' gas desorption with time is similar to the decay of gas emission in the boreholes drilled in coal seam, and they could both be described by the following equation:

$$Q_t = \frac{u_0}{b_v}(1 - e^{-b_v t})b_v \tag{4.23}$$

where b_v is the coefficient describing the decay of gas desorption rate with time, and the other signs are the same as above.

Kissell et al. (USA) [28] believe that the process of gas desorption can be described by diffusion equation and the accumulative desorption amount at early time of this process is in direct proportion to the square root of time. Such relationship can be used to calculate coal's gas leakage amount when coring in slurry medium. The specific approach is: plot the collected samples' cumulative desorption amount (in air medium) and use square root of desorption time in rectangular coordinate system, then employ the approach of straight-line extrapolation to get back to starting time of gas diffusion of coal sample in the slurry medium. Kissell thinks that starting time of gas diffusion for coal core in slurry medium depends on the type of circulating fluid used in core boring. For air or wet gas, the diffusion starts when the core barrel penetrating the coal seam and the leakage time of coal gas is the sum of coring time, lifting time and the time before samples are sealed in canister and measuring gas desorption. For drilling slurry, the gas diffusion starts when coal core is lifted at the half depth of drill. For that case, the leakage time of coal gas can be defined as a sum of the half lifting time and the time before samples are taken out, sealed in canister and measuring gas desorption. Therefore, Kissell established the industrial standards for the determination of coal seam gas content, namely the USBM desorption method, which is recognized worldwide.

Smith and Williams et al. (USA) [29, 30] aim at some technical flaws of USBM desorption method and put forward a method to calculate the gas leakage amount of coring process in slurry medium, which is known as the Smith-Williams desorption method. Smith-Williams desorption method can determine the gas content of coal seam by using drilling cuttings, gathered from the circulating fluid during drilling. The desorbed gas from coal cuttings is collected with a procedure similar to the USBM method.

Chinese scholar Zhaofeng Wang [14] simulated coal samples' desorption process in water and slurry medium using coal samples of different metamorphic degrees, different particle size, different gas pressure (content) and medium pressure. He determined the relationship between coal gas desorption amount with time when medium pressure decreases linearly as well as studying the gas desorption law in water and slurry medium, and finally establishing a new method calculating coal samples' gas leakage amount when drill hole sampling is in water and slurry medium. By fitting the measured data of gas desorption from simulated coring and drill-lifting process, he showed that the gas desorption process of coal core in slurry medium abides the following rule:

$$Q_t = u_0[\frac{(1+t)^{1+n_2} - 1}{1+n_2}] \tag{4.24}$$

where n_2 is a coefficient.

The form of Eq. (4.24) is identical to Eq. (4.18) which is for air medium, but the physical process and physical significance they express are completely different. In air medium, the index $(1 + n_2) < 1$, indicating that the gas desorption in coal is a decaying process; however, in Eq. (4.24) the index $(1+n_2) > 1$, indicating that in the process of drill lifting and coring, the gas desorption of coal core is an accelerating process on the overall trend.

4.2.4 Test Methods of Gas Desorption Properties

The laboratory test of coal desorption properties is performed by measuring the gas desorption process of the instantaneously decompressed coal sample under the adsorption equilibrium state. The schematic diagram of the experimental setup is shown in Fig. 4.16, and the laboratory testing process is as follows:

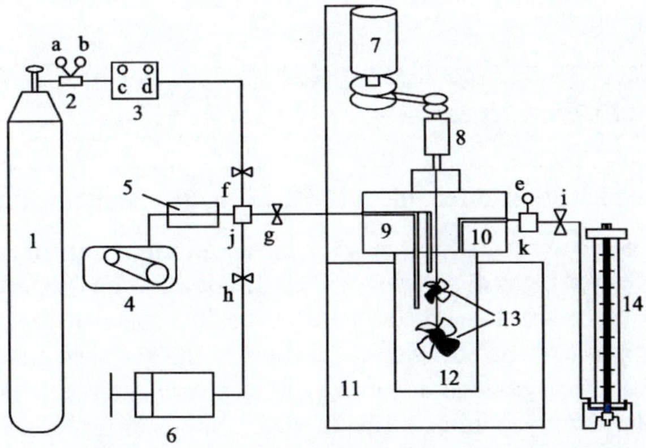

Fig. 4.16 Schematic diagram of the gas desorption experimental setup. 1—High pressure CH_4 cylinder; 2—Pressure reducing valve; 3—Reference container; 4—Vacuum pump; 5—Compound vacuum gauge; 6—Constant-flux pump; 7—Blender motor; 8—Stirring apparatus; 9—Water inlet and gas intake; 10—Gas outlet; 11—Thermostatic oil bath; 12—Coal sample container; 13—Agitating vane; 14—Desorption cylinder; a to e—Pressure gauges; f to i—Valves; j to k—side outlet ell.

(1) Pre-treatment of coal sample: The prepared coal sample (about 50 g) is packed into a coal sample container, and the container can be compacted as much as possible to reduce the volume of dead space inside. Cover the sample with degreasing cotton and then seal the container.
(2) Vacuum degassing of coal sample: Open the constant temperature oil bath, prepare the vacuum pump, set the oil bath temperature to (60 ± 1) °C. Open the valve of coal sample container to degass the coal sample for 2 h until the vacuum degree stabilizes at 4 Pa.
(3) After degassing, adjust the constant temperature water bath to (30 ± 1) °C, open the high-pressure gas cylinder valve and reference container valve, so that the high-pressure gas cylinders and gas container are linked. Charge the gas to the reference container to the gas pressure of about 1.2 times the ideal adsorption equilibrium pressure of the sample container, then close the high-pressure gas cylinder valve and link the reference container to the coal container. The equilibrium time of methane adsorption to coal sample is about 6–48 h (different particle size of coal sample needs different adsorption time), until the adsorption equilibrium state is achieved.
(4) Determination of gas desorption process of coal sample: First of all, ensure that the stopwatch and desorption cylinder are ready. Measure and record the air temperature at atmospheric pressure. Then open the coal container valve, let the free gas in coal container escape to the atmosphere, when the pressure indicator value of coal container decreases to zero, connect the valve with the desorption cylinder immediately and start the time according to the experimental purpose to set the time interval. Read and record the gas desorption amount in desorption cylinder. The measured gas desorption should be converted to a volume in a standard state.

4.2.5 *Application of Gas Desorption Properties of Coal in gas content determination*

1. Basic principles of the direct method of gas content determination

Gas desorption behavior in coal is that when the equilibrium state of coal is changed (such as pressure drop), gas desorption will begin for a certain period of time until a new dynamic equilibrium state is established. When the coal is in the original in-situ state, the gas pressure of the coal sample is equal to the gas pressure of the original coal body and the free gas and adsorbed gas in a coal sample is in the dynamic equilibrium state. During mining, strips of coal at the coal face are exposed to the atmosphere at the normal atmospheric pressure. The reduction of pressure causes the equilibrium state of the free gas and the adsorbed gas in the coal to change. The adsorbed gas in the coal begins to desorb until the gas pressure in the coal is equal to the atmospheric pressure so that a new dynamic equilibrium is achieved.

The direct method for original gas content measurement in coal seam is proposed based on the above principles. First, using the gas desorption theory, the amount of desorbed gas and the residual gas of the coal sample are measured. Then, the lost amount of gas is calculated based on the gas desorption and the time interval it took to remove the sample from its insitu adsorbed state to sealing the sample in the container. The sum of the lost gas amount, desorbed gas amount and residual gas amount is the original gas content of the coal seam.

2. Test process of the direct method of gas content determination

The test process of direct method first needs to select the appropriate site for gas content determination, by drilling a borehole the coal samples are collected from the deep portion of coal seam and immediately sealed up in the container, then an on-site 2 h test is applied to determine the desorbed gas content (X_j). A suitable empirical formula needs to be selected to calculate the lost gas amount (X_s) that occurred before sealing the coal sample into the coal sample container. When all of these are done, the coal sample container is taken back to the laboratory to determine the residual gas content (X_c). And the sum of the lost gas amount, desorbed gas amount and residual gas amount is the coal seam gas content (X), mathematically summarized by:

$$X = X_s + X_j + X_c \tag{4.25}$$

where the unit of the gas content is m³/t.

3. Calculating process of the direct method of gas content determination

1) Volume correction for the 2 h gas desorption test conducted underground

As mentioned above, the desorbed gas content in the sealed container is measured underground, however, the directly measured data, namely the underground 2 h gas desorption volume measurement needs to be corrected into the standard state volume before the gas desorption amount per unit mass can be calculated. The method for volume correction can be performed using Eq. (4.26):

$$V_{t_0} = \frac{273.2(p_0 - 0.00981h_w - p_t)V_t}{101.3(273.2 + T_w)} \tag{4.26}$$

where: V_{t_0} is the gas volume in the standard state, cm³; V_t is desorbed gas volume in the desorption cylinder at the time t, cm³; p_0 is the atmospheric pressure, kPa; T_w is the water temperature inside the cylinder, °C; h_w is the height of water column in cylinder, mm; p_t is the saturated vapor pressure of water at the time t, kPa.

2) Gas volume conversions for the negative-pressure desorption in the laboratory

(1) Negative-pressure gas desorption before pulverization: using the vacuum degassing device shown in Fig. 4.17, coal samples after 2 h desorption underground are degassed at room temperature and constant temperature

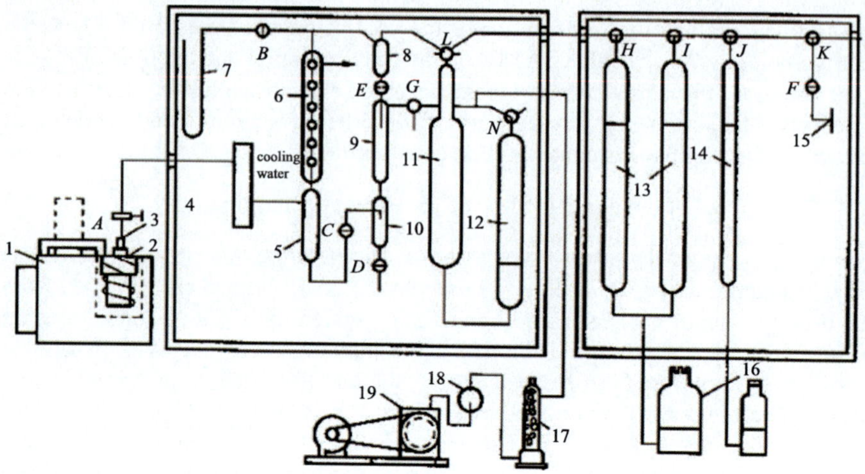

Fig. 4.17 Schematic diagram of the vacuum degassing device. 1—Thermostatic water bath; 2—Airtight sample container; 3—Puncture needle; 4—Tube filter; 5—Water-collected bottle; 6—Cooling pipe; 7—Mercury vacuum gauge; 8—Water-interdicted bottle; 9—Water adsorption tube I; 10—Water adsorption tube II; 11—Water adsorption bottle; 12—Vacuum bottle; 13—The big burette; 14—The small burette; 15—Gas flowing pipe; 16—Leveling bottle; 17—Drying tube; 18—Separation device; 19—Vacuum pump; *A*—Screw clamp; *B* to *F*—Unidirectional screw clamp; G to K—Three-way piston; L to M - Three-way piston of 120°

of 95–100 °C respectively, until the desorbed gas volume is less than 10 cm^3 per 30 min, and then analyze the gas composition with a gas chromatograph.

(2) Negative-pressure gas desorption after comminution: once the negative-pressure gas desorption before comminution is finished, quickly transfer the coal sample from the container to the ball mill tank and seal. Grind the coal sample until the weight percentage of the coal particles below 0.25 mm in size is more than 80%. Use the vacuum degassing device shown in Fig. 4.16 for the negative-pressure gas desorption after comminution. After that, to convert the gas desorption volumes before/after comminution to the standard state use Eq. (4.27):

$$V_{Tn0} = \frac{273.2(p_0 - 0.0167T_C - p_{T_n}) \times V_{Tn}}{101.3(273.2 + T_n)} \tag{4.27}$$

where, V_{Tn0} is the gas volume in the standard state, cm^3; V_{Tn} is gas volume in the cylinder at the laboratory temperature T_n and the atmospheric pressure p_1, cm^3; T_n is the laboratory temperature, °C; T_C is the barometer temperature, °C; p_{T_n} is the saturated vapor pressure of saturated salt solution at the laboratory temperature, kPa.

3) Estimation method for lost gas content

At present, the methods of $\sqrt{t}$ and power function are commonly used in China for estimating the lost gas amount, as well as other validated methods.

(1) The $\sqrt{t}$ method:
This method takes the straight line relationship between V_{t_0} and $\sqrt{t}$ at the beginning of the coal samples exposure to determine the lost gas content, namely:

$$V_{t_0} = c + d\sqrt{t} \tag{4.28}$$

where c and d are constants.
When $t = 0, V_{t_0} = c$, and the value of c is the amount of lost gas. Before the calculation of the c value, you should plot the data first with $\sqrt{t}$ versus V_{t_0} and select out the data points that show obvious linear relationship. Then determine the value of c using the least squares method, which is the lost gas content.

(2) Power function method

Convert the measured(t, V_t)data to the data of desorption rate of ($\frac{t_i+t_{i-1}}{2}$, q_t), then by fitting the desorption rate data using Eq. (4.29) find out q_0 and n.

$$q_t = q_0 \cdot (1+t)^{-n} \tag{4.29}$$

where q_t is the gas desorption rate at time of t, cm^3/min; q_0 is the gas desorption rate when $t = 0$, cm^3/min; t is the gas desorption time including the sampling time T_0, min; n is the attenuation coefficient of gas desorption rate, $0 < n < 1$.
The lost gas amount of coal sample is calculated using Eq. (4.30):

$$V_s = q_0\left[\frac{(1+t_0)^{1-n} - 1}{1-n}\right] \tag{4.30}$$

where V_s is the lost gas amount of coal sample, cm^3; t_0 is exposure time of coal sample, min.

4) Calculation of the natural gas composition in coal seam

The natural gas composition of the coal seam is analyzed using the gas composition data obtained from the negative-pressure desorbed gas or the underground desorbed gas before comminution. Defining the concentrations of the different gases obtained by gas chromatographic analysis as $C(O_2)$, $C(N_2)$, $C(CH_4)$ and $C(CO_2)$ respectively, using equations below, the airless base concentrations of the various gas components, namely the natural gas composition of coal seam, can be calculated:

$$A(N_2) = \frac{C(N_2) - 3.57C(O_2)}{100 - 4.57C(O_2)} \times 100 \tag{4.31}$$

$$A(CH_4) = \frac{C(CH_4)}{100 - 4.57C(O_2)} \times 100 \quad (4.32)$$

$$A(CO_2) = \frac{C(CO_2)}{100 - 4.57C(O_2)} \times 100 \quad (4.33)$$

where, $A(N_2)$, $A(CH_4)$ and $A(CO_2)$ are the airless base concentrations of the various gas components, %.

5) For the gas containing desorption gas, loss of gas and outgassed gas, their volumes can be converted into the volume of airless base coal seam gas with the Eq. (4.34):

$$V_i = \frac{V_{Tn0}(100 - 4.57C_{O_2})}{100} \quad (4.34)$$

where, C_{O_2} is the oxygen concentration at STP, %.

6) Calculation of the gas content

There are two ways to describe the gas content of coal. One is the gas content on air dry basis (raw coal), and the other is the gas content on dry ash free basis. Dry ash free basis refers to the air dry basis weight minus moisture and ash free:

$$X_i = \frac{\sum V_i}{m} \quad (4.35)$$

where X_i is the gas content of coal sample, cm^3/g; m is mass of coal samples (air dry basis and dry ash free basis respectively), g; V_i is the gas volumes, cm^3.

4.3 Gas Diffusion in Coal

4.3.1 Physical Process of Gas Diffusion in Coal

Gas diffusion, derived from the random motion of particles in a phase, is a concentration equilibrium process that matter migrates from high concentration to low concentration due to the random molecular motion [31]. Usually, we regard the coal as micro-porous body to discuss the diffusion process of gas in coal particles, which belongs to the gas diffusion in porous medium. Many phenomena in mining practices, such as the gas emission from coal during the coal mining, the emission of gas from the pulverized coal during the outburst development process and the gas desorption indexes of drilling cuttings that can be used to predict the coal and gas outburst, are all based on the gas diffusion in coal.

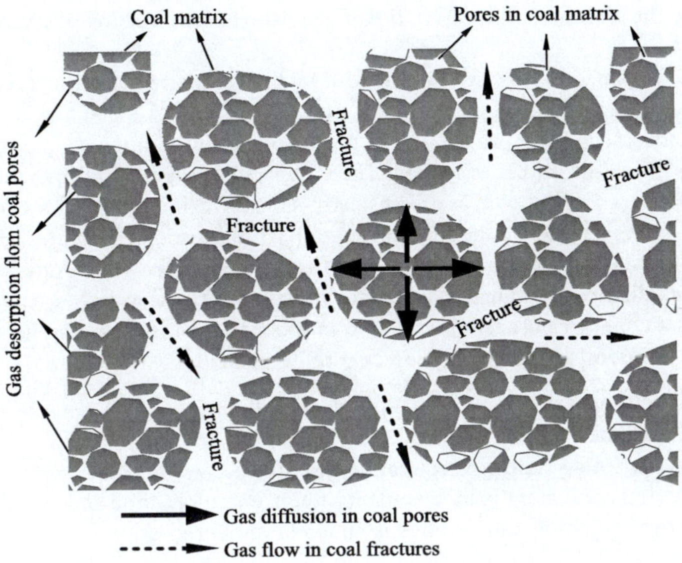

Fig. 4.18 Gas migration process in coal

The gas migration in coal is a quite complex process. According to the molecular movement theory, it is performed instantly when the gas molecules are adsorbed and/or desorbed by the coal pore surface [32]. But actually, the gas migration in coal needs some time due to the various resistance caused when the gas diffuses out of all kinds of pores in coal matrix and flows through the fractures in coal. While studying the gas migration behavior in coal, the dual porosity media model is generally accepted around the world, in which coals are regarded as the porous media consisting of coal matrix and fractures [33, 34]. As shown in Fig. 4.18 [35], the gas migration in coal can be simplified to the following consecutive processes: Firstly, the adsorbed gas on the coal pore surface is desorbed, and then, the free gas diffuses from the pores in coal matrix to the fractures; finally, gas flows out of coal bodies.

The diffusion we usually talk about refers to the process that the gas in coal migrates from the coal matrix pores to the fractures in coal. In order to describe and quantify the diffusion process from a mathematical point-of-view, the following equation is used [31, 36, 37]:

$$J = -D\frac{\partial C}{\partial x} \tag{4.36}$$

where J is the diffusive flux, kg/(m^2 s); D is the diffusion coefficient or diffusivity, and the value characterizes the diffusion "rate", m^2/s; c is the concentration, kg/kg; x is the diffusion distance, m; $\frac{\partial c}{\partial x}$ is the concentration gradient, and the negative sign means that the direction of diffusive flux is opposite to the direction of concentration gradient.

The above equation was first put forward by the German physiologist Adolf Fick and hence the name as Fick's law or Fick's first law. The Fick's first law is a formula used in a macroscopic approach and is not involved in the microscopic motion of molecules inside the diffusion system. The Fick's law is suitable for both the steady-state diffusion process as well as the unsteady-state diffusion process. Due to the fact that diffusion is caused by the random motion of microscopic particles, therefore, solutions to Fick's diffusion equation are often related to statistical functions, such as Gaussian distribution function and error function [31].

Coal is a complex porous material and in order to facilitate solving its diffusion complexity, the following assumptions are often made:

(1) Coal cuttings are composed of spherical particles;
(2) Coal particles are homogeneous and isotropic;
(3) Gas flow follows the principle of continuity;
(4) Diffusion coefficient is independent of concentration, time and coordinates;
(5) Gas desorption of coal cuttings is an isothermal process.

4.3.2 Mathematical Models of Coal Gas Diffusion

Most researchers believe that the gas diffusion movement in coal particles conforms to the classical Fick diffusion law, however the Fick's law only links the diffusion flux to the concentration but does not provide an equation for solving the concentration evolution. In general, the concentration of components in the diffusion problem is related to time and space, namely $C = C(x, t)$, where x represents the position in the one-dimensional direction. Therefore, an equation for solving $C(x, t)$ must contain differentials of t and x, respectively, that is, the partial differential equation must be used to describe the variation of concentration C with t and x. To this end, we also need to introduce another equation that connects the flux and the concentration, which is the mass conservation equation:

$$\frac{\partial C}{\partial t} = -\frac{\partial J(x)}{\partial x} \tag{4.37}$$

Combining Fick's first law with the conservation equation of mass, we obtain:

$$\frac{\partial C}{\partial t} = D\frac{\partial^2 C}{\partial x^2} \tag{4.38}$$

If the coal particles are assumed to be small spheres, then the spherical coordinates of Eq. (4.38) are as follows:

$$\frac{\partial C}{\partial t} = D\left(\frac{\partial^2 C}{\partial r^2} + \frac{2}{r}\frac{\partial C}{\partial r}\right) \tag{4.39}$$

where t is the diffusion time, s; r is the radius of the sphere coal particle, m.

Equations (4.38) and (4.39) are the most basic diffusion equations, and there are analytical solutions under many initial and boundary conditions. Most of the more complex diffusion problems can also be transformed into the above equation. In addition, different boundary conditions will also result in various analytical solutions from the Fick's model. Among the various boundary conditions, the most commonly used boundary is the Dirichlet boundary condition, which is also known as the first boundary condition. This example can be used to solve the analytical solution of the diffusion equation.

The Dirichlet boundary condition gives the value of the function at the boundary, which is ideal for describing the initial gas diffusion law of slight destruction coal type.

If $u = Cr$, Eq. (4.39) can be simplified as:

$$\frac{\partial u}{\partial t} = D\frac{\partial^2 u}{\partial r^2} \tag{4.40}$$

The initial conditions and boundary conditions are:

$$\begin{cases} u = 0(r = 0, t > 0) \\ u = r_1 C_1 (r = r_1, t > 0) \\ u = r C_0 (0 < r < r_0, t = 0) \end{cases} \tag{4.41}$$

where C_0 is the initial gas concentration at the adsorption equilibrium state, kg/kg; C_1 is the gas concentration on the surface of coal particles, kg/kg; r_1 is the outer radius of the coal particles, m.

Therefore, the diffusion equation is transformed into a second order parabolic partial differential equation with nonhomogeneous boundary condition. Using the separation variable method, Eq. (4.42) can be obtained [36]:

$$\begin{cases} C = C_1 - (C_0 - C_1)\frac{2r_1}{\pi r}\sum\limits_{n=1}^{\infty}\left[\frac{(-1)^n}{n}e^{-(\frac{n\pi}{r})^2 Dt}\text{Sin}\frac{n\pi r}{r_1}\right] \\ C|_{r=0} = C_1 - 2(C_0 - C_1)\sum\limits_{n=1}^{\infty}\left[(-1)^n e^{-(\frac{n\pi}{r})^2 Dt}\right] \end{cases} \tag{4.42}$$

Integrating the above equation, the total amount of diffusing substance leaving the sphere is:

$$\frac{Q_t}{Q_\infty} = 1 - \sum_{n=1}^{\infty}\left[\frac{6}{(n\pi)^2}\mathrm{e}^{-\left(\frac{n\pi}{r_1}\right)^2 Dt}\right] \tag{4.43}$$

where Q_∞ is the accumulated gas desorption amount when t approaches ∞, $Q_\infty = 4\pi r_1^2(C_1 - C_0)/3$, mL/g.

The above expression is an infinite series, and can be calculated by the MATLAB programming. From the calculated results, it is found that when $n \geq 7$ the diffusion coefficient D becomes stable.

However, Eq. (4.42) is too complex and not suitable for practical engineering. Therefore, some scholars used the error function to solve the analytical solution of the diffusion equation, and obtained the following expressions [31, 36]:

$$\begin{cases} C = C_0 - (C_0 - C_1)\dfrac{r_1}{r}\displaystyle\sum_{n=0}^{\infty}\left[erfc\dfrac{(2n+1)r_1 - r}{\sqrt{4Dt}} - erfc\dfrac{(2n+1)r_1 + r}{\sqrt{4Dt}}\right] \\ C|_{r=0} = C_0 - (C_0 - C_1)\dfrac{2r}{\sqrt{\pi Dt}}\displaystyle\sum_{n=0}^{\infty} e^{-(2n+1)^2 r_1^2/4Dt} \end{cases} \tag{4.44}$$

where $\mathrm{erfc}(z)$ is the complementary error function.

Integrating the Eq. (4.43), we can obtain:

$$\frac{Q_t}{Q_\infty} = \frac{6\sqrt{Dt}}{r_1}\left[\frac{1}{\sqrt{\pi}} + 2\sum_{n=1}^{\infty}\mathrm{ierfc}\left(\frac{nr_1}{\sqrt{Dt}}\right)\right] - \frac{3Dt}{r_1^2} \tag{4.45}$$

where $erfc(z)$ is the integrated error function.

Generally, the gas diffusion coefficients of coal are very small, ranging from 10^{-10} to 10^{-17} m^2/s [36], therefore, when the gas diffusing time is less than 10 min and $\frac{Q_t}{Q_\infty} < 0.5$, the above expression can be further simplified as:

$$\frac{Q_t}{Q_\infty} = \frac{6}{r_1}\sqrt{\frac{Dt}{\pi}} \tag{4.46}$$

By transforming the above formula, Eq. (4.47) can be obtained:

$$Q_t = \frac{6Q_\infty}{r_1}\sqrt{\frac{Dt}{\pi}} = K_1\sqrt{t} \tag{4.47}$$

where K_1 is the commonly used gas desorption index of drilling cuttings, mL/(min$^{0.5}$ g).

Equation (4.47) is the theoretical basis for determining desorption-based gas content, as well as the drilling cuttings gas desorption index for coal and gas outburst prediction.

4.3.3 The Desorption Index of Drilling Cuttings K_1

The desorption index of drilling cuttings K_1 is a reflection of coal desorption characteristics and physical structure characteristics, which is mainly used in the risk prediction for a coal and gas outbursts at the working face and/or a regional validation.

1. Test principle

The desorption index of drilling cuttings K_1 was proposed by the Chongqing Research Institute of China Coal Technology Engineering Group. When the gas adsorption equilibrium is achieved in coal, the sudden exposure of granular coal samples to the atmosphere will lead to a rapid gas desorption of coal. The gas desorption of coal sample can be expressed by Eq. (4.47). The physical significance of the coefficient K_1 in this equation is the total gas desorption amount per gram of coal sample within 1 min of being exposed to the atmosphere. It comprehensively reflects the amount of coal gas content and the attenuation degree of the desorption speed.

If Q_i is the cumulative volume of gas desorption from the time t_0 when the coal sample is exposed to the atmosphere, then:

$$Q_i = K_1\sqrt{t_i} - W \tag{4.48}$$

where W is the lost gas amount per gram of coal before the time t_0, mL/g.

If $X_i = \sqrt{t_i}$, then:

$$K_1 = \frac{\sum_{i=1}^{10}(Q_i - \overline{Q})(X_i - \overline{X})}{\sum_{i=1}^{10}(X_i - \overline{X})^2} \tag{4.49}$$

$$\overline{Q} = \sum_{i=1}^{10} Q_i/10 \tag{4.50}$$

$$\overline{X} = \sum_{i=1}^{10} X_i/10 \tag{4.51}$$

Therefore, if the coal containing gas is instantly exposed to the atmosphere or instrument similar to the atmospheric environmental conditions, then by measuring the gas desorption volume or desorption speed per gram of coal at regular time intervals, the desorption index of drilling cuttings can be obtained. The relationship between the measured data and the coal exposure duration follow the isovolumetric, isobaric, volume-varied and pressure-varied principles.

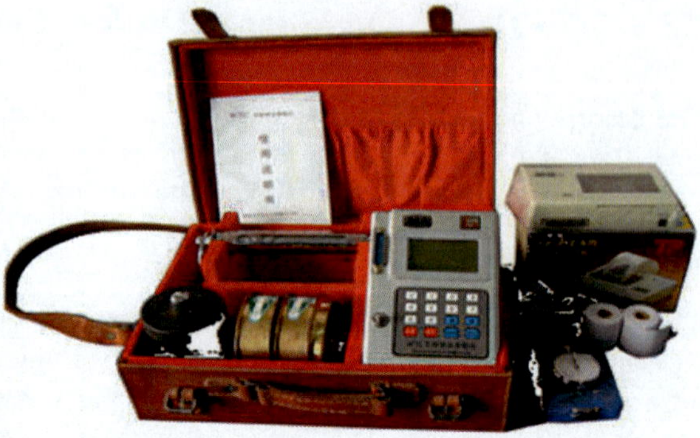

Fig. 4.19 WTC type gas outburst prediction instrument

2. Test methods

Use the WTC type gas outburst prediction instrument, which is shown in Fig. 4.19 to determine the gas desorption index K_1 of drilling cuttings. Start the timer when the borehole is drilled to the pre-calculated position. Recover the drilled coal and obtain a 10 g of the coal sample sieved to a particle size of 1–3 mm. Inserted the sample into the coal sample bottle of the WTC type gas outburst prediction instrument. After 1 min, seal the coal sample bottle and wait a moment. When the instrument carries out the gas desorption test, enter the exposure duration and length of the borehole and the instrument would calculate the value of K_1.

3. Factors influencing the gas desorption index of drilling cuttings

(1) Exposure duration of coal sample
The longer the exposure of coal samples, the longer is the desorption time of coal samples. From Eq. (4.47) and the physical significance of K_1, the exposure time has no effect on K_1. However, different exposure time leads to the various graph scatter fitting time and range, and the final result may be different [38]. Through the laboratory research, Shao [39] showed that when the exposure time was 3 min, the fitting K_1 was 13.37–21.07% smaller than when the exposure time was 1 min.

(2) Particle size of coal sample
Under the same conditions, the smaller the particle size of the coal sample, the greater the drilling cuttings gas desorption index K_1. But the degree of influence on the measurement result is gradually weakened with the increase of the particle size. When the particle size is larger than 1 mm, the impact of size variation on K_1 greatly decreases [40].

(3) Destruction type of coal sample
The higher the damage of coal, the larger the gas desorption rate. The gas desorption rate of coal in the coal seam with the risk of coal and gas outburst is higher than that in the coal seam with no risk of coal and gas outburst.

(4) Gas pressure in coal seam
The drilling cuttings gas desorption index K_1 increases exponentially with the increase of the gas pressure. In addition to the above factors, the determination of the drilling cuttings gas desorption index K_1 is also affected by human factors, sampling methods, cuttings mixed factors and other aspects [40].

References

1. Yang, R. T., & Saunders, J. T. (1985). Adsorption of gases on coals and heattreated coals at elevated temperature and pressure: 1. Adsorption from hydrogen and methane as single gases. *Fuel, 64*(5), 616–620.
2. Zwietering, P., & Van Krevelen, D. W. (1954). Chemical structure and properties of coal IV-Pore structure. *Fuel, 33*(3), 331–337.
3. Moffat, D. H., & Weale, K. E. (1955). Sorption by coal of methane at high pressures. *Fuel, 34*(4), 449–462.
4. Gu, T. R., & Zhu, B. Y. (2001). *Surface chemistry*. Science press. (in Chinese).
5. Cui, Y. J., Zhang, Q. L., & Yang, X. L. (2003). The adsorption properties of coal and the variation of the equivalent adsorption heat. *Natural Gas Industry, 23*(4), 130–131. (in Chinese).
6. Wang, F. (2016). *Kinetics characteristic of methane adsorption/desorption in coal and its application in the rapid determination of methane parameters*. China University of Mining and Technology. (in Chinese).
7. Cheng, Y. P., Jiang, H. N., & Zhang, X. L. (2017). Effects of coal rank on physicochemical properties of coal and on methane adsorption. *Journal of Coal Science and Engineering*.
8. Cheng, Y. P., Wang, H. F., Wang, L., et al. (2012). *Theories and engineering applications on coal mine gas control*. China University of Mining and Technology Press. (in Chinese).
9. Wang, L. G. (2013). *Experiment studies on displacing deep coalbed $CH_4$by gas injection and characteristic trace after displacement*. China University of Mining and Technology. (in Chinese).
10. Zhong, L. W., & Zhang, X. M. (1990). The adsorption capacity of coal and the relation to the metamorphic degree and components of coal. *Coal Geology and Exploration, 4*, 29–35. (in Chinese).
11. Yu, B. F., & Wang, Y. A. (2000). *Technical manual for coal mine gas disaster prevention and utilization*. China Coal Industry Publishing House. (in Chinese).
12. Yu, Q. X. (2012). *Coal mine gas control*. China University of Mining and Technology Press. (in Chinese).
13. Zhang, L., He, X. Q., & Nie, B. S. (2000). Research on the adsorption process of coal on the gas. *Mining Safty and Environmental Protection, 27*(6), 1–2. (in Chinese).
14. Wang, Z. F. (2001). *Study on the cas desorption law of coal in the medium of air, water and mud and its application*. China University of Mining and Technology. (in Chinese).
15. Yang, Q. L. (1987). Experimental study on the gas desorption law in coal dust. *Safety in Coal Mines, 18*(2), 9–16. (in Chinese).
16. Wang, Z. F. (1998). Study on the method of determining the gas permeability coefficient in coal seam through the permeability of coal particles. *Safety in Coal Mines, 29*(6), 3–5. (in Chinese).
17. Joubert, J. I., Grein, C. T., & Bienstock, D. (1973). Sorption of methane in moist coal. *Fuel, 52*(3), 181–185.
18. Levine, J. R., & Johnson, P. (1993). High pressure microbalance sorption studies. In *International Coalbed Methane Symposium* (pp. 187–195).

19. Clarkson, C. R., & Bustin, R. M. (2000). Binary gas adsorption/desorption isotherms: Effect of moisture and coal composition upon carbon dioxide selectivity over methane. *International Journal of Coal Geology, 42*(4), 241–271.
20. Chen, X. J. (2013). *Impact of injected water on kinetics characteristics of methane desorption in coal*. China University of Mining and Technology. (in Chinese).
21. Barrer, R. M. (1951). *Diffusion in and through solids*. Cambridge University Press.
22. Winter, K., & Janas, H. (1959). Gas emission characteristics of coal and methods of determining the desorbable gas content by means of desorbometers. In *XIV International Conference of Coal Mine Safety Research*.
23. А.Э.ПЕТРОСЯН. (1983). *Coal mine gas emission*. China Coal Industry Publishing House. (in Chinese).
24. Airey, E. M. (1968). Gas emission from broken coal. An experimental and theoretical investigation. *International Journal of Rock Mechanics and Mining Sciences and Geomechanics Abstracts, 5*(6), 475–494.
25. Bolt, B. A., & Jinnes, J. A. (1959). Diffusion of Carbon dioxide from coal. *Fuel, 38*, 333–337.
26. Wang, Y. A., & Yang, S. J. (1980). Some characteristics of coal seam with the risk of coal and gas outburst. *International Journal of Coal Science and Technology, 01*, 47–53. (in Chinese).
27. Sun, C. X. (1983). *Study on the desorbed coal gas emission and the gas desorption characteristics of coal in the coal seam with risk of coal and gas outburst*. Chongqing Research Institute. (in Chinese).
28. Kissell, F. N., McCulloch, C. M., & Elder, C. H. (1973). The direct method of determining methane content of coalbeds for ventilation design. US Bur. Mines. Rep. Invest, 7767: 17.
29. Williams, D. M., & Smith, F. L. (1984). Diffusion models for gas production from coals. *Fuel, 63*, 251–255.
30. Smith, D. M., & Williams, F. L. (1981). New technique for determining the methane content of coal. In *Proceedings of the Intersociety Energy Conversion Engineering Conference*.
31. Zhang, Y. (2008). *Geochemical kinetics*. Princeton University Press.
32. Wu, S. Y. (2005). *Research of methane-coalbed coupling movement theory and its application——Gas solid Coupling movement theory with adsorption*. Northeastern University. (in Chinese).
33. Guo, H., Cheng, Y., Ren, T., et al. (2016). Pulverization characteristics of coal from a strong outburst-prone coal seam and their impact on gas desorption and diffusion properties. *Journal of Natural Gas Science and Engineering, 33*, 867–878.
34. Liu, Q. Q., Cheng, Y. P., Zhou, H. X., et al. (2015). A mathematical model of coupled gas flow and coal deformation with gas diffusion and klinkenberg effects. *Rock Mechanics and Rock Engineering, 48*(3), 1163–1180.
35. Guo, H., Cheng, Y., Yuan, L., et al. (2016). Unsteady-state diffusion of gas in coals and its relationship with coal pore structure. *Energy Fuels*.
36. Yang, Q. L., & Wang, Y. A. (1986). Theory of methane diffusion from coal cuttings and its application. *Journl of China Coal Society, 8*(3), 87–94. (in Chinese).
37. Yang, Q. L., & Wang, Y. (1988). Mathematical simulation of the radial methane flow in spherical coal grains. *Journal of China University of Mining and Technology, 3*, 58–64. (in Chinese).
38. Shi, G. S., & Wei, F. Q. (2014). Theoretical analysis of factors influencing the gas desorption index K1 of drilling cutting. *Journal of Safety and Environment, 14*(5), 8–10. (in Chinese).
39. Shao, J. (1991). Discuss on the gas desorption index K1 of drilling cutting. *Safety in Coal Mines, 3*, 34–39. (in Chinese).
40. Zhao, X. S., & Liu, S. (2002). Influencing factor on measure error of the gas desorption index K1 of drilling cutting. *Mining Safety and Environmental Protection, 29*(2), 3–5. (in Chinese).

Chapter 5
Foundation of Coal Mechanics

Learning Outcomes

- Stress concept, differences between internal force and external force;
- The description method of the stress state under different conditions and the method to calculate the stress state using the Mohr stress circle;
- Strain, principal strain and volumetric strain;
- Frequently used strength criterion;
- Concept of the effective stress and its solution method.

5.1 Stress State

5.1.1 Concept of Stress

The basic concept of force in coal mechanics consists of external force and stress. The external force is generated from the outside of coal seam and can be further divided into two types, surface force and body force. Surface force is applied to the coal surface, such as gas pressure, coal support force and the contact forces between different coal blocks. Body force, such as gravity and inertia force, is continuously distributed within the coal and is also called a mass force.

Coal deforms when the external force is applied, and this deformation will further generate interaction among different coal parts and surfaces within the coal body. This interaction is usually named as internal force. It is well known that the interaction forces exist within the coal mass even when the coal is not affected by the external force. In coal mechanics, the internal force can also be treated as the "additional interaction force" among different internal parts of coal, which is caused by the external force and has the effect of restoring the static equilibrium state.

When trying to work out some method to calculate whether the internal force acting on a small area of coal (also known as stress) will exceed coal strength or not, intuitively it will be safer to use the sum of internal forces over a larger area. On

Y. Cheng et al., *Coal Mechanics*,
https://doi.org/10.1007/978-981-16-3895-4_5

the small scale, significant variation of internal forces and coal properties can exist due to the heterogeneity of materials such as coal. On the basis of microstructure analysis and image analysis, rock/coal heterogeneity can statistically describe its average strength that can be used when estimating strength of a larger piece of coal. However, on the larger scale, again coal is not continuous. As described before in Chap. 3 on the macroscale coal seam consists of many blocks and pieces of variable properties that need to be considered.

The mechanical quantity used to describe the concentration of internal force in any part of coal under the effect of external force is called the stress. Note that to satisfy the static equilibrium, any force is opposed with an equal reaction force in the opposite direction.

The concept of stress is used to describe the intensity of internal forces distribution in a body under the influence of a set of applied surface forces. The idea is quantified by defining the state of stress at a point in a body in terms of the forces acting on an orthogonally oriented area.

To explain the concept of stress, we assume that through the point *P* there is a plane *C* cutting the coal into two parts, part *A* and part *B*, as shown in Fig. 5.1. If we move away the part *B*, the acting force between part *A* and part *B* can be expressed by the internal force distribution on the plane *mn*. By selecting a small area ΔA including the point *P* from the plane *mn*, we can define the internal force vector as Δp, thus, the average magnitude of the resultant internal force is $\Delta p/\Delta A$. If we

Fig. 5.1 Concept of stress

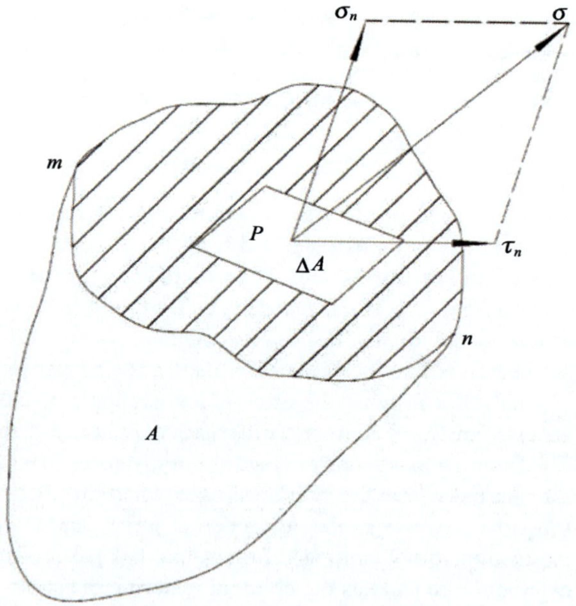

shrink ΔA infinitely small to the point P, then under the condition that internal force is continuously distributed, the value of $\Delta p/\Delta A$ tends to a fixed limit named σ, that can be expressed by:

$$\sigma = \lim_{\Delta A \to 0} \frac{\Delta p}{\Delta A} \tag{5.1}$$

This limit value of σ is called the stress at the point P on the plane mn, which is a vector. As ΔA is a scalar, the direction of σ should be consistent with that of Δp. The stress σ can be divided into two components, one is parallel to the outward normal direction of ΔA and it is called normal stress σ_n, and the other is named shear stress τ_n that is parallel to the tangential direction of ΔA. The normal stress σ_n and shear stress τ_n can be defined as

$$\sigma_n = \lim_{\Delta A \to 0} \frac{\Delta p_n}{\Delta A} \tag{5.2}$$

$$\tau_n = \lim_{\Delta A \to 0} \frac{\Delta p_t}{\Delta A} \tag{5.3}$$

where Δp_n and Δp_t represent the normal and tangential components of the internal force vector on the plane mn, respectively.

5.1.2 Specification of Stress State at a Point

In the previous section, we have discussed the concept of stress. As the plane mn and the point P is selected arbitrarily, it can be concluded that the magnitude and the direction of stress σ is not only related to the position of point P but also related to the direction of the plane mn. However, since there is an infinite number of planes with infinite variation in direction that intercept the point P, stress σ is different for each different plane intercepting point P. We can determine the stress state of point P only if we know the stress vector on all planes in any direction at the point P, which is surely impossible. Therefore, how to determine stress state at a point is the topic of the following section.

The state of stress at a point refers to the summation of the stress conditions on different planes through this point. As shown in Fig. 5.2, to study the stress state of a random point P in coal, we select an extremely small regular hexahedron (along the x, y, z axes respectively) around the point P, which is called the micro-body. The sides of the micro-body are parallel to the corresponding coordinate axis, the boundary lengths of the micro-body are dimensionless and named dx, dy and dz, respectively. As the selected micro-body is very small, it can be assumed that the stress on the six planes is uniformly distributed. Thus, the stress on the six planes

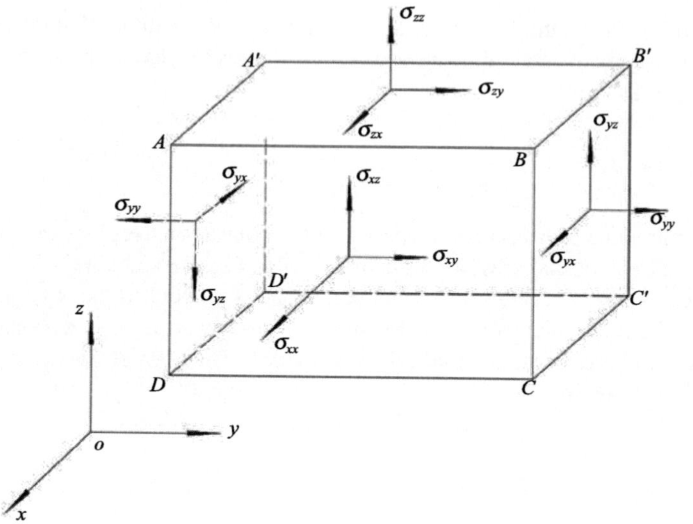

Fig. 5.2 Specification of the stress state at a point

can be expressed by using a set of stress vectors of the central point of these planes, these stress vectors include a normal stress and two tangential stresses.

When the micro-body is infinitely small, the stress condition of the hexahedron can be treated as the stress condition of point *P*. Since the stress vector on each surface can be broken down according to the directions of the three coordinate axes, thus, there are nine stress components on the three pairs of surfaces perpendicular to each other in the micro-body, and it is convenient to write the nine stress components in the form of a stress matrix, defined by

$$\begin{pmatrix} \sigma_{xx} & \sigma_{xy} & \sigma_{xz} \\ \sigma_{yx} & \sigma_{yy} & \sigma_{yz} \\ \sigma_{zx} & \sigma_{zy} & \sigma_{zz} \end{pmatrix}$$

Naming of the nine stress components should obey the following rules reference:

(1) The stress component with two same subscript letters is normal stress, otherwise, it is tangential stress.
(2) For the tangential stress, the first subscript letter represents the normal direction of the plane, and the second subscript letter represents the coordinate that is parallel to the tangential stress.

At the same time, a rule for the plus-minus sign of the stress components are as follows: For the normal stress, the sign for the compressive stress is plus and for the tensile stress is minus; For the tangential stress, if the outward normal direction of the plane is the same with the positive direction of the coordinate, then when the direction of the tangential stress is the same as the positive direction of the coordinate (plus), otherwise the sign of the tangential stress is minus.

Based on the above rule, it can be concluded that there are three normal stresses and six tangential stresses for the point P. However, according to the theorem of tangential stress equivalence, the tangential stresses acting on two planes perpendicular to each other and perpendicular to the intersection line are equal, namely have the same magnitude and ± sign), actually there are only three independent tangential stress components and the subscript letters of the tangential stress are interchangeable. For a certain point in coal, if we know three normal stress components of $\sigma_{xx}, \sigma_{yy}, \sigma_{zz}$ and three independent tangential stress components of $\sigma_{xy}, \sigma_{xz}, \sigma_{yz}$, the normal and tangential stress on any plane at this point can be calculated. Therefore, only six independent stress components are required to define the complete stress state at a point.

5.1.3 *Plane Stress State and the Mohr Stress Circle*

In the previous section, based on the coal under three-dimensional stress, we discussed the stress state at a point *P* in the three-dimensional space, with nine stress components parallel to the coordinate planes at the point *P*. Suppose that if we have already known the six independent stress components of point *P*, how can we determine the stress state of an arbitrary plane at a point? When considering the actual requirement of coal mechanics this section mainly discusses the plane stress state and two methods for determining the state of plane stress using an analytical and graphical methods. This is because many mechanical characteristics of experiments concerning coal-containing gas are conducted under pseudo-triaxial conditions.

The characteristic of the plane problem is that the external force (including surface force and body force) on the coal and its stress have no connection with one of the coordinate direction. Plane problem can be further categorized into plane strain problem and plane stress problem reference.

Plane strain problem

Consider the scale of a specific coal in a certain direction (usually defined as the *z* direction) which is far longer than that of the other two directions, and the cross section with the external forces perpendicular to the *z* axis don't change along the *z* axis reference. When ignoring the edge effects, coal deformation is the same in any plane perpendicular to the *z* axis. As shown in Fig. 5.3, the stress distribution condition around the coal seam gas drainage borehole is a typical example of the plane strain problem. Assuming *u*, *v*, *w* as the displacement components of the micro-body in the coordinate direction of *x*, *y*, *z* respectively, and because the scale in z direction is very large, the displacement in z direction can be neglected. Since the coal deformations only appear in the plane perpendicular to the z axis, both the *u* and *v* displacement have no relation with *z*. For the plane problem, the displacements have the characteristics as follows:

Fig. 5.3 Illustration of the plane strain problem

$$\begin{cases} u = u(x, y) \\ v = v(x, y) \\ w = 0 \end{cases} \tag{5.4}$$

According to Hooke's law, the stress components of any point obey the rule $\sigma_{zx} = \sigma_{xz} = \sigma_{zy} = \sigma_{yz} = 0$ in the plane perpendicular to the z axis. However, σ_{zz} isn't equal to zero due to the constraint of deformation in z axis, thus, the stress condition of any point in that plane is:

$$\begin{pmatrix} \sigma_{xx} & \sigma_{xy} & 0 \\ \sigma_{yx} & \sigma_{yy} & 0 \\ 0 & 0 & \sigma_{zz} \end{pmatrix}$$

In fact, σ_{zz} isn't an independent quantity, but can be calculated by σ_{xx} and σ_{yy}. Therefore, there are only three independent stress components for the plane strain problem, i.e., σ_{xx}, σ_{yy} and σ_{xy}.

Plane stress problem: As shown in Fig. 5.4, when the dimension of the coal body along one direction (usually z coordinate) is far shorter than the other two directions, namely the coal is a thin and uniform slab. Then, if all the external loads are applied on the middle plane or the plane parallel to the middle plane, and these external loads are symmetrical to the middle plane, the stress state can be simplified as the plane stress problem.

For the plane stress problem, as the slab is thin and the external forces distribute uniformly over the thickness, it is reasonable to assume that the stress is also distributed uniformly over the thickness. Therefore, there is $\sigma_{zz} = \sigma_{zx} = \sigma_{zy} = 0$ for any small area perpendicular to the z coordinate axis. Meanwhile, according to the equivalent law of shear stress, we also have $\sigma_{xz} = \sigma_{yz} = 0$. Thus, for the plane stress problem, the stress state at any point in the plane is

Fig. 5.4 Illustration of the plane stress problem

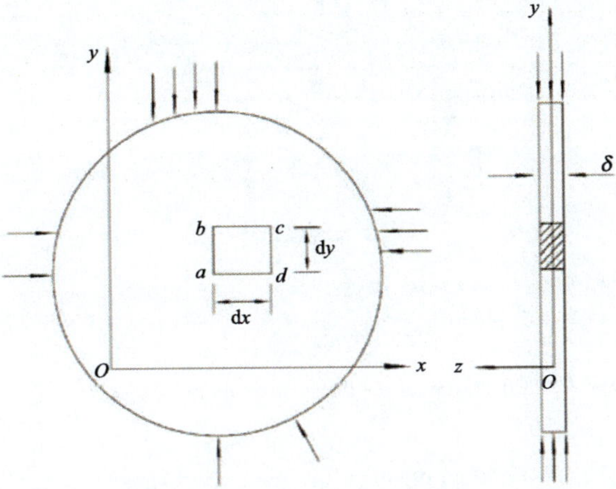

$$\begin{pmatrix} \sigma_{xx} & \sigma_{xy} & 0 \\ \sigma_{yx} & \sigma_{yy} & 0 \\ 0 & 0 & 0 \end{pmatrix}$$

The summary of the differences between plane stress problem and plane strain problem is listed in Table 5.1.

It can be found from Table 5.1 that there are only three independent stress components for both plane strain problem and plane stress problem, i.e., σ_{xx}, σ_{yy} and σ_{xy}.

Table 5.1 Summary of the differences between plane stress problem and plane strain problem

Description	Plane stress problem		Plane strain problem	
	Unknown quantities	Known quantities	Unknown quantities	Known quantities
Displacement	u, v	$w \neq 0$	u, v	$w = 0$
Strain	$\varepsilon_{xx}, \varepsilon_{yy}, \varepsilon_{xy}$	$\varepsilon_{yz} = \varepsilon_{xz} = 0$ $\varepsilon_{zz} = -\frac{\mu}{E}(\sigma_{xx} + \sigma_{yy})$	$\varepsilon_{xx}, \varepsilon_{yy}, \varepsilon_{xy}$	$\varepsilon_{yz} = \varepsilon_{xz} = \varepsilon_{zz} = 0$
Stress	$\sigma_{xx}, \sigma_{yy}, \sigma_{xy}$	$\sigma_{yz} = \sigma_{xz} = \sigma_{zz} = 0$	$\sigma_{xx}, \sigma_{yy}, \sigma_{xy}$	$\sigma_{yz} = \sigma_{xz} = 0$ $\sigma_{zz} = -\mu(\sigma_{xx} + \sigma_{yy})$
External force	The acting surfaces of body force surface forces are parallel to the plane *Oxy*, and the external force distributes uniformly along the plate thickness		The acting surfaces of body force surface forces are parallel to the plane *Oxy*, and the external force doesn't change in *z* axis	
Shape	The size in *z* axis is far less than that of the plate (thin slab with uniform thickness)		The assumed size along *z* axis is far bigger than that in the plane *Oxy* (long cylinder with uniform cross section)	

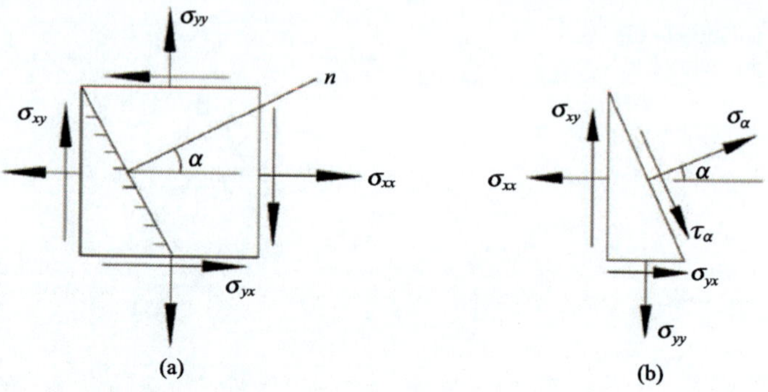

Fig. 5.5 Illustration of the stress on an inclined section

Thus, for a clear description, we only discuss the two dimensional stress state of the plane stress problem.

For the plane stress problem, there are two methods to calculate the stress state at a point, namely the analytical method and graphical method. The analytical method is based on solving of multiple sets of equations, which is not convenient thus, we only discuss the graphical method in the book.

The stress state of the plane stress unit is illustrated in Fig. 5.5, where we select an oblique section named α to introduce the graphical method. The exterior normal direction of this section is at α degrees angle to the x-axis.

Based on the idea of cross-section method, the plane stress unit can be sliced along the α oblique section. If we set the area of the oblique section as dA, then take the lower left part of plane stress unit as the study object and according to the stress equilibrium theory, we have

$$\sigma_{xx}\mathrm{d}A\cos\alpha - \sigma_{yx}(\mathrm{d}A\sin\alpha) - (\sigma_{\alpha}\cos\alpha)\mathrm{d}A - (\tau_{\alpha}\sin\alpha)\mathrm{d}A = 0 \tag{5.5a}$$

$$\sigma_{yy}\mathrm{d}A\sin\alpha - \sigma_{xy}(\mathrm{d}A\cos\alpha) - (\sigma_{\alpha}\sin\alpha)\mathrm{d}A + (\tau_{\alpha}\cos\alpha)\mathrm{d}A = 0 \tag{5.5b}$$

By solving Eqs. (5.5a) and (5.5b), we obtain:

$$\sigma_{\alpha} = \frac{\sigma_{xx} + \sigma_{yy}}{2} + \frac{\sigma_{xx} - \sigma_{yy}}{2}\cos 2\alpha - \sigma_{xy}\sin 2\alpha \tag{5.6}$$

$$\tau_{\alpha} = \frac{\sigma_{xx} - \sigma_{yy}}{2}\sin 2\alpha + \sigma_{xy}\cos 2\alpha \tag{5.7}$$

Combining Eqs. (5.6) and (5.7) yields

$$(\sigma_{\alpha} - \frac{\sigma_{xx} + \sigma_{yy}}{2})^2 + \tau_{\alpha}^2 = (\frac{\sigma_{xx} - \sigma_{yy}}{2})^2 + \sigma_{xy}^2 \tag{5.8}$$

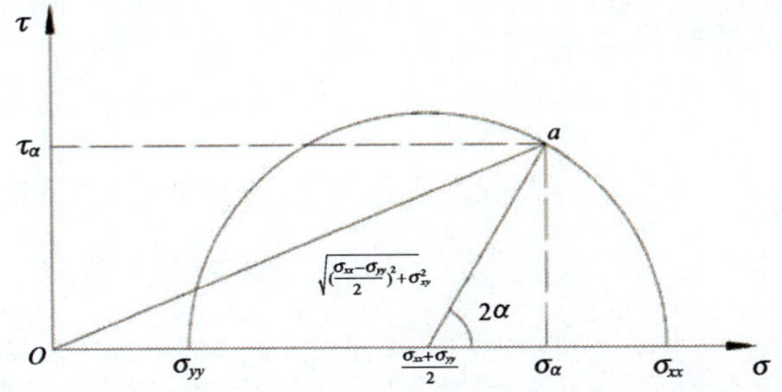

Fig. 5.6 Illustration of the Mohr stress circle

According to Eq. (5.8), it can be concluded that the locus of stress (σ_α, τ_α) of the α oblique section is a circle on the stress plane (σ, τ). The coordinate of the center of the circle is ($\frac{\sigma_{xx}+\sigma_{yy}}{2}$, 0), and radius is $\sqrt{(\frac{\sigma_{xx}-\sigma_{yy}}{2})^2+\sigma_{xy}^2}$. As shown in Fig. 5.6, this circle is known as the stress circle or Mohr's circle, which is firstly presented by Mohr in 1914. By using the Mohr's circle obtained from the stress state of plane stress unit, we can obtain the stress state of any oblique section of the unit.

5.1.4 Principal Stresses and Their Directions

The stress state of arbitrary micro plane over a point in coal can be completely determined by three direction cosines n_1, n_2, n_3 of the outer normal direction $\vec{n}$ and the six independent stress components σ_{ij}. When the micro plane rotates, its outward normal direction will rotate simultaneously, thus, the normal stress σ_n and the τ_n will change correspondingly (both in magnitude and direction). During the continuous rotation of this micro plane as well as changing its outer normal direction, there must exist a special situation where the shear stress components of the micro plane vanish and only normal stress exists. This special plane on which only normal stress exists is defined as the principal plane, and its outward normal direction is defined as the principal direction, and the normal stress along the principal direction is named the principal stress. In an arbitrary point of the stressed coal, there are three such reference directions representing directions of three principal stresses and three principal directions.

As shown in Fig. 5.7, the plane *ABC* is defined as the principal plane, and the principal stress on it is σ_n. If the six independent components are already known, the components of the principal stress σ_n on the three coordinate axis can be respectively expressed as:

$$\sigma_n n_1 = \sigma_{11} n_1 + \sigma_{21} n_2 + \sigma_{31} n_3 \tag{5.9a}$$

$$\sigma_n n_2 = \sigma_{12} n_1 + \sigma_{22} n_2 + \sigma_{32} n_3 \tag{5.9b}$$

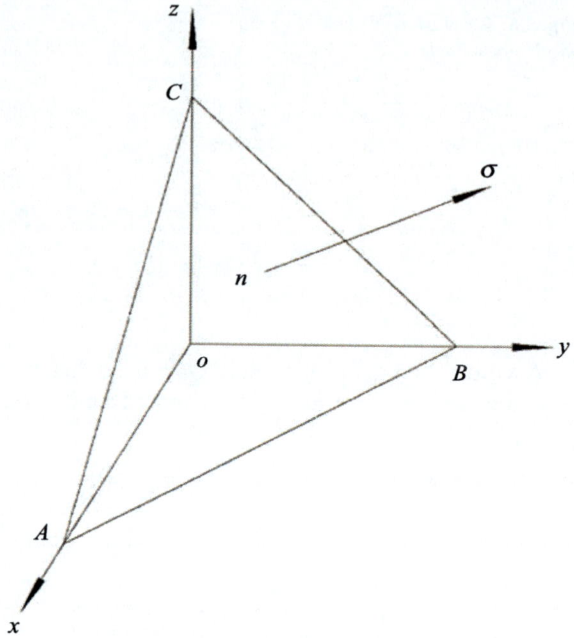

Fig. 5.7 Illustration of the principal stresses

$$\sigma_n n_3 = \sigma_{13} n_1 + \sigma_{23} n_2 + \sigma_{33} n_3 \tag{5.9c}$$

The above equation can be transformed into:

$$\begin{cases} (\sigma_{11} - \sigma_n) n_1 + \sigma_{21} n_2 + \sigma_{31} n_3 = 0 \\ \sigma_{12} n_1 + (\sigma_{22} - \sigma_n) n_2 + \sigma_{32} n_3 = 0 \\ \sigma_{13} n_1 + \sigma_{23} n_2 + (\sigma_{33} - \sigma_n) n_3 = 0 \end{cases} \tag{5.10}$$

If Eq. (5.10) represents the three dimensional homogeneous linear algebraic equations with six known independent components and the direction cosines n_1, n_2 and n_3 they obey the following relation:

$$n_1^2 + n_2^2 + n_3^2 = 1 \tag{5.11}$$

Because n_1, n_2 and n_3 can't be zero at the same time, the coefficient determinant of Eq. (5.10) must be equal to zero if we want to get a nonzero solution:

$$\begin{vmatrix} \sigma_{11} - \sigma_n & \sigma_{21} & \sigma_{31} \\ \sigma_{12} & \sigma_{22} - \sigma_n & \sigma_{32} \\ \sigma_{13} & \sigma_{23} & \sigma_{33} - \sigma_n \end{vmatrix} = 0 \tag{5.12}$$

Expanding the determinant, we have

$$\sigma_n^3 - (\sigma_{11} + \sigma_{22} + \sigma_{33})\sigma_n^2 + (\sigma_{11}\sigma_{22} + \sigma_{22}\sigma_{33} + \sigma_{33}\sigma_{11} - \sigma_{12}^2 - \sigma_{23}^2 - \sigma_{31}^2)\sigma_n$$
$$- (\sigma_{11}\sigma_{22}\sigma_{33} + 2\sigma_{12}\sigma_{23}\sigma_{31} - \sigma_{11}\sigma_{23}^2 - \sigma_{22}\sigma_{31}^2 - \sigma_{33}\sigma_{12}^2) = 0 \quad (5.13)$$

Defining the quantities I_1, I_2 and I_3 as the first, second and third stress invariants, their expressions are as follow:

$$\begin{aligned} I_1 &= \sigma_{11} + \sigma_{22} + \sigma_{33} \\ I_2 &= \sigma_{11}\sigma_{22} + \sigma_{22}\sigma_{33} + \sigma_{33}\sigma_{11} - \sigma_{12}^2 - \sigma_{23}^2 - \sigma_{31}^2 \\ I_3 &= \sigma_{11}\sigma_{22}\sigma_{33} + 2\sigma_{12}\sigma_{23}\sigma_{31} - \sigma_{11}\sigma_{23}^2 - \sigma_{22}\sigma_{31}^2 - \sigma_{33}\sigma_{12}^2 \end{aligned} \quad (5.14)$$

Then, Eq. (5.13) can be simplified as

$$\sigma_n^3 - I_1\sigma_n^2 + I_2\sigma_n - I_3 = 0 \quad (5.15)$$

Equation (5.15) is a cubic equation of σ_n, which is called the characteristic equation of stress state. By solving Eq. (5.15) we can obtain three real roots which describe the three principal stresses (σ_1, σ_2 and σ_3). The three groups of direction cosines of the three principal stresses correspond to the three groups of principal planes. As the principal stresses are only a consequence of the loading mode and its magnitude, they are independent of any selected coordinate system, thus, the coefficients I_1, I_2 and I_3 in Eq. (5.15) are also independent of any selected coordinates. For the confirmed stress state, I_1, I_2 and I_3 become constant quantities defining the principal stresses.

Overall, σ_1 is the maximum principal stress, σ_2 is the intermediate principal stress and σ_3 is the minimum principal stress. These principal stresses are unique and exist only in one particular orientation where all shear stresses vanish. Directions of these three principal stresses ($\sigma_1, \sigma_2, \sigma_3$) are perpendicular to each other.

5.2 Strain State

In this section, some deformation laws of coal are introduced, such as the basic concept of deformation and strain, geometric equation, principal strain and volumetric strain.

5.2.1 Displacement and Strain Concepts

In Sect. 5.1, stress concepts and the state in coal bearing the external force are discussed, and in this section we examine displacements and strain in coal under the effect of external force.

Under external forces, the relative positions of points within the coal body change, namely the points in coal where displacements occur. If the displacements occur, but the relative position of each point doesn't change, namely the relative displacements keep constant through the initial state to the final state during external loading, then only rigid displacements and rotations in coal body occur. This kind of displacement is called rigid displacement. Otherwise, if the relative position of each point changes, namely the displacement induces changes in coal shape, then it is called the coal deformation or strain. In this section, we only discuss the deformation of stressed coal, namely the strain of coal bearing external force, rather than the rigid displacement.

As shown in Fig. 5.8, a coal body generates elastic deformation when the external force is applied, the solid line contour is the shape of coal at the initial state and the dashed contour is the shape of coal after deformation. Selecting a point *A* in the coal body, and defining its position after deformation as the point *A*', the displacement of point *A* can be defined as u, v and w in the *x*, *y*, *z* direction. As discussed above, coal deformation will induce change of relative position to the points in coal and the displacements of different points within the coal would vary, thus the displacements of any point can be expressed by the space coordinate function of the displacement components:

$$u = u(x, y, z) \quad v = v(x, y, z) \quad w = w(x, y, z) \tag{5.16}$$

Fig. 5.8 Illustration of the coal deformation and the displacement of a single point *A*

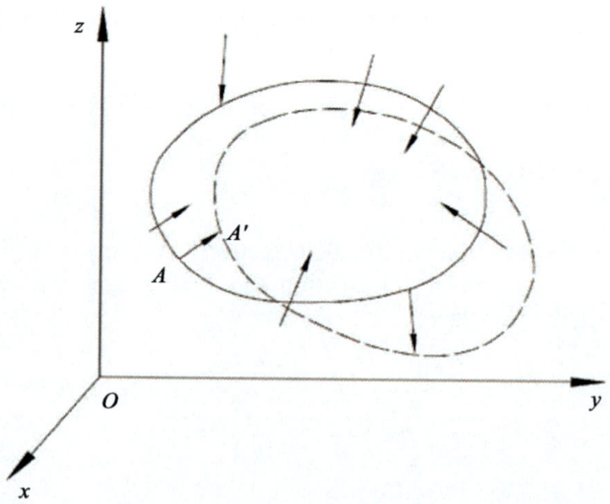

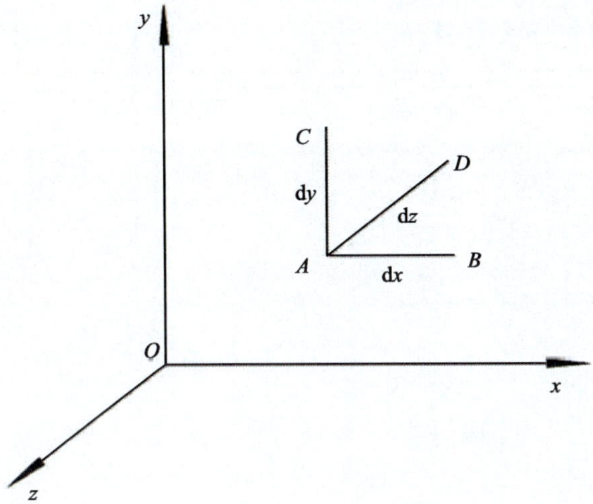

Fig. 5.9 Strain state of a point

The coal body follows the continuity assumption in the elastic deformation stage. The displacement component of any point in coal can be defined as the single valued continuous function. If we can confirm the displacements of points in coal, the strain state of the internal points can be determined.

As shown in Fig. 5.9, in order to describe the deformation characteristic of any point of a coal body, taking point *A* we select three orthogonal infinitesimal sections *AB*, *AC* and *AD*, and their length is dx, dy and dz, respectively. As the relative positions of the points change due to the external force induced deformation, the length and included angle of the three infinitesimal sections will change. This defines the relative length change of the infinitesimal section as the normal strain of point *A* (ε), while also defining the relative included angle change of the infinitesimal section as the shearing strain of point *A* (γ).

The normal strain of point *A* in the *x*, *y*, and *z* directions can be expressed by ε_x, ε_y and ε_z, respectively. In coal mechanics, we define the normal strain as positive in compression and negative in expansion. For the shear strain, the change of the included angle between *AB* and *AC* is defined as γ_{xy}, which for *AC* and *AD* is γ_{yz} and for *AD* and *AB* is γ_{zx}. In coal mechanics, we define the shear strain as positive when the included angle decreases, and negative when the included angle increases.

5.2.2 *Geometric Equation*

The concept of displacement and strain has been introduced in the previous section, and in this section the relationship between the strain component and displacement component is discussed. As shown in Fig. 5.10, a hexahedral micro-body at point *A* is selected from a coal body, and the length of the three edges *AB*, *AC* and *AD*

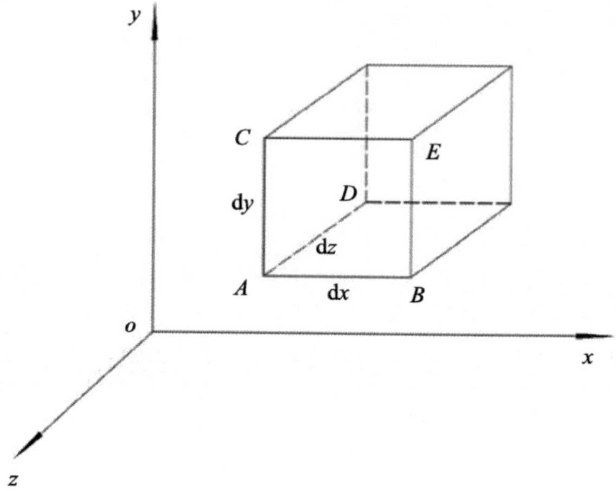

Fig. 5.10 Micro-body bearing external force

is $\mathrm{d}x$, $\mathrm{d}y$ and $\mathrm{d}z$, respectively. Based on the deformation characteristics of micro-body under external force, the fundamental assumptions concerning the geometric equations under small deformation conditions will be examined first. The strain of different points in coal is usually non-uniform, the selected micro-body is very small and the deformation is limited to the infinitesimal deformation, therefore, we can make the following assumptions:

(1) The surface of the micro-body is planar before and after deformation;
(2) The edge of the micro-body is a straight line before and after deformation.
(3) The lines parallel to each other keep to be parallel before and after deformation.

To simplify the discussion about the micro-body deformation, based on the above-mentioned assumptions, we discuss the deformation of the micro-body projection first, then we can determine the deformation of micro-body based on the projection deformation. Figure 5.11 illustrates the projection on the micro-body on the *x*, *y* plane. The coordinates of points *A*, *B* and *C* at the initial state (before the deformation of micro-body) are:

$$A(x, y) \quad B(x + \mathrm{d}x, y) \quad C(x, y + \mathrm{d}y)$$

Due to the micro-body deformation, points *A*, *B* and *C* move to *A'*, *B'* and *C'*, and their coordinates change to:

$$\begin{aligned}
&A'\ [x + u(x, y), y + v(x, y)]\\
&B'\ [x + \mathrm{d}x + u(x + \mathrm{d}x, y), y + v(x + \mathrm{d}x, y)]\\
&C'\ [x + u(x, y + \mathrm{d}y), y + \mathrm{d}y + v(x, y + \mathrm{d}y)]
\end{aligned}$$

Expanding: $u(x + \mathrm{d}x, y)$, $v(x + \mathrm{d}x, y)$, $u(x, y + \mathrm{d}y)$ and $v(x, y + \mathrm{d}y)$ by using the Taylor series expansion method, and ignoring the influence of the higher order

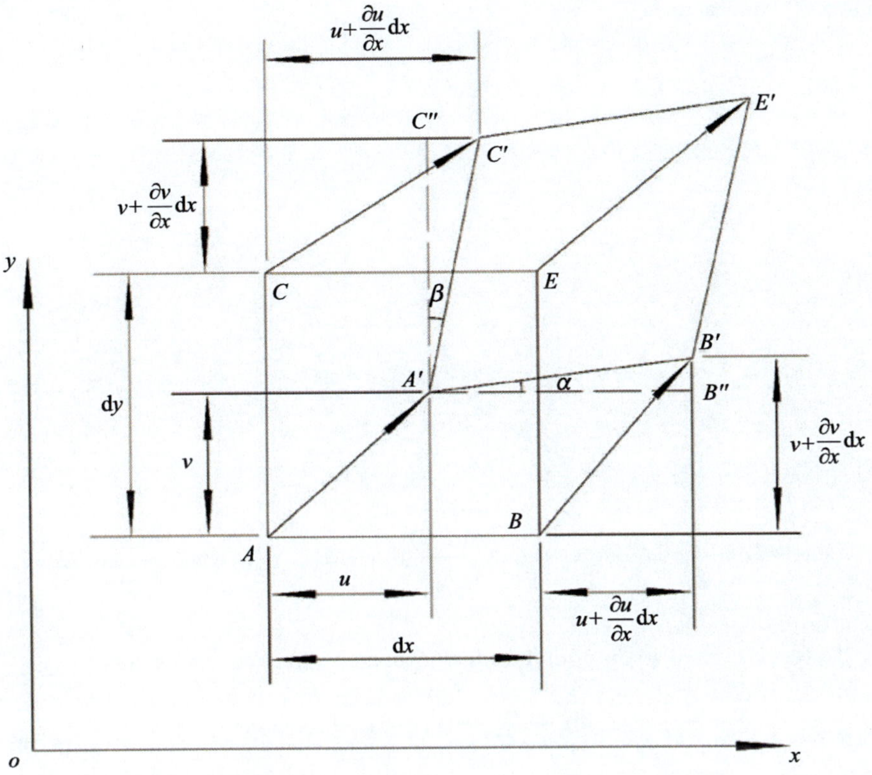

Fig. 5.11 Projection deformation of the micro-body

term, we have:

$$
\begin{aligned}
u(x + \mathrm{d}x, y) &= u(x, y) + \frac{\partial u(x, y)}{\partial x}\mathrm{d}x \\
v(x + \mathrm{d}x, y) &= v(x, y) + \frac{\partial v(x, y)}{\partial x}\mathrm{d}x \\
u(x, y + \mathrm{d}y) &= u(x, y) + \frac{\partial u(x, y)}{\partial y}\mathrm{d}y \\
v(x, y + \mathrm{d}y) &= v(x, y) + \frac{\partial v(x, y)}{\partial y}\mathrm{d}y
\end{aligned}
\tag{5.17}
$$

According to Fig. 5.11, the projection of the edge AB after deformation is $A'B'$, and the projection of the edge $A'B'$ on the x-direction is $A'B''$. At the infinitesimal deformation condition, the length of $A'B'$ is almost equal to that of $A'B''$, namely:

Fig. 5.12 Principal strain

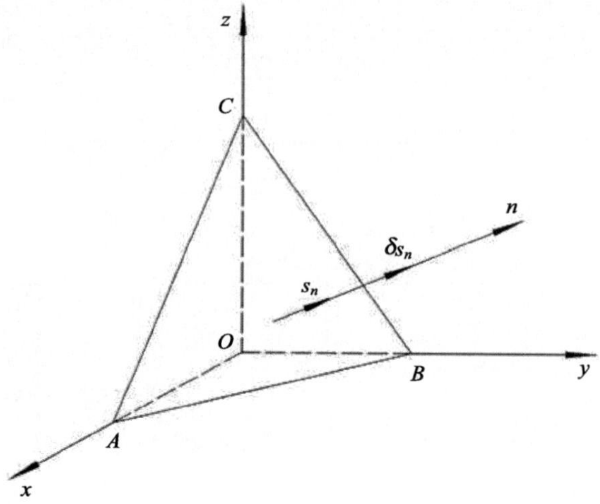

$$l_{A'B'} \approx l_{A'B''} = [\mathrm{dx} + u(x, y) + \frac{\partial u(x, y)}{\partial x}\mathrm{dx}] - u(x, y) = \mathrm{dx} + \frac{\partial u(x, y)}{\partial x}\mathrm{dx} \tag{5.18}$$

Thus, the normal strain ε_x of the edge *AB* in the *x* direction can be defined as

$$\varepsilon_x = \frac{[\mathrm{dx} + \frac{\partial u(x,y)}{\partial x}\mathrm{dx}] - \mathrm{dx}}{\mathrm{dx}} = \frac{\partial u(x, y)}{\partial x} \tag{5.19}$$

Similarly, the normal strain ε_y, ε_z of the edge *AB* in the *y* direction and *z* direction can be defined as:

$$\varepsilon_y = \frac{\partial v(x, y)}{\partial y} \quad \varepsilon_z = \frac{\partial w(x, y)}{\partial z} \tag{5.20}$$

In the space coordinates, the normal strain can be defined as:

$$\varepsilon_x = \frac{\partial u(x, y, z)}{\partial x} \quad \varepsilon_y = \frac{\partial v(x, y, z)}{\partial y} \quad \varepsilon_z = \frac{\partial w(x, y, z)}{\partial z} \tag{5.21}$$

The above equations define the relationship between the normal strain components and displacement components of any points in coal. As the displacement and strain have both geometric sense, the above-mentioned equations are usually called geometric equations or Cauchy's equations.

5.2.3 Principal Strain and Volumetric Strain

Using similar reasoning to those discussed in the stress state section, the plane on which the shear strain is equal to zero is the principal plane. The normal direction of the principal plane is defined as the principal strain direction, and the normal strain perpendicular to the principal plane is the principal strain. There are three principal planes at a point and therefore three principal strain directions and three corresponding principal strains ε_1, ε_2 and ε_3.

Calculation method of ε_1, ε_2 and ε_3: As shown in Fig. 5.12, we can find a vector s_n along the normal direction of the plane *ABC*. During the deformation process, the direction of the vector remains unchanged while the increment of the vector change is δs_n. As the direction of s_n and δs_n are the same, the components of them are proportional to the other:

$$\frac{\delta s_n}{s_n} = \frac{\delta s_x}{s_x} = \frac{\delta s_y}{s_y} = \frac{\delta s_z}{s_z} \tag{5.22}$$

where s_x, s_y, s_z, δs_x, δs_y and δs_z are the components of s_n and δs_n in x, y and z coordinates.

According to the definition of strain:

$$\varepsilon_n = \frac{\delta s_n}{s_n} \tag{5.23}$$

Thus:

$$\delta s_x = \varepsilon_n s_x \quad \delta s_y = \varepsilon_n s_y \quad \delta s_z = \varepsilon_n s_z \tag{5.24}$$

In addition, we have

$$\begin{aligned} \delta s_x &= \varepsilon_x s_x + \varepsilon_{xy} s_y + \varepsilon_{xz} s_z \\ \delta s_y &= \varepsilon_{xy} s_x + \varepsilon_y s_y + \varepsilon_{yz} s_z \\ \delta s_x &= \varepsilon_{xz} s_x + \varepsilon_{yz} s_y + \varepsilon_z s_z \end{aligned} \tag{5.25}$$

Equation (5.25) can be transformed into

$$\varepsilon_n^3 - I_1 \varepsilon_n^2 + I_2 \varepsilon_n - I_3 = 0 \tag{5.26}$$

where,

$$I_1 = \varepsilon_{xx} + \varepsilon_{yy} + \varepsilon_{zz} \tag{5.27a}$$

$$I_2 = (\varepsilon_{xx}\varepsilon_{yy} + \varepsilon_{yy}\varepsilon_{zz} + \varepsilon_{zz}\varepsilon_{xx}) - (\varepsilon_{xy}^2 + \varepsilon_{yz}^2 + \varepsilon_{zx}^2) \tag{5.27b}$$

$$I_3 = \varepsilon_{xx}\varepsilon_{yy}\varepsilon_{zz} + 2\varepsilon_{xy}\varepsilon_{yz}\varepsilon_{zx} - (\varepsilon_{xx}\varepsilon_{yz}^2 + \varepsilon_{yy}\varepsilon_{zx}^2 + \varepsilon_{zx}\varepsilon_{zx}^2) \quad (5.27c)$$

By solving Eq. (5.26) we can obtain three real roots, namely the principal strains ε_1, ε_2 and ε_3. And because when the strain state is fixed, ε_1, ε_2 and ε_3 will not change with the coordinate direction. Thus, I_1, I_2 and I_3 are invariants, which are usually named as the first, the second and the third invariant.

If we take a micro-body (cuboid) around the point A, whose length of the three sides is dx, dy and dz, thus, its volume is dxdydz. When the coal generates tiny deformation induced by external force, the strain components of point A are assumed as ε_{xx}, ε_{xy}, ε_{yy}, ε_{xy}, ε_{zz} and ε_{zx}. Thus, after the deformation which turns the tiny cuboid into oblique parallelepiped, the length of the three sides of the oblique parallelepiped changes into $(1+\varepsilon_{xx})\mathrm{d}x$, $(1+\varepsilon_{yy})\mathrm{d}y$ and $(1+\varepsilon_{zz})\mathrm{d}z$. Under the small deformation condition, the volume change of the micro-body is induced by the length change of the three sides. Thus, the volume of the oblique parallelepiped can also be calculated by the product of the lengths of three sides, namely:

$$v = (1+\varepsilon_{xx})\mathrm{d}x(1+\varepsilon_{yy})\mathrm{d}y(1+\varepsilon_{zz})\mathrm{d}z \quad (5.28)$$

The change in unit volume is called the volumetric strain:

$$\varepsilon_v = \frac{(1+\varepsilon_{xx})\mathrm{d}x(1+\varepsilon_{yy})\mathrm{d}y(1+\varepsilon_{zz})\mathrm{d}z - \mathrm{d}x\mathrm{d}y\mathrm{d}z}{\mathrm{d}x\mathrm{d}y\mathrm{d}z} \quad (5.29)$$

By omitting the higher order term, Eq. (5.29) can be simplified as

$$\varepsilon_v = \varepsilon_{xx} + \varepsilon_{yy} + \varepsilon_{zz} \quad (5.30)$$

If all the three surfaces of the micro-body are the principal planes of the point, Eq. (5.30) can be transformed to

$$\varepsilon_v = \varepsilon_1 + \varepsilon_2 + \varepsilon_3 \quad (5.31)$$

Equation (5.30) is the commonly used calculation formula for volumetric strain.

5.3 Strength Criterion

Strength criterion is also known as the failure criterion, which defines rock/coal failure under the effect of a stress or a combined stress. Due to the variable rock properties, for different kinds of rocks their failure characteristics can be quite different. Various strength criteria have been established according to the different characteristics of rock failure, but so far, the universally applicable strength criterion has not been found. Among the various rock strength criteria, the Mohr–Coulomb strength criterion, D-P (Drucker-Prager) strength criterion and Griffith strength criterion are of the most representative and widely used.

5.3.1 Mohr–Coulomb Strength Criterion

The Mohr–Coulomb (MC) strength criterion is the most widely used method to describe rock/coal failure in rock mechanics and coal mechanics. It was first proposed by C.A. Coulomb as the "friction" criterion in 1773. Then, he speculated that rock strength which is also known as the abrasion resistance was equal to the rock's cohesive force resisting the shear friction and the friction force generated by the normal force of the shear surface. Afterwards, Coulomb's theory was further developed by Mohr, who pointed out that the properties of materials themselves are also a function of stress.

The Mohr–Coulomb failure criterion is established on the statistical analysis of the uniaxial/ triaxial compression test results. The basic idea of the Mohr–Coulomb failure criterion is that rock failure does not occur under simple stress state, but appears under the combined effects of different normal stress and shear stress. The failure of rock materials subjected to shear and normal stresses belongs to the shear failure mode. The Mohr–Coulomb failure criterion can be intuitively described by the Mohr limit stress circle.

As shown in Fig. 5.13, in the τ-σ coordinate system, the strength envelope of Mohr–Coulomb failure criterion is expressed as

$$\varphi \quad \tau_f = c + \sigma_n \tan \varphi \tag{5.32}$$

where τ_f is the ultimate shear strength under the effect of the normal stress σ_n, MPa; c is the cohesion of rock, MPa; φ is the angle of internal friction.

Under the stress state shown in Fig. 5.13, the normal stress and shear stress of a plane in the coal can be determined by the stress circle made of the principal

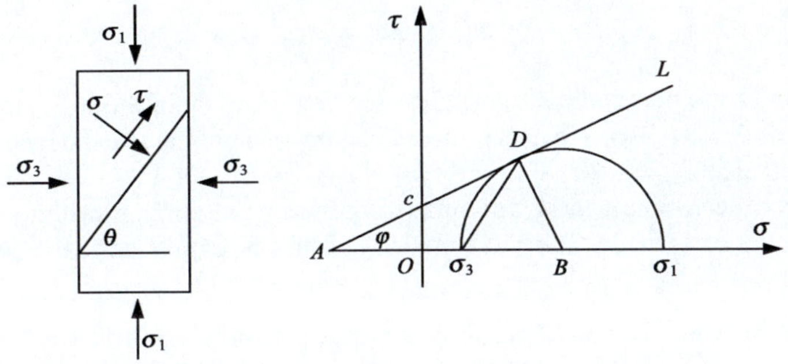

Fig. 5.13 Mohr–Coulomb strength criterion

stresses σ_1 and σ_3. If the stress circle is located beneath the strength envelope *AL*, the corresponding stress conditions given by of the principal stresses σ_1 and σ_3 do not reach the ultimate strength of the material, and the material will not fail. However, if the stress circle is over the strength envelope *AL*, the corresponding stress conditions given by of the principal stresses σ_1 and σ_3 exceed the ultimate strength of the material, and the material will fail. Otherwise, if the stress circle just touches the strength envelope *AL*, the material is in the state of limit equilibrium, and the shear failure may occur. Moreover, it can also be noticed that the Mohr–Coulomb strength envelope is the trace line composed of the failure points for a particular rock at different limit stress circle.

In Eq. (5.32), the cohesion and the angle of internal friction are the two most widely used parameters representing the strength characteristics of rock. Cohesion is the mutual attraction between the internal (adjacent) parts of the same material, representing the molecular attraction force existing among the molecules within the material. The angle of internal friction, that appeared in the Coulomb's theory for the first time, is also an important component of rock shear strength. The cohesion and the internal friction angle are both indirectly measurable parameters, which can be calculated by the statistical analysis of the uniaxial/triaxial compression test results.

Mohr–Coulomb criterion assumes the failure of rock material belongs to shear failure, but actually the failure of rock is the process of crack extension, growth and connection, and the final failure of rock involves more than just shear failure. The Mohr–Coulomb failure criterion can only provide the strength condition of rock failure, but cannot describe the process of failure occurrence and development, which is also the limitation of this criterion.

5.3.2 *Drucker–Prager Strength Criterion*

The Drucker–Prager (D-P) failure criterion considers that the failure of material is decided by the first invariant of the stress tensor and the second invariant of the stress deviator tensor, which can be expressed as:

$$aI_1 + \sqrt{J_2} = k \tag{5.33}$$

where a and k are material constants. I_1 is the first invariant of the stress tensor and is defined as $I_1 = \sigma_1 + \sigma_2 + \sigma_3$. J_2 is the second invariant of the stress deviator tensor and is defined as $J_2 = \frac{1}{6}\left[(\sigma_1 - \sigma_2)^2 + (\sigma_2 - \sigma_3)^2 + (\sigma_1 - \sigma_3)^2\right]$.

In order to determine a and k, a comparison between D-P strength criterion and MC strength criterion is made. The MC criterion can also be expressed in the alternative form:

$$f(\sigma_1, \sigma_3) = \frac{\sigma_1 - \sigma_3}{2} + \frac{(\sigma_1 + \sigma_3)\sin\psi}{2} - C\cos\psi \tag{5.34}$$

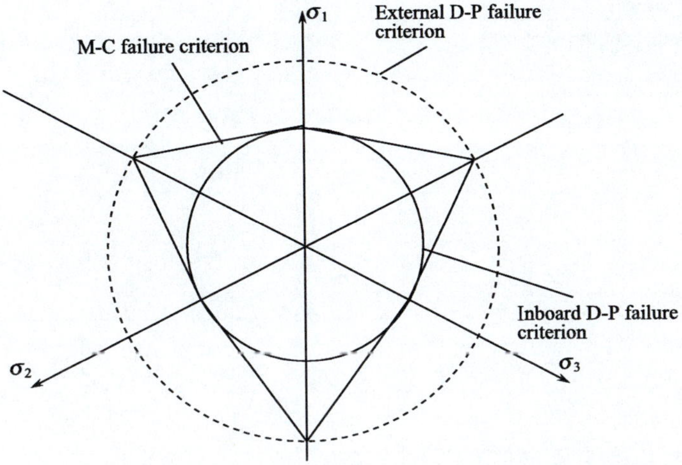

Fig. 5.14 Failure criterion in the π-plane

As shown in Fig. 5.14, MC strength criterion in the principal stress space is a hexagonal pyramid, and which is an irregular hexagon in the π-Plane.

When the conical point of the failure plane of D-P strength criterion is in a superposition with the conical point of the failure plane of MC strength criterion, the transversal line of the failure plane of D-P strength criterion in the π-Plane is defined as the D-P circle.

When the external vertexes of the hexagons of D-P strength criterion are in a superposition with the external vertexes of the hexagons of MC strength criterion, the relationship between a, k (of D-P strength criterion) and c, φ (of M-C strength criterion) can be defined as:

$$a = \frac{2\sin\varphi}{\sqrt{3}(3-\sin\varphi)}, \quad k = \frac{6c\cos\varphi}{\sqrt{3}(3-\sin\varphi)}$$

When the internal vertexes of the hexagons of D-P and M-C strength criteria are in superposition, a, k, c and φ have the following relationship:

$$a = \frac{2\sin\varphi}{\sqrt{3}(3+\sin\varphi)}, \quad k = \frac{6c\cos\varphi}{\sqrt{3}(3+\sin\varphi)}$$

In the case of plane strain, a, k, c and φ have the following relationship:

$$a = \frac{2\tan\varphi}{\sqrt{9+12\tan^2\varphi}}, \quad k = \frac{3c}{\sqrt{9+12\tan\varphi}}$$

The advantages of the D-P failure criterion are the influence of hydrostatic pressure and the second invariant of the stress deviator tensor has been taken into account.

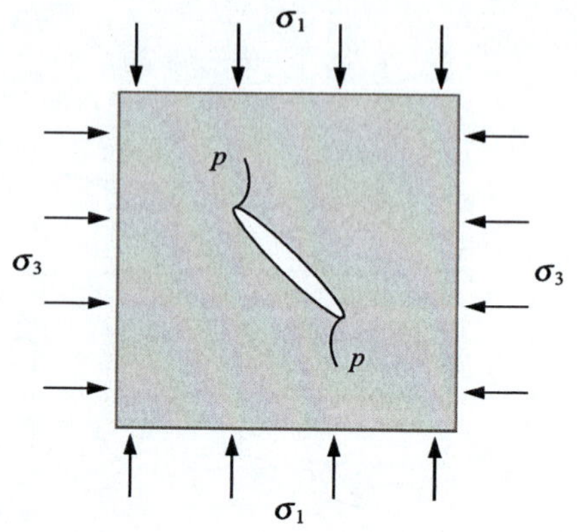

Fig. 5.15 Griffith crack model of plane compression

5.3.3 *Griffith's Strength Criterion*

Fracture mechanics was developed during World War I by English aeronautical engineer, A. A. Griffith, to explain the failure of brittle materials. Griffith was concerned with the energetics of fracture, and considered the energy changes associated with incremental crack extension.

In 1920, Griffith indicated that the breakage of brittle materials (like steel and glass) was induced by the tension stress concentration around the micro-crack (Griffith crack) tip, as shown in Fig. 5.15. And in 1924, the Griffith's strength criterion was finally proposed. Griffith's strength criterion is a pure theoretical criterion. Though Griffith's strength criterion can't provide appropriate interpretation of macroscopical crack development and expansion in rock material, but as the expansion condition of the micro-crack in rock material, namely the initial condition for the increase of rock material damage, this criterion is of good applicability. Griffith's strength criterion theoretically explains the crack expanding phenomenon in rock, which greatly promotes the development of the fracture mechanics.

In order to further analyze the stress distribution of a material with cracks, some researchers deduced the Griffith's strength criterion by using the stress solutions of elliptical holes, which is

$$\begin{cases} \dfrac{(\sigma_1-\sigma_3)^2}{\sigma_1+\sigma_3}=8\sigma_t \ (\sigma_1+3\sigma_3\geq 0) \\ \sigma_1=-\sigma_3 \ (\sigma_1+3\sigma_3\leq 0) \end{cases} \tag{5.35}$$

where σ_t is the maximum tension stress around the crack tip, MPa.

Griffith's strength criterion is deduced according to the research results of brittle materials such as steel and glass, thus it is very suitable for the studies of brittle rock failure. However, though Griffith's theory can provide the theoretical interpretation of crack development and expansion inside the rock, but can't provide appropriate interpretation of macroscopical crack development and expansion of rock. Thus, Griffith's strength criterion is not suitable for application in engineering practice.

5.4 The Effective Stress

Presence of a freely moving fluid in a porous rock can affect its mechanical response, and in turn, change of the porous media's mechanical characteristics will also affect the fluid flow properties. The branch of mechanics that describe the mechanical response and volume response of the pore elastic medium is called poroelasticity theory [1]. This phenomenon also refers to the fluid–solid coupling effect which has received extensive attention in many engineering fields.

The study on fluid–solid coupling is first noted by Terzaghi [2] in 1923 in his researches about surface subsidence, the famous one-dimensional effective stress principle was also proposed in that research. After that, in 1936 Renbdulic applied the effective stress principle of Terzaghi into three-dimensions. These studies concerning the effective stress principle provided theoretical basis for the poroelasticity theory and the earliest single-porosity poroelasticity theory was proposed by Biot [3, 4] in 1935 and 1941. With the continuous development of poroelasticity theory, the classical single-porosity poroelasticity theory, dual-poroelasticity theory, tri-poroelasticity theory and multi-porous poroelasticity theory are gradually formed.

In the poroelasticity theory, the effective stress principle plays an important role in the description of the mechanical response of pore medium to the pore fluid pressure action [5–13].

$$\sigma_{ij}^{e} = \sigma_{ij} - \beta\delta_{ij}p \tag{5.36}$$

where σ_{ij}^{e} is the effective stress; σ_{ij} is the total stress (positive in compression); β is the effective stress coefficient; δ_{ij} is the Kronecker delta tensor; p is the fluid pressure, MPa.

Although the law of single-pore effective stress for a single-pore medium is relatively perfect, application of this law for the multiple pore medium such as fractured rock mass and coal is limited. From experimental studies concerning the permeability of dual porosity coal, it is noticed that in the dual media of pore fracture, the fluid pressure in pore and fracture is not consistent (see Fig. 5.16), while the mechanical properties of pore and fracture are also different, therefore the effect of fluid on medium's skeleton of pore/fracture will vary as well.

To better describe the mechanical response from pore fracture dual porosity medium under gas pressure, the effective stress law for dual porosity media is introduced into the permeability model as follows [13–16]:

$$\sigma_{ij}^{e} = \sigma_{ij} - (\beta_f p_f + \beta_m p_m)\delta_{ij} \tag{5.37}$$

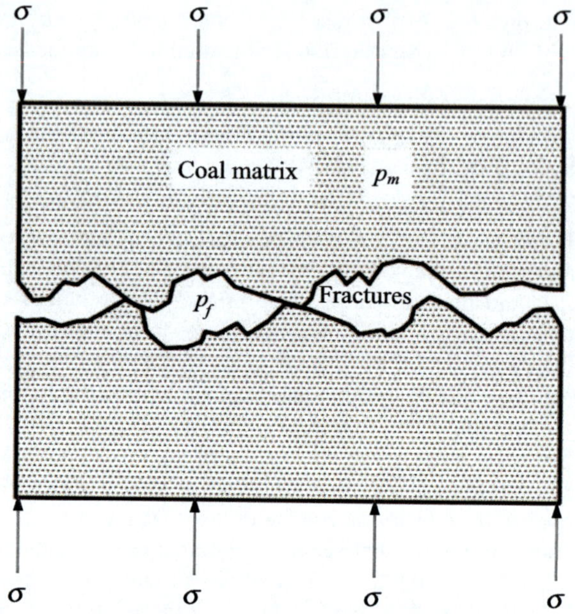

Fig. 5.16 Illustration of the effective stress

where β_f and β_m are effective stress coefficients for the fractures and matrix, respectively; p_f and p_m are the fluid pressures in the coal fractures and matrix pores, respectively, MPa.

The effective stress coefficients for the fractures (β_f) and matrix (β_m) in Eq. (5.37) can be calculated as [13]:

$$\beta_f = 1 - \frac{K}{K_m} \tag{5.38}$$

$$\beta_m = \frac{K}{K_m} - \frac{K}{K_s} \tag{5.39}$$

where K is the bulk modulus of coal, MPa; K_m is the bulk modulus of coal matrix, MPa; K_s is the bulk modulus of coal skeleton, MPa.

At present, the bulk modulus of coal skeleton is difficult to be measured directly, and its value is usually obtained by theoretical calculation according to its physical significance. The three bulk modulus mentioned above can be calculated using the following expressions [17]:

$$K = \frac{E}{3(1 - 2\mu)} \tag{5.40}$$

$$K_m = \frac{E_m}{3(1 - 2\mu)} \tag{5.41}$$

$$K_s = \frac{K_m}{1 - 3\phi_m(1-\mu)/[2(1-2\mu)]} \tag{5.42}$$

where E is the Young's modulus of coal, MPa; E_m is the Young's modulus of the coal matrix, MPa; μ is the Poisson's ratio of the coal; ϕ_m is the matrix porosity of coal.

References

1. Detournay, E., & Cheng, A. H. D. (1993). Fundamentals of poroelasticity, comprehensive rock engineering: principles, practice and projects[J]. *Analysis and Design Method, 140*(1), 113–171.
2. Terzaghi K. (1923) Die berechnung der durchlassigkeitsziffer des tones aus dem verlauf der hydrodynamischen spannungserscheinungen[J]. *Sitzungsberichte der Akademie der Wissenschaften in Wien, MathematischNaturwissenschaftliche Klasse, Abteilung IIa, 132*, 125–138.
3. Biot M. A. (1935). Le problèmc de la consolidation des matières argileuses sous une charge[C]. *Annales de la Société Scientifique de Bruxelles*, série B, *55*, 110–113.
4. Maurice, A. B. (1941). General theory of three-dimensional consolidation. *Journal of Applied Physics, 12*(2), 155–164.
5. James, R. R., & Michael, P. C. (1976). Some basic stress diffusion solutions for fluid-saturated elastic porous media with compressible constituents. *Reviews of Geophysics, 14*(2), 227–241.
6. John, R. (1985). Effect of pore fluid diffusion on deformation and failure of rock. *Mechanics of Geomaterials*, 315–347.
7. Zhang, J. (2013). Borehole stability analysis accounting for anisotropies in drilling to weak bedding planes [J]. *International Journal of Rock Mechanics and Mining Sciences, 60*,160–170.
8. Zhang, J., Bai, M., & Roegiers, J. C. (2006). On drilling directions for optimizing horizontal well stability using a dual-porosity poroelastic approach. *Journal of Petroleum Science and Engineering, 53*(1), 61–76.
9. Zhang, J., & Roegiers, J. C. (2015). Double porosity finite element method for borehole modeling. *Rock Mechanics and Rock Engineering, 38*(31), 217–242.
10. Zhang, J., Bai, M., & Roegiers, J. C. (2003). Dual-porosity poroelastic analyses of wellbore stability. *International Journal of Rock Mechanics and Mining Sciences, 40*, 473–483.
11. Zhang, J. (2002). *Dual-porosity approach to wellbore stability in naturally fractured reservoirs*. University of Oklahoma.
12. Liu, J. S., Chen, Z. W., Derek, E., et al. (2011). Interactions of multiple processes during CBM extraction: A critical review. *International Journal of Coal Geology, 87*(3), 175–189.
13. Chen, M., & Chen, Z. D. (1999). Effective stress laws for multi-porosity media. *Applied Mathematics and Mechanics, 20*(11), 1207–1213.
14. Kagan, T., & Yavuz, M. C. (1995). Effective stress principle for saturated fractured porous media. *Water Resources Research, 31*(12), 3103–3106.
15. Arkady, G., & Roger, B. (2000). Flow of coal-bed methane to a gallery. *Transport in Porous Media, 41*(1), 1–16.
16. Zhang, J., Bai, M., & Roegiers, J. C. (2004). Dual-porosity elastoplastic analyses of non-isothermal one dimensional consolidation. *Geotechnical and Geological Engineering, 22*(4), 589–610.
17. Willi, P., Eva, G., & Gabriela, T. (2006). Elasticity of porous ceramics—A critical study of modulus–porosity relations. *Journal of the European Ceramic Society, 26*(7), 1085–1097.

Chapter 6
Strength and Deformation Characteristics of Coal Containing Gas

Learning Outcomes

- The principle, apparatus, and basic requirements of conventional compression tests and confining pressure unloading tests for coal containing gas, along with the coal sample preparation methods;
- The methods to calculate the cohesion, internal friction angle, elasticity modulus, and Poisson's ratio of coal;
- The influencing factors of coal strength and deformation characteristics;
- The macroscopic damage features and microcosmic damage mechanism of coal containing gas.

6.1 Mechanical Tests of Coal Containing Gas

Coal strength plays a significant role in various mining related engineering activities, including the evaluation of coal and gas outburst dangers, pillar design, hydraulic fracturing design (involving enhanced gas drainage), coalface support, coal seam CO_2 sequestration, and other activities.

The primary method for determining coal strength is by conducting a series of uniaxial/triaxial compressive tests in the laboratory, whose procedure has been standardized by the International Society for Rock Mechanics (ISRM). Because the ISRM suggested method provides us with a relatively straightforward standard to determine coal strength, many coal strength factors have been studied. Poulsen et al. [1] reported that coal loses significant strength when it becomes water saturated due to its absorption of water. Ranjith et al. [2] investigated the weakening effect of CO_2 on Australian black coal by conducting acoustic emission and uniaxial tests, which have also been studied by a number of researchers (for example, [3–5]). Medhurst and Brown [6] conducted a series of triaxial compression tests on samples with different diameters to investigate scale effects on the mechanical behavior of coal, and based on these experimental results, Poulsen and Adhikary [7] developed and calibrated

Y. Cheng et al., *Coal Mechanics*,
https://doi.org/10.1007/978-981-16-3895-4_6

a numerical Bonded Particle Model to study the scale effect on coal strength. Li et al. [8] demonstrated that coal strength first increases and then decreases with increase in the loading rate during triaxial compression tests. Recently, due to innovations in technologies for deep underground mining at great depths (>1000 m) or tunneling underneath mountains, effects of high confining pressure and temperature on coal (rock) strength have drawn a lot of attention [9–13]. Based on studies in excavation engineering, Kaiser and Kim [14] reported that the strength characteristics of massive rock in the direct vicinity of excavation differ significantly from that which is remote with higher confinement. Furthermore, Xu et al. [15] reported that temperature generates a significant weakening effect on the compressive strength of coal.

In addition to the above-mentioned factors on coal (rock) strength properties, it should be noted that stress path can also significantly impact coal (rock) strength. Underground coal or rock is under a three-dimensional stress equilibrium condition. However, excavation of an underground mine disturbs the original in situ stress state, leading to stress redistribution around the underground cavity. With rock removal, the stress redistribution is characterized by the unloading process, under which the mechanical properties of rock (coal) are different from that of the loading process as the confining pressure is kept constant [16, 17]. Moreover, the unloading-induced stress release and its related deformation relaxation during excavation can drastically influence the stability of underground openings and might induce rock destruction or rock burst under the high in -situ stress conditions [18, 19]. Ding et al. [20] investigated the mechanical behavior of sandstone under unloading conditions and found that its load bearing ability decreased gradually due to unloading of the confining pressure. Chen et al. [21] reported that the volume deformation of halite under unloading conditions was greater than that under the triaxial loading conditions. To study rock burst processes, which usually occur on the underground excavation surface, He et al. [22] conducted single-face dynamic unloading tests on limestone under true-triaxial conditions and showed that the acoustic emission (AE) energy release increases rapidly from the onset of unloading to sample failure.

6.1.1 Testing Facility and Experimental Principles

The mechanical properties of materials are mainly determined using an ordinary compression testing machine or a stiff compression testing machine. As the ordinary compression testing machine cannot provide sufficient stiffness for rock or concrete compression tests, when the loading force reaches its peak strength, the material will rapidly disintegrate. However, the stiff compression testing machine can provide sufficient stiffness for rock and concrete compression tests, and is usually used to obtain the complete stress–strain curve of coal.

Figure 6.1 shows a photograph of the coupling test system for adsorption–permeability–mechanics at the National Engineering Research Center for Coal and Gas Control, China University of Mining and Technology. This system can be used to

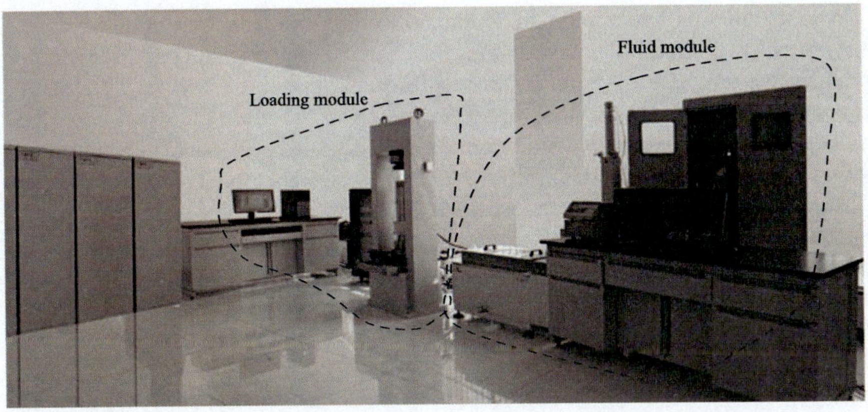

Fig. 6.1 Photograph of the coupling test system for adsorption–permeability–mechanics

conduct uniaxial/triaxial compression tests of coal/rock materials as well as permeability tests with various gases. It consists of loading module, fluid module, and data acquisition device. It has a variety of loading control modes, which can be converted during the experiments. A schematic of the test system is shown in Fig. 6.2.

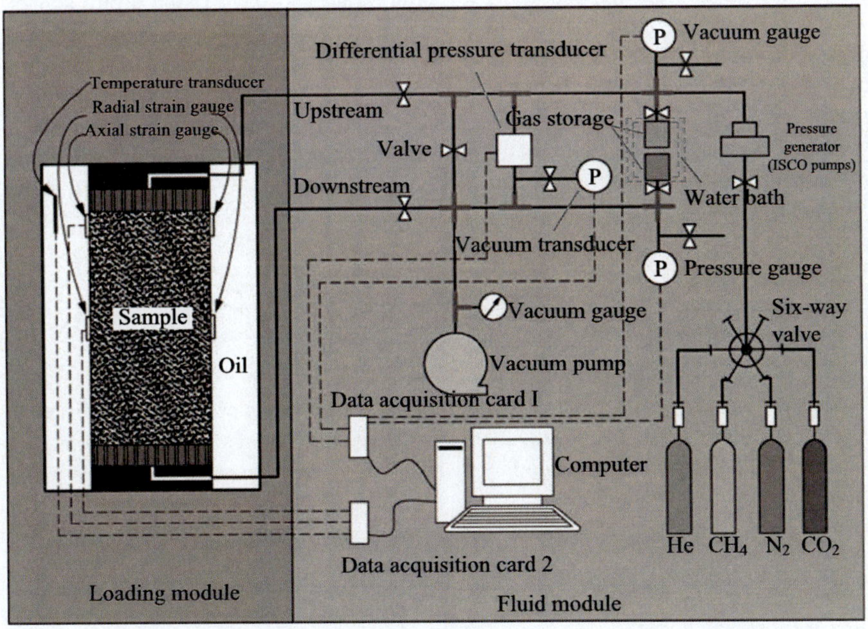

Fig. 6.2 Schematic diagram of the coupling test system

The loading module comprises a uniaxial loading frame, a triaxial pressure cell, a servo system, two oil hydraulic pumps (one for axial loading and one for confining pressure), a temperature control unit, and a loading control device. The loading control modes can be used during experiments and include the axial stress-controlled mode, radial stress-controlled mode, axial piston displacement-controlled mode, and radial piston displacement-controlled mode. The maximum axial load of the system can reach 600 kN (control precision: $\leq\pm1\%$), the effective range of the axial piston displacement is 0–230 mm (control precision: $\leq\pm0.5\%$), the maximum confining pressure can reach 60 MPa (control precision: $\leq\pm1\%$), and the effective range of the radial piston displacement is 0–230 mm (control precision: $\leq\pm0.5\%$). The adjustment range of the temperature control unit to the triaxial pressure cell ranges from ambient temperature to 90 °C (control precision: $\leq\pm0.2\%$).

The fluid module consists of a high-precision plunger pump, vacuum pump, gas storage reservoir, thermostatic water bath, gas source, and pressure-resistant pipe system. The maximum gas pressure provided by the plunger pump is 40 MPa (control precision: $\leq\pm1\%$) and can be operated in either pressure control mode or flow control mode. The designed maximum tolerable pressure of the pipeline system is 100 MPa, which is suitable for permeability tests using various gases such as CH_4, CO_2, N_2, He, and O_2. The gas storage reservoir and the main pipe system are soaked in the thermostatic water bath (control precision of temperature: $\leq\pm0.2$ °C).

The data acquisition device consists of a computer, data acquisition cards, and different types of sensors, including an axial strain sensor, a radial strain sensor, a loading sensor, a temperature sensor, a pressure sensor, a differential pressure transducer, and a vacuum gauge. During the test, real-time collection and recording of all types of experimental data can be realized by coordinating the computer with the sensors. Ideally, each sensor should not exceed its maximum measurement range, which is 200 kPa, 8 mm, and 4 mm for the differential pressure transducer, axial strain sensor, and radial strain sensor, respectively.

6.1.2 Basic Requirements for Strength Tests of Coal Containing Gas

Coal strength usually refers to the maximum tolerable stress of coal under different loading conditions, and is an indicative parameter rather than an inherent parameter of coal. Coal strength is influenced by many factors when measured experimentally, including:

(1) The shape and size of a coal sample, whereby a larger test sample generally corresponds to a higher strength value. There is a large difference in coal strength among cube-, rectangle-, and cylindrical-shaped samples.

(2) The dimensions of a coal sample. The obtained strength of coal samples with different dimension (height/diameter ratio or length/width ratio) usually differs.

(3) The loading rate, which is proportional to coal strength in the field of statics, whereby a higher loading rate results in a higher strength value.
(4) The mechanical stress path of the test.
(5) The types of gases that are saturated during the test, and/or different gas pressures.
(6) The moisture content of coal samples and experimental temperature.

The International Society for Rock Mechanics (ISRM) and the Standardization Administration of the People's Republic of China have invested considerable effort in developing a succession of standards for the coal samples used in strength tests, including those for sampling, shape, size, loading rate, moisture content, and temperature. According to these standards, the coal samples used in strength tests should be in accordance with the following criteria:

(1) The diameter and the height/diameter ratio of the standard coal sample are (50^{+5}_{-2}) mm and (2 ± 0.2), respectively. The influence of the dimension scale on coal strength should be considered if it diverges from the standard value. If it is impossible for the coring cylinder sample, a cuboid sample with dimensions of 50 mm × 50 mm × 100 mm can also be used.
(2) The machining accuracy of a coal sample should be in accordance with the Chinese National Standard GB/T 50266-2013 ("***Standard for test methods of engineering rock mass***"), which states that (i) the non-parallelism and roughness of the two ends of a coal sample should be <0.005 mm and <0.02 mm, respectively, (ii) the diameter error along the coal sample height should be <0.3 mm, (iii) the two ends of a coal sample should be perpendicular to the sample's axes, and (iv) the maximum deviation should be <0.25°.
(3) When coal samples are tested with a natural water content, samples are placed in a dryer at room temperature for 1–2 d with a small amount of water at the bottom to maintain humidity (the samples should not be touched with the water). When testing dry coal samples, samples should be first dried at 105–110 °C for 24 h.

When conducting a mechanical test, a sample is placed inside the triaxial pressure cell on the base plate, as shown in Fig. 6.3. To complete this, the dents on the sample surface must first be coated with silica gel. After the silica gel solidifies, the coal sample is placed between the top and bottom pedestals and wrapped with a heat-shrinking tube (rubber sleeve). This tube is heated with a heat gun to shrink it against the sample and pedestals. Then, each end of the heat-shrunk tube is covered with a rubber belt and further fastened with ferrules. Subsequently, the axial and radial strain sensors are set (Fig. 6.3), and the sample is placed into the triaxial pressure cell. Finally, the data collection interfaces and gas pipes are linked, and the airtightness of the entire system is checked.

Fig. 6.3 Photograph of a sample with the sensors

6.1.3 Coal Sample Preparation Methods

At present, the commonly used sampling methods in coalmines mainly include the drilling core method and lump coal sampling method. The former one uses a drilling machine and core tube to drill the coal cores. Then, the standard coal sample can be obtained using a rock-cutting machine and a rock grinder. In comparison, the lump coal sampling method requires a larger, more integral coal block from an underground coalmine. First, a rock drill-core is used to obtain coal cores, and then a rock-cutting machine and rock grinder are employed to make the standard coal samples. However, the drilling process of both methods produces a large amount of vibration; thus, these methods are only applicable to hard coal with a good integrity.

The raw coal samples considered in this book were prepared according to the lump coal sampling method. Figure 6.4 displays the main stages of the sample preparation process. First, large coal blocks (50 cm long, 25–30 cm wide, and 16–20 cm high) were collected from an underground coalmine. It was ensured that each coal block had a clear layer and stripes (Fig. 6.4a). Before the coring commenced, a flat plane was made on a coal block along the side perpendicular to the direction of the bedding plane. Then, the coal block was fixed and drilling proceeded to obtain cylindrical

Fig. 6.4 Stages of the lump coal sampling method, showing a a coal block b coring of a coal sample, c a coal block after coring, and d standard coal samples

coal samples (diameter of 50 mm; length of >100 mm) (Fig. 6.4b). Finally, each cylindrical sample was cut and polished to the standard sample size of Φ 50 mm × 100 mm. It was ensured that each prepared coal sample had a smooth surface and was marked with labels on the side surface. The samples were then wrapped with preservative film and stored for experiments.

Owing to the loose structure of deformed coal after stress release, it is extremely difficult to obtain standard coal samples using the above-mentioned methods. The macrostructure of deformed coal can be primarily classified into fragmentation structure, fractionation structure, powder particle structure, and mylonitic structure [23, 24]. However, regardless of the type of deformed coal, it is nearly impossible to prepare a Φ 50 mm × 100 mm coal sample by coring; therefore, it is necessary to use a reconstituted coal sample to conduct studies on soft coal mechanics and deformation properties. Reconstituted coal has been widely used in experimental investigations of the mechanics, deformation, and permeability characteristics of coal. For instance, Jasinge et al. [25] considered the anisotropy and strong discreteness of raw coal, and used cement as an additive to make reconstituted coal samples with various ingredients under a range of molding pressures. The authors studied the mechanics, deformation, and permeability characteristics of their reconstituted coal samples and compared with those of raw coal, and finally suggested the application prospects of reconstituted coal with respect to research concerning CO_2 geological sequestration. Peng et al. [26] investigated the effects of gas on the mechanical properties of reconstituted coal, and determined the influencing mechanism of gas on coal and gas outbursts. The research group of Yin [27, 28] used reconstituted coal to conduct extensive experiments, including the mechanical properties of coal containing gas, permeability evolution during the complete stress–strain process, and the effect of geo-stress on the permeability characteristics of outburst coal. Considering that reconstituted coal has been applied in many experiments, Xu et al. [29] investigated the fractal characteristics of pores in reconstituted coal with different coal particle sizes. Cao [30] analyzed the difference in failure modes between reconstituted coal and raw coal during triaxial compression tests from the perspective of mesoscopic damage mechanics as well as the influence of deformation/damage on the permeability rate. In an earlier study by the author's research team, reconstituted coal was also used to conduct experimental investigations on the damage and permeability variations during the unloading process, thus obtaining the relationship between microscopic damage and permeability evolution in stress-relief coal. Their research results were applied to the optimization of borehole arrangement in the gas extraction of protected coal seams.

Table 6.1 summarizes and classifies the methods for preparing reconstituted coal samples. It can be seen that there are large differences in the particle size, molding pressure, molding time, and additives of different coal sample preparation methods, which reflect the lack of a standard for the preparation of reconstituted coal samples. In addition, compared with raw coal, there are significant differences in the adsorption characteristics and stress sensitivity of conventionally made reconstituted coal.

Table 6.1 Summary of the methods for preparing reconstituted coal samples

No.	Particle size	Molding pressure (MPa)	Molding time (h)	Loading mode	Additive	Source
1	<1 mm	9	24	Displacement control	Water	Refs. [24, 31]
2	0.1–1 mm	5.7	~24	N/A	Water, cement	Refs. [32, 33]
3	<1 mm	5	3–26	N/A	Water, cement	Ref. [23]
4	<0.8 mm	N/A	N/A	N/A	Water, cement	Ref. [34]
5	0.2–0.25	100	0.5	N/A	Water	Refs. [35, 36]
6	40–60 mesh	100	N/A	N/A	Water	Refs. [27, 28, 37–39]
7	40–60 mesh	100	0.35	N/A	Water	Ref. [30]
8	0.18–4 mesh	4	N/A	N/A	N/A	Ref. [40]

Unlike hard coal, it is difficult to prepare a standard coal sample using soft coal; therefore, it would be very useful to explore a new sampling method that does not disturb the structure of soft coal.

Reconstituted coal samples are traditionally made in a laboratory by the following four steps:

(1) Small coal blocks are crushed and sieved using a milling machine. A vibrating screen is then used to select the small coal particles (<1 mm).
(2) These fine coal particles are mixed with distilled water, and the mixture is placed into a specially designed steel mold with a diameter of 50 mm (Fig. 6.5a).
(3) The stress-controlled mode is used to compress the coal particles at a loading rate of 300 N/s. The compression is maintained for 3 h when the load reaches 200 kN. The coal sample is then extruded from the steel mold and trimmed to an approximate length of 100 mm using a diamond cutter. Subsequently, the sample is polished to a reasonable uniformity and smoothness in accordance with ISRM standards [41].
(4) The reconstituted sample is dried in a vacuum oven at 60 °C for 24 h. Then, the sample is sealed and ready for testing. Figure 6.5b shows some reconstituted coal samples.

A useful tip for the preparation of reconstituted samples is as follows. Because a mixed coal sample is consistent in particle size and moisture content, the first reconstituted coal sample can be used to calculate the weight of the wet coal sample that is required to produce the standard reconstituted coal sample (size: Φ 50 mm $\times$ 100 mm). Therefore, by properly weighing a mixed wet coal sample of a certain weight, it is possible to produce reconstituted coal with a height of exactly 100 mm, which could reduce the work stage of using a cutting machine.

Meanwhile, to obtain deformed soft coal with a complete structure and standard size, an air-operated soft coal sampling device has been developed. As shown in Fig. 6.6, the sampling device is composed of an air pump, pressure regulating valve, manual gas circuit reversing valve, check valve, pipe diameter adaptor, three-way

(a) (b)

Fig. 6.5 **a** Mold and **b** reconstituted coal samples

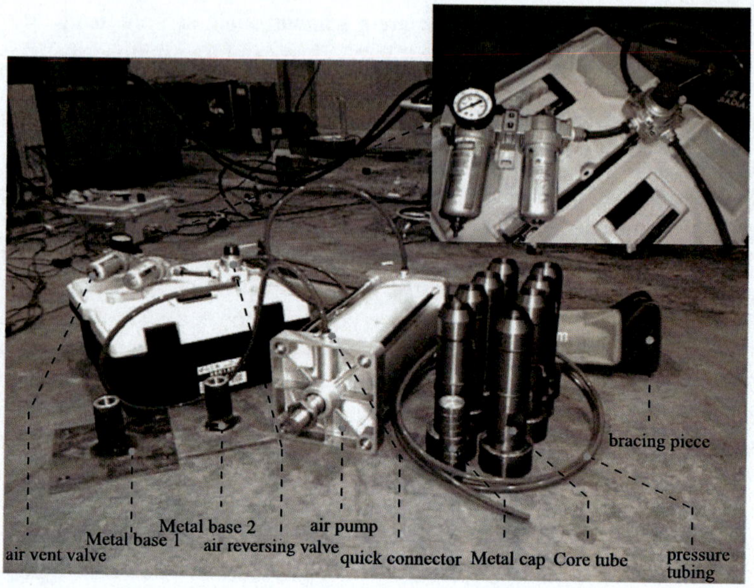

Fig. 6.6 Photograph of the air-operated soft coal sampling device

valve, pressure line, quick plug, core tube, reverse support strut, and support base. This device uses the underground pressured air as a power source and has no electrical equipment; thus, it can be quickly installed, and is easy and safe to use.

As shown in Fig. 6.7, when using the air-operated soft coal sampling device, the sampling location should be chosen in a narrow coal roadway or drill field so that the support component of the device can take effect safely. During the sampling process, the air pump is driven by pressured air, which can provide a thrust force of up to 5000 kg. With the support of the reverse support strut, the core tube is pressed into a soft coal seam. Subsequently, the manual gas circuit reversing valve is switched to change the gas inlet direction, and the piston of the air pump works backwards to pull the core tube out of the coal. For the convenience of demolding, the core tube was designed to be detachable. It has a length, inner diameter, and outer diameter of 250 mm, 50 mm, and 55 mm, respectively, and the front end has a chamfer design. The linkage between the core tube and tube base has a threaded connection, as does the linkage between the tube base and air pump. After on-site sampling, cooperating with the special clamp (bearing capacity: 100–150 MPa), undertaking a stiff compression test, and performing the stress recovery method, a Φ 50 mm cylindrical soft coal sample with slight disturbance can be prepared.

Once a coal sample is transported to the laboratory, it needs to be wetted from the top before compression. Then, by cooperating with the special clamping die, stiff compression testing machine, and stress recovery method, the reconstituted sample is compressed. The molding pressure is set to 100 MPa, and the molding time is set to 12 h. After demolding, the reconstituted sample should be placed in a vacuum

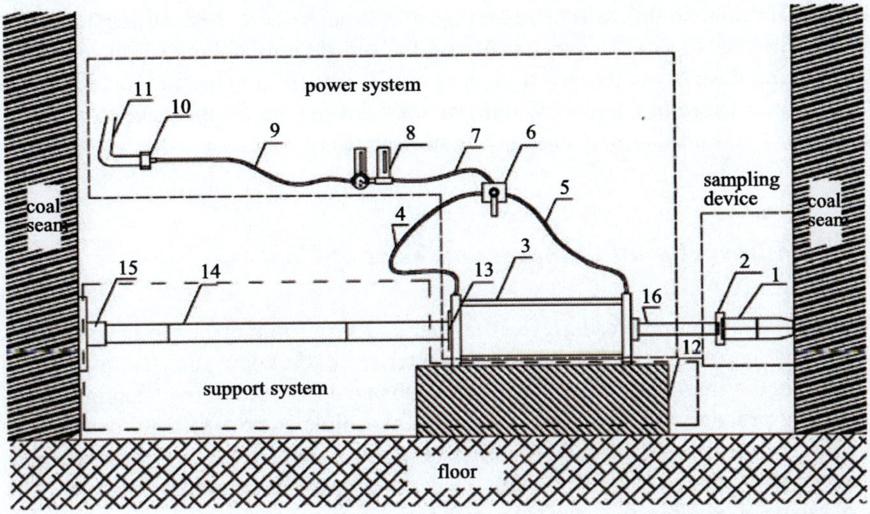

Fig. 6.7 Schematic showing the underground operation of the air-operated soft coal sampling device

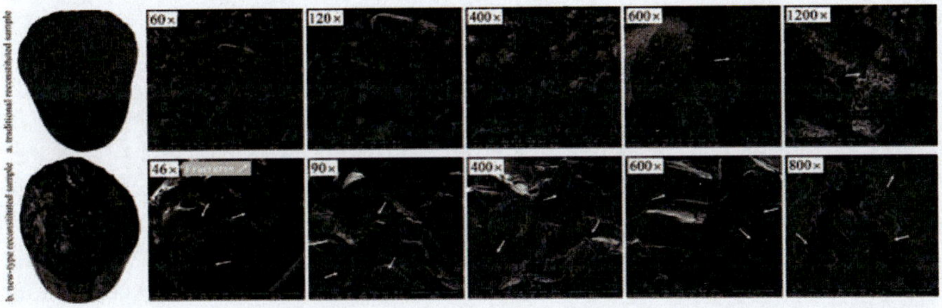

Fig. 6.8 Photomicrographs of a prepared slightly disturbed soft coal sample

drying oven at 60 °C for 24 h. After drying, the sample is wrapped with a heat shrink tube, and then cut and polished into a proper size of Φ 50 mm × 100 mm, as shown in Fig. 6.8. Through this method, the prepared slightly disturbed soft coal sample can maintain the structural characteristics of soft coal.

6.2 Strength Characteristics of Coal Containing Gas

Coal strength is the ability of coal to resist the damage caused by external forces, that is, the maximum tolerable stress of coal under different loading conditions. Because of the various load forms, coal strength mainly includes the uniaxial compression strength, triaxial compression strength, tensile strength, and shear strength. As the tensile strength of coal is very small and the tensile stress should be avoided in engineering design, the tensile strength of coal is difficult to measure and has been less studied. Therefore, in this section, we only discuss the compressive strength of coal, which has been widely used in engineering design.

6.2.1 Conventional Compression Tests of Coal

The conventional compression test is the basic and most common method for studying the mechanical properties of coal. The characteristic of the conventional compression test is that the confining pressure remains unchanged (for the uniaxial condition, the confining pressure is set to zero) until the coal sample is damaged. By conducting a series of conventional compression tests with different confining pressures, the uniaxial strength and triaxial strength (peak stress at various confining pressures) of coal can be obtained, which can also provide the basis for the unloading point choice of the pre-peak confining pressure unloading test.

The uniaxial compression strength of coal refers to the ratio of the uniaxial damage load to the area of the load-bearing plane for a standard coal sample under laboratory conditions, namely:

$$\sigma_c = F/A \tag{6.1}$$

where σ_c is the uniaxial strength (Pa), F is the axial damage load (N), and A is the area of the load-bearing plane; for a cylindrical sample, it is the area of the top/bottom circle (m^2).

The triaxial strength of coal is the maximum load that coal can bear under the effect of the triaxial compression condition until damage appears. When conducting a triaxial compression test, a coal sample needs to bear both a uniaxial load and lateral pressure. There are two loading modes for the triaxial compression test: the true triaxial test and pseudo triaxial test. As shown in Fig. 6.9c, for the true triaxial test, the coal sample should be a cubic specimen, and the stress state is $\sigma_1 > \sigma_2 > \sigma_3$ (where σ_1 is the maximum principal stress, and σ_2 and σ_3 are the lateral principal stresses). However, the implementation of this loading mode is complicated, and the requirements for the instruments are also high. Thus, the true triaxial loading mode is seldom used to determine the coal strength properties. The commonly applied triaxial compression test is the pseudo-triaxial test used in the loading mode, as shown in Fig. 6.9b. The sample can be cubic or cylindric. Because the lateral principal stress is provided by the hydraulic pressure, the lateral stresses in different directions are the same; therefore, the lateral stress is also called the confining pressure. In this loading mode, the hydraulic oil does not create friction on the surface of the sample, and the confining pressure can be applied evenly to the coal sample. Additionally, it should be noted that when conducting the pseudo triaxial loading mode, most triaxial presses cannot apply axial load separately; thus, the axial load data determined in the

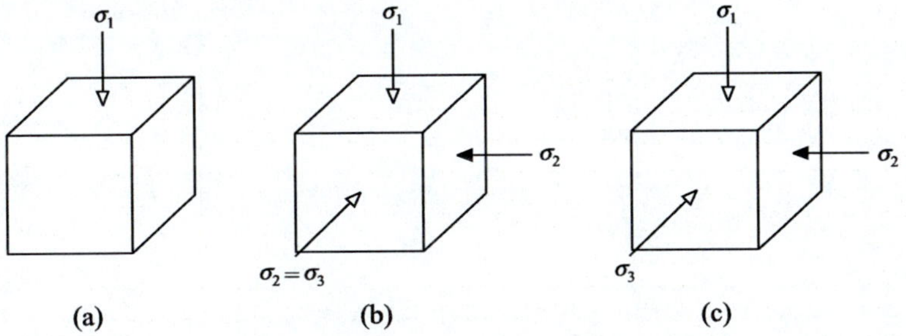

Fig. 6.9 Cubic specimen under **a** uniaxial stress, **b** traditional triaxial stress, in which the two lateral confining stresses are equal, and **c** true triaxial stress, in which all three principal stresses are possibly different

experiment belong to the deviatoric stress, namely the difference between the axial pressure and the confining pressure.

The loading method and loading rate of the conventional compression test are considered in this book. The specific test methods are as follows:

(1) Restricted by the loading mode of the test system, an axial load of <100 N should first be applied to the sample after it is installed to ensure that the head of the sample is aligned with the press. The hydrostatic strength (confining pressure) is then loaded at a rate of 10 N/s to some predetermined values (i.e., 2, 4, 6, and 8 in the confining pressure-controlled mode).
(2) Before failure occurs, the axial strength is loaded at a rate of 50 N/s (in the axial load-controlled mode) while keeping the confining pressure constant. After the coal undergoes plastic deformation, the loading mode should be transferred to the displacement-controlled mode at a loading rate of 5 mm/min. The loading should be stopped until the coal fails or the deformation reaches the maximum scale value of the strain gauge.
(3) It should be noted that when performing the conventional triaxial compression test of coal containing gas, the gas adsorption of coal needs to reach the equilibrium state.

The deviatoric stress–strain curves of conventional reconstituted coal samples, slightly disturbed soft coal samples, and raw coal samples obtained under conventional compression conditions are shown in Figs. 6.10, 6.11, and 6.12, respectively.

The peak stress and peak point strain (axial strain, radial strain, and volume strain) of the conventional triaxial compression test for three types of coal samples under different confining pressures are summarized in Table 6.2. As shown in Figs. 6.10, 6.11, and 6.12, and Table 6.2, the peak strength and peak point strain increased with increasing confining pressure, indicating that the ability of the coal samples to resist damage increased. Moreover, it can also be observed that strain softening occurred with an increase in confining pressure.

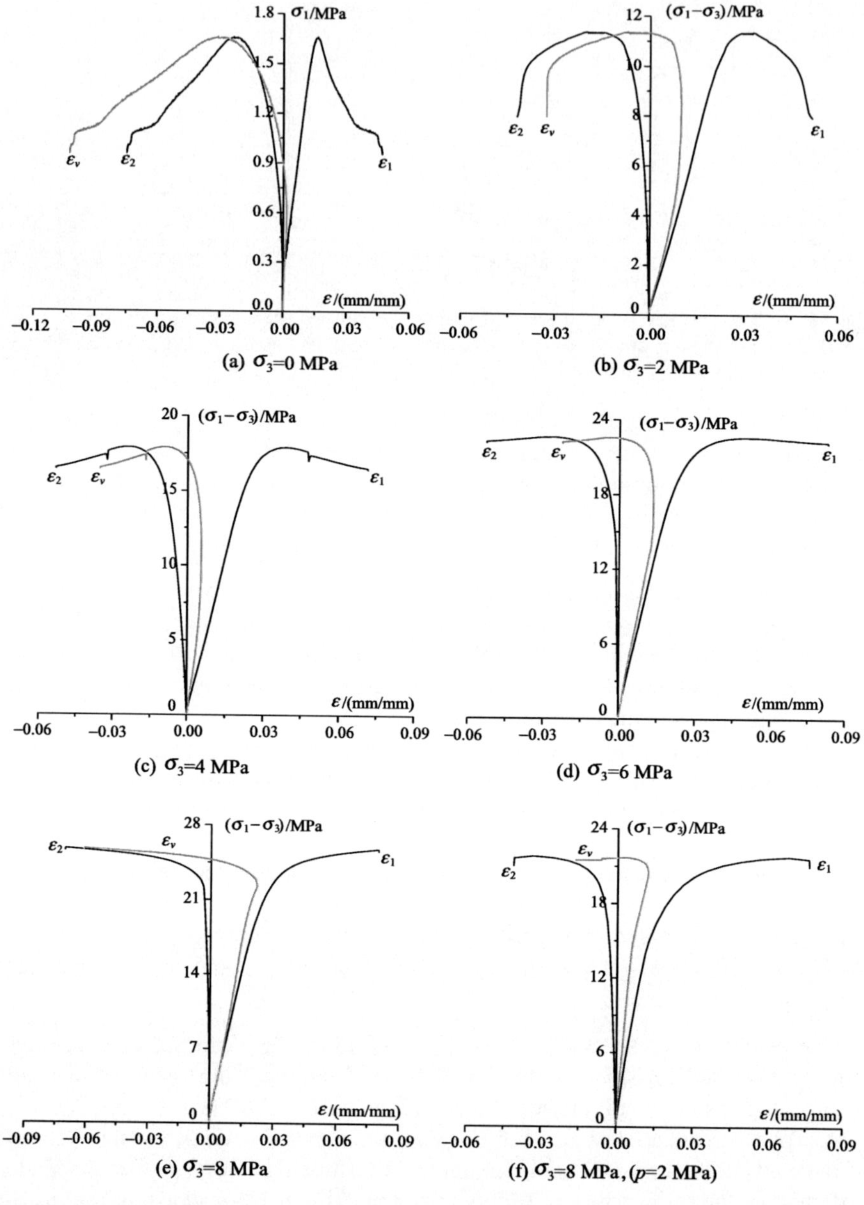

Fig. 6.10 Deviatoric stress–strain curves of conventional reconstituted coal samples obtained under conventional compression conditions

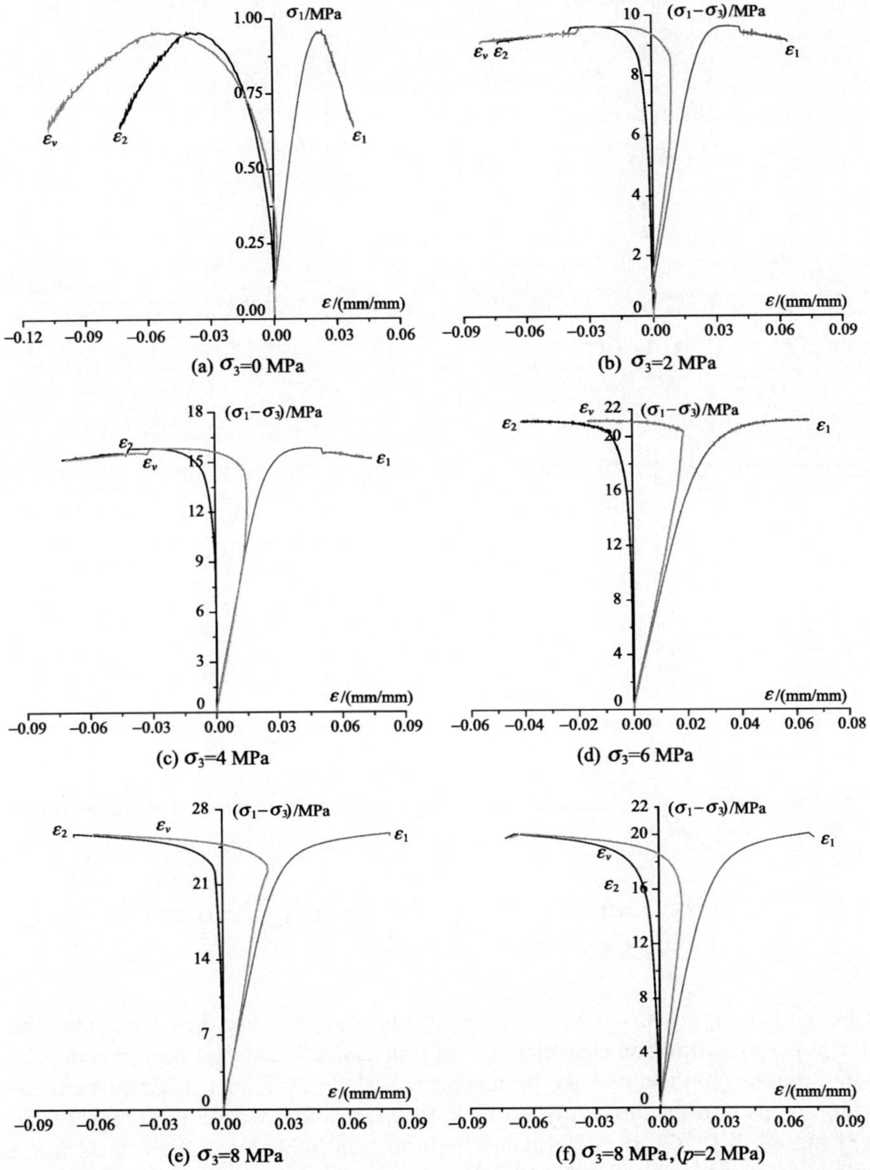

Fig. 6.11 Deviatoric stress–strain curves of slightly disturbed soft coal samples obtained under conventional compression conditions

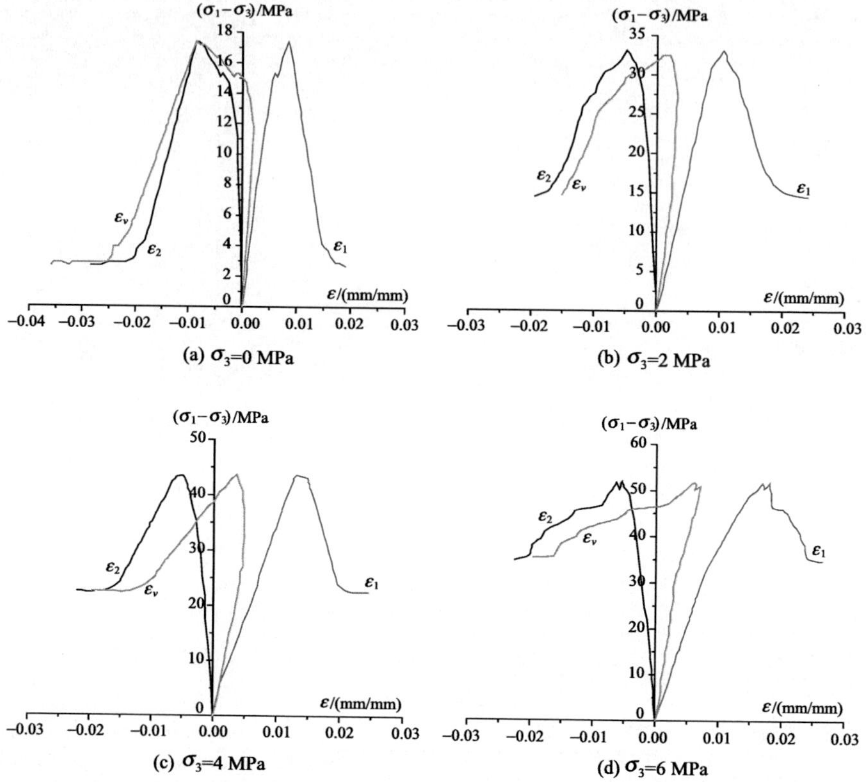

Fig. 6.12 Deviatoric stress–strain curves of raw coal samples obtained under conventional compression conditions

6.2.2 *Pre-peak Confining Pressure Unloading Tests of Coal Containing Gas*

Through loading tests, scholars have conducted a series of studies concerning the strength and deformation characteristics of coal and rock, and many research results with guiding significance have been achieved. However, under different mechanical stress paths of loading and unloading, the mechanical properties of engineering rock masses differ significantly, and the deformation and failure of rock mass due to unloading are widely distributed. With the further study of rock mechanics, the deformation and failure of rock mass under actual conditions have attracted increasing attention, and many scholars have indicated that research on the deformation/failure characteristics of rock mass under unloading conditions is of great importance [42]. Consequently, unloading experiments have received considerable attention in the field of rock mechanics. The damage to the rock mass under unloading conditions was much more serious than that under loading conditions. The damage degree of

Table 6.2 Conventional compression test results of three types of coal samples

No.	Sample label	Confining pressure (MPa)	Peak strength (MPa)	Peak point strain ($\times 10^{-3}$)		
				Axial	Radial	Volumetric
1	RC-1	0	1.66	16.3	−23.7	−31.1
2	RC-2	2	13.72	30.8	−16.8	−2.8
3	RC-3	4	21.90	38.8	−24.1	−9.4
4	RC-4	6	28.57	50.7	−26.9	−3.1
5	RC-5	8	36.14	97.2	−53.5	−9.8
6	RC-6	8 (2)	29.75	67.8	−33.9	0.0
7	NSC-1	0	0.95	24.0	−40.5	−57.0
8	NSC-2	2	11.64	35.9	−28.0	−20.1
9	NSC-3	4	19.56	53.2	−44.1	−35.0
10	NSC-4	6	27.17	64.6	−40.0	−15.4
11	NSC-5	8	33.76	79.6	−70.1	−60.6
12	NSC-6	8 (2)	28.08	71.1	−68.1	−65.1
13	NC-1	0	17.37	8.4	−8.2	−8.6
14	NC-2	2	33.10	10.6	−4.9	−0.5
15	NC-3	4	43.65	13.0	−5.4	3.6
16	NC-4	6	51.72	18.0	−6.4	6.0

Note 8 (2) means that the confining pressure was 8 MPa and the gas pressure was 2 MPa; RC represents conventional reconstituted coal samples; NSC represents slightly disturbed soft coal samples; and NC represents raw coal samples

the pre-peak confining pressure unloading test was larger than that of the post-peak confining pressure unloading test, and the damage degrees of both the pre/post-peak confining pressure unloading tests were larger than those of the loading tests. The authors concluded that the rock capacity was significantly expanded and exhibited brittle failure characteristics during the unloading process, in which the shear strength (c) decreased and the value of φ increased.

It is well known that after excavation or borehole drilling, the stress condition of the surrounding rock changes, forming a new secondary stress distribution that exhibits the characteristics of tangential loading and radial unloading [43]. It should be noted that the axial load considered in the laboratory refers to the maximum principal stress on site, which is the tangential stress for the coal around a borehole, whereas the confining load considered in the laboratory refers to the minimum principal stress on site, which is the radial stress for the coal around a borehole.

According to the evolution of the actual stress state around a borehole, the loading axial pressure (maximum principal stress) and unloading confining pressure (minimum principal stress) were chosen as the mechanical path in this book. The mechanical path of loading axial pressure–unloading confining pressure can be further divided into three categories: the loading rate of axial pressure that is ① equal to the unloading rate of confining pressure, ② less than the unloading rate of confining pressure, and ③ larger than the unloading rate of confining pressure. Considering the asymmetry characteristic of engineering problems, we choose the "loading rate of axial pressure that is larger than the unloading rate of confining pressure" as the mechanical path of the test in this book. Additionally, in order to obtain more details about the test, especially after the coal sample is damaged, the unloading point is set before the peak strength of coal, that is, the test belongs to the pre-peak confining pressure unloading test.

The specific steps of the pre-peak confining pressure unloading test are as follows:

(1) Restricted by the loading mode of the test system, an axial load of <100 N should first be applied to the sample after it is installed to ensure that the head of the sample is aligned with the press. The hydrostatic strength (confining pressure) is then loaded at a rate of 10 N/s to some predetermined values (i.e., 2, 4, 6, and 8 MPa in the confining pressure-controlled mode).
(2) The axial stress-controlled mode is used to load the axial stress at a rate of 50 N/s until 80% of the peak strength is reached (which is determined by a conventional triaxial compression test).
(3) The axial strength is continuously loaded at a rate of 50 N/s, and the confining pressure is unloaded at a rate of 10 N/s. The loading should be stopped until the coal fails or the deformation reaches the maximum scale value of the strain gauge.
(4) It should be noted that when performing the confining pressure unloading test of coal containing gas, the gas adsorption of coal needs to reach the equilibrium state.

Figures 6.13 and 6.14 show the deviatoric stress–strain curves of conventional reconstituted coal samples and slightly disturbed coal samples, respectively, which were obtained under confining pressures of 2, 4, 6, and 8 MPa, as well as under a confining pressure of 8 MPa and a gas pressure of 2 MPa. The horizontal dotted line in each plot represents the starting point for confining pressure unloading. The stress state at the beginning of unloading and the stress state when failure occurs are summarized in Table 6.3.

6.2.3 *Analysis of the Strength Characteristics of Coal Containing Gas*

1. Influence of unloading confining pressure on coal strength

The internal friction angle [44] and cohesion of a rock are two important parameters that are widely used in the engineering field to describe the strength characteristics of rocks. Cohesion is the mutual attraction between the internal adjacent parts of the same substance, and represents the molecular force existing among the molecules of the same substance. The internal friction angle first appeared in Coulomb's theory,

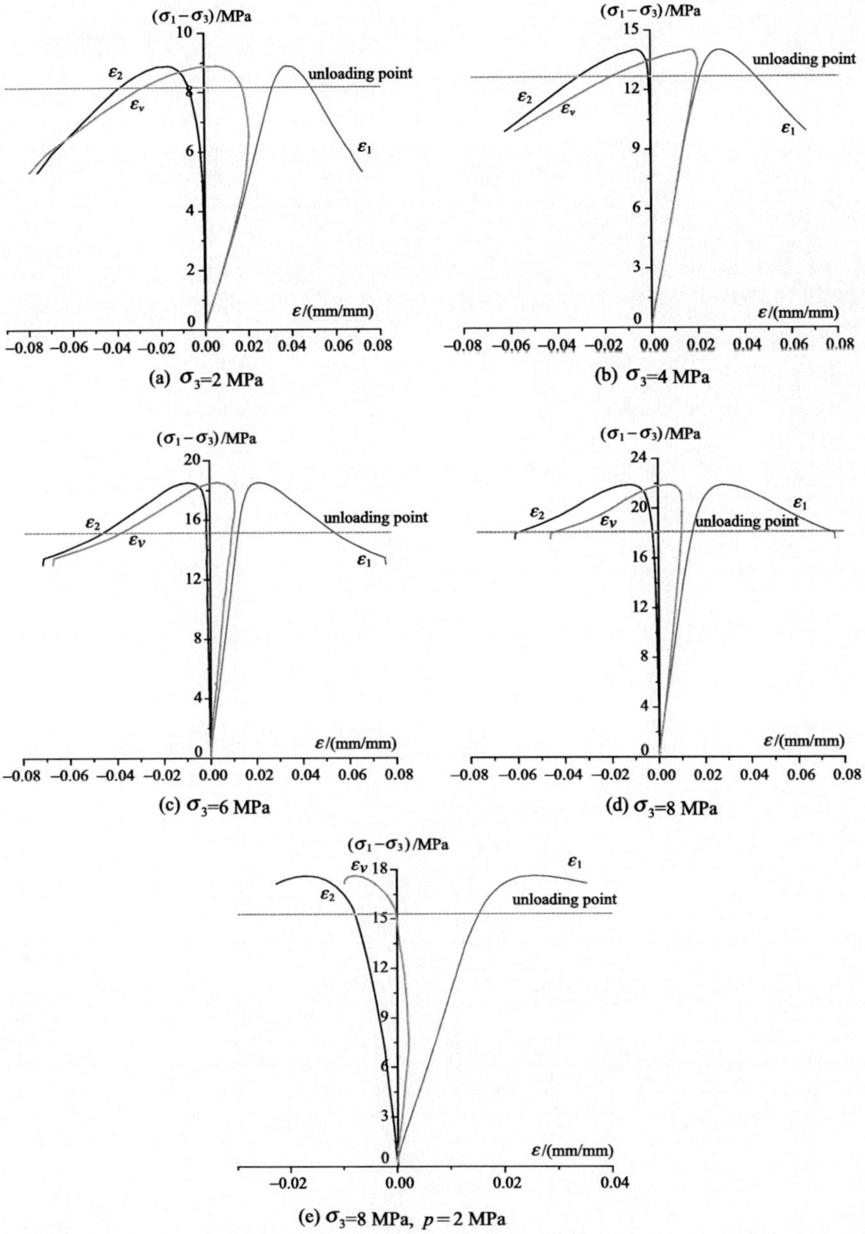

Fig. 6.13 Deviatoric stress–strain curves of conventional reconstituted coal samples obtained under conventional compression conditions

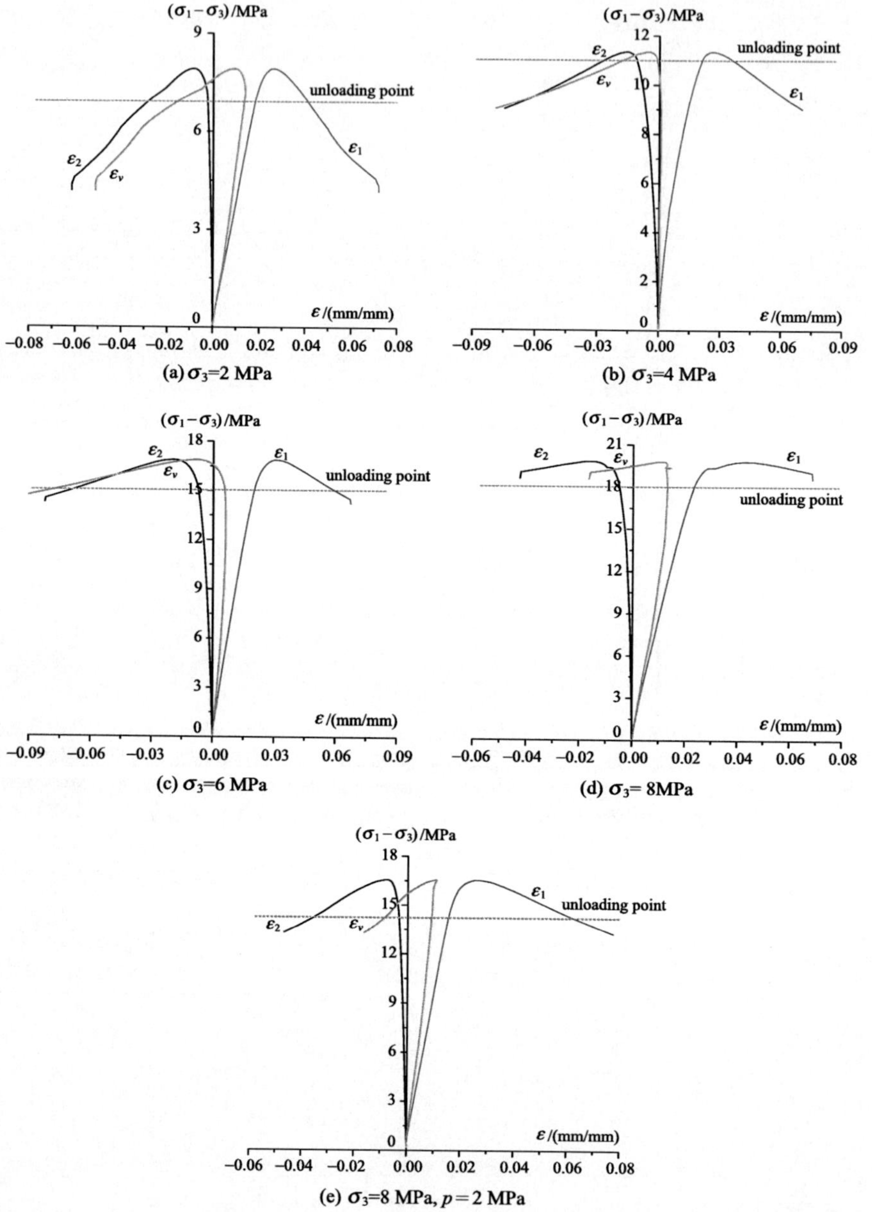

Fig. 6.14 Deviatoric stress–strain curves of slightly disturbed reconstituted coal samples obtained under conventional compression conditions

Table 6.3 Stress state summary of two types of reconstituted coal samples in pre-peak confining pressure unloading tests

No.	Sample label	Stress state before confining pressure unloading		Stress state at failure	
		σ_3	$\sigma_1 - \sigma_3$	σ_3	$\sigma_1 - \sigma_3$
1	RC-7	2	8.13	1.69	8.89
2	RC-8	4	12.70	3.02	14.02
3	RC-9	6	15.86	4.40	18.52
4	RC-10	8	18.06	5.75	21.91
5	RC-11	8 (2)	15.47	6.57	17.63
6	NSC-7	2	6.97	1.61	7.93
7	NSC-8	4	11.07	3.66	11.35
8	NSC-9	6	15.08	4.74	16.88
9	NSC-10	8	18.07	6.31	19.82
10	NSC-11	8 (2)	14.45	6.60	16.55

Note 8 (2) means that the confining pressure was 8 MPa and the gas pressure was 2 MPa; RC represents conventional reconstituted coal samples; NSC represents slightly disturbed soft coal samples

which is also an important component of the shear strength of a rock. The cohesion and internal friction angle of a rock are both indirectly measurable parameters, which can be calculated by statistical analysis of the triaxial compression test results. For an accurate regression analysis of the compression results, the Mohr–Coulomb criterion is translated into the following principal stress form [45]:

$$\sigma_1 = \psi + \zeta\sigma_3 \tag{6.2}$$

where ψ represents cohesion (Eq. (6.3)), and ζ represents the internal friction angle (Eq. (6.4)):

$$\psi = \frac{2c\cos\varphi}{1 - \sin\varphi} \tag{6.3}$$

$$\zeta = \tan^2\left(45^\circ + \frac{\varphi}{2}\right) \tag{6.4}$$

Because coal failure under uniaxial compression belongs to shear dilatancy, the uniaxial compressive strength is not entirely induced by shear failure [46, 47], and its value is significantly lower than the triaxial compressive strength. Thus, in this research, to reflect the real effect of confining pressure [48] on the triaxial compressive strength of coal samples as much as possible, the uniaxial compression strength was not incorporated into the regressive calculations of shear strength parameters. The Mohr–Coulomb circles obtained from the conventional compression condition are displayed in Fig. 6.15.

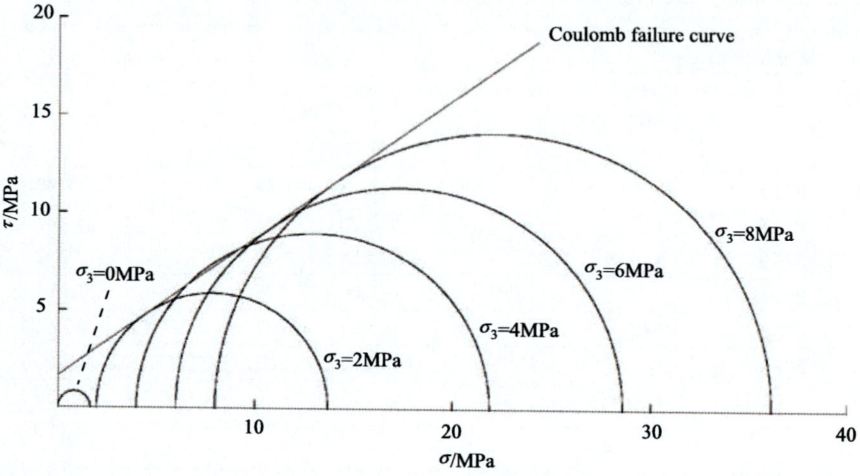

Fig. 6.15 Mohr–Coulomb circles obtained under the conventional compression condition

By using the least square method to fit the data of conventional coal loading tests and pre-peak unloading tests, the slope (ζ) and intercept (ψ) of the straight line can be obtained. Then, the cohesion and internal friction angle of coal for both tests can be deduced through the physical meanings of ζ and ψ. Figure 6.16 displays the fitted lines obtained from the conventional loading tests and pre-peak unloading tests.

From Fig. 6.16, the relationships between σ_1 and σ_3 for a conventional reconstituted coal sample under conventional compression tests and confining pressure unloading tests can be expressed as follows.

Loading:

$$\sigma_1 = 6.6 + 3.7\sigma_3 \tag{6.5}$$

Unloading:

$$\sigma_1 = 3.91 + 4.21\sigma_3 \tag{6.6}$$

The relationships between σ_1 and σ_3 for a slightly disturbed reconstituted coal sample under conventional compression tests and confining pressure unloading tests can be expressed as follows.

Loading:

$$\sigma_1 = 4.54 + 3.7\sigma_3 \tag{6.7}$$

Unloading:

$$\sigma_1 = 3.18 + 3.65\sigma_3 \tag{6.8}$$

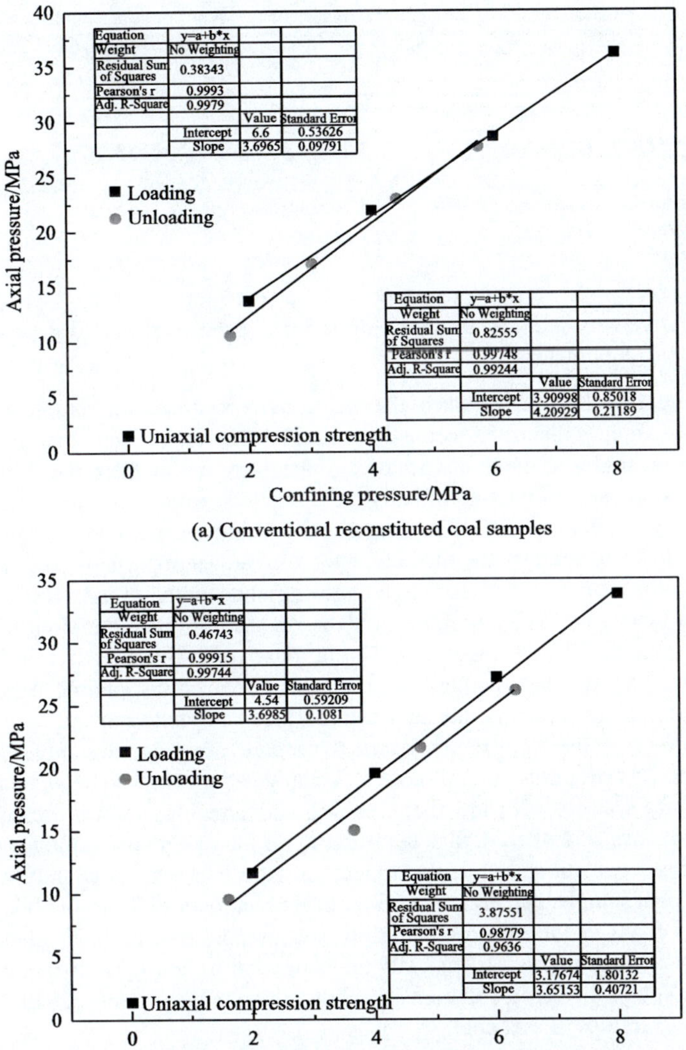

(a) Conventional reconstituted coal samples

(b) Slightly disturbed reconstituted coal samples

Fig. 6.16 Relationship curves of σ_1 and σ_3 obtained from loading and unloading tests

Table 6.4 summarizes the linear regression results and strength parameters of the two types of coal samples.

The R^2 values of the conventional reconstituted coal sample (loading and unloading) were very high (>0.99), as was that for the slightly disturbed soft coal sample in the loading test, whereas that of the unloading test was 0.964. These results indicate that the strength parameters of the two coal samples obtained by regression

Table 6.4 Computation results of strength parameters of two types of coal samples

Sample type	Stress path	ψ (MPa)	ζ	Cohesion (MPa)	Internal friction angle (°)	R^2
Conventional reconstituted coal sample	Loading	6.60	3.70	1.72	35.06	0.998
	Unloading	3.91	4.21	0.95	38.03	0.992
Slightly disturbed soft coal sample	Loading	4.54	3.70	1.18	35.06	0.997
	Unloading	3.18	3.65	0.83	34.74	0.964

calculation are very reliable. Further comparison of these results allows us to draw the following conclusions:

(1) Comparing the loading data of the traditionally reconstituted sample with those of the slightly disturbed reconstituted sample, it can be found that the internal friction angles of the two types of samples were nearly the same, whereas the cohesion value of the traditionally reconstituted sample was larger than that of the slightly disturbed sample. The cohesion value depends on the original cohesive force formed by the molecular force in adjacent parts of coal, while the magnitude of the internal friction angle depends on the friction resistance and interlocking action between the coal matrix. According to the Mohr–Coulomb criterion, the shear strength of coal is composed of the internal friction force and cohesion. At present, within the geo-stress interval corresponding to the current mining depth in China, the shear strength of coal is dominated by cohesion. Therefore, pulverizing coal particles to prepare reconstituted samples damages the matrix properties of soft coal and changes the cohesion between adjacent parts of soft coal. Meanwhile, crushing and screening reduce the equivalent matrix size of soft coal, thus resulting in an increased cohesion of the reconstituted coal sample after re-compaction. However, for the slightly disturbed soft coal sample, because the structural characteristics of the coal body would have changed little during sampling and preparation, the equivalent matrix scale of this sample was basically unchanged; thus, the strength parameters of the slightly disturbed soft coal sample were more consistent with the strength characteristics of soft coal.

(2) Comparing the loading/unloading data of the traditionally reconstituted sample with those of the slightly disturbed reconstituted sample, it is observed that the cohesion values obtained from the loading tests were higher than those obtained from the unloading tests; hence, there was a significant degradation effect of unloading on coal strength. The cohesion values of the traditionally reconstituted and slightly disturbed soft samples reduced by 44.77% and 29.66%, respectively, between the loading and unloading tests. For the traditionally reconstituted sample, the internal friction angle in the unloading test exceeded that of the loading test, while the reverse was true for the slightly disturbed soft coal sample. The possible cause for this phenomenon is that shear failure was more intense during unloading for the traditionally reconstituted sample,

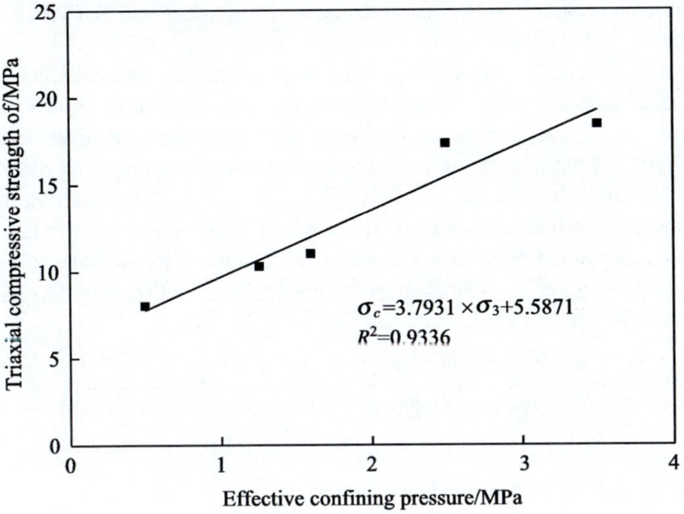

Fig. 6.17 Triaxial compressive strength of coal with increasing effective confining pressure

which would have increased the roughness of the fracture surface and the internal friction angle [49]. For the slightly disturbed soft coal sample, the unloading process may have been accompanied by the pulverization of coal matrix blocks, and the pulverized coal particles could have acted as a lubricant on the fracture surface, thus leading to a reduction in the internal friction angle.

2. Influence of methane on coal strength

It is well known that methane (CH_4) has a significant influence on coal strength. As shown in Fig. 6.17, the compressive strength of coal containing gas decreased with increasing gas pressure when conducting a triaxial compression test under a constant confining pressure [50]. Even when the effective confining pressure remained constant, the coal strength also decreased with increasing gas pressure. In an uniaxial compression test, the uniaxial compressive strength of coal containing gas decreased with an increase in gas pressure. Additionally, under the same pressure, the stronger the adsorption properties of coal, the lower the compressive strength of the gas in coal.

The influence of methane on coal strength can be divided into two aspects. On the one hand, according to the principle of effective stress, the free gas in coal fractures changes the effective confining pressure; thus, the compressive strength of coal varies with the change in confining pressure. According to the Mohr–Coulomb criterion, the triaxial compressive strength of coal (σ_c) has a linear relationship with the confining pressure (σ_3):

$$\sigma_c = \sigma_{cu} + n_\sigma \sigma_3 \tag{6.9}$$

where σ_{cu} is the uniaxial compressive strength at full shear failure (MPa), and n_σ is a coefficient.

On the other hand, the adsorbed gas in coal generates a nonmechanical reducing effect on coal strength. Many researchers have made different interpretations of nonmechanical effects. For example, Wang [51] reported that adsorbed gas can wedge the microfractures of coal, thus reducing cohesion and strength. Astakhov et al. [52] pointed out that adsorbed gas generates a capillary condensation effect in microfractures, which reduces coal strength. The influence of methane on coal strength was explained by Yao and Zhou [53] based on Renhbinder's theory and Griffith's fracture criterion. The fracture toughness of a coal fracture (K_{Ic}) is closely related to its surface free energy (E) as follows:

$$K_{\mathrm{Ic}} = \sqrt{\frac{2\psi E}{\pi c_l}} \tag{6.10}$$

where c_l is the initial length of a fracture (m).

After coal adsorbs methane, the surface free energy of coal decreases. This makes it easier for coal fractures to expand, and leads to an overall decrease in coal strength. The higher the gas pressure, the more gas that is adsorbs by coal, thus resulting in a larger reduction in the surface free energy and strength of the affected coal.

He and Zhou [51] considered that gas is absorbed onto the new surface of a crack after it is formed; hence, the surface free energy induced by the adsorption of methane does not act on the generation of a new fracture surface, and the energy is released in the form of adsorption heat. The authors reported that the adsorbed methane will form an adsorbed film layer, whose lubrication action will decrease the friction on the surface of coal fractures, thus reducing the strength of the affected coal.

As there have been many studies concerning the mechanical properties of coal and its mechanical response, this book focuses on the deterioration of coal cohesion by gas and the calculation methods of the coal strength parameters (cohesion and Poisson's ratio) under different gas pressures based on the mechanical test data of coal without gas.

Figure 6.18 presents a scatter plot of the effect of gas on the peak strength of coal. The test data in the figure are classified into the following three groups: a confining pressure of 6 MPa, a confining pressure of 8 MPa, and a confining pressure of 8 MPa at gas pressure of 2 MPa.

From the relationship between the peak stress and confining pressure in Fig. 6.18, the peak strength of the conventional triaxial loading test exceeded that of the pre-peak confining pressure unloading test. Regardless of the type of test, the peak strengths of the samples were ranked as: a confining pressure of 8 MPa > a confining pressure of 8 MPa at a gas pressure of 2 MPa > a confining pressure of 6 MPa. It is known that gas affects the strength of coal in two ways: one is that it changes the effective confining pressure, and the other is the nonmechanical effect induced by gas adsorption. Both of these effects cause the strength of coal to deteriorate. According to the effective stress principle, if the effective stress coefficient is set to

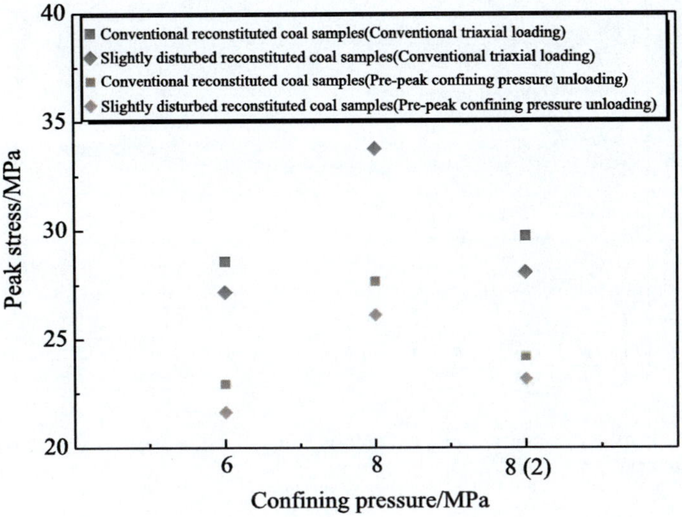

Fig. 6.18 Influence of gas on the peak strength of coal

1, the effective confining pressure of the gas-containing coal should be 6 MPa. Thus, when further affected by the nonmechanical effect, the peak strength at a confining pressure of 8 MPa and gas pressure of 2 MPa should be less than that at a confining pressure of 6 MPa, which is inconsistent with the test results, indicating that the effective stress coefficient should be <1. However, because the difference between the values of the two peak strengths is not very large, the effect of increasing the confining pressure on the peak strength is obvious; therefore, the effective stress coefficient can be set to 1 if the requirements are not too strict.

From the above analysis, it can be observed that the existence of gas will have a greater impact on the peak strength of coal. Next, we make a quantitative analysis of the impact of methane on coal strength, and then seek a method to use non-gas-based mechanical test data to calculate the coal strength parameters (cohesion and Poisson's ratio) under different gas pressures.

Equation (6.11) defines the relationship between the peak strength and confining pressure of coal without the influence of gas. According to the effective stress principle, when the gas pressure is p, if the nonmechanical effect of the gas is not considered, the relationship between the peak strength of coal and the confining pressure is expressed as follows [26]:

$$\sigma_1 - p = \psi' + \zeta'(\sigma_3 - p) \tag{6.11}$$

Considering both the mechanical and nonmechanical effects of gas, the relationship between the peak strength of coal and the confining pressure should be in accordance with Eq. (6.12) [26]:

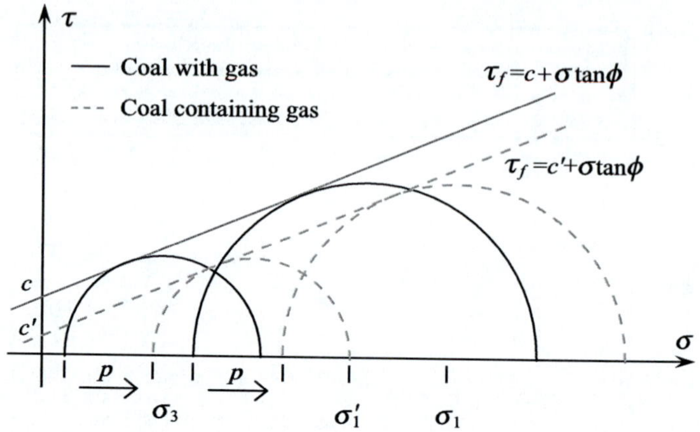

Fig. 6.19 Schematic diagram showing the influence of gas on the deterioration of the strength parameters of coal

$$\sigma_1 - p = \psi' + \zeta'(\sigma_3 - \eta p) \tag{6.12}$$

where η is a comprehensive parameter reflecting the coal adsorption capacity and gas adsorption effect, the value of which depends on the gases that the coal sample adsorbs. The largest η value is for CO_2, followed by N_2 (slightly greater than 1) and CH_4 (1.07–1.19) (under the test conditions, please refer to the literature [26]). For the situation where different coal samples adsorb the same gas, the values of η are also different, which is affected by the coal adsorption capacity. The larger the value of η, the stronger the nonmechanical effect of the adsorbed gas, and the greater the influence on the strength of coal.

In the following, the Mohr stress circle is used to explain the effect of gas on the deterioration of the cohesion and internal friction angle of coal (see Fig. 6.19). For convenience, the nonmechanical effect of gas is neglected.

As shown in Fig. 6.19, when a sample of coal containing gas (gas pressure is p) is in the peak stress state, the confining pressure and axial pressure are σ_3 and σ_1', respectively. According to the principle of effective stress, the effective stress acting on the radial and axial directions of the coal is equal to $\sigma_3 - p$ and $\sigma_1' - p$, respectively, which is equivalent to the Mohr stress circle offsetting to the left side of p, while the diameter is constant. During this process, the Mohr–Coulomb strength envelope is also equivalent to the offset to the left, with the slope remaining constant (i.e., the internal friction angle does not change) and the cohesion term increasing from c' to c. Hence, when only the effect of gas mechanics is considered, this indicates that only the cohesion of coal deteriorates, whereas the internal friction angle does not change.

The above analysis also explains the engineering phenomenon of why a coal body is hardened (the bearing capacity increases) when the coal seam gas is extracted. This means of interpretation is only for the purpose of easy understanding, when using

Table 6.5 Peak strength and deformation characteristics of raw coal containing gas and tectonic coal samples

Sample	No.	Confining pressure (MPa)	Gas pressure (MPa)	Peak strength (MPa)	Elastic modulus (MPa)		Poisson's ratio	
					Measured value	Average value	Measured value	Average value
Natural coal samples	YM-2–1	2	1	24.70	3638	3840	0.28	0.26
	YM-4–1	4	1	37.60	4075		0.28	
	YM-4–3	4	3	35.10	3588		0.24	
	YM-6–3	6	3	23.20	4059		0.24	

non-gas-bearing coal mechanics test data to calculate the coal strength parameters under different gas pressures, the Mohr stress should be converted to the right. It is important to note that when the confining pressure is σ_3, the difference in peak strength between coal containing gas and coal without gas is $\sigma_1 - \sigma_1'$, rather than $\sigma_1 - p$. In addition, when considering the nonmechanical action of gas, which is equivalent to a further reduction in the diameter of the circle of molar stress as it moves to the left, the friction angle of coal will change owing to the change in slope of the Mohr–Coulomb strength envelope.

With data from raw coal samples, we will use the above method to calculate the influence of gas on the peak strength of coal and to quantitatively evaluate the weakening effect of free gas on coal strength. Table 6.5 shows the peak strength and deformation parameters of gas-containing coal samples with different gas pressures and confining pressures.

As mentioned above, the effect of gas on the peak strength of coal mainly includes two aspects: the mechanical effect of free gas, and the nonmechanical effect of adsorbed gas. The mechanical effect of free gas can be evaluated based on the principle of effective stress, with the internal gas being set as the internal force. Coal will produce volume expansion after gas adsorption, which weakens coal cohesion and strength. At the same time, gas adsorbed on the surface of the crack between the coal matrix forms an adsorption film. The adsorption film lubricates and reduces the resistance of the coal body, which also leads to a decrease in coal strength.

According to Griffith's criterion and the Mohr–Coulomb criterion, the relationship between the uniaxial compressive strength and the surface tension of a coal with a crack length of L_c can be expressed as follows:

$$\sigma_c = \sqrt{\frac{2E\gamma_0}{\pi L_c}} = 2c\sqrt{\frac{1+\sin\varphi}{1-\sin\varphi}} \tag{6.13}$$

After gas adsorption, assuming that the elastic modulus and crack length change are not significant, and that the internal friction angle also changes little, the decrease in coal strength is mainly due to a sorption-induced decrease of the surface tension and free energy. The uniaxial compressive strength of coal adsorbed gas can be expressed as:

$$\sigma_{cp} = \sqrt{\frac{2E\gamma}{\pi L_c}} = 2c_p\sqrt{\frac{1+\sin\varphi}{1-\sin\varphi}} \tag{6.14}$$

According to Eqs. (6.13) and (6.14), the relationship between the uniaxial compressive strength of coal before and after gas adsorption can be expressed as:

$$\frac{\sigma_{cp}}{\sigma_c} = \frac{c_p}{c} = \sqrt{\frac{\gamma}{\gamma_0}} = \sqrt{1-\frac{\Delta\gamma}{\gamma_0}} \tag{6.15}$$

According to Eq. (6.15), the cohesion of coal after gas adsorption can be expressed as:

$$c_p = c\sqrt{1-\frac{\Delta\gamma}{\gamma_0}} \tag{6.16}$$

According to the Mohr–Coulomb criterion in Eq. (6.16), the relationship between the shear stress and the normal stress in coal containing both free gas and adsorbed gas can be expressed as:

$$\begin{cases} \left[\sigma - \frac{(\sigma_1+\sigma_3-2p)}{2}\right]^2 + \tau^2 = \left[\frac{(\sigma_1-\sigma_3)}{2}\right]^2 \\ \tau = c\sqrt{1-\frac{\Delta\gamma}{\gamma_0}} + \sigma\tan\varphi \end{cases} \tag{6.17}$$

When the influences of free gas and adsorbed gas are both considered in coal strength, the center and radius of the ultimate stress circle for coal generating failure are $\left(\frac{\sigma_1+\sigma_3-2p}{2}, 0\right)$ and $\frac{\sigma_1-\sigma_3}{2}$, respectively, and the destruction envelope also changes. The mechanical effect of free gas mainly changes the effective stress environment of coal, mainly because the effective stress of a coal body decreases once the coal within it contains free gas; hence, the stress circle of the coal body remains the same size, but the center of the coal body moves to the left for a distance of p. This will lead to a stress circle being located closer to the failure envelope, thus promoting the failure of coal. In contrast, the nonmechanical effect of adsorbed gas mainly changes the internal structure of coal, reducing its internal mechanical parameters, and then changing the failure envelope of coal. This allows the envelope approach to the Mohr stress circle and promotes the destruction of coal. Figure 6.20 provides a specific graphical explanation.

As shown in Fig. 6.20, the affected forms of free gas and adsorbed gas can be identified, but we cannot quantitatively evaluate the extent of the two types of gases in the process of coal damage; thus, further quantitative analyses are required.

In addition to the form of shear stress and the form of normal stress, the Mohr–Coulomb criterion can also be expressed as the relationship between the maximum

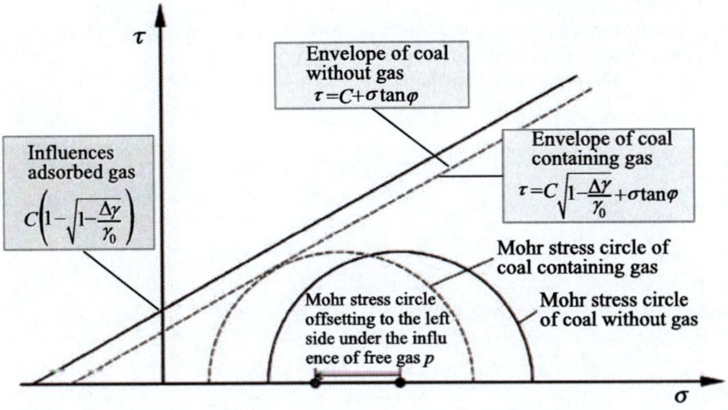

Fig. 6.20 Relationships between the normal stress and shear stress of coal

principal stress and the minimum principal stress as follows:

$$\sigma_1 = \frac{1+\sin\varphi}{1-\sin\varphi}\sigma_3 + 2c\sqrt{\frac{1+\sin\varphi}{1-\sin\varphi}} - \frac{2\sin\varphi}{1-\sin\varphi}p - 2c\sqrt{\frac{1+\sin\varphi}{1-\sin\varphi}}\left(1-\sqrt{1-\frac{\Delta\gamma}{\gamma_0}}\right) \tag{6.18}$$

The relationships between the maximum and minimum principal stresses in the three cases of coal without gas, coal containing free gas, and coal containing free gas and adsorbed gas are shown in Fig. 6.21.

Equation (6.2) is the relationship between the maximum and minimum principal stresses of gas-free coal, and Eq. (6.18) is the relationship between the maximum and minimum principal stresses of the coal containing gas. By comparing Eq. (6.2) and Eq. (6.18), it can be found that the difference between the two cases is in the

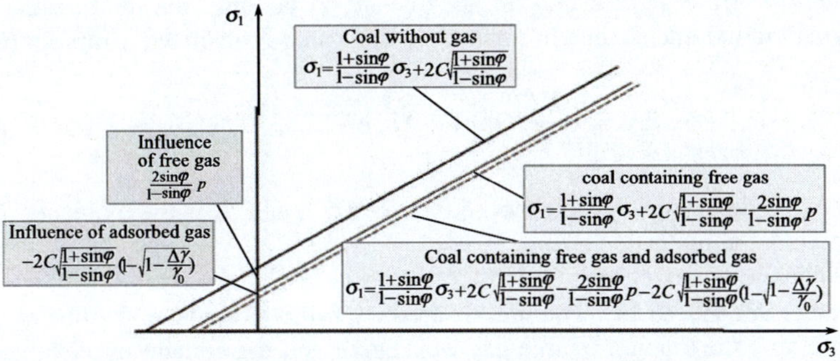

Fig. 6.21 Relationships between the maximum and minimum principal stresses of coal

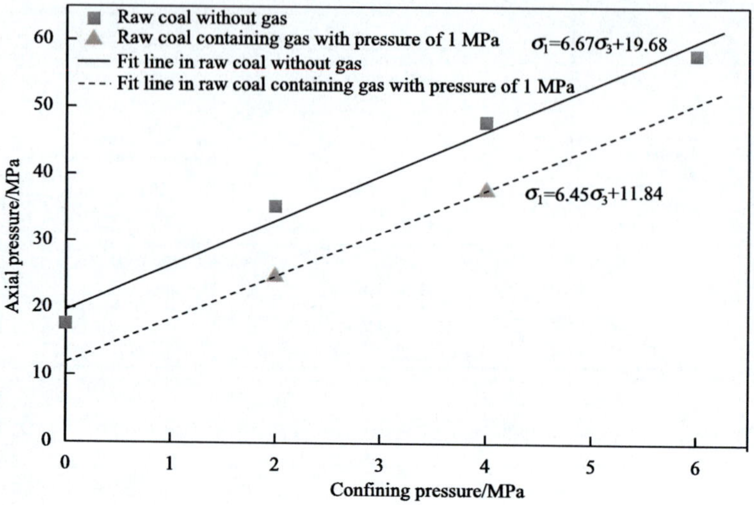

Fig. 6.22 Influence of gas on the peak strength of coal

latter two terms, i.e., $-\frac{2\sin\varphi}{1-\sin\varphi}p$ and $-2c\sqrt{\frac{1+\sin\varphi}{1-\sin\varphi}}\left(1-\sqrt{1-\frac{\Delta\gamma}{\gamma_0}}\right)$, which represent the free gas and adsorption gas terms, respectively. In this book, the obtained internal friction angle of coal is taken as being between 30° and 50°, i.e., $0 < \sin\varphi < 1$. In this case, the free gas and adsorbed gas terms are both negative; hence, the free gas and adsorbed gas both reduce the ability of coal to resist the external force.

When the cohesion and friction angles of gas-free coal samples are already known, the degree of weakening of the free gas on the strength of coal containing gas can be quantitatively evaluated according to Eq. (6.18); however, as the initial surface tension (γ_0) is unknown, and whose value is difficult to obtain directly, Eq. (6.18) cannot be used to directly evaluate the effect of the adsorbed gas on the weakening of the strength of coal. The fitting method can be used to solve the weakening degree of the free gas and adsorbed gas in coal samples, as shown in Fig. 6.22.

After coal adsorbs gas, both the surface tension and surface free energy change. As the values of surface tension and surface free energy per unit area are the same (only in different units), the change in surface tension can be calculated using Eq. (6.19):

$$\Delta\gamma = \frac{RT}{N_A S_0}\ln\left[1+\left(b_0 e^{-\frac{\Delta H_{ad}}{RT}}/\sqrt{T}\right)p\right] \tag{6.19}$$

where N_A is the Avogadro constant (6.023×10^{23}), and S_0 is the molecular cross-sectional area of methane (1.78×10^{-19}).

The change in surface tension at 30 °C can be obtained by substituting the relevant parameters into Eq. (6.19). The initial surface tension of coal can then be obtained by combining the amount of free gas weakening and the amount of adsorbed gas. The details are summarized in Table 6.6.

Table 6.6 Weakening effect of gas on coal strength and surface tension

Coal type	Gas pressure (MPa)	Weakening effect of free gas on coal strength (MPa)	Weakening rate of free gas (confining pressure: 3 MPa) (%)	Weakening effect of adsorbed gas on coal strength (MPa)	Weakening rate of adsorbed gas (confining pressure: 3 MPa) (%)	Variation of surface tension (N/m)	Initial surface tension (N/m)
Natural coal	1	5.67	14.32	2.77	7.00	0.0204	0.0776

According to Table 6.6, when the gas pressure is 1 MPa, the weakening effect of free gas on coal strength is 5.67 MPa, and the weakening effect of adsorbed gas on coal strength is 2.77 MPa. Thus, the weakening effect of free gas on coal strength is 2.1 times that of adsorbed gas. The initial surface tension of natural coal is between 0.0673 and 0.0776 N/m (average of 0.0724 N/m).

6.3 Deformation Characteristics of Coal Containing Gas

6.3.1 Analysis of the Deformation Characteristics of Coal Containing Gas

The stress–strain curve reflects the deformation characteristics of a coal sample under different stress conditions. The deformation parameters such as the elastic modulus and Poisson's ratio can be calculated by calculating the data obtained from the stress–strain curve. The elastic modulus and Poisson's ratio of the uniaxial test are calculated as follows:

$$E = \sigma_1/\varepsilon_1 \tag{6.20}$$

$$\mu = -\varepsilon_3/\varepsilon_1 \tag{6.21}$$

where ε_1 is the axial strain, and ε_3 is the radial strain.

The formulae for calculating the elastic modulus and Poisson's ratio for the triaxial test are as follows:

$$E = \frac{(\sigma_1 - 2\mu\sigma_3)}{\varepsilon_1} \tag{6.22}$$

$$\mu = \frac{(B\sigma_1 - \sigma_3)}{[\sigma_3(2B - 1) - \sigma_1]} \tag{6.23}$$

where $B = \varepsilon_3/\varepsilon_1$.

In general, the elastic modulus and Poisson's ratio in both uniaxial and triaxial stress states can be obtained by selecting the straight-line segment data of the stress–strain curves of the axial and radial strains. However, in the specific calculation process, the choice of data points may have a significant impact on the calculation results. To eliminate the errors caused by manual selection, we calculated the deformation modulus and Poisson's ratio for the entire stress–strain process. To further analyze the difference in the deformation characteristics of two types of coal samples and the influence of loading modes on the deformation characteristics of coal, the deformation parameters of the same stress state are plotted with the evolution of the axial strain in Figs. 6.23 and 6.24.

Figure 6.23 shows the relationships between Poisson's ratio (obtained from conventional triaxial loading tests) and the axial strain, and the deformation modulus and the axial strain. The main points are as follows:

(1) The evolution characteristics of the conventional reconstituted coal samples and slightly disturbed soft coal samples were relatively close; in both cases, the deformation modulus exhibited the characteristics of an initial rapid decline, followed by a stable period or slow decline, and then a final rapid decline.

(2) In each evolutionary stage of the deformation modulus, the most important concern is the steady state/slow descent phase because this corresponds to the elastic deformation stage of the coal sample. A comparison of the results obtained at different confining pressures reveals that the elastic modulus of the conventional reconstituted coal sample was larger than that of slightly disturbed soft coal sample under the same confining pressure. When the peak intensity was reached, the slightly disturbed soft coal sample produced a larger axial strain. For both samples, the elastic modulus presented an increasing trend with increasing confining pressure. This is because coal usually contains many internal defects (e.g., intertwined fractures), and high confining pressure is conducive to closing such defects, thus increasing the stiffness and elastic modulus of coal.

(3) In the elastic deformation stage, the deformation modulus was basically stable, which suggests that the coal sample had the obvious characteristics of a linear elastic medium. In contrast, a slow descent implies that the coal sample had the characteristics of a nonlinear elastic medium. It can be concluded that, the slightly disturbed soft coal sample exhibited strong nonlinear elastic medium characteristics at a confining pressure of >2 MPa, whereas conventional reconstituted coal began to exhibit strong nonlinear elastic medium characteristics at a confining pressure of >4 MPa. With respect to the rate of decrease of the deformation modulus, that of the slightly disturbed soft coal sample was higher than that of the conventional reconstituted coal sample under different confining pressure conditions, thus indicating that the nonlinear elastic medium characteristics of the former were stronger than those of the latter. The main reason for this is that slightly disturbed soft coal samples retain the characteristics of unevenness and microfracture development of on-site soft coal.

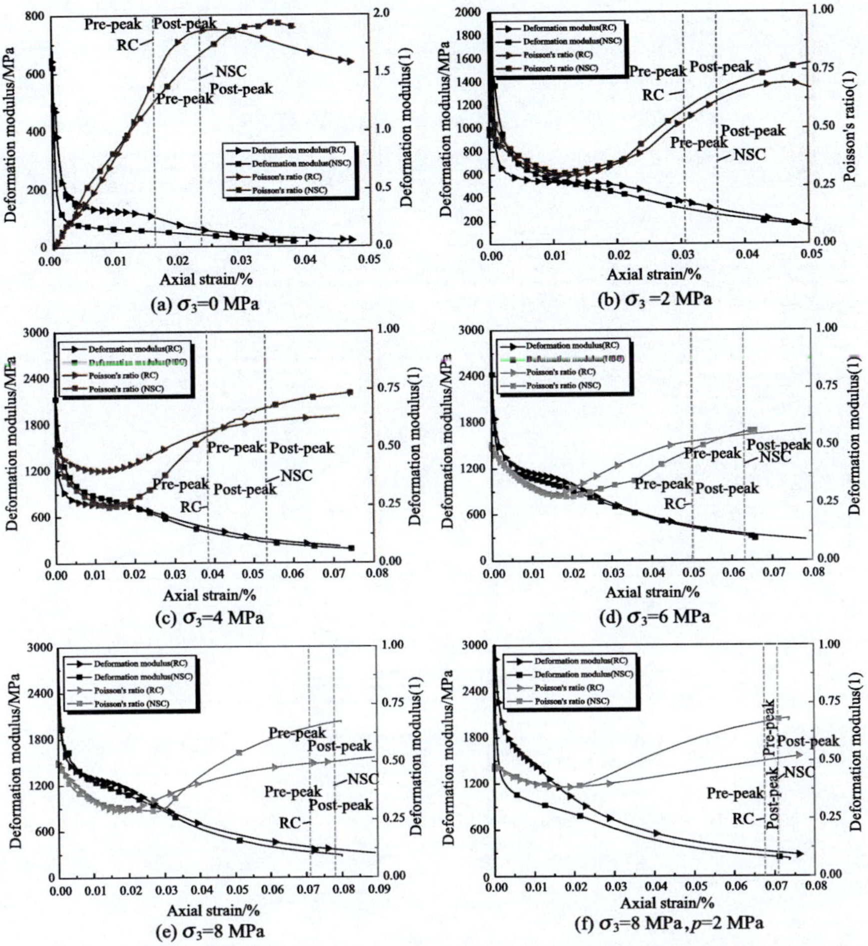

Fig. 6.23 Poisson's ratio–axial strain curves and deformation modulus–axial strain curves for coal samples under conventional compression with different confining pressures

(4) After a substantially steady or slow descent stage, the deformation modulus of the coal samples generally underwent an obvious downward process. This was mainly due to the rapid decrease in the modulus of elasticity after the yielding of the coal sample, that is, the coal sample underwent a yield stage before the peak strength had been reached. Comparing the duration of the yield stage under different confining pressures (between the beginning of the rapid decrease and the dashed line in Fig. 6.23), it can be found that the duration of the yield stage gradually increased with an increase in the confining pressure; hence, an increased confining pressure implies that the failure of coal requires a more plastic deformation.

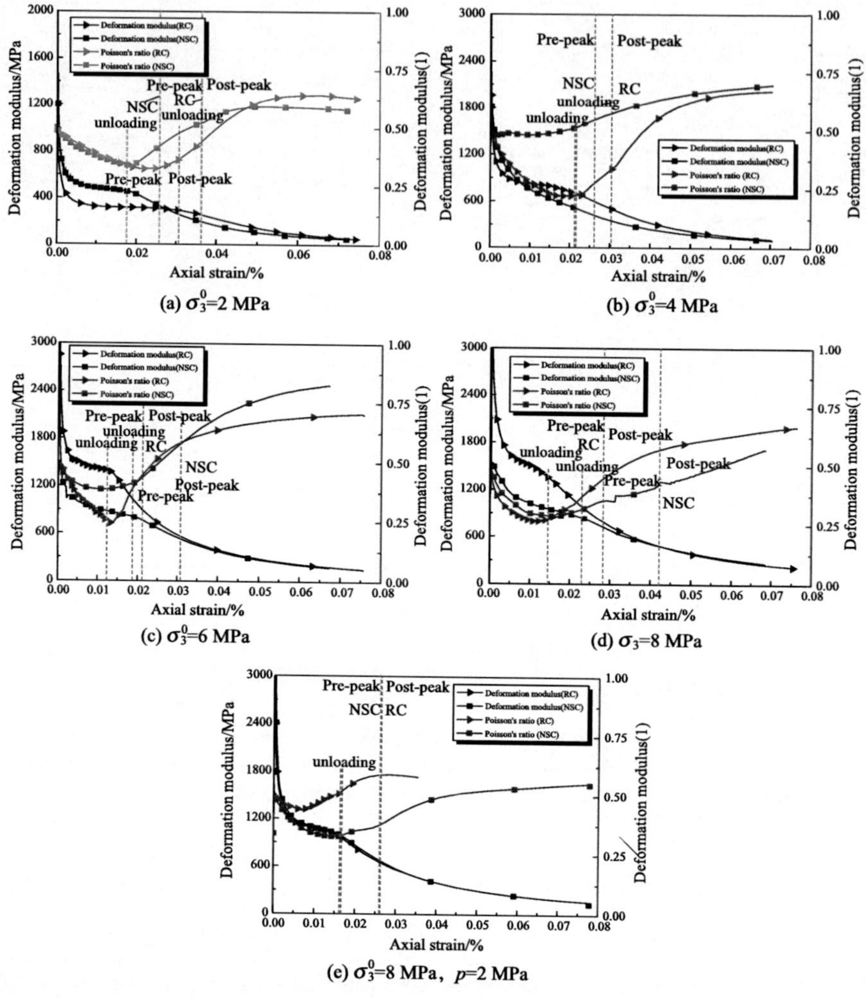

Fig. 6.24 Poisson's ratio–axial strain curves and deformation modulus–axial strain curves for coal samples under pre-peak confining pressure unloading conditions

(5) Except the coal sample and confining pressure conditions in Fig. 6.23a, the evolution of Poisson's ratio in other cases could be divided into three stages: the initial declining stage, the yield failure rising stage, and the final stage of slow rise and trend of stabilization. It is important to note that the Poisson's ratio of coal is the ratio of the absolute value of the radial strain to the axial strain during uniaxial tension or compression (in this book, it refers to compression). Poisson's ratio is an elastic parameter reflecting the material deformation characteristics, and its value should be between 0 and 0.5. At the same time, Poisson's ratio is the most common mechanical parameter, which is easily

obtained through the conventional mechanical test. Its application has also been developed, such that it can reveal the deformation and failure behavior of coal to some extent. When approaching the peak strength of coal, Poisson's ratio could exceed 0.5, and showed a slow increasing trend, which was due to the development of small fractures in the coal sample. During the continuous axial loading, tiny primary fractures would have continued to expand and extend until penetration and failure, thus resulting in a significant increase in radial deformation. At this point, Poisson's ratio no longer provided a general parameter of the coal deformation parameter; it also included the expansion of the fracture deformation. During unloading, the post-peak "stabilizing value" of Poisson's ratio, which was greater than the test results of conventional triaxial loading, related to the expansion, extension, and transfixion of small cracks in the coal sample, which were much more intense than those of the conventional triaxial loading during the process of confining pressure unloading as well as axial pressure loading. As a result, the increase in radial deformation was significant, and the Poisson's ratio at this moment no longer provided a general parameter of the coal deformation parameter, but also included the expansion of the fracture deformation in coal.

Figure 6.24 presents the relationships between the deformation modulus and the axial strain, and Poisson's ratio and the axial strain in the pre-peak confining pressure unloading tests of coal samples. The main points are as follows:

(1) For the pre-peak confining pressure unloading test, both the deformation modulus of the conventional reconstituted coal sample and slightly disturbed soft coal sample exhibited the characteristics of an initial rapid decline, followed by a basically stable or slowed decline phase, and then a final rapid decline. The deformation modulus evolution characteristics at the loading stage were consistent with the results of the conventional triaxial test. That is, under the same confining pressure condition, the elastic modulus of conventional reconstituted coal sample exceeded that of slightly disturbed soft coal sample. The latter presented strong nonlinear elastic medium characteristics at a confining pressure of >2 MPa, whereas the nonlinear elastic medium characteristics of the conventional reconstituted coal sample only became significant at a confining pressure of >4 MPa.

(2) In Fig. 6.4, "unload (RC)" and "unload (NSC)" represent the unloading points of the conventional reconstituted coal and slightly disturbed soft coal samples, respectively. It can be seen from Fig. 6.24 that, regardless of the type of sample, both deformation modulus values started to decrease rapidly from the unloading point. In addition, the rate of decrease was obviously faster than that of the conventional triaxial loading test under the same conditions. The duration of the yield stage reflected by the deformation modulus was significantly smaller than that of the conventional triaxial loading test, thus indicating that the failure of coal samples during the pre-peak confining pressure unloading condition was more sudden than that of the conventional loading condition.

(3) For the conventional reconstituted coal and slightly disturbed soft coal samples, the evolution characteristics of Poisson's ratio could also be divided into three stages: the initial declining stage, the rising stage before and after the unloading point, and the final stage of the slow rise and trend of stabilization. Similar to the conventional triaxial loading results, Poisson's ratio exceeded 0.5 and showed a slow increasing trend when approaching the peak strength. However, when unloading, the post-peak "stabilizing value" of Poisson's ratio was larger than the conventional triaxial loading test results. This is because there were many small fractures in the coal sample, and during the continuous loading of the axial pressure and unloading of confining pressure, the tiny fractures that continued to expand and extend were much more violent than those under conventional triaxial loading conditions.

Table 6.7 summarizes the elastic modulus and Poisson's ratio values of the two

Table 6.7 Deformation parameters of the conventional triaxial compression test and the pre-peak confining pressure unloading test

No.	Sample label	Confining pressure (MPa)	Elastic modulus (MPa)	Poisson's ratio
1	RC-1	0	128	0.45
2	RC-2	2	536	0.30
3	RC-3	4	767	0.40
4	RC-4	6	1141	0.32
5	RC-5	8	1294	0.34
6	RC-6	8 (2)	1130	0.39
7	RC-7	2	323	0.39
8	RC-8	4	806	0.24
9	RC-9	6	1427	0.31
10	RC-10	8	1507	0.27
11	RC-11	8 (2)	1075	0.46
12	NSC-1	0	62	0.46
13	NSC-2	2	522	0.31
14	NSC-3	4	769	0.26
15	NSC-4	6	991	0.29
16	NSC-5	8	1147	0.31
17	NSC-6	8 (2)	919	0.40
18	NSC-7	2	484	0.38
19	NSC-8	4	681	0.48
20	NSC-9	6	958	0.39
21	NSC-10	8	1110	0.34
22	NSC-11	8 (2)	1055	0.33

Note 8 (2) represents a confining pressure of 8 MPa and gas pressure of 2 MPa; RC represents the conventional reconstituted coal samples; NSC represents the slightly disturbed soft coal samples types of coal samples considering the elastic phase of the conventional triaxial compression test and the loading phase of the pre-peak confining pressure unloading test.

6.3.2 *Influence of Gas on Coal Deformation*

Figure 6.25 shows the scatter plot of the influence of gas on the peak strength, elastic modulus, and Poisson's ratio of coal (data from Table 6.7). The test data in the figure are classified according to the confining pressure: 6, 8, and 8 MPa at a gas pressure of 2 MPa.

In the previous chapter, it was concluded that the elastic modulus increased with increasing confining pressure, and that Poisson's ratio decreased with increasing confining pressure. Therefore, for the three cases, the peak strength was the greatest and Poisson's ratio was lowest at a confining pressure of 8 MPa. This implies that the elastic modulus of coal decreases as the gas pressure increases under a constant confining pressure. The possible reason for this can be analyzed from two aspects: the effective stress and the nonmechanical effects of gas adsorption. The nonmechanical effect of gas adsorption can be fixed in the triaxial compression tests of coal without gas or with a constant gas pressure. Generally, a coal body contains many internal defects, such as intertwined fractures. A high confining pressure is favorable for the compaction of the defects in a coal body, thus increasing the stiffness and elastic modulus of the coal body. However, the internal defects and degree of compaction of coal vary widely. For tight coal, owing to its small internal dimension, the compaction effect of the confining pressure on the defect is very small, and the elastic modulus is less affected by the confining pressure, which is consistent with the experimental results of tight rocks such as marble [54, 55]. For a coal body with an uneven distribution of internal defects, fractures are compacted and closed under different confining pressures. This changes the normal stress on the internal fracture surface because the internal deformation of a coal sample is affected by the internal friction on the fracture plane, and the elastic modulus is related to the confining pressure.

At a confining pressure of 8 MPa and gas pressure of 2 MPa, the elastic modulus was slightly smaller than that when the confining pressure was 6 MPa (Fig. 6.25), while the Poisson's ratio of the former was slightly higher than that of the latter. These results suggest that the nonmechanical effects of adsorbed gas also affect the elasticity modulus and Poisson's ratio. This may be because adsorbed gas can wedge within the microcracks of a coal body, thereby weakening the compaction effect of the confining pressure on the microcracks. This phenomenon results in a decrease in the stiffness and elastic modulus of the coal body, and an increase in Poisson's ratio.

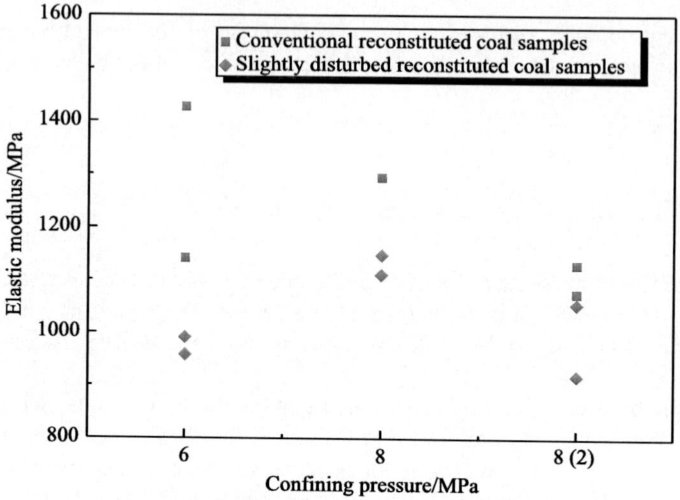

(a) Confining pressure versus elastic modulus

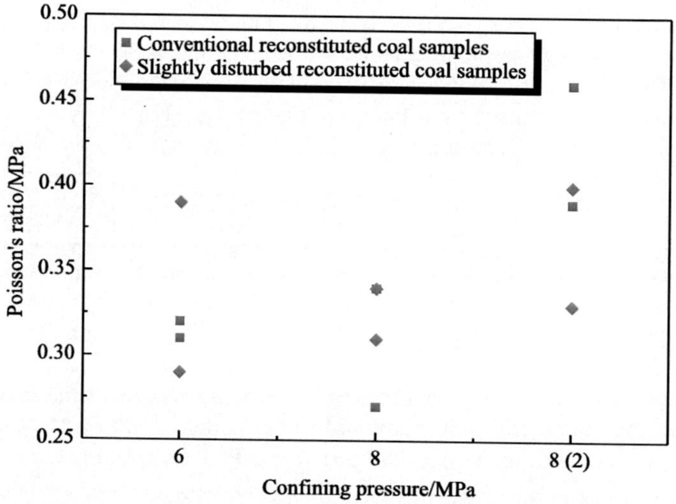

(b) Confining pressure versus Poisson's ratio

Fig. 6.25 Influence of gas on the elastic modulus and Poisson's ratio of coal

6.4 Constitutive Equation of Coal Containing Gas

6.4.1 Analysis of the Constitutive Relationship Characteristics of Coal

The stress–strain relationship of coal is very complicated, and includes the effects of anisotropy, nonlinearity, elastoplasticity, brittleness, ductility, creepage, and conversion from brittleness to ductility. However, until now, the proposed constitutive models can only simulate the main characteristics of a certain type of coal (or rock) under certain loading conditions, and a constitutive model that can fully represent the constitutive properties of various types of coal (or rock) under any loading conditions is still lacking. The stress–strain curve of coal presents different constitutive relationship characteristics according to its properties and different loading modes. As shown in Fig. 6.26, Miller divided 28 types of rock into six types based on their pre-peak stress–strain curves through a uniaxial compression test.

These six types of stress–strain curves have the following characteristics [56]:

(1) The stress–strain curve of type I is a straight line or approximate straight line, and the plastic stage is not obvious until the sample fails suddenly. Materials with such properties can be classified as "elastomers".

(2) The stress–strain curve of type II approximates a straight line at low stress levels; however, when the stress increases to a certain level, the stress–strain curve shows de-flexibility characteristics (the slope of the line gradually decreases with the increase in stress) until the sample fails. Materials with such properties can be classified as "elastoplastic bodies".

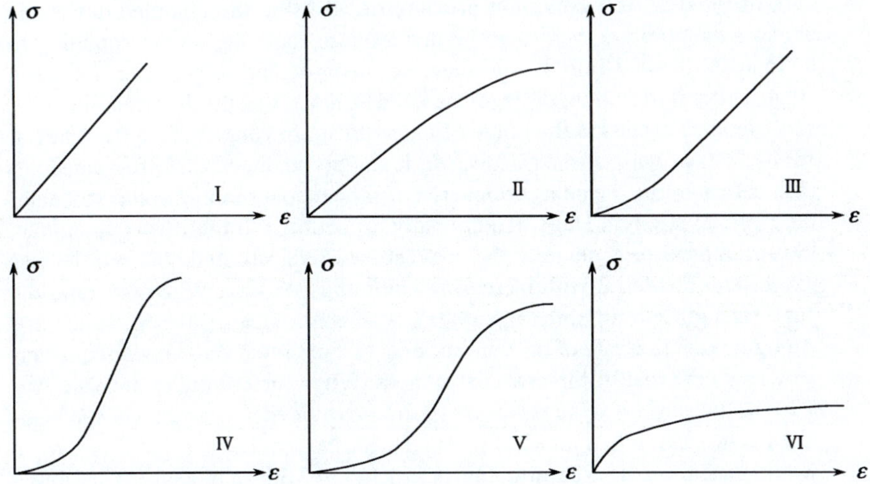

Fig. 6.26 Typical pre-peak stress–strain curves of rock

(3) The stress–strain curve of type III shows upsweep characteristics at low stress levels; when the stress increases to a certain level, the stress–strain curve gradually converts to a straight line until the sample fails. Materials with such properties can be classified as "plastic elastomers".
(4) The stress–strain curve of type IV also presents upsweep characteristics at low stress levels; however, when the stress increases to a certain level, the stress–strain curve gradually becomes a straight line, and then exhibits de-flexibility characteristics with a further increase in stress until the sample fails. Materials with such properties can be classified as "plastic–elastic–plastic bodies".
(5) Although the stress–strain curves of types V and IV are similar, the former has a smoother slope. Materials with such properties can also be classified as "plastic–elastic–plastic bodies".
(6) The stress–strain curve of type VI is a straight line during the early loading period, and then shows inelastic and creepage characteristics. Materials with such properties can be classified as "elastic–viscous bodies".

Based on an overview of the constitutive relationship of a given rock mass, we analyze the constitutive relationship characteristics of different types of coal under various loading modes (conventional triaxial loading and pre-peak confining pressure unloading). Subsequently, we further analyze the influence of the confining pressure (or initial confining pressure) on the constitutive relationship. The deviatoric stress–strain curves for coal samples obtained under conventional compression and confining pressure unloading conditions are illustrated in Fig. 6.27.

By comparing the deviatoric stress–axial strain curves of two types of coal samples, the following characteristics can be found:

(1) For the conventional reconstituted coal sample, the deviatoric stress–axial strain curve was similar to that of the raw coal sample at confining pressure of 0 MPa. At a confining pressure of >0 MPa, the obtained deviatoric stress–axial strain curves could be divided into four stages corresponding to an increased axial strain: elastic stage, strain hardening stage, strain softening stage, and residual strength stage. It should be noted that the deformation of the coal samples exceeded the range of the axial strain gauge before the residual strength stage was reached. However, it can be predicted that after considerable deformation, the conventional reconstituted coal sample would still enter the residual strength stage. Additionally, in contrast to the raw coal sample, the strain hardening stage of the conventional reconstituted coal sample was longer and increased with increasing confining pressure. When reaching the peak strength (strain softening stage), with the increase in axial strain, the strength reduction rate of the conventional reconstituted sample was considerably less than that of the raw coal sample. When the confining pressure was 8 MPa, the deviatoric stress–axial strain curve of the conventional reconstituted coal sample had an ideal plastic stage, thus indicating that this sample would generate a large deformation before failure with an increase in confining pressure.

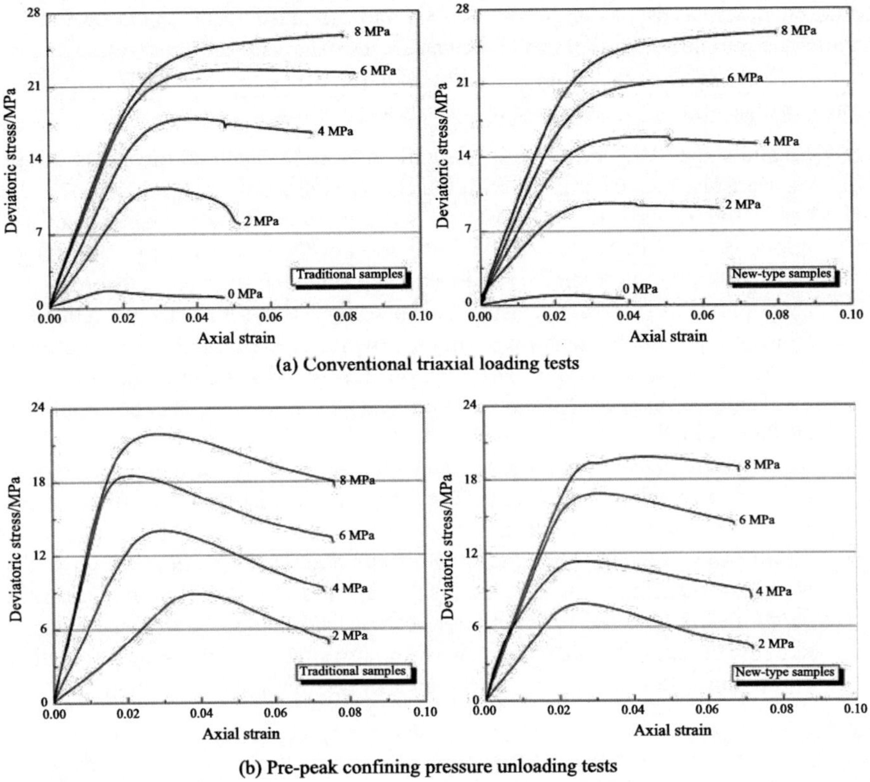

Fig. 6.27 Deviatoric stress–axial strain curves of coal samples

(2) The deviatoric stress–axial strain curve of the slightly disturbed soft coal sample was similar to that of the conventional reconstituted coal sample. However, the slightly disturbed soft coal sample experienced a long strain-softening stage under a low confining pressure condition (e.g., 2 MPa), which suggests that the plasticity of this sample was higher than that of the conventional reconstituted coal sample.

(3) A comparison of the deviatoric stress–axial strain curves under different loading modes reveals that, with increasing axial strain, the reduced rates of both coal strengths in the stage of strain softening were higher than those of the loading condition. Hence, the two types of coal exhibited more "brittle" characteristics under a confining pressure unloading condition compared with a loading condition. This is consistent with the results of other scholars' comparative tests of loading and unloading with rocks, where the brittle failure features were more obvious during unloading.

From the deviatoric stress–axial strain curves of the conventional compression tests and the pre-peak confining pressure unloading tests, the studied coal samples could be classified as elastic–plastic bodies, and the plastic stage could be further subdivided into three stages. For the convenience of analysis, the stress–strain curve characteristics of the coal samples are illustrated in Fig. 6.28. The basic nouns and concepts of the plastic constitutive relationship are introduced as follows:

(1) Elastic stage OA: The curve is basically a straight line, and the stress–strain relationship is in accordance with the elastic constitutive relationship.
(2) Strain hardening stage AC: Point A is the yield limit and upper limit of the elastic stage, which is also the starting point of the plastic deformation stage. This is because there can be unrecoverable plastic deformation ε_p (the recoverable deformation is called elastic deformation ε_e) if the axial stress is unloaded from point B (unloading the axial pressure while the confining pressure remains constant). From point A, the axial stress increases with increasing strain; thus, this stage is called the "strain hardening stage", which can also be simplified as "hardening".
(3) Ideal plastic stage CD: This stage is characterized by an increasing strain, while the stress does not increase. The stress–strain relationship in this segment follows the ideal plastic constitutive relationship.
(4) Strain softening stage DE: This stage is characterized by a decrease in stress with increasing strain until the coal is damaged (under high confining pressure conditions, where the residual strength can be maintained to some extent). Because the axial stress decreases with increasing strain in this stage, it is called the "strain softening stage", which can also be simplified as "softening".

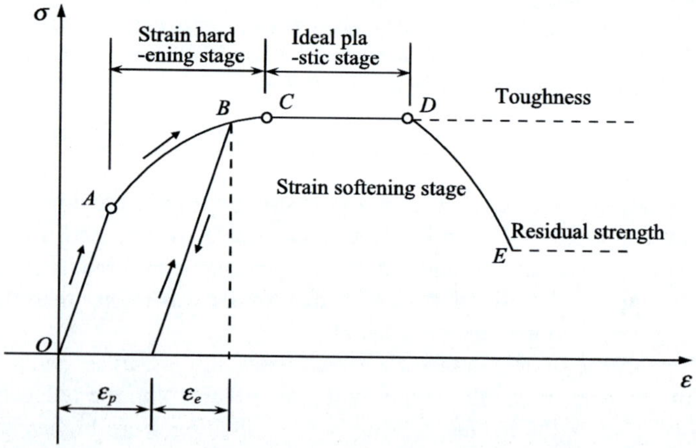

Fig. 6.28 Schematic of a stress–strain curve and the corresponding constitutive relationship of coal

6.4.2 Linear Elastic Stage

The coal body consists of matrix blocks and fractures, and the fractures consist of orthogonal butt cleats and face cleats. The coal matrix contains a pore system, which is the adsorption space for gas, over 90% of which exists on the micropore surface (as a representative of the internal surface of coal) in the adsorption state. The pore system of coal forms a diffusion channel, while the fracture system forms a seepage channel. The development of coal body fractures is affected by factors such as the coal forming environment, coal metamorphism, stratum stress, and the geological structure. In the process of coal mining or gas extraction, coal fractures are also affected by mining stress and gas pressure among other factors.

The influence of gas on coal strength not only includes the mechanical effect of the effective stress, but also the nonmechanical effect of the coal matrix adsorption expansion deformation. Meanwhile, in the process of gas desorption and flow, the temperature of coal will decrease because gas desorption needs to absorb a large amount of heat, thus inducing the thermal expansion effect of the coal matrix, which can be described as follows:

$$d\varepsilon_{mt} = \eta dT \tag{6.24}$$

where ε_{mt} is the thermal expansion deformation of the coal matrix, and η is the thermal expansivity of coal (K^{-1}).

The deformation of the coal matrix induced by gas adsorption and temperature change is expressed as Eq. (6.25):

$$\varepsilon_m = \frac{3a\rho RT}{EV_{mol}} \ln(1 + bp) + \eta(T - T_0) \tag{6.25}$$

where b is the adsorption constant. Its relationship with temperature T can be expressed as:

$$b = \frac{b_c}{\sqrt{T}} \exp\left(\frac{\Delta H}{RT}\right) \tag{6.26}$$

The deformation of the coal matrix influences the coal volume and fracture volume. When studying coal permeability, it is generally assumed that the deformation of the coal matrix completely acts on fractures, which significantly overestimates the effects of matrix deformation. Because the volume of coal does not change, this assumption is not in line with the actual situation. Therefore, the influence of the coal matrix deformation on the fracture and coal volume should have a coefficient of <1.

Figure 6.29 presents the ideal abstract model of coal, which shows that the coal body is not cut into completely independent matrix blocks, but connected by rock bridges to block the deformation of the coal matrix. Karacan [57] and Dawson

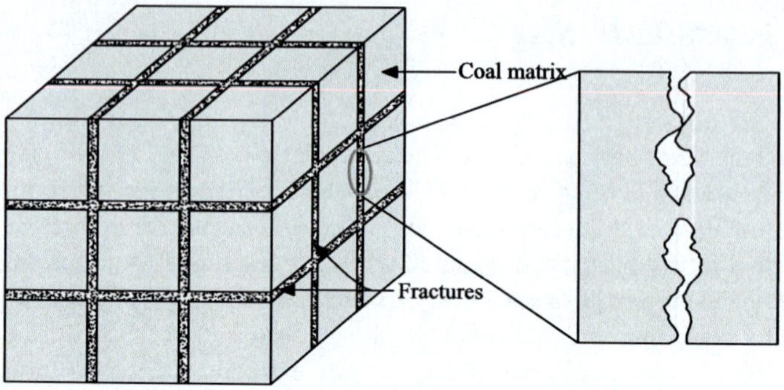

Fig. 6.29 Schematic diagram of the structures of coal and fractures

et al. [58] found that there is a large amount of inorganic mineral material in coal fractures, including of kaolin, pyrite and illite. These mineral fillings constitute the macroporous structure, which can allow gas to pass quickly without inducing expansion. Karacan [57, 59] reported that each part of a coal body can produce corresponding deformation and conduct self-adjustment when the coal generates deformation induced by adsorption. Thus, it can be concluded that only a part of the coal matrix deformation can influence the coal fractures. We introduce a factor f_m (a coefficient of the coal matrix influencing the fracture deformation of a coal body) to describe the effect of coal matrix deformation on fracture deformation. The f_m value ranges from 0 to 1, and mainly depends on the fracture distribution characteristics of coal and fracture filling properties; thus, f_m is a fixed value for a given coal body.

The contribution of the coal matrix deformation to the bulk volume and fracture volume can be expressed as follows:

$$\mathrm{d}V_{mb} = (1 - f_m)\mathrm{d}V_m = (1 - f_m)V_m(\mathrm{d}\varepsilon_s + \mathrm{d}\varepsilon_T) \tag{6.27}$$

$$\mathrm{d}V_{mf} = f_m\mathrm{d}V_m = f_m V_m(\mathrm{d}\varepsilon_s + \mathrm{d}\varepsilon_T) \tag{6.28}$$

where V_{mf} is the fracture volume change due to the deformation of the coal matrix (m^3), V_{mb} is the coal volume change due to the coal matrix deformation (m^3), and V_m is the volume of the coal matrix (m^3).

For coal at the linear elastic stage of OA, the constitutive relationship is consistent with Hooke's law. As the adsorption-induced deformation effect is similar to the thermal expansion effect, the constitutive relationship of coal gas at the linear elastic stage can refer to the thermal elastic constitutive equation of porous media:

$$\begin{aligned}\Delta\varepsilon_{ij} = \frac{1}{2G}\Delta\sigma_{ij} - \left(\frac{1}{6G} - \frac{1}{9K}\right)\Delta\sigma_{kk}\delta_{ij} + \frac{\beta}{3K}\Delta P\delta_{ij} \\ - (1 - f_m)\left(\frac{\Delta\varepsilon_{sw}}{3} + \frac{\alpha_T}{3}\Delta T\right)\delta_{ij}\end{aligned} \tag{6.29}$$

where K is the bulk modulus of the studied coal $\left(K = \frac{E}{3(1-2\mu)}\ \text{MPa}\right)$; G is the shear modulus $\left(G = \frac{E}{2(1+\mu)}\ \text{MPa}\right)$; μ is Poisson's ratio; β is the Biot coefficient of the effective stress ($\beta = 1 - K/K_m$); and K_m is the bulk modulus of the coal matrix (MPa).

6.4.3 Nonlinear Elastoplastic Stage

After coal generates dilatation, it enters the nonlinear elastoplastic stage, and the constitutive relationship can be described by the nonlinear Hooke's law [60].

When the volumetric strain of coal reaches dilatation point A, damage begins to appear, which causes internal cracks to generate, develop, and converge. As the loading continues, the damage of the coal body gradually increases and failure occurs at the peak strength point C. The damage variable D_a is defined as Eq. (6.30):

$$D_a = \frac{\varepsilon_{vG} - \varepsilon_v}{\varepsilon_{vG}} \tag{6.30}$$

where ε_{vG} is the volumetric strain at dilation point A.

Then, the elastic modulus E_D at the nonlinear elastoplastic stage is a function of the damage variable D_a:

$$E_D = f(D_a) \tag{6.31}$$

Research has shown that the elastic modulus at the nonlinear elastoplastic stage is a quadratic function of the volumetric strain; thus, Eq. (6.30) can be transformed into Eq. (6.32):

$$E_D = a_E D_a^2 + b_E D_a + E_A \tag{6.32}$$

where E_A is the elastic modulus of coal at dilatation point A, which is equal to the elastic modulus at the elastic stage (MPa). a_E and b_E are the coefficients obtained by fitting the experimental data.

When $D_a = 0$ coal is not damaged, and E_D is equal to E_A. The constitutive equation of coal at the nonlinear elastoplastic stage can be defined as follows:

$$\begin{aligned}\Delta\varepsilon_{ij} = {} & \frac{1}{2G_D}\Delta\sigma_{ij} - \left(\frac{1}{6G_D} - \frac{1}{9K_D}\right)\Delta\sigma_{kk}\delta_{ij} + \left(\frac{1}{3K} - \frac{1}{3K_m}\right)\Delta P\delta_{ij} \\ & - (1 - f_m)\left(\frac{\Delta\varepsilon_{sw}}{3} + \frac{\alpha_T}{3}\Delta T\right)\delta_{ij}\end{aligned} \tag{6.33}$$

where K_D is the bulk modulus of coal at the nonlinear elastoplastic stage (MPa), and G_D is the shear modulus of coal at the nonlinear elastoplastic stage (MPa).

6.4.4 Ideal Plastic Stage

At the ideal plastic stage, the stress of coal is kept constant, and plastic flow occurs. Thus, the constitutive equation of coal in the complete stress–strain process can be defined as Eq. (6.34):

$$\begin{gathered}\Delta\varepsilon_{ij} = \frac{1}{2G}\Delta\sigma_{ij} - \left(\frac{1}{6G} - \frac{1}{9K}\right)\Delta\sigma_{kk}\delta_{ij} + \left(\frac{1}{3K} - \frac{1}{3K_m}\right)\Delta P\delta_{ij} \\ - (1 - f_m)\left(\frac{\Delta\varepsilon_{sw}}{3} + \frac{\alpha_T}{3}\Delta T\right)\delta_{ij} \\ \begin{cases} E = E_A, & 0 \le \varepsilon_1 \le \varepsilon_y; \\ E = a_E D^2 + b_E D + E_G, & \varepsilon_y \le \varepsilon_1 \le \varepsilon_c; \\ E = \frac{\sigma_c - \sigma_p}{\varepsilon_c - \varepsilon_p}, & \varepsilon_c \le \varepsilon_1 \le \varepsilon_p; \\ \text{Plastic flow}, & \varepsilon_1 > \varepsilon_p. \end{cases}\end{gathered} \tag{6.34}$$

6.4.5 Strain Softening Stage

The linear negative elastic modulus can be used to handle the constitutive equation at the strain-softening stage DE. At this stage, the elastic modulus can be defined as follows:

$$E_d = \frac{\sigma_c - \sigma_p}{\varepsilon_c - \varepsilon_p} \tag{6.35}$$

where σ_c is the compressive strength of coal (MPa); σ_p is the residual strength of coal (Mpa); ε_c is the axial strain when the coal stress reaches its compressive strength; and ε_p is the axial strain at the lowest point E of the strain softening stage.

The constitutive equation of coal at the strain-softening stage can be defined by Eq. (6.36):

$$\begin{gathered}\Delta\varepsilon_{ij} = \frac{1}{2G_d}\Delta\sigma_{ij} - \left(\frac{1}{6G_d} - \frac{1}{9K_d}\right)\Delta\sigma_{kk}\delta_{ij} + \left(\frac{1}{3K} - \frac{1}{3K_m}\right)\Delta P\delta_{ij} \\ - (1 - f_m)\left(\frac{\Delta\varepsilon_{sw}}{3} + \frac{\alpha_T}{3}\Delta T\right)\delta_{ij}\end{gathered} \tag{6.36}$$

where K_d is the bulk modulus of coal at the strain-softening stage (MPa), and G_d is the shear modulus of coal at the strain-softening stage (MPa).

6.5 Macroscopic Damage Features of Coal Containing Gas

As the fracture evolution and damage dilation mechanism of coal will be discussed based on the C.D. Martin fracture strain model in Sect. 6.6, in this section only the macroscopic damage features of coal containing gas are discussed. As shown in Fig. 6.30, coal samples exhibit different damage features under different confining pressure conditions:

(1) Tensile failure under uniaxial conditions (Fig. 6.30a): Under the effect of axial load, a coal sample will produce tensile stress along the transverse direction, which is the result of Poisson's effect. When the transverse tensile stress exceeds the tensile limit of the coal sample, the coal sample produces irregular vertical cracks, which are considered as tensile failure.
(2) Shear failure at a low confining pressure (Fig. 6.30b): The confining pressure has an inhibitory effect on the generation of vertical fractures. Thus, when a coal sample fails, the generated fractures are often along the weak structure surface, which has a certain angle (usually <45°) to the loading direction. This type of damage is caused by the shear stress of the damaged surface, which exceeds the limit, and is therefore regarded as shear failure.
(3) Damage dilatation failure at a high confining pressure (Fig. 6.30c): If the confining pressure is continuously increased, the compression deformation of a coal sample gradually changes from brittleness to ductility. During the failure process, many microcracks are produced and gradually cut through to form multiple failure surfaces. At the same time, the volume expansion of the coal sample is obvious and the shape of the waist drum is macroscopically presented. Accordingly, it is characterized as a damage dilatation failure (multiple shear failure).

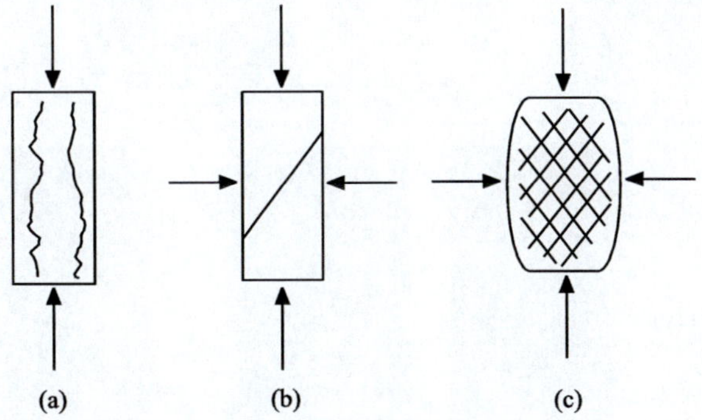

Fig. 6.30 Three common types of macroscopic failure of coal samples: **a** tensile failure, **b** shear failure, and **c** damage dilatation failure (multiple shear failure)

It should be noted that the above mentioned "low confining pressure" and "high confining pressure" are closely related to the coal strength properties, whereby a higher confining pressure is required for the damage dilation of higher strength coal samples, and vice versa.

Figure 6.31 illustrates the shear failure of a raw coal sample and the damage dilation failure of a slightly disturbed coal sample. It can be observed that the failure mode of the raw coal sample followed typical tensile failure at a confining pressure of 2 MPa and gas pressure of 1 MPa, or at a confining pressure of 4 MPa and gas pressure of 3 MPa. In contrast, the failure mode of the raw coal sample was typical shear failure at a confining pressure of 4 MPa and gas pressure of 1 MPa, or at a confining pressure of 6 MPa and gas pressure of 3 MPa. The failure mode of the slightly disturbed soft coal sample was damage dilatation failure (multiple shear failure). The asymmetric dilatation of this sample was mainly because its density was hardly compressed uniformly during the compression process.

The stress–strain curves of the conventional reconstituted coal and slightly disturbed soft coal samples in Fig. 6.27 reveal that:

(1) In conventional triaxial loading tests, both of the two types of coal samples showed the plastic strain softening property, and with an increase in confining pressure, the strain softening property gradually decreased. For both types of sample, the post-peak strain softening property gradually transformed to

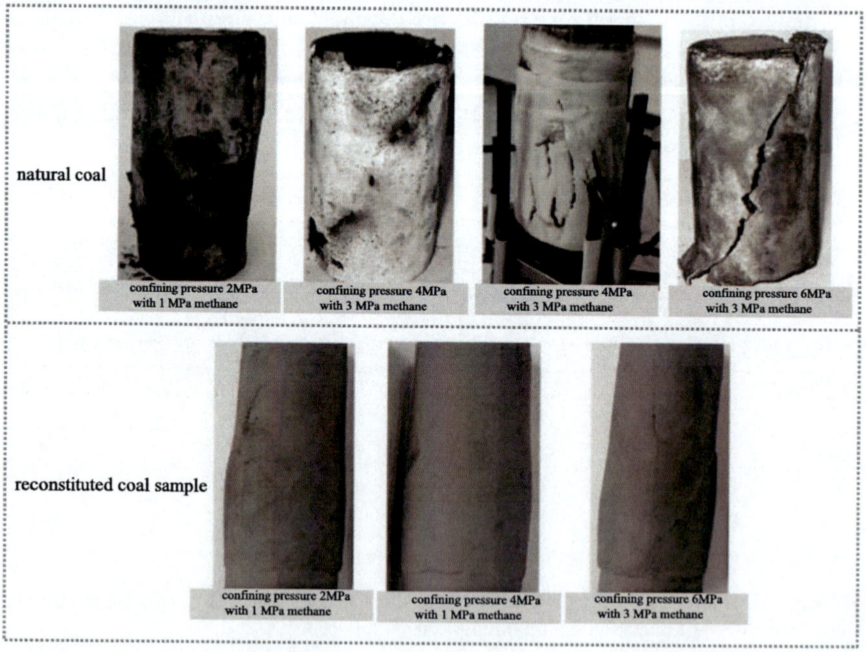

Fig. 6.31 Damaged test coal samples

ductility, with obvious ductility at a confining pressure of 8 MPa. In the pre-peak confining pressure unloading tests, the two types of coal samples also exhibited plastic strain-softening properties. Although the post-peak plastic strain-softening property also gradually decreased with an increase in the initial confining pressure, the plastic strain-softening property was still obvious. The stress state transformation of the surrounding rock in an underground coalmine (i.e., chamber excavations such as working face mining, roadway tunneling, and borehole drilling) belongs to confining pressure unloading, and the original geo-stress level will exceed the confining pressure of a test. Thus, the plastic strain-softening property of soft coal after failure is the main deformation behavior of the surrounding rock and chamber in an underground coalmine.

(2) As shown in Fig. 6.31, coal samples generated an obvious volume expansion before/after failure. Regardless of whether the conventional triaxial test or the pre-peak confining pressure unloading test was used, the volumetric strain of the coal samples underwent the process of compression first, and then experienced gradual expansion. In addition, the volume deformation also changed suddenly around the peak strength point, exhibiting the characteristics: before the peak strength point, the volume deformation was small; after this point, the volume deformation increased rapidly. The confining pressure had a significant effect on the volume expansion of coal after failure, and the volume strain curve of the coal sample after the peak strength point gradually slowed down with an increase in confining pressure. By comparing the volumetric strain characteristics of the conventional triaxial test with those of the pre-peak confining pressure unloading test, it was found that the volumetric strain curve of the conventional triaxial test was slightly less steep than that of the pre-peak confining pressure unloading test after the peak strength point, which indicates that volume expansion was more obvious under the confining pressure unloading condition. By comparing the volumetric strain characteristics of the conventional reconstituted coal sample with those of the slightly disturbed soft coal sample, it can also be observed that the volumetric strain curve of the former was slightly less steep than that of the latter after the peak strength point, thus implying that the volume expansion of the slightly disturbed soft sample was more obvious (Fig. 6.32).

6.6 Failure Mechanism and Strength Theory of Coal Containing Gas

In Sect. 5.3 of Chap. 5, the representative rock strength criteria were reviewed and evaluated. The Mohr–Coulomb criterion has made great achievements in the field of rock mechanics and engineering. However, this criterion ignores the inelastic stage of a rock before the peak strength, which is different from the constitutive characteristics of soft coal. The Griffith criterion is also unsuitable for practical engineering because of its limitations. In addition, the use of these strength criteria can only obtain the

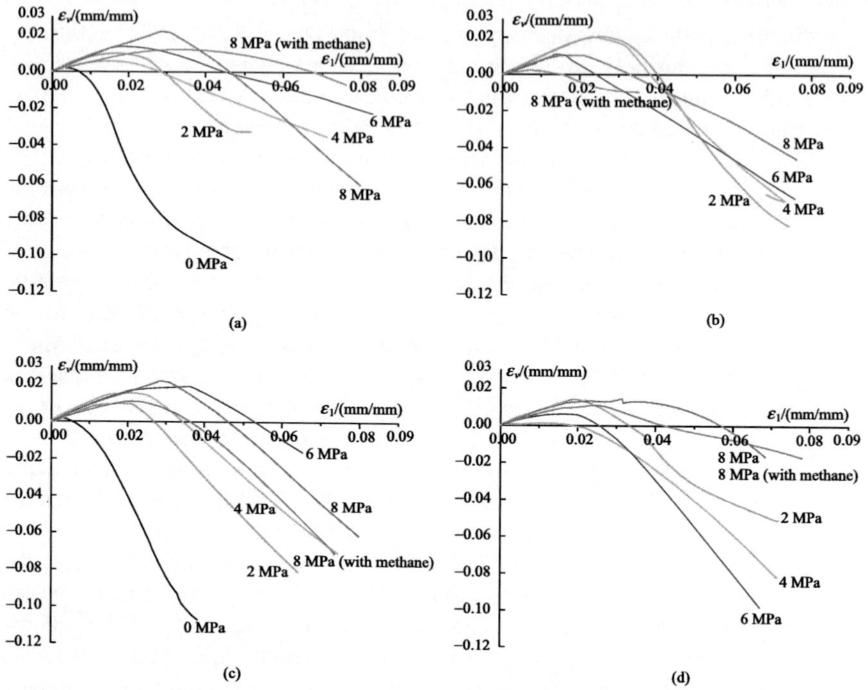

Fig. 6.32 Relationship between volume strain and axial strain for a **a** reconstituted sample (conventional triaxial compression), **b** reconstituted sample (pre-peak confining pressure unloading), **c** slightly disturbed soft sample (conventional triaxial compression), and **d** slightly disturbed soft sample (pre-peak confining pressure unloading)

final state of coal damage, but cannot quantitatively describe the gradual damage process of soft coal.

6.6.1 C. D. Martin Fracture Strain Model

Under the long-term of external loading, there are various defects in the natural coal body, such as microcracks and cleats. From the perspective of mechanics, the failure of coal and rock is due to the expansion and connection of many microfractures, which gradually form a macroscopic fracture after the change of the stress environment. Therefore, the damage state of coal/rock mass is closely related to the evolution of microfractures, and the corresponding yield and damage are the macroscopic results of the continuous development of microscopic fractures [61]. To quantitatively describe the law of fracture evolution during the process of coal failure, many scholars have used various methods to carry out experimental observations, such as the dynamic observation of coal surface fracture evolution by using optical microscopy [62], monitoring of the acoustic emission signals using the acoustic emission monitoring system during coal failure [63, 64], and real-time computer tomography (CT) scans of the coal damage process by using an industrial CT machine [65, 66]. In recent years, the theory of fracture evolution has also made great progress. For instance, Kemeny et al. presented a shear slip model of natural fractures [67], and Sammis presented a micro-hole stress concentration model [68]; however, the fracture strain model proposed by Martin is still the most widely applied [69]. This section introduces the idea of the fracture strain model and discusses how it can be used to study the coal damage dilation mechanism.

As shown in Fig. 6.33, according to the difference in the microfracture evolution state of rock samples under different load levels, the pre-peak stress–strain curve can be divided into four stages:

(1) Stage I is the compression closure stage of the primary fractures, and the characteristic stress corresponding to the end of this stage is called the fracture closure stress (σ_{cc}).
(2) Stage II represents the elastic deformation stage. The characteristic stress corresponding to the end point of this stage is called the fracture initiation stress (or initial dilatation stress, σ_{ci}). It is not only the starting point of the expansion

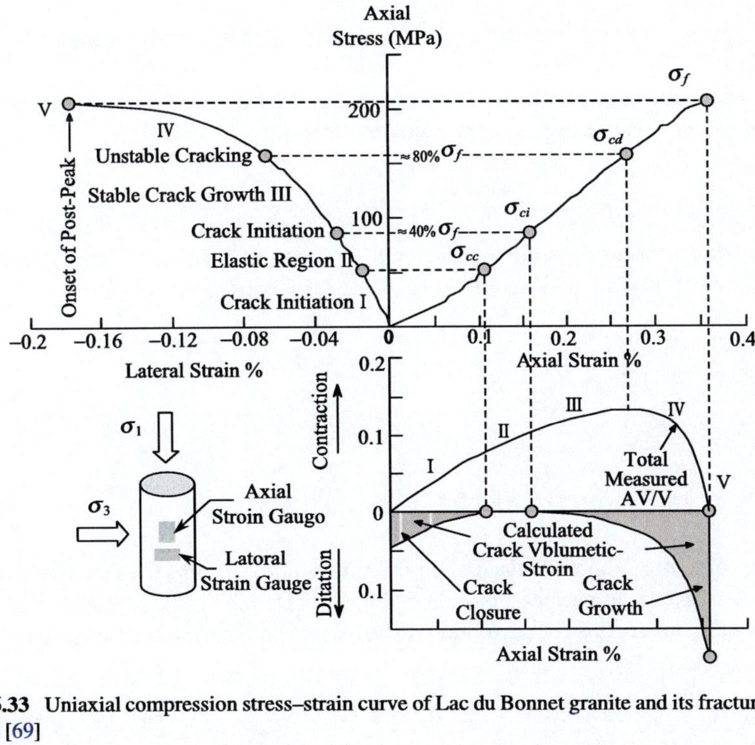

Fig. 6.33 Uniaxial compression stress–strain curve of Lac du Bonnet granite and its fracture strain curve [69]

of fractures, but also the symbol of fracture dilation, which is also called the dilation point of the stress–strain curve.

(3) Stage III is the stage of fracture stability expansion. The characteristic stress corresponding to the end point of this stage is called the damage/dilatation stress, σ_{cd}. It not only indicates that coal begins to yield, but also corresponds to the maximum value of volume compression (the maximum value of volumetric strain).

(4) Stage IV is the unsteady expansion stage of the fracture, and the characteristic stress corresponding to the end of the stage is called the peak stress, σ_f.

After the introduction of the four characteristic stresses in the process of progressive failure of coal and rock mass, the definition and calculation method of fracture strain will be introduced [69]. Fracture strain refers to the change in axial and radial deformation of a rock caused by the initial crack initiation and expansion of the rock under the action of the load, including the axial fracture strain and radial fracture strain [70]. In practical applications, we are more concerned about the fracture volume strain $\left(\varepsilon_v^f\right)$, which cannot be directly measured, but can be calculated using the following method:

$$\varepsilon_v^f = \varepsilon_v - \varepsilon_v^g \tag{6.37}$$

where ε_v is the volume strain of coal, and ε_v^g is the elastic volume strain of the coal skeleton.

In triaxial mechanics, the volumetric strain of coal cannot be directly measured, but can be calculated using an approximate method:

$$\varepsilon_v = \varepsilon_1 + 2\varepsilon_2 \tag{6.38}$$

The elastic volume strain of the coal skeleton can be calculated using the generalized Hooke's law. Under the pseudo-triaxial condition, we have:

$$\varepsilon_v^g = \frac{1-2\mu}{E}(\sigma_1 + 2\sigma_3) \tag{6.39}$$

Then:

$$\varepsilon_v^f = \varepsilon_1 + 2\varepsilon_2 - \frac{1-2\mu}{E}(\sigma_1 + 2\sigma_3) \tag{6.40}$$

After obtaining the formula for the fracture volume strain, the fracture volume strain of coal can be obtained based on the results of the triaxial mechanical test. Furthermore, the fracture strain model can be used to analyze the fracture propagation law for a coal body.

6.6.2 *Damage Dilatation Progressive Failure Mechanism of Coal*

Figure 6.34 displays the deviatoric stress–strain curve of a slightly disturbed coal sample under an initial confining pressure of 6 MPa. Based on the calculated law of fracture volume evolution, the mechanism of coal damage dilation progressive failure can be discussed. According to the definition of the characteristic stress of the fracture strain model, the deviatoric stress–strain curve and fracture strain curve can be divided into five stages. Accordingly, the damage dilatation progressive failure process of the slightly disturbed coal sample in Fig. 6.34 can be described by the corresponding characteristic stress of each stage [71]:

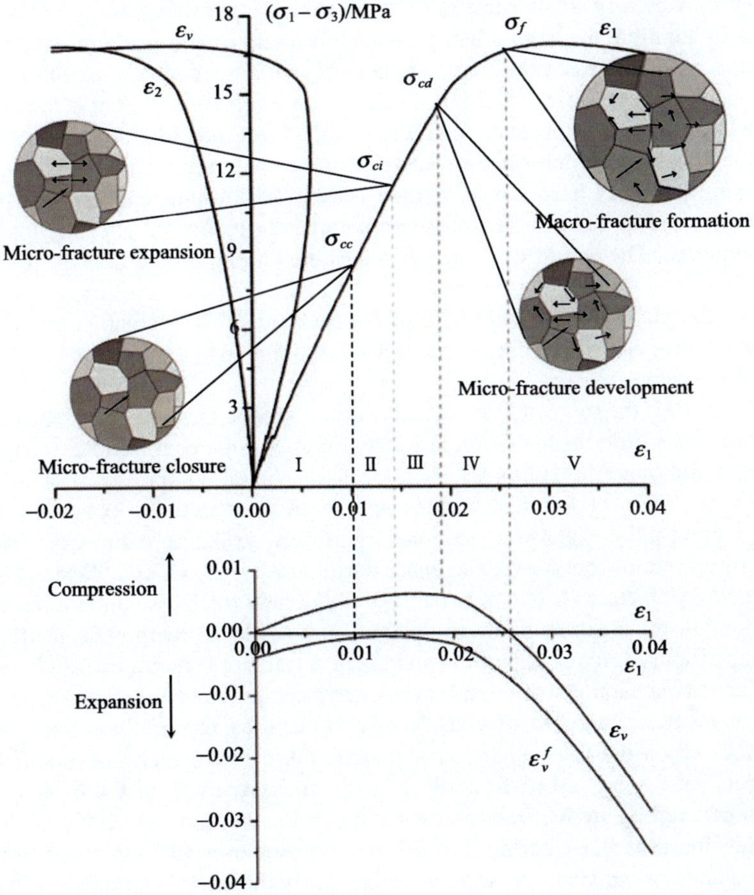

Fig. 6.34 Deviatoric stress–strain curve and fracture strain curve of a pre-peak confining pressure unloading test

(1) Stage I is the stage of fracture compaction: In this stage, the volume strain and fracture volume strain of the coal sample both increased (positive compression), indicating that the volume of the coal sample was compressed. Except the cracks that are parallel to the axial stress, most of the primary cracks were compacted. The structure and properties of coal samples do not change irreversibly at this stage, and the duration of this stage depends on the density of the primary fractures and their initial orientation. As shown in Fig. 6.34, the coal sample experienced a considerable period of fracture compaction, which is consistent with the loose structure of soft coal. The end of the phase is determined by the fracture closure stress, σ_{cc}, which is the axial stress corresponding to the end of the rapid increase in the fracture volume.

(2) Stage II is the stage of elastic deformation: In this stage, the volume strain of the coal sample continued to increase, and the change of the fracture volumetric strain was very small. Most of the primary fractures were compacted in stage I. In addition, the axial loading in stage II could only cause a relative degree of slipping between the fractures (possibly with a small number of closed or open fissures), which would be insufficient to produce a lot of new fractures or a large volume of primary crack expansion. Therefore, fractures did not affect the linear elastic deformation behavior of the coal sample in stage II. The coal sample did not have a macroscopic irreversible mechanical response in this stage, and its deformation could be completely recovered after the load was removed. The end of this stage is determined based on the fracture initiation stress, σ_{ci}. Research has demonstrated that the characteristic stress of stage II is generally approximately 40% of the peak stress [70], which is equal to the axial stress corresponding to the inflection point at which the fracture volume strain begins to decrease.

(3) Stage III is the stage of fracture stable development. Once the load reached σ_{ci}, there was little further increase in the volume of the coal sample. It suggests that, although the compression deformation (axial compression) of the coal sample was still dominant at this stage, its advantages to the expansion deformation (radial expansion) were not significant. At the same time, the fracture volume strain continued to decrease during this stage, which indicates that the primary fractures began to expand, and that new fractures would have developed in the weak structure of the coal matrix blocks. During stage III, both the primary fracture expansion and the new fracture generation were basically carried out parallel to the axial (maximum principal stress direction) at random. However, mesoscopic information reveals that the length of these fractures did not exceed the scale of the coal matrix; these microcracks were still independent of each other. According to the stress analysis of the oval fracture conducted by Inglis, fracture formation and expansion are still in the stable development phase during stage III [61, 69]. However, fracture generation and expansion could have caused a deterioration of the cohesion of the coal sample because the development of fractures in stage III would have caused the macroscopic irreversible mechanical response of the coal sample. This means that the deformation of the coal sample was permanent even when the load was

removed. Although this permanent deformation would be much less than the total deformation of this stage, it would be sufficient to affect the macroscopic mechanical properties of coal. The end point of stage III is determined by the damage dilation stress, σ_{cd}, which is the starting point of the dilatation of the coal sample and the maximum point of the volume compression of the coal sample. Other researchers have indicated that this characteristic stress is generally approximately 80% of the peak stress [70], which is equal to the axial stress corresponding to the point at which the coal volume strain is the largest.

(4) Stage IV is the unsteady expansion stage of a fracture; the starting point of this stage is the damage dilation stress, σ_{cd}. When the characteristic stress level reached σ_{cd}, the volume expansion of the coal sample exceeded the compression deformation, and the coal volume strain curve shifted from the maximum point to the negative direction; hence, the coal sample entered the phase of dilation and yield. The fracture volume strain during stage IV rapidly reduced, which indicates that the development of the fracture underwent a qualitative change. At that moment, any new fractures were insufficient to absorb the energy released by fracture expansion; thus, fracture expansion accelerated and then entered the unsteady extension stage. During this stage, the expansion and formation of fractures continuously increased, and the weak facets of the structure were destroyed. The fractures in the coal sample converged and penetrated, and the fracture density and fracture length were multiplied to form many irreversible deformations. Moreover, in this process, the fracture activity in the coal sample showed the characteristic of a gradually decreasing fracture and gradually increasing fracture slip, as well as the trend of forming potential macroscopic fractures. Finally, it should be noted that if the axial load level exceeds σ_{cd}, even when the stress state does not change, the coal sample will be destroyed after a certain period of time. Therefore, the damage dilation stress is also called the long-term strength. The end of this stage corresponds to peak stress, σ_f.

(5) Stage V is the post-peak failure stage, and is the moment when the axial load reaches the peak strength and begins to decrease. During this stage, the volumetric strain and fracture volume strain of the coal sample showed decreasing trends, the coal sample softened, and dilatation accelerated. As a result, the coal sample would have been cut up by macroscopic fractures, and the rupture surfaces in the coal would have slip, which would finally lead to the overall destruction of the coal sample or a decrease in the bearing capacity.

Through the analysis of the coal volume strain and the fracture volume strain, it was found that the process of damage dilatation progressive failure on the coal sample was also the process of (i) the first gradual compaction of internal fractures, (ii) the original crack expansion induced by the tip stress concentration, (iii) new fracture generation, development, connection, and (iv) the formation of a macro-damaged surface.

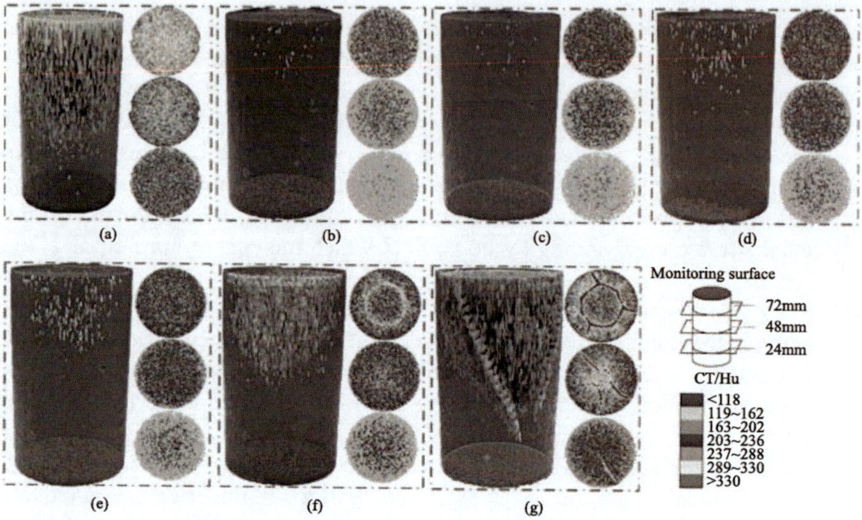

Fig. 6.35 Damage-fracture evolution of coal under the fixed axial displacement–confining pressure unloading condition, under **a** the initial condition, **b** a confining pressure of 8 MPa, **c** a confining pressure of 6 MPa, **d** a confining pressure of 4.8 MPa, **e** a confining pressure of 3.6 MPa, **f** a confining pressure of 2.4 MPa, and **g** a confining pressure of 0.8 MPa

To study the damage and fracture evolution of a coal sample during the unloading process, we carried out a CT scanning experiment of coal damage fracture under a fixed axial displacement–confining pressure unloading condition. As shown in Fig. 6.35, the results of CT scanning were reconstructed using a numerical technique, and the fracture evolution of the coal sample during the experiment was obtained. As shown in Fig. 6.35, there were many microfractures in the coal sample, even when it was not loaded (Fig. 6.35a). As the loading progressed, most of the microcracks in the coal sample were closed (Fig. 6.35b). With the unloading process, when the confining pressure was reduced from 8 to 4.8 MPa, the fracture evolution was mainly due to the expansion of the microfractures. The expansion of the microfractures started from the central part of the core and gradually developed into the peripheral part of the core (Fig. 6.35b–d). When the confining pressure was reduced to 3.6 MPa, the fractures developed stably. Although they remained discrete, they had a dominant direction, and the cone was concentrated in the upper part of the coal sample (Fig. 6.35e). When the confining pressure was reduced to 2.4 MPa, the macroscopic fractures converged to form a conical failure surface (Fig. 6.35g). When the confining pressure was further reduced to 0.8 MPa, the failure surface continued to develop, eventually forming an inverted conical failure plane (Fig. 6.35g).

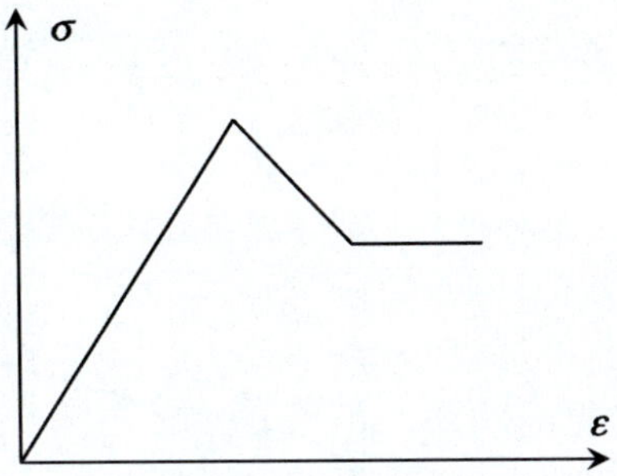

Fig. 6.36 Typical complete stress–strain curve

6.6.3 Damage Dilatation Yield Criterion of Coal

At present, the most widely used rock failure criteria when conducting mechanical analyses or numerical simulation for rock engineering studies include the Mohr–Coulomb criterion, Drucker–Prager criterion, and Hoek–Brown criterion, which are all based on the ideal elastoplastic constitutive model and strain-softening model (as shown in Fig. 6.36). When using these failure criteria and constitutive models, the non-elastic deformation generated before the peak strength is usually treated as elastic deformation; that is, once the rock exceeds the elastic stage, it will immediately enter the yield stage. In addition, entering the yield phase is equivalent to destruction. This simplified method is different from the constitutive relationship and damage dilation mechanism obtained from previous experiments. However, as these rock failure guidelines are concise, clear, and easy to use, they still have strong vitality and deserve consideration.

From the above damage dilation mechanism, it can be found that when the stress level of coal exceeds the damage dilation stress σ_{cd}, coal will generate irreversible permanent deformation as well as damage dilatation. Thus, the damage dilation stress can be used to replace the peak strength, and to build the damage dilation yield criterion with the same form of the Mohr–Coulomb strength criterion as follows:

$$\sigma_{cd} = \psi' + \zeta'\sigma_3 \tag{6.41}$$

ψ' and ζ' also can be defined as:

$$\psi' = \frac{2c' \cos \varphi'}{1 - \sin \varphi'} \tag{6.42}$$

$$\zeta' = \tan^2\left(45^\circ + \frac{\varphi'}{2}\right) \tag{6.43}$$

where c' is the damage dilation cohesion of the studied coal (MPa), and φ' is the damage dilation internal friction angle of the studied coal (°).

As shown in Fig. 6.37, the damage dilation yield criterion can be interpreted as

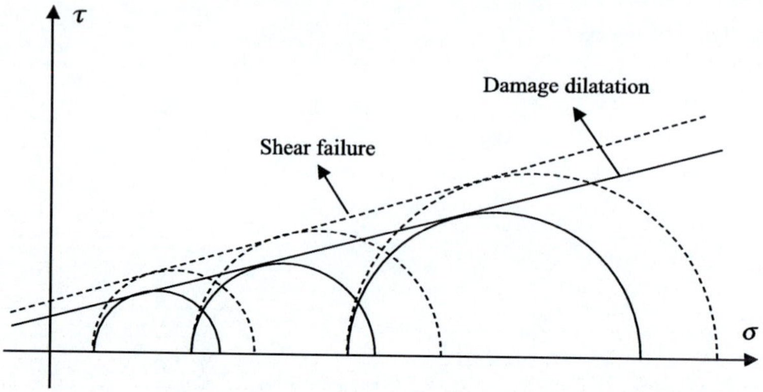

Fig. 6.37 Damage dilatation yield criterion

follows: the minimum principal stress remains constant while the maximum principal stress decreases, and the strength envelope of the damage dilation yield criterion is the tangent of a new set of stress circles.

According to the definition and calculation method in Sect. 6.6.1, the peak strength σ_f and damage dilation stress σ_{cd} of the conventional reconstituted coal samples and slightly disturbed soft coal samples in conventional triaxial tests and pre-peak confining pressure unloading tests are summarized in Table 6.8.

The peak strength (σ_f) and damage dilation stress (σ_{cd}) of the both types of coal samples in conventional triaxial tests and confining pressure unloading tests are illustrated in Fig. 6.38.

Using the same data reduction method described in Sect. 6.3.1, the relationship between the damage dilation stress and confining pressure can be calculated by the minimum linear quadratic fitting (using the Mohr–Coulomb criterion function to represent the principal stress, whereby the uniaxial strength is still excluded). The obtained relationships for different types of coal samples under different experimental conditions can be expressed as:

Conventional reconstituted coal sample in a conventional triaxial test:

$$\sigma_{cd} = 2.47 + 3.4\sigma_3 \tag{6.44}$$

Conventional reconstituted coal sample in a pre-peak confining pressure unloading test:

$$\sigma_{cd} = 1.19 + 3.75\sigma_3 \tag{6.45}$$

Table 6.8 List of damage dilatation stress and peak strength

No.	Sample label	Confining pressure (MPa)	σ_{cd} (MPa)	σ_f (MPa)
1	RC-1	0	0.54	1.66
2	RC-2	2	10.02	13.72
3	RC-3	4	15.05	21.90
4	RC-4	6	22.68	28.57
5	RC-5	8	30.15	36.14
6	RC-6	8 (2)	28.44	29.75
7	RC-7	2	8.60	10.58
8	RC-8	3.67	15.05	17.04
9	RC-9	5.62	22.37	22.92
10	RC-10	6.99	27.31	27.66
11	RC-11	8 (2)	15.53	24.20
12	NSC-1	0	0.25	0.95
13	NSC-2	2	9.97	11.64
14	NSC-3	4	16.80	19.56
15	NSC-4	6	26.33	27.17
16	NSC-5	8	30.25	33.76
17	NSC-6	8 (2)	22.76	28.08
18	NSC-7	1.94	9.15	9.54
19	NSC-8	4	11.98	15.01
20	NSC-9	6	18.27	21.62
21	NSC-10	8	25.97	26.13
22	NSC-11	6.69 (2)	22.32	23.15

Note 8 (2) means that the confining pressure was 8 MPa and the gas pressure was 2 MPa; RC represents conventional reconstituted coal samples; NSC represents slightly disturbed soft coal samples

Slight disturbed soft coal sample in a conventional triaxial test:

$$\sigma_{cd} = 2.25 + 3.52\sigma_3 \tag{6.46}$$

Slight disturbed soft coal sample in a pre-peak confining pressure unloading test:

$$\sigma_{cd} = 2.34 + 2.81\sigma_3 \tag{6.47}$$

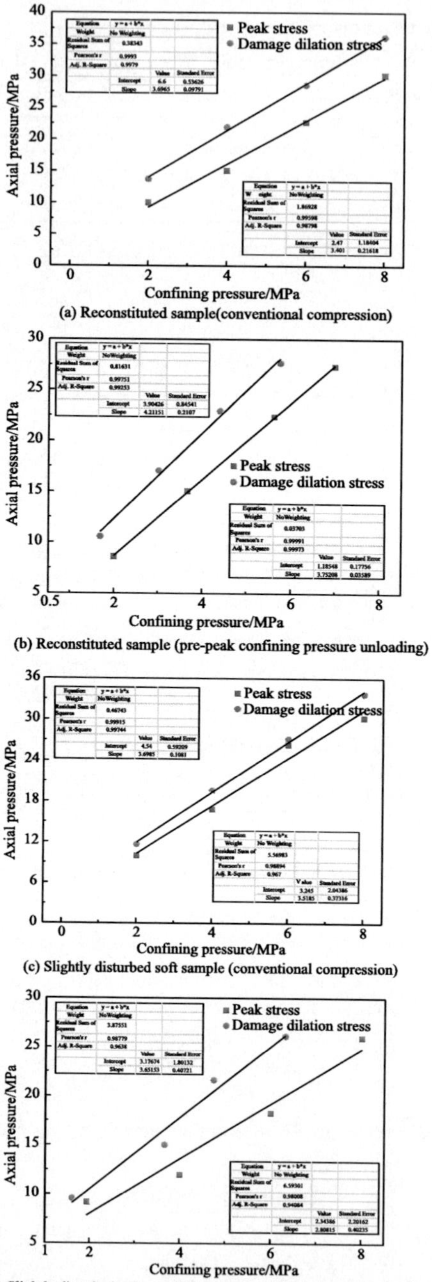

(a) Reconstituted sample(conventional compression)

(b) Reconstituted sample (pre-peak confining pressure unloading)

(c) Slightly disturbed soft sample (conventional compression)

(d) Slightly disturbed soft sample (pre-peak confining pressure unloading)

Fig. 6.38 Relationship curves of characteristic stress and confining pressure

References

1. Poulsen, B. A., Shen, B., Williams, D. J., et al. (2014). Strength reduction on saturation of coal and coal measures rocks with implications for coal pillar strength. *International Journal of Rock Mechanics and Mining Sciences, 71*, 41–52.
2. Ranjith, P. G., Jasinge, D., Choi, S. K., et al. (2010). The effect of CO_2 saturation on mechanical properties of Australian black coal using acoustic emission. *Fuel, 89*, 2110–2117.
3. Viete, D. R., & Ranjith, P. G. (2006). The effect of CO_2 on the geomechanical and permeability behaviour of brown coal: implications for coal seam CO_2 sequestration. *International Journal of Coal Geology, 66*, 204–216.
4. Ranjith, P. G., & Perera, M. S. A. (2012). Effects of cleat performance on strength reduction of coal in CO_2 sequestration. *Energy, 45*, 1069–1075.
5. Perera, M. S. A., Ranjith, P. G., & Viete, D. R. (2013). Effects of gaseous and super-critical carbon dioxide saturation on the mechanical properties of bituminous coal from the Southern Sydney Basin. *Applied Energy, 110*, 73–81.
6. Medhurst, T. P., & Brown, E. T. (1998). A study of the mechanical behaviour of coal for pillar design. *International Journal of Rock Mechanics and Mining Sciences, 35*, 1087–1105.
7. Poulsen, B., & Adhikary, D. (2013). A numerical study of the scale effect in coal strength. *International Journal of Rock Mechanics and Mining Sciences, 63*, 62–71.
8. Li, H. T., Jiang, C. X., Jiang, Y. D., Wang, H. W., & Liu, H. B. (2015). Mechanical behavior and mechanism analysis of coal samples based on loading rate effect. *Journal of China University of Mining and Technology, 44*, 430–436 (in Chinese).
9. Brotóns, V., Tomás, R., Ivorra, S., et al. (2013). Temperature influence on the physical and mechanical properties of a porous rock: San Julian's calcarenite. *Engineering Geology, 167*, 117–127.
10. Renshaw, C. E., & Schulson, E. M. (2007). Limits on rock strength under high confinement. *Earth and Planetary Science Letters, 258*, 307–314.
11. Alshayea, N. A., Khan, K., & Abduljauwad, S. N. (2000). Effects of confining pressure and temperature on mixed-mode (I–II) fracture toughness of a limestone rock. *International Journal of Rock Mechanics and Mining Sciences, 37*, 629–643.
12. Haimson, B., & Chang, C. (2000). A new true triaxial cell for testing mechanical properties of rock, and its use to determine rock strength and deformability of Westerly granite. *International Journal of Rock Mechanics and Mining Sciences, 37*, 285–296.
13. Cai, M., & Kaiser, P. K. (2014). In-situ rock spalling strength near excavation boundaries. *Rock Mechanics and Rock Engineering, 47*, 659–675.
14. Kaiser, P. K., & Kim, B. (2014). Characterization of strength of intact brittle rock considering confinement-dependent failure processes. *Rock Mechanics and Rock Engineering, 48*, 107–119.
15. Xu, J., Zhang, D., Peng, S. J., Liu, D., & Wang, L. (2011). Experimental research on influence of temperature on mechanical properties of coal containing methane. *Chinese Journal of Rock Mechanics and Engineering, 30*, 2730–2735 (in Chinese).
16. Zhou, X., Zhang, Y., & Ha, Q. (2008). Real-time computerized tomography (CT) experiments on limestone damage evolution during unloading. *Theoretical and Applied Fracture Mechanics, 50*, 49–56.
17. Wu, G., & Zhang, L. (2004). Studying unloading failure characteristics of a rock mass using the disturbed state concept. *International Journal of Rock Mechanics and Mining Sciences, 41*, 419–425.
18. Tao, M., Li, X., & Wu, C. (2012). Characteristics of the unloading process of rocks under high initial stress. *Computers and Geotechnics, 45*, 83–92.
19. Zhao, X., Wang, J., Cai, M., et al. (2014). Influence of unloading rate on the strainburst characteristics of Beishan granite under true-triaxial unloading conditions. *Rock Mechanics and Rock Engineering, 47*, 467–483.
20. Ding, Q. L., Ju, F., Mao, X. B., et al. (2016). Experimental investigation of the mechanical behavior in unloading conditions of sandstone after high-temperature treatment. *Rock Mechanics and Rock Engineering, 49*, 2641–2653.

21. Chen, J., Jiang, D., Ren, S., et al. (2016). Comparison of the characteristics of rock salt exposed to loading and unloading of confining pressures. *Acta Geotechnica, 11*, 221–230.
22. He, M., Miao, J., & Feng, J. (2010). Rock burst process of limestone and its acoustic emission characteristics under true-triaxial unloading conditions. *International Journal of Rock Mechanics and Mining Sciences, 47*, 286–298.
23. Ranjith, P. G., Shao, S. S., Viete, D. R., et al. (2012). Carbon dioxide storage in coal: reconstituted coal as a structurally homogeneous substitute for coal. *International Journal of Coal Preparation and Utilization, 32*, 265–275.
24. Jasinge, D., Ranjith, P. G., & Choi, S. K. (2011). Effects of effective stress changes on permeability of latrobe valley brown coal. *Fuel, 90*, 1292–1300.
25. Wang, C. G., Zhang, J. D., Zang, Y. X., et al. (2021). Time-dependent coal permeability: Impact of gas transport from coal cleats to matrices. *Journal of Natural Gas Science and Engineering, 88*, 103806.
26. Xu, J., Zhang, D., Peng, S., et al. (2011). Experimental research on influence of temperature on mechanical properties of coal containing methane. *Chinese Journal of Rock Mechanics and Engineering, 30*, 2730–2735 (in Chinese).
27. Li, X., Yin, G., Zhao, H., et al. (2010). Experimental study of mechanical properties of outburst coal containing gas under triaxial compression. *Chinese Journal of Rock Mechanics and Engineering, 29*, 3350–3358 (in Chinese).
28. Jiang, C., Yin, G., Li, X., et al. (2010). Experimental study of gas permeability of outburst coal briquettes in complete stressstrain process. *Chinese Journal of Rock Mechanics and Engineering, 29*(3), 482–483 (in Chinese).
29. Xu, J., Liu, D., Peng, S., et al. (2010). Experimental research on influence of particle diameter on coal and gas outburst. *Chinese Journal of Rock Mechanics and Engineering, 29*, 1231–1237 (in Chinese).
30. Cao, S., Yong, L. I., Ping, G., et al. (2010). Comparative research on permeability characteristics in complete stress-strain process of briquettes and coal samples. *Chinese Journal of Rock Mechanics and Engineering, 29*(5), 899–906 (in Chinese).
31. Jasinge, D., Ranjith, P. G., Choi, X., et al. (2012). Investigation of the influence of coal swelling on permeability characteristics using natural brown coal and reconstituted brown coal specimens. *Energy, 39*, 303–309.
32. Jasinge, D., Ranjith, P. G., Choi, X., et al. (2009). Mechanical properties of reconstituted Australian black coal. *Journal of Geotechnical and Geoenvironmental Engineering, 135*, 980–985.
33. Jasinge, D., Ranjith, P. G., Kodikara, J. et al. (2007). A comparison of stress strain behaviour of reconstituted and natural black coal. In *11th ISRM Congress*. International Society for Rock Mechanics.
34. Jasinge, D., Ranjith, P. G., & Choi, S. (2011). Development of a reconstituted brown coal material using cement as a binder. *Environmental Earth Sciences, 64*, 631–641.
35. Chen, H. D., Cheng, Y. P., Zhou, H. X., et al. (2013). Damage and permeability development in coal during unloading. *Rock Mechanics and Rock Engineering, 46*(6), 1–14.
36. Chen, H. D. (2013). *Damage and permeability development of unloading coal body during mining the protective coal seam*. Xuzhou: China University of Mining and Technology (in Chinese).
37. Wang, J., Shao, T., & Zhao, H. (2011). Experimental study of effect of gas on mechanical properties of outburst coal. *Journal of Mining and Safety Engineering, 3*, 12.
38. Yin, G., Huang, Q., Zhang, D., et al. (2010). Test study of gas seepage characteristics of gas-bearing coal specimen during process of deformation and failure in geostress field. *Chinese Journal of Rock Mechanics and Engineering, 29*(2), 336–343 (in Chinese).
39. Yin, G. Z., Jiang, C. B., Li, X. Q., et al. (2011). An experimental study of gas permeabilities of outburst and nonoutburst coals under complete stress-strain process. *Rock and Soil Mechanics, 32*(6), 1613–1619.
40. Jiang, X. U., Qi, L. U., Xin, W. U., et al. (2011). The fractal characteristics of the pore and development of briquettes with different coal particle sizes. *Journal of Chongqing University, 9*, 013 (in Chinese).

41. Christiansson, R., & Hudson, J. (2003). ISRM suggested methods for rock stress estimation—Part 4: Quality control of rock stress estimation. *International Journal of Rock Mechanics and Mining Sciences, 40*, 1021–1025.
42. Cai, M. (2008). Influence of intermediate principal stress on rock fracturing and strength near excavation boundaries—Insight from numerical modeling. *International Journal of Rock Mechanics and Mining Sciences, 45*, 763–772.
43. Zhang, J., Bai, M., & Roegiers, J. C. (2003). Dual-porosity poroelastic analyses of wellbore stability. *International Journal of Rock Mechanics and Mining Sciences, 40*, 473–483.
44. Zhao, J. (2000). Applicability of Mohr-Coulomb and Hoek-Brown strength criteria to the dynamic strength of brittle rock. *International Journal of Rock Mechanics and Mining Sciences, 37*, 1115–1121.
45. Li, H., Xia, C., Yan, Z., et al. (2007). Study on marble unloading mechanical properties of Jinping hydropower station under high geostress conditions. *Chinese Journal of Rock Mechanics and Engineering, 26*, 2104–2109 (in Chinese).
46. Su, C., Zhai, X., Li, Y., et al. (2006). Study on deformation and strength of coal samples in triaxial compression. *Yanshilixue Yu Gongcheng Xuebao/Chinese Journal of Rock Mechanics and Engineering, 25*, 2963–2968 (in Chinese).
47. Liu, Q. S., Liu, K. D., Zhu, J. B., et al. (2014). Study of mechanical properties of raw coal under high stress with triaxial compression. *Chinese Journal of Rock Mechanics and Engineering, 33*, 24–34 (in Chinese).
48. You, M. (2014). Effect of confining pressure on strength scattering of rock specimen. *Chinese Journal of Rock Mechanics and Engineering, 33*, 929–937 (in Chinese).
49. Huang, R. Q., & Huang, D. (2014). Evolution of rock cracks under unloading Condition. *Rock Mechanics and Rock Engineering, 47*, 453–466.
50. Meng, L. (2013). *Research on coal damage characteristics and gas migration law of coal containing gas*. China University of Mining and Technology (in Chinese).
51. He, X. Q., Nie, B. S., Wang, E. Y., et al. (2007). Electromagnetic emission forecasting technology of coal or rock dynamic disasters in mine. *Journal of China Coal Society, 32*(1), 56–59 (in Chinese).
52. Astakhov, A. V., Khazov, S. P., & Ékonomova, L. N. (1992). Effect of capillary condensate of methane on electrophysical and strength properties of coal. *Journal of Mining Science, 28*, 231–234.
53. Yao, Y. P., & Zhou, S. N. (1988). The mechanical property of coal containing gas. *Journal of China University of Mining and Technology, 1*, 1–7 (in Chinese).
54. Robinson, L. (1959). The effect of pore and confining pressure on the failure process in sedimentary rock. In *The 3rd US Symposium on Rock Mechanics* (*USRMS*). American Rock Mechanics Association.
55. He, Y. N., Tao, L. J., & Cao, W. G. (2010). *Brief course of rock mechanics*. China University of Mining and Technology Press (in Chinese).
56. Cai, M. F. (2002). *Rock mechanics and rock engineering*. Science Press (in Chinese).
57. Karacan, C. Ö. (2007). Swelling-induced volumetric strains internal to a stressed coal associated with CO_2 sorption. *International Journal of Coal Geology, 72*, 209–220.
58. Dawson, G. K. W., Golding, S. D., Esterle, J. S., et al. (2012). Occurrence of minerals within fractures and matrix of selected Bowen and Ruhr Basin coals. *International Journal of Coal Geology, 94*, 150–166.
59. Karacan, C. Ö. (2003). Heterogeneous sorption and swelling in a confined and stressed coal during CO_2 injection. *Energy and Fuels, 17*, 1595–1608.
60. Huang, W., Shen, M., & Zhang, Q. (2010). Study of unloading dilatancy property of rock and its constitutive model under high confining pressure. *Chinese Journal of Rock Mechanics and Engineering, 29*(S2), 3475–3481.
61. Hu, Y. H. (2008). *Study on mechanical properties of granites under high pressure conditions and its constitutive models*. Institute of Rock and Soil Mechanics Chinese Academy of Sciences (in Chinese)

62. Liu, Y. B. (2009). *Study on the deformation and damage rules of gas-filled coal base on meso-mechanical experiments*. Chongqing University (in Chinese).
63. Zhao, H., & Yin, G. (2011). Study of acoustic emission characteristics and damage equation of coal containing gas. *Rock and Soil Mechanics, 32*(3), 667–671.
64. Gang, W. U., & Zhen, Y. Z. (1998). Acoustic emission character of rock materials failure during various stress states. *Chinese Journal of Geotechnical Engineering-Chinese Edition, 20*(2), 82–85 (in Chinese).
65. Li, T. C., & Lü, H. B. (2010). CT real-time scanning tests on single crack propagation under triaxial compression. *Chinese Journal of Rock Mechanics and Engineering, 29*(2), 289–296.
66. Cheng, Z., Zuo, Y. Z., & Ding, H. S. (2011). Application of CT technology in geotechnical mechanics. *Journal of Yangtze River Scientific Research Institute, 28*(3), 33–38 (in Chinese).
67. Kemeny, J. M., & Cook, N. G. W. (1986). *Crack models for the failure of rocks in compression* (pp. 1–18). Lawrence Berkeley Lab.
68. Sammis, C., & Ashby, M. (1986). The failure of brittle porous solids under compressive stress states. *Acta Metallurgica, 34*, 511–526.
69. Martin, C. D. (1993). *The strength of massive Lac du Bonnet granite around underground openings*. University of Manitoba Manitoba.
70. Huang, S. L. (2008). *Study on mechanical model of brittle rock under high stress condition and its engineering applications*. Institute of Rock and Soil Mechanics. Chinese Academy of Sciences (in Chinese)
71. Liu, Q. Q. (2015). *Damage and permeability evolution mechanism of dual-porosity coal under multiple stress paths and its application*. China University of Mining and Technology (in Chinese).

Chapter 7
Seepage Properties and Permeability Evolution Model of Coal

Learning Outcomes

- Basic concepts, theoretical basis, and derivation process of coal permeability;
- Theoretical basis and derivation of the classical permeability model;
- Test instrument, test principle, and methods of coal permeability;
- Competitive effect of swelling deformation and effective stress on permeability evolution;
- Evolution law of coal permeability under the effect of stress perturbation.

7.1 Basic Concepts and Theoretical Basis of Coal Permeability

7.1.1 *Basic Concepts of Coal Permeability*

Gas drainage technologies are the main means of lowering the gas content of coal seams and eliminating the gas hazard of coalmines. As China has a huge reservoir of CBM resources, the extraction and utilization of coal seam gas resources can realize the significant triple benefits of ensuring the safe mining of coal resources, promoting the clean and efficient use of coalmine gas, and protecting the environment. The key factor in controlling coalmine gas drainage or CBM recovery is coal permeability.

Seepage is an important attribute of coal seams, which reflects the ability of a coal body to allow fluid to pass through it. Permeability (k) is a parameter that characterizes the seepage properties of coal, and includes the absolute permeability, phase permeability, and relative permeability. The physical meanings of the three physical qualities are summarized as follows:

(1) If there is only one phase fluid existing in the pore/fracture system of coal, the ability of the coal body to allow the fluid to pass through is called the absolute permeability.

Y. Cheng et al., *Coal Mechanics*,
https://doi.org/10.1007/978-981-16-3895-4_7

(2) If there are several phase fluids existing in the pore/fracture system of coal, then the ability of the coal body allowing each phase fluid to pass through is called the phase permeability (otherwise known as the effective permeability).
(3) The ratio of the phase permeability to the absolute permeability is called the relative permeability.

In this book, the content concerning the permeability of coal seams mainly focuses on the absolute permeability. Hence, in the absence of special instructions, the absolute permeability of a coal seam is simplified as the "coal seam permeability". Permeability is an inherent property of porous media and is only related to the skeleton properties of porous media, such as the particle size, particle size distribution, particle packing, specific surface area, tortuosity, and porosity.

According to Darcy's law, the fluid velocity is proportional to the pressure gradient:

$$v = \frac{Q}{A} = KJ = K\frac{\Delta h}{l} \tag{7.1}$$

where v is the fluid velocity, m/s; Q is the fluid flow of porous media with a cross-sectional area A per unit of time, m^3/s; A is the cross-sectional area of the porous material, m^2; K is the seepage coefficient, m/s; J is the hydraulic gradient along the flow direction; Δh is the head difference along the flow direction, m; and l is the corresponding flow path of the head difference Δh, m.

The seepage coefficient is not only related to the physical properties of a given porous media, but also the properties of a given fluid. However, using the seepage coefficient to reflect the penetration of porous media is far from satisfactory, and it is hoped that the influence of porous media and fluid on permeability can be separated. To achieve this, Eq. (7.2) is proposed:

$$K = \frac{k\rho g}{\mu} \tag{7.2}$$

where k is the permeability of coal, m^2; ρ is the fluid density, kg/m^3; and μ is the fluid dynamic viscosity, Pa s.

Considering that the fluid flow direction is opposite to the pressure gradient, by substituting Eq. (7.2) into Eq. (7.1) we obtain Eq. (7.3):

$$v = -\frac{k}{\mu}\frac{\partial p}{\partial x} \tag{7.3}$$

where $\partial p/\partial x$ is the pressure gradient along the flow direction, Pa/m; p is the fluid pressure, Pa; and the minus sign indicates that the fluid flow direction is opposite to the pressure gradient.

The permeability parameter is widely used in underground engineering, petroleum and gas exploitation, and CBM recovery. The frequently-used units of permeability are "Darcy" (D) or "millidarcy" (mD), whose physical meaning can be described as follows. Under the viscous flow condition, when the pore space of a porous medium with a length of 1 cm and a cross-sectional area of 1 cm^2 is completely filled with a single-phase fluid with a viscosity of 10^{-3} Pa s (i.e., the viscosity of water at 20.2 °C), the permeability of this porous medium will be 1 D if a fluid flow of 1 cm^3/s can be determined under a pressure difference of 1 atm (101,325 Pa).

In China, the permeability coefficient of a coal seam (λ) is often used to characterize the difficulty of gas (mainly CH_4) flowing in a coal seam. As the coal seam gas is a type of compressible fluid, the flow velocity should be converted into the value under the standard condition; thus, introducing the isothermal condition formula of $pv = p_n v_n$ into Eq. (7.3), we obtain Eq. (7.4):

$$v_n = -\frac{k}{2\mu p_n}\frac{\mathrm{d}p^2}{\mathrm{d}x} = -\lambda\frac{\mathrm{d}p^2}{\mathrm{d}x} \tag{7.4}$$

where λ is the permeability coefficient of a coal seam, $m^2/(MPa^2\ d)$; p_n is atmospheric pressure under the standard condition, 0.101325 MPa; and v_n is the flow velocity of gas under the standard condition, m/d.

The physical significance of the permeability coefficient represents the daily amount of gas (temperature in Celsius (°C) under the standard condition) flowing through a coal block with a length of 1 m and coal face area of 1 m^2, under the conditions of a square pressure difference of 1 MPa^2 between two sides of a 1 m^3 coal block. The conversion relationship between permeability and the permeability coefficient is shown in Table 7.1.

The field-measured values of the permeability coefficient and the converted permeability for some of China's coalmines are summarized in Table 7.2.

Owing to the importance of coal permeability in coal seam gas drainage and CBM recovery, Chinese and overseas scholars have carried out many investigations. From the perspective of the research scale, studies on coal permeability can be divided into two categories: one is the measurement of the permeability coefficient at the field scale, and the other is the research on the physical simulation and permeability model at the laboratory scale. The permeability coefficient is an indirect measurable parameter of coal seam, and the most commonly used method is the "drilling flow method" proposed by Zhou [1–4]. In the field test, the "drilling flow method" can be further subdivided into the following three methods, and the specific choice of method should be based on the influence scope of the disturbance zone [2].

(1) In general, the one-way flow method is suitable for coal seams with a high-strength structure and many developed fractures; however, only when the

Table 7.1 Conversion relationship between permeability and the permeability coefficient

Parameter	Permeability				Permeability coefficient λ
	D	mD	m^2	μm^2	$m^2/(MPa^2\ d)$
Value	1	1000	9.869233×10^{-13}	0.9869233	40,000

Table 7.2 Permeability coefficient and converted permeability for some coalmines in China

Coalmine	Coal seam	λ (m^2/(MPa^2 d)	*k*/mD	Coalmine	Coal seam	λ (m^2/(MPa^2 d))	*k*/mD
Fushun Longfeng Mine	Ben seam	140–150	3.5–3.75	Baishaliwangmiao Mine	6	0.12–0.32	(0.3–0.8) × 10^{-2}
Fushun Shengli Mine	Ben seam	29.6–36.8	0.74–0.92	Baishatanjiachong Mine	6	0.24–0.47	(0.6–1.18) × 10^{-2}
Yangquan No. 1 Mine	3	0.019	0.475 × 10^{-3}	Lianshaolixin Mine	4	1.10	0.0275
Beipiao Taiji Mine	10	0.0028–0.004	(0.7–1) × 10^{-4}	Liuzhihuachu Mine	7	0.0178	0.445 × 10^{-3}
Beipiao Taiji Mine	4	0.006	0.16 × 10^{-3}	Liuzhidayong Mine	7	0.0862	2.15 × 10^{-3}
Beipiao Taiji Mine	3	0.0144	0.36 × 10^{-3}	Baotou Hetangou Mine	G	11.3–17.4	0.28–0.44
Beipiao Sanbao Mine	9B	0.039	0.975 × 10^{-3}	Hebi No. 6 Mine	Damei	1.2–1.8	0.03–0.045
Jiaozuo zhucun Mine	Damei	0.55–3.6	0.013–0.09	Hegang Nanshan Mine	15	30.8–62.4	0.77–1.56
Zhongliangshanbei Mine	K1	0.64–0.68	(1.61–1.7) × 10^{-2}	Anyang Longshan Mine	Damei	0.7–1.0	0.018–0.025
Tianfu moxinpo Mine	9	0.042–0.14	(1.06–3.5) × 10^{-3}	Shenyang Hongleng Mine	12	0.02–0.06	(0.5–1.5) × 10^{-3}
Huainan Xie No. 1 Mine	B11	0.092	2.3 × 10^{-3}	Pingdingshan No.8 Mine	Yi_{15}	0.076	1.9 × 10^{-3}

(continued)

Table 7.2 (continued)

Coalmine	Coal seam	λ ($m^2/(MPa^2$ d)	*k*/mD	Coalmine	Coal seam	λ ($m^2/(MPa^2$ d))	*k*/mD
Huainan Xie No. 2 Mine	C13	0.135	3.37×10^{-3}	Huaibei Luling Mine	Houmei	0.028	0.7×10^{-3}
Furongbaijiao Mine	K3	0.20	5×10^{-3}	Shizuishan No. 1 Mine		0.087	2.18×10^{-3}
Hongwei Tanjiachong	6	0.472	1.18×10^{-2}	Shenyang Hongleng Mine	12	0.014	3.5×10^{-4}
Huainan Pan No.1 Mine	C13	0.01135	2.8375×10^{-4}	Huainan Lier Mine	B8	0.1717	4.29×10^{-3}
Huainan Xinzhuangzi Mine	B4	0.0445	1.11×10^{-3}	Huainan Xinzhuangzi Mine	B7	0.0586	1.47×10^{-3}

exposed area of coal seams is sufficiently large can it meet the requirements of the hypothesis of the one-way flow theory.

(2) When the permeability of a coal seam is mainly related to microfractures, the actual state of the gas flow field is consistent with the basic assumption of the radial flow theory. In this case, the radial flow method is more suitable.

(3) For a soft coal seam, its strength is low, the scale of the coal matrix block is small, and it is greatly affected by the disturbance of boreholes; hence, the spherical flow method is more appropriate for such situations.

At the laboratory scale, physical simulation experiments have been generally used to study the influencing factors of coal permeability from various perspectives. For example, Harpalani [5, 6], and Mazumder et al. [7] investigated the evolution law of coal permeability under the combined action of effective stress and sorption-induced deformation by experimentally simulating the gas pressure change in the CBM recovery process. Chen et al. [8, 9] and Pan et al. [10] used adsorbing and non-adsorbing gases to test the effective stress coefficient of coal samples, and further discussed the single impact of effective stress on the evolution of coal permeability. Seidle and Huitt [11], Durucan and Huitt [12], Robertson [13, 14], Van Bergen [15], Ranjith [16], and Battistutta [17] studied the swelling deformation characteristics of coal induced by gas adsorption as well as its influence on coal permeability. Robertson developed an optical measuring device to accurately determine the adsorption swelling of coal. By using the device, the swelling deformation induced by different gases such as CH_4, CO_2, N_2, and He were tested, and through the experimental results, three common permeability models were corrected. Considering the stress states of in situ coal reservoirs, Mitra [5], Liu [18], and Perera [19] tested the change of coal permeability with pore pressure and confining pressure under the uniaxial strain condition, and then corrected common permeability models based on their experimental results. Klinkenberg [20], Wu and Karsten [21] studied the influence of the Klinkenberg effect on the permeability of low-permeability fractured coal, while Dawson and Eeterle [22] determined the relationships between the permeability of differently ranked coal samples and their cleat system (cleat spacing, cleat height, and strip characteristics of coal samples). In addition, Perera [23] conducted an experiment to reveal the influence of temperature on coal permeability.

In the 1990s, Xian, Zhou, and Jiang et al. [24–27] discussed the relationships between coal permeability and the factors of geo-stress, pore pressure, temperature, and electric field parameters, and established the corresponding empirical formula and theoretical model. Lin and Zhou [28] studied the relationship between pore pressure, permeability, and sorption-induced deformation under a constant confining pressure condition, and indicated that exponential relationships existed between pore pressure and permeability as well as pore pressure and sorption-induced deformation. Yang et al. [29] and Hu et al. [30] experimentally investigated the effect of gas adsorption and coal deformation on coal permeability, and obtained a parabolic trend between coal permeability and pore pressure. Qin et al. [31] and Fu et al. [32] tested the evolution of coal permeability with a change in confining pressure under a constant effective stress condition or a constant gas (CH_4) pressure condition, and established mathematical prediction models of the variation of coal permeability with the surface density, occurrence, and aperture evolution of fractures, respectively. Sun and Ling [33] studied the change law of coal permeability during the deformation process by changing the confining pressure or pore pressure, and obtained an empirical equation of coal permeability variation with the confining pressure or pore pressure. Wu [34] and Li et al. [35] established a permeability model considering the adsorption deformation of the coal skeleton. Yin et al. [36–41] and Xu et al. [42–44] performed many experimental studies on the permeability of coal containing gas under different mechanical paths, and also analyzed the influence of effective stress, adsorption deformation, coal damage, temperature, and other factors on coal permeability. Peng et al. [45] constructed a permeability model considering the scale effect of coal, while Xie et al. [46] established a theoretical expression of coal permeability under the coupling effect of abutment pressure, pore pressure, and sorption-induced deformation. Chen et al. [47, 48] performed CT scanning tests and permeability experiments using reconstituted coal specimens subjected to the unloading stress path, which was similar to the protective seam mining process, and also studied the coal damage and permeability evolution characteristics during the unloading process. In addition, Pan [49] tested the evolution of coal permeability during the unloading process using raw coal samples with different bedding plane directions (vertical, parallel, and oblique crossing), and then obtained a mining-enhanced permeability model that showed a sudden change characteristic of coal permeability.

From the above review concerning the current research status of physical simulation experiments of coal permeability, it can be found that researchers have always performed their test conditions on the specified engineering background of a given study area. Among them, scholars have mainly considered the evolution of coal reservoir permeability during CBM recovery as the research object, focusing on the effects of adsorption-induced deformation and effective stress on the reservoir permeability under the original stress state. Apart from the above-mentioned research hotspot, Chinese scholars are also interested in the evolution characteristics of coal permeability under the mining-induced unloading condition. Based on the research findings regarding the stress evolution law during actual underground engineering, Chinese scholars have performed a number of physical simulation experiments under different loading/unloading stress paths in laboratories.

7.1.2 Theoretical Basis of Permeability Evolution

In 1880, Seelheim proposed that the permeability of porous media is related to the square value of some characteristic parameters of the fracture (pore) diameter [50]. Since then, many empirical and semi-empirical equations have been established to predict the permeability of porous media. Among them, the most famous is the Kozeny–Carman (KC) equation, which was proposed by Kozeny [51] in 1927 and later modified by Carman [52]. Because the fracture (pore) structures and connectivity of porous media such as coal, petroleum, and gas reservoirs are very complex, it is generally very difficult to establish a strict mathematical model to accurately describe the relationships between permeability and fracture (pore) features. Hence, the KC equation is still widely used in many fields, including underground seepage, petroleum and gas exploration, chemical engineering, biochemistry, and electrochemistry.

In the following section, the main experimental basis and derivation process of the KC equation are introduced. As shown in Fig. 7.1, the water seepage test involves filling the bottom of a cylindrical container with fine compacted sand to a certain thickness. Then, water enters the container from the top, flows through the fine sand layer under the action of gravity, and finally flows out from the bottom.

The seepage process of water through the fine sand layer is schematically illustrated in Fig. 7.2, whereby water flows through the penetrated fractures under the action of gravity. To describe the migration characteristics of water in fractures, the fracture structure of porous media and the physical process of water seepage need to be further determined and simplified. At present, the most widely used simplified model is the capillary bundle model, in which the fractures are regarded as a cylindrical seepage region, and the surrounding material is the skeleton of porous media without seepage activity (Fig. 7.3).

In the case of existing multiple fractures, the simplified physical model of a capillary bundle is presented in Fig. 7.4.

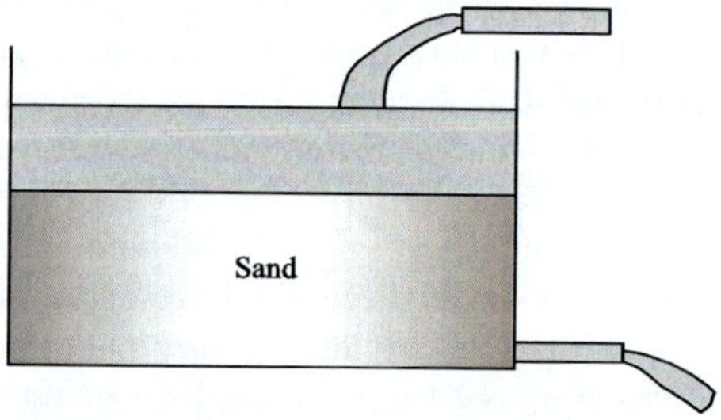

Fig. 7.1 Water seepage test in fine sand layer

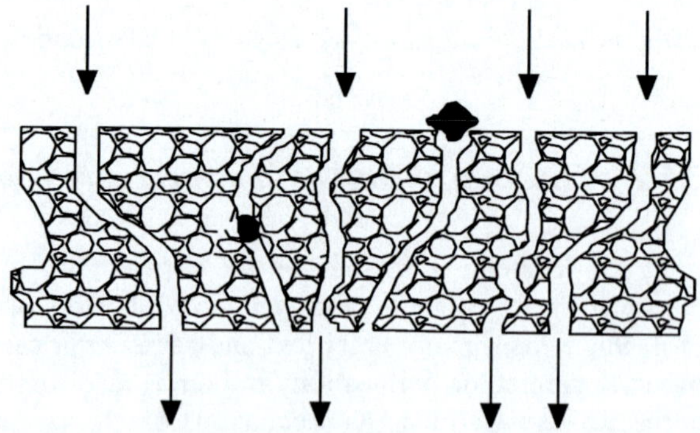

Fig. 7.2 Simplified physical model of a water seepage test through a fine sand layer

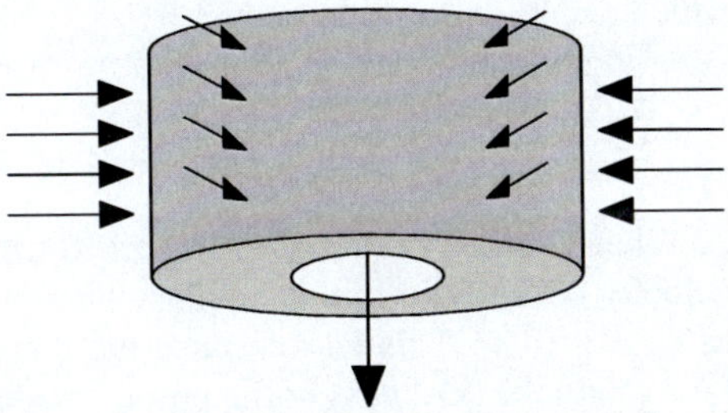

Fig. 7.3 Three-dimensional capillary bundle model

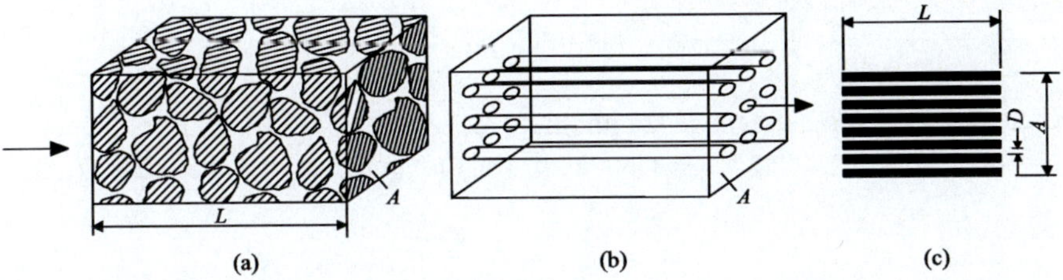

Fig. 7.4 Simplified physical model of porous media. **a** Porous media, **b** simplified three-dimensional capillary bundle model, and **c** simplified two-dimensional capillary bundle model

For the convenience of deriving the KC equation, the three-dimensional capillary bundle model (Fig. 7.4b) is further simplified into the two-dimensional capillary bundle model (Fig. 7.4c), where L is the thickness (length) of the porous media layer, D is the diameter of the capillary bundle, A is the top and bottom area of the porous media layer, and A' is the top and bottom area of the n group capillary bundle. With the above settings, the fracture porosity (ϕ_f) of the porous media layer can be expressed as Eq. (7.5):

$$\phi_f = \frac{A'}{A} \tag{7.5}$$

where ϕ_f is the fracture porosity of the porous media, %.

Assuming Q is the water quantity flowing out from the bottom of the porous media layer per unit of time, we obtain Eq. (7.6):

$$Q = Av = A'v' \tag{7.6}$$

where v is the average velocity of water flowing through the bottom surface of the porous media, m/s, and v' is the average velocity of water flowing through the bottom surface of the capillary bundle, m/s.

Hence, it can be determined that:

$$v' = \frac{Av}{A'}, \tag{7.7}$$

which can be expressed as:

$$v' = \frac{v}{\phi_f} \tag{7.8}$$

When Carman [52] tried to correct Kozeny's equation, he realized that the actual velocity of the fluid in the capillary bundle was greater than the results calculated using Eq. (7.8). This was because the seepage length in the capillary bundle was larger than the thickness of the porous media layer (see Fig. 7.2); therefore,

$$v' = \frac{vl'}{\phi_f l} \tag{7.9}$$

where l' is the average seepage length, m.

Meanwhile, according to the concept of fracture porosity, we obtain Eq. (7.10):

$$\phi_f = \frac{V_f}{V} = \frac{V_f}{V_f + V_s} \tag{7.10}$$

where V is the total volume of the porous medium, m^3, V_f is the volume of the fractures and pores, m^3, and V_s is the skeleton volume of the porous medium, m^3.

According to the relationships among the above three parameters, the following relationships can be derived:

$$V_f = V\phi_f \tag{7.11}$$

$$V_s = V(1 - \phi_f) \tag{7.12}$$

$$V_f + V_s = \frac{V_s}{(1 - \phi_f)} \tag{7.13}$$

$$V_f = \frac{\phi_f V_s}{(1 - \phi_f)} \tag{7.14}$$

Defining S as the surface area of the capillary bundle, we obtain the follows:

$$S = n\pi D l' \tag{7.15}$$

$$V_f = \frac{n\pi D^2 l'}{4} = \frac{SD}{4} \tag{7.16}$$

$$D = \frac{4\phi_f V_s}{S(1 - \phi_f)} \tag{7.17}$$

where n is the number of capillary bundles, and D is the diameter of the capillary bundle, m.

Because S is also the surface area of the solid skeleton, then,

$$S = \pi d_s^2 \tag{7.18}$$

$$V_s = \frac{\pi d_s^3}{6} \tag{7.19}$$

where d_s is the equivalent mean diameter of the skeleton particles, m.

Substituting Eqs. (7.18) and (7.19) into Eq. (7.17), we obtain Eq. (7.20):

$$D = \frac{2\phi_f d_s}{3(1 - \phi_f)} \tag{7.20}$$

In addition, according to the Poiseuille equation, for a porous material layer with a thickness of l, when the pressure difference between the top and bottom is Δp, we obtain Eq. (7.21) [20]:

$$\frac{\Delta p}{l} = -32\frac{\mu v'}{D^2} \tag{7.21}$$

Substituting Eq. (7.9) into Eq. (7.21), and transforming Eq. (7.21) into the form of differential equations, we obtain Eq. (7.22):

$$\frac{\mathrm{d}p}{\mathrm{d}x} = -32\frac{\mu v l'}{\phi_f l D^2} \tag{7.22}$$

Substituting Eq. (7.20) into Eq. (7.22) yields the following:

$$\frac{\mathrm{d}p}{\mathrm{d}x} = -72\frac{\mu v l'(1 - \phi_f)^2}{l d_s^2 \phi_f^3} \tag{7.23}$$

After mathematical treatment, we obtain Eq. (7.24):

$$v = -\frac{\frac{l d_s^2 \phi_f^3}{72 l'(1-\phi_f)^2}}{\mu}\frac{\mathrm{d}p}{\mathrm{d}x} \tag{7.24}$$

According to Darcy's law, it can be determined that

$$v = -\frac{k}{\mu}\frac{\mathrm{d}p}{\mathrm{d}x} \tag{7.25}$$

Hence,

$$k = \frac{l d_s^2 \phi_f^3}{72 l'(1-\phi_f)^2} \tag{7.26}$$

For the permeability evolution caused by the gas–solid coupling effect, the parameters of l' and d_s are almost unchanged, if there are no newly created fractures.

$$\frac{k}{k_0} = \frac{\frac{l d_s^2 \phi_f^3}{72 l'(1-\phi_f)^2}}{\frac{l d_s^2 \phi_{f0}^3}{72 l'(1-\phi_{f0})^2}} = \left(\frac{\phi_f}{\phi_{f0}}\right)^3 \frac{(1-\phi_{f0})^2}{(1-\phi_f)^2} \tag{7.27}$$

Because the fracture porosity is considerably less than 1, Eq. (7.27) can be further simplified into Eq. (7.28):

$$\frac{k}{k_0} = \left(\frac{\phi_f}{\phi_{f0}}\right)^3 \tag{7.28}$$

Equation (7.28) is the theoretical basis for the derivation of the permeability evolution caused by the gas–solid coupling effect, which is known as the "Cubic law".

7.2 Typical Models of Permeability Evolution

Chinese and overseas scholars place a different emphasis on the physical simulation tests of coal permeability, which is also reflected in research on coal permeability models. As shown in Fig. 7.5, the coal reservoir is located in the original stress state during the CBM exploitation process, and does not experience a complex conversion of the stress state. The deformation caused by the change in effective stress is caused by elastic deformation; therefore, the permeability of coal is controlled by the opening and closure of the original fractures under the influence of effective stress and sorption-induced coal deformation. Hence, overseas scholars have established many permeability models to describe the dynamic evolution process of coal permeability during CBM exploitation. However, when Chinese scholars try to establish the evolution models of coal permeability, the conversion of the complex stress state must be considered because plastic deformation will occur in the coal body, and even enter the post-peak failure stage, under the effect of mining disturbance stress. Unfortunately, there is a lack of ideal theoretical tools to handle the relationship between the mechanical response and volumetric response during the extension of original fractures or the formation of new fractures. Therefore, in general, the coal permeability models proposed by Chinese scholars are usually obtained by fitting test data with some functional relationships.

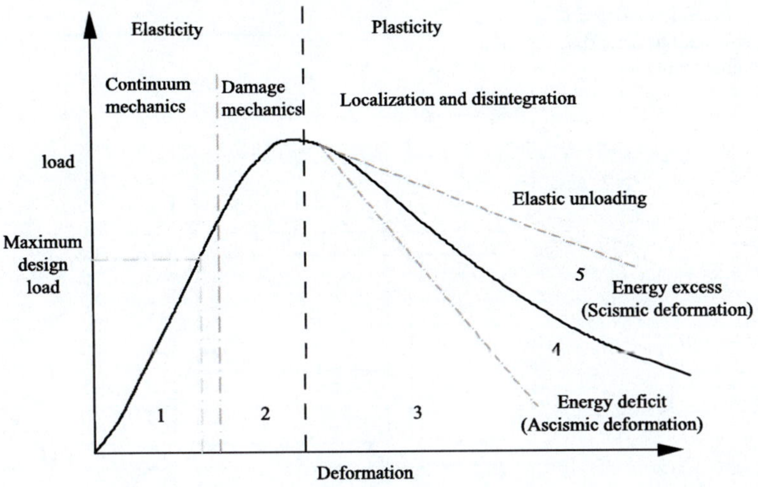

Fig. 7.5 Typical complete stress–strain curve [56]

Coal is a complex dual-porosity material with abundant pores and fractures. To establish a permeability model, and thus quantitatively analyze the effects of sorption-induced swelling deformation and effective stress on the evolution of coal permeability, the physical structure of the coal body needs to be simplified. At present, the matchstick model and cubic geometry model are the most commonly accepted simplified models (Fig. 7.6) [53–55].

In theoretical research and engineering practice, fracture porosity is often used to describe the development degree of coal fractures. The flow of gas in a coal seam is in accordance with Darcy's law, and fracture porosity is also often associated with coal permeability in research.

For the convenience of formula derivation, the cubic geometry model shown in Fig. 7.6b is transformed into the schematic diagram in Fig. 7.7, where *a* represents

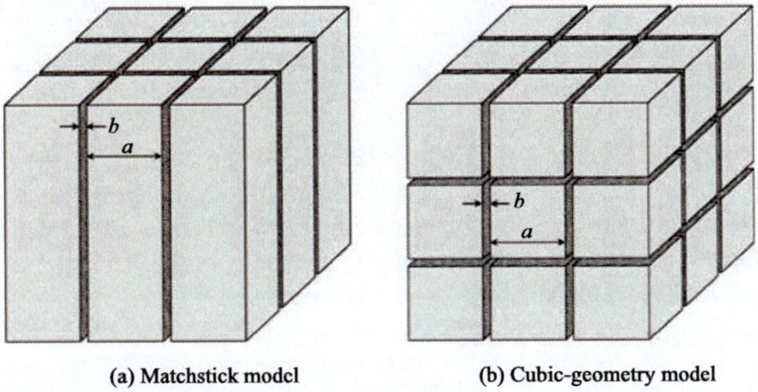

Fig. 7.6 Simplified models of the physical structure of coal

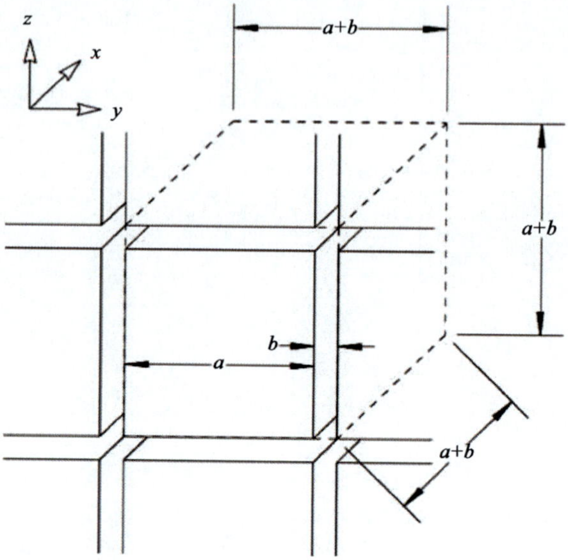

Fig. 7.7 Schematic diagram for the calculation using the cubic geometry model

the length of the side of the coal matrix block, and b represents the width of the coal fractures. The fracture porosity (ϕ_f) is the ratio of the fracture volume of a coal body to the volume (apparent volume) of coal within it:

$$\phi_f = \frac{V_f}{V_t} = \frac{V_t - a^3}{V_t} = \frac{(a+b)^3 - a^3}{(a+b)^3} = \frac{3a^2b + 3ab^2 + b^3}{a^3 + 3a^2b + 3ab^2 + b^3} \tag{7.29}$$

Owing to the relation of $b \ll a$, we can obtain Eq. (7.30):

$$\phi_f = \frac{3a^2b}{a^3 + 3a^2b} = \frac{3b}{a + 3b} \tag{7.30}$$

Similarly, because $b \ll a$, Eq. (7.30) can be further simplified to Eq. (7.31):

$$\phi_f = \frac{3b}{a} \tag{7.31}$$

For $b \ll a$, the influence of the fracture intersection on gas flow can be neglected. Assuming that the direction of gas flow in the coal body is along the X direction (perpendicular to the surface), the corresponding gas flow can be considered to flow in a smooth flat rectangular tube. The laminar flow velocity of the gas in the rectangular tube is expressed as follows:

$$v = \frac{b^2}{12\mu}\frac{\Delta p}{L} \tag{7.32}$$

where Δp is the pressure difference between the inlet and outlet of a fracture, MPa; and L is the length of the rectangular tube, m.

For a rectangular tube, the relationship between the flow rate and flow velocity is given by Eq. (7.33):

$$q = vA_f = vab \tag{7.33}$$

By substituting Eq. (7.32) into Eq. (7.33), we obtain Eq. (7.34):

$$q = \frac{ab^3}{12\mu}\frac{\Delta p}{L} \tag{7.34}$$

In the cubic geometry model, the flow in a certain direction is controlled by two sets of fractures, which are mutually perpendicular. Assuming that there are n matrix blocks in a cross-section of coal, the flow rate of the gas through the cross section should be twice that of the gas through the flat plate, namely:

$$Q = \frac{2nab^3}{12\mu}\frac{\Delta p}{L} = \frac{nab^3}{6\mu}\frac{\Delta p}{L} \tag{7.35}$$

According to Darcy's law, the flow rate of the gas through the cross-section of coal is expressed as:

$$Q = \frac{kA}{\mu}\frac{\Delta p}{L} \tag{7.36}$$

where A is the sectional area, which can be expressed as:

$$A = n(a+b)^2 = n\left(a^2 + 2ab + b^2\right) \tag{7.37}$$

Owing to the relationship $b \ll a$, $2ab + b^2$ can be neglected; thus, we obtain Eq. (7.38):

$$A = na^2 \tag{7.38}$$

Substituting Eq. (7.38) into Eq. (7.36) yields:

$$Q = \frac{kna^2}{\mu}\frac{\Delta p}{L} \tag{7.39}$$

Then, by combing Eqs. (7.35) and (7.39), we obtain:

$$\frac{kna^2}{\mu}\frac{\Delta p}{L} = \frac{nab^3}{6\mu}\frac{\Delta p}{L} \tag{7.40}$$

$$k = \frac{b^3}{6a} \tag{7.41}$$

Then, substituting Eq. (7.31) into Eq. (7.41) yields:

$$k = \frac{a^2}{162}\phi_f^3 \tag{7.42}$$

Equation (7.42) is the relationship between the permeability and porosity of coal in the cubic geometry model.

For the matchstick model, fracture porosity can be obtained using a similar derivation method as:

$$\phi_f = \frac{2b}{a} \tag{7.43}$$

Then, the permeability of the matchstick model is expressed as Eq. (7.44):

$$k = \frac{b^3}{12a} = \frac{a^2}{96}\phi_f^3 \tag{7.44}$$

Equation (7.44) is the relationship between coal permeability and fracture porosity in the matchstick geometry model.

According to Eqs. (7.42) and (7.44), we obtain:

$$\frac{k}{k_0} = \left(\frac{\phi_f}{\phi_{f0}}\right)^3\left(\frac{a}{a_0}\right)^2 \tag{7.45}$$

Because the sorption-induced coal matrix deformation is very small ($a \approx a_0$), the square term in Eq. (7.44) is approximately equal to 1, and Eq. (7.45) can be simplified as:

$$\frac{k}{k_0} = \left(\frac{\phi_f}{\phi_{f0}}\right)^3 \tag{7.46}$$

Equation (7.46) is the commonly used cubic law in coal permeability studies. From the functional relationship between coal permeability and fracture porosity obtained from the derivation of the cubic geometry and matchstick geometry, it is observed that the permeability and fracture porosity still meet the cubic relationship, which indicates that simplifying the physical structure of coal based on the cubic model or matchstick model is feasible with respect to the relationship between coal permeability and fracture porosity.

Corresponding to the physical simplified models of coal, two boundary conditions are also established by simplifying the stress state of coal reservoirs, namely, the uniaxial strain condition and the triaxial stress condition [18, 57].

(1) Under the uniaxial strain condition, the lateral strain is zero, while the vertical strain is not zero. Taking CBM exploitation as an example, with decreasing gas pressure, deformation will occur in both the vertical and horizontal directions of a coal seam. Because the horizontal scale of a coal seam is much larger than its vertical scale, from the viewpoint of strain, the horizontal deformation can be neglected (i.e., the lateral strain tends to zero).

(2) The stress state of the triaxial stress condition is easy to understand. It indicates that a coal body suffers stress from three mutually perpendicular directions.

Under these two simplified conditions, scholars have established many coal permeability models. The representative permeability models are introduced in the following subsections.

7.2.1 Palmer–Mansoori Model

Using CBM exploitation as the research background, Palmer and Mansoori [58] assumed that the underground coal seam is located in the stress state of uniaxial strain and constant vertical stress, and regarded adsorption-induced swelling as thermal expansion. In addition, the authors suggested that the fracture porosity change is the result of sorption-induced deformation and effective stress. Based on these assumptions and the poroelasticity theory, they developed a permeability model (the Palmer–Mansoori (PM) model) considering the desorption-induced shrinkage of coal and the fracture compression effect.

In the PM model, coal is composed of fractures and coal matrix. Thus, the volume of coal blocks (V_t) and fracture porosity (ϕ_f) can be expressed as Eqs. (7.47) and (7.48), respectively:

$$V_t = V_f + V_g \tag{7.47}$$

$$\phi_f = \frac{V_f}{V_t} \tag{7.48}$$

where V_f is the volume of the coal fracture, m^3; and V_g is the volume of the coal matrix, m^3.

Under the effects of external stress, the strain of the coal fractures can be expressed as:

$$\mathrm{d}\varepsilon_f = \frac{\mathrm{d}V_f}{V_f} = \frac{\mathrm{d}V_t}{V_f} - \frac{\mathrm{d}V_g}{V_f} = \frac{\mathrm{d}V_t/V_t}{V_f/V_t} - \frac{\mathrm{d}V_g/V_g}{V_f/V_g} = \frac{\mathrm{d}\varepsilon_t}{\phi_f} - \left(\frac{1-\phi_f}{\phi_f}\right)\mathrm{d}\varepsilon_g \tag{7.49}$$

where ε_t is the bulk strain of the coal, and ε_g is the strain of the coal matrix.

Assuming that the strain is infinitesimal, then under uniaxial strain conditions, we obtain Eq. (7.50) [58]:

$$-\mathrm{d}\phi_f = \left[\frac{1}{M} - (1-\phi_f) f\gamma\right](\mathrm{d}\sigma - \mathrm{d}p) + \left[\frac{K}{M} - (1-\phi_f)\right]\gamma \mathrm{d}p - \left[\frac{K}{M} - (1-\phi_f)\right]\alpha_T \mathrm{d}T \quad (7.50)$$

where M is the constrained axial modulus $M = \frac{(1-\upsilon)E}{(1+\upsilon)(1-2\upsilon)}$, MPa; K is the bulk modulus $\frac{K}{M} = \frac{1}{3}\left[\frac{1+\upsilon}{1-\upsilon}\right]$, MPa; E is the elastic modulus, MPa; f is a coefficient ranging from 0 to 1; υ is Poisson's ratio; $\mathrm{d}T$ is the temperature gradient, °C; α_T is the thermal expansion coefficient, °C; and γ is the compressibility factor of the coal matrix, Pa^{-1}.

For porosity $\phi_f \ll 1$ (actually, the porosity of coal is generally only 0.04–1.50%), as is the case in coal seams, there is no change in overburden stress ($\mathrm{d}\sigma = 0$). Hence, Eq. (7.50) can be simplified as:

$$-\mathrm{d}\phi_f = -\frac{1}{M}\mathrm{d}p + \left[\frac{K}{M} + f - 1\right]\gamma \mathrm{d}p - \left[\frac{K}{M} - 1\right]\alpha_T \mathrm{d}T \quad (7.51)$$

In Eq. (7.51), the term $\alpha_T \mathrm{d}T$ represents thermal expansion/contraction. If the temperature drops, the medium shrinks and the fractures increase in width. This is analogous to matrix shrinkage, where the fracture width increases as gas desorbs during pressure drawdown. The volumetric strain of coal is associated with gas adsorption, and desorption can be modeled using a Langmuir-type equation. According to the principle of similitude, the volumetric strain increment can be expressed as:

$$\alpha_T \mathrm{d}T \equiv \frac{\mathrm{d}}{\mathrm{d}p}\left(\frac{\varepsilon_l \beta p}{1+\beta p}\right)\mathrm{d}p \quad (7.52)$$

where ε_l is the ultimate expansion strain of the coal matrix, and β is a coefficient.

Thus, Eq. (7.51) can be rewritten as Eq. (7.53) or Eq. (7.54):

$$-\mathrm{d}\phi_f = -\frac{1}{M}\mathrm{d}p + \left[\frac{K}{M} + f - 1\right]\gamma \mathrm{d}p - \left[\frac{K}{M} - 1\right]\frac{\mathrm{d}}{\mathrm{d}p}\left(\frac{\varepsilon_l \beta p}{1+\beta p}\right)\mathrm{d}p \quad (7.53)$$

$$\left.\begin{aligned} \mathrm{d}\phi_f &= c_m \mathrm{d}p + \varepsilon_l\left[\tfrac{K}{M} - 1\right]\tfrac{\mathrm{d}}{\mathrm{d}p}\left(\tfrac{\beta p}{1+\beta p}\right)\mathrm{d}p \\ c_m &= \tfrac{1}{M} - \left[\tfrac{K}{M} + f - 1\right]\gamma \end{aligned}\right\} \quad (7.54)$$

After integrating Eq. (7.54) and dividing by ϕ_{f0}, we obtain Eq. (7.55):

$$\frac{\phi_f}{\phi_{f0}} = 1 + \frac{c_m}{\phi_{f0}}(p - p_0) + \frac{\varepsilon_l}{\phi_{f0}}\left(\frac{K}{M} - 1\right)\left(\frac{\beta p}{1+\beta p} - \frac{\beta p_0}{1+\beta p_0}\right) \quad (7.55)$$

Then, substituting Eq. (7.31) into Eq. (7.41) yields:

$$\frac{k}{k_0} = \left[1 + \frac{c_m}{\phi_{f0}}(p - p_0) + \frac{\varepsilon_l}{\phi_{f0}}\left(\frac{K}{M} - 1\right)\left(\frac{\beta p}{1 + \beta p} - \frac{\beta p_0}{1 + \beta p_0}\right)\right]^3 \tag{7.56}$$

In 2009, Palmer et al. [59] revised the original model to account for the influence of the fracture spreading orientation of coal as:

$$c_m = \frac{g}{M} - \left[\frac{K}{M} + f - 1\right]\gamma \tag{7.57}$$

where g is a geometric term related to the orientation of the coal fracture system, and ranges from 0 to 1.

7.2.2 Shi–Durucan Model

Similar to the PM model, Shi and Durucan [60, 61] also used CBM exploitation as a research background to develop the Shi–Durucan (SD) model. However, the authors suggested that coal permeability is mainly controlled by the horizontal stress under uniaxial strain conditions.

Assuming that coal is a homogenous, isotropic, thermoelastic porous medium, the stress–strain relationship of coal can be expressed as Eq. (7.58):

$$\left.\begin{aligned} \Delta\sigma_{ij} &= 2G\Delta\varepsilon_{ij} + \lambda\Delta\varepsilon_v\delta_{ij} + \left(\lambda + \tfrac{2}{3}G\right)\alpha_T\Delta T\delta_{ij} \\ \Delta\varepsilon_v &= \Delta\varepsilon_{xx} + \Delta\varepsilon_{yy} + \Delta\varepsilon_{zz} \end{aligned}\right\} \tag{7.58}$$

where G is the shear modulus; λ is the Lame constant; δ_{ij} is the Kronecker delta (when $i = j$ then $\delta_{ij} = 1$, otherwise $\delta_{ij} = 0$); and ε_v is the volumetric strain of coal.

In a non-isothermal hot-pore elastomer, a temperature drop results in skeleton shrinkage, leading to an increase in the porosity of the porous medium, which is analogous to the desorption-induced matrix shrinkage in coal seams. By replacing the thermal contraction term with the matrix shrinkage term, the constitutive relationship for an isothermal gas desorption process of coal can be written as:

$$\Delta\sigma_{ij} = 2G\Delta\varepsilon_{ij} + \lambda\Delta\varepsilon_v\delta_{ij} + \left(\lambda + \frac{2}{3}G\right)\Delta\varepsilon_s\delta_{ij} \tag{7.59}$$

where ε_s is the shrinkage volumetric strain of the coal matrix induced by gas desorption in the coal.

Because the shear stress components are unaffected by thermal contraction/matrix shrinkage, we only discuss the constitutive relationship of the three normal stress components, that is,

$$\left.\begin{aligned}\Delta\sigma_{xx} &= 2G\Delta\varepsilon_{xx} + \lambda\Delta\varepsilon_v + \left(\lambda + \tfrac{2}{3}G\right)\Delta\varepsilon_s \\ \Delta\sigma_{yy} &= 2G\Delta\varepsilon_{yy} + \lambda\Delta\varepsilon_v + \left(\lambda + \tfrac{2}{3}G\right)\Delta\varepsilon_s \\ \Delta\sigma_{zz} &= 2G\Delta\varepsilon_{zz} + \lambda\Delta\varepsilon_v + \left(\lambda + \tfrac{2}{3}G\right)\Delta\varepsilon_s\end{aligned}\right\} \tag{7.60}$$

Under uniaxial strain conditions (i.e., $\Delta\varepsilon_{xx} = \Delta\varepsilon_{yy} = 0$),

$$\Delta\sigma_{xx} = \Delta\sigma_{yy} = \frac{\upsilon}{1-\upsilon}\Delta\sigma_{zz} + \frac{E\Delta\varepsilon_s}{3(1-\upsilon)} \tag{7.61}$$

Assuming that the overburden stress of the coal seam remains unchanged during the CBM exploitation process, then $\Delta\sigma_{zz} = -p$. By substituting this into Eq. (7.61), the change in the effective horizontal stress is expressed as:

$$\sigma - \sigma_0 = -\frac{\upsilon}{1-\upsilon}(p - p_0) + \frac{E\varepsilon_l}{3(1-\upsilon)}\left(\frac{p}{p+P_L} - \frac{p_0}{p_0+P_L}\right) \tag{7.62}$$

The relationship between fracture permeability and stress can be described by an exponential equation as [62]:

$$k = k_0 \mathrm{e}^{-3c_f(\sigma-\sigma_0)} \tag{7.63}$$

where c_f is the volume compressibility of fracture.

Substituting Eq. (7.62) into Eq. (7.63), the coal fracture permeability can be described by Eq. (7.64):

$$\frac{k}{k_0} = \mathrm{e}^{3c_f\left(\frac{\upsilon}{1-\upsilon}(p-p_0) - \frac{E\varepsilon_l}{3(1-\upsilon)}\left(\frac{p}{p+P_L} - \frac{p_0}{p_0+P_L}\right)\right)} \tag{7.64}$$

7.2.3 *Robertson–Christiansen Model*

Robertson and Christiansen [14, 55] analyzed the measurement of permeability data under variable stress conditions in the laboratory and developed a permeability model: the Robertson–Christiansen (RC) model. This model was derived for cubic geometry considering the impacts of stress and sorption-induced swelling/shrinkage on coal permeability.

As shown in Eq. (7.41), permeability is a cubic function of the fracture width, and depends more on the fracture width than the matrix width of coal. Hence, a simplified assumption is that only the change of fracture width needs to be considered. By taking the derivative of Eq. (7.41) with respect to the pressure results in Eq. (7.65), we obtain:

$$\frac{\mathrm{d}k}{\mathrm{d}p} = \frac{3b^2}{12a}\frac{\mathrm{d}b}{\mathrm{d}p} \tag{7.65}$$

where the term $\mathrm{d}b/\mathrm{d}p$ is the change in the fracture width due to an infinitesimal change in stress (pressure).

The change in the fracture width can be divided into three different cases as follows:

(1) The change in a fracture caused by stress compressing the fracture space, Δb_f. According to the principle of effective stress, the relationship between the effective stress σ_e, total stress σ, and pore pressure is given by:

$$\sigma_e = \sigma - \beta p \tag{7.66}$$

where β is an effective stress coefficient, which is usually taken as 1, thus yielding:

$$\sigma_e = \sigma - p \tag{7.67}$$

Hence, the change in the effective stress is equal to the difference between the change in stress and the change in pore pressure:

$$\Delta\sigma_e = (\sigma - \sigma_0) - (p - p_0) \tag{7.68}$$

The fracture compressibility is defined as the rate of change in porosity with stress as follows:

$$c_f \equiv \frac{-1}{\phi_0}\frac{\Delta\phi}{\Delta\sigma} \tag{7.69}$$

Substituting Eq. (7.43) and Eq. (7.67) into Eq. (7.69), we obtain:

$$\Delta b_f = -b_0 c_f[(\sigma - \sigma_0) - (p - p_0)] \tag{7.70}$$

(2) The change in a fracture caused by elastic compression of stress to the matrix, Δb_m. For porous materials (e.g., coal), a change in pore pressure has a significant impact on the fracture width owing to the compressibility of the matrix blocks. In the case of coal matrix blocks, each block is predominantly surrounded by pore space and is held in place by a limited number of contact points with adjacent blocks. Hence, the contact between the matrix blocks can be ignored, and the coal blocks can be abstracted into a cube. Therefore, changes in pore pressure would appear to have a much greater impact on the elastic compressibility of the matrix blocks than changes in the confining pressure (external stress). To simplify the resulting equations, we assume that changes in the fracture width due to the mechanical elastic compression of the matrix blocks are (i) solely a function of pore pressure changes, and (ii) independent of changes in overburden or confining pressure. In addition, the volumetric mechanical strain, ε_{vm}, is defined as the change in volume with respect to the original volume as:

$$-\varepsilon_{vm} = \frac{V - V_0}{V_0} = \frac{p - p_0}{K} = \frac{3(1 - 2\upsilon)(p - p_0)}{E} \tag{7.71}$$

For an isotropic homogeneous medium, the linear strain (ε_{lm}) is given by Eq. (7.72):

$$\varepsilon_{lm} = \frac{1}{3}\varepsilon_{vm} = -\frac{(1 - 2\upsilon)(p - p_0)}{E} \tag{7.72}$$

The linear strain (ε_{lm}) is defined as:

$$\varepsilon_{lm} \equiv \frac{\Delta a}{a_0} = \frac{a - a_0}{a_0} \tag{7.73}$$

A positive change in the length of a matrix block results in an equal but negative change in the fracture width, namely,

$$\Delta a = -\Delta b \tag{7.74}$$

Hence, the change in the fracture width (Δb_m) caused by the pore pressure compression matrix can be expressed as:

$$\Delta b_m = \frac{a_0(1 - 2\upsilon)(p - p_0)}{E} \tag{7.75}$$

(3) The change in a fracture caused by sorption-induced deformation, Δb_s. The linear strain caused by the sorption of gas, ε_{ls}, can be defined in terms of the difference between two points on the sorption-induced strain curve modeled using the Langmuir-type equation:

$$\varepsilon_{ls} = \frac{\varepsilon_l p}{p + P_L} - \frac{\varepsilon_l p_0}{p_0 + P_L} = \frac{\varepsilon_l P_L}{(p + P_L)(p_0 + P_L)}(p - p_0) \tag{7.76}$$

The change in the width of coal matrix blocks, Δa_s, can be defined as:

$$\Delta a_s = a_0 \varepsilon_{ls} = \frac{a_0 \varepsilon_l P_L}{(p + P_L)(p_0 + P_L)}(p - p_0) \tag{7.77}$$

However, $\Delta b_s = -\Delta a_s$, and substituting this into the above equation results in Eq. (7.78), which relates the change in the fracture width with respect to the change in pressure due to the sorptive elastic properties of coal:

$$\Delta b_s = \frac{-a_0 \varepsilon_l P_L}{(p + P_L)(p_0 + P_L)}(p - p_0) \tag{7.78}$$

To summarize, the total change in the fracture width, Δb, can be expressed as:

$$\Delta b = -b_0 c_f[(\sigma - \sigma_0) - (p - p_0)] + \frac{a_0(1-2\upsilon)(p-p_0)}{E} - \frac{a_0 \varepsilon_l P_L}{(p+P_L)(p_0+P_L)}(p-p_0) \tag{7.79}$$

When the change in Δb is very small, Eq. (7.65) can be rewritten as:

$$\frac{\Delta k}{\Delta p} = \frac{3b^2}{12a}\frac{\Delta b}{\Delta p} \tag{7.80}$$

Substituting Eq. (7.79) into Eq. (7.80) yields:

$$\frac{\Delta k}{\Delta p} = -\frac{3b_0^3}{12a}c_f\left(\frac{\sigma-\sigma_0}{\Delta p} - 1\right) + \frac{3b_0^3}{12a}\frac{a_0}{b_0}\left(\frac{(1-2\upsilon)}{E} - \frac{\varepsilon_l P_L}{(p+P_L)(p_0+P_L)}\right) \tag{7.81}$$

Substituting $a/b = 3/\phi$ and $b^3/(12a) = k$ into Eq. (7.81) gives:

$$\frac{\Delta k}{\Delta p} = 3k\left[-c_f\left(\frac{\sigma-\sigma_0}{\Delta p} - 1\right) + \frac{3}{\phi_0}\left(\frac{(1-2\upsilon)}{E} - \frac{\varepsilon_l P_L}{(p+P_L)(p_0+P_L)}\right)\right] \tag{7.82}$$

Then, by rearranging Eq. (7.82) as $\Delta p \to 0$, we obtain:

$$\frac{dk}{3k} = -c_f[(\sigma-\sigma_0) - dp] + \frac{3}{\phi_0}\left(\frac{(1-2\upsilon)}{E} - \frac{\varepsilon_l P_L}{(p+P_L)(p_0+P_L)}\right)\mathrm{d}p \tag{7.83}$$

By integrating the above equation results in, we obtain:

$$\frac{k}{k_0} = \mathrm{e}^{3\left\{-c_f[(\sigma-\sigma_0)-(p-p_0)] + \frac{3}{\phi_0}\left[\frac{(1-2\upsilon)}{E}(p-p_0) - \frac{\varepsilon_l P_L}{(p_0+P_L)}\ln\left(\frac{p+P_L}{p_0+P_L}\right)\right]\right\}} \tag{7.84}$$

The fracture compressibility, c_f, is not necessarily constant, but is often a function of the effective stress, namely,

$$c_f = \frac{c_0}{\alpha(\sigma_e - \sigma_{e0})}\left[1 - \mathrm{e}^{-\alpha(\sigma_e - \sigma_{e0})}\right] \tag{7.85}$$

where c_0 is the initial fracture compressibility, and α is the fracture compressibility change rate.

Hence, considering the influence of the stress impact on the fracture compressibility, Eq. (7.84) can be rewritten as:

$$\frac{k}{k_0} = \mathrm{e}^{3\left\{-c_0 \frac{1-e^{-\alpha[(\sigma-\sigma_0)-(p-p_0)]}}{\alpha} + \frac{3}{\phi_0}\left[\frac{(1-2\upsilon)}{E}(p-p_0) - \frac{\varepsilon_l P_L}{(p_0+P_L)} \ln\left(\frac{p+P_L}{p_0+P_L}\right)\right]\right\}} \tag{7.86}$$

7.2.4 *Evaluation of Classical Permeability Models*

The previous chapters introduced the theoretical basis of permeability evolution and three popular permeability models. From the three aspects of basic concepts, basic assumptions, and the following applications, this section comprehensively evaluates the PM, SD, and RC permeability models, which are listed in Table 7.3.

First, the basic concepts and basic assumptions of the three permeability models are introduced and analyzed. As shown in Fig. 7.8, these permeability models have the following main characteristics:

(1) PM model: This strain-based model assumes conditions of uniaxial strain and constant vertical stress and follows a rigorous mechanical derivation process. In this model, the physical structure of coal blocks is simplified to the matchstick model.

(2) SD model: This model is similar to the PM model, and assumes conditions of uniaxial strain and constant vertical stress. The physical structure of coal blocks is also simplified to the matchstick model; however, this model is stress-based rather than strain-based.

Table 7.3 Widely used permeability models

Model	Formula description
PM model	$\frac{k}{k_0} = \left[1 + \frac{c_m}{\phi_{f0}}(p - p_0) + \frac{\varepsilon_l}{\phi_{f0}}\left(\frac{K}{M} - 1\right)\left(\frac{\beta p}{1+\beta p} - \frac{\beta p_0}{1+\beta p_0}\right)\right]^3$ $c_m = \frac{1}{M} - \left(\frac{K}{M} + f - 1\right)\gamma,\ M = \frac{E(1-v)}{(1+v)(1-2v)}$
SD model	$\frac{k}{k_0} = \exp\left\{3C_f\left[\frac{v}{1-v}(p - p_0) - \frac{\varepsilon_l}{3}\frac{E}{1-v}\left(\frac{p}{P_L+p} - \frac{p_0}{P_L+p_0}\right)\right]\right\}$
RC model	$\frac{k}{k_0} = \exp\left\{-3c_0\frac{1-\mathrm{e}^{-\alpha[(\sigma-\sigma_0)-(p-p_0)]}}{\alpha} + \frac{9}{\phi_0}\left[\frac{(1-2\upsilon)}{E}(p - p_0) - \frac{\varepsilon_l P_L}{(p_0+P_L)}\ln\left(\frac{p+P_L}{p_0+P_L}\right)\right]\right\}$

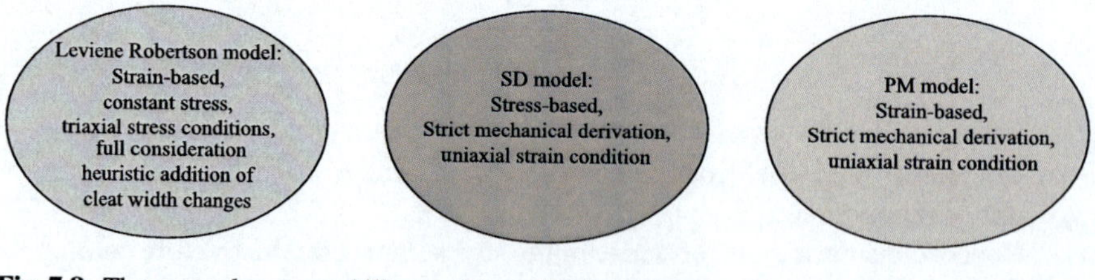

Fig. 7.8 Three popular permeability models and their features

(3) RC model: This strain-based model is derived for the cubic geometry model under a hydrostatic confining pressure condition and follows a rigorous mechanical derivation process.

From the above characteristics, some conclusions can be summarized:

(1) The three typical permeability models follow the rigorous mechanical derivation process, and satisfy the theoretical basis of permeability evolution, whether the coal structure is simplified as the matchstick model or cubic geometry model.
(2) The PM and RC models are strain-based, whereas the SD model is stress-based. From Sect. 7.1.2, it can be inferred that the relationship between permeability and porosity meets the cubic law. The strain-based model is derived from the mechanical relationship between stress and the porosity change due to the effects of sorption-induced deformation and effective stress. The impact of effective stress can be derived either by the generalized Hooke's law of strain form or by the generalized Hooke's law of stress form. If the sorption-induced deformation effect is deduced by stress, the sorption-induced strain should be transformed into stress. Strain-based relationships and stress-based relationships can transform into each other. Establishing the relationship between porosity and strain is a more intuitive logic, while using the former method is relatively simple.
(3) The PM and SD models both assume uniaxial strain and constant vertical stress conditions, while the RC model assumes hydrostatic pressure conditions. The two assumptions are different from the actual stress state, which also simplifies the complexity of the derivation of coal permeability. The uniaxial strain and constant vertical stress assumptions are consistent with the stress state of coal during CBM exploitation, while the hydrostatic pressure assumptions can be converted into a three-dimensional unequal stress state through appropriate changes.

By analyzing the similarities and differences of the three typical permeability models, it is noted that the essence of each model is to characterize the porosity and permeability changes during the variations of effective stress and sorption-induced deformation, namely, the evolution of porosity and permeability under gas–solid coupling effects. With the development of permeability theory in recent years, many permeability models have been established. However, it is undeniable that most permeability models are based on the above three permeability models.

There are still many aspects worthy of further consideration and research with respect to the accuracy and applicability of popular permeability models. The results of the above permeability models vary under different engineering conditions or test conditions. As shown in Fig. 7.9, the permeability data from the San Juan Basin could be best matched by the PM model. However, Fig. 7.10 reveals that the SD model agreed well with the experimental results, while the prediction of the PM model exhibited a relatively large deviation.

In addition, as displayed in Fig. 7.11, according to the experimental data and

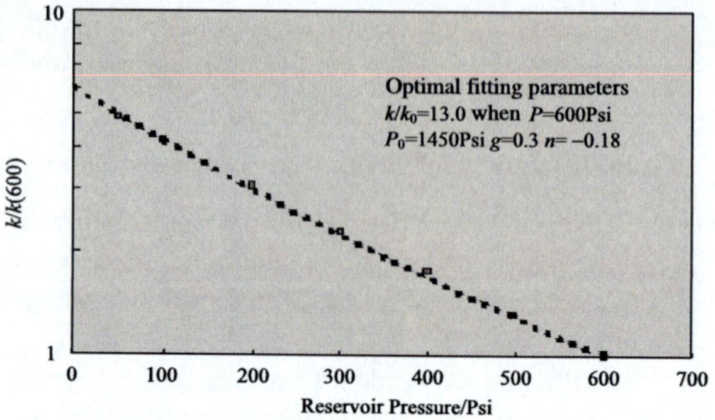

Fig. 7.9 Permeability data of the San Juan Basin matched by the PM model [59, 63]

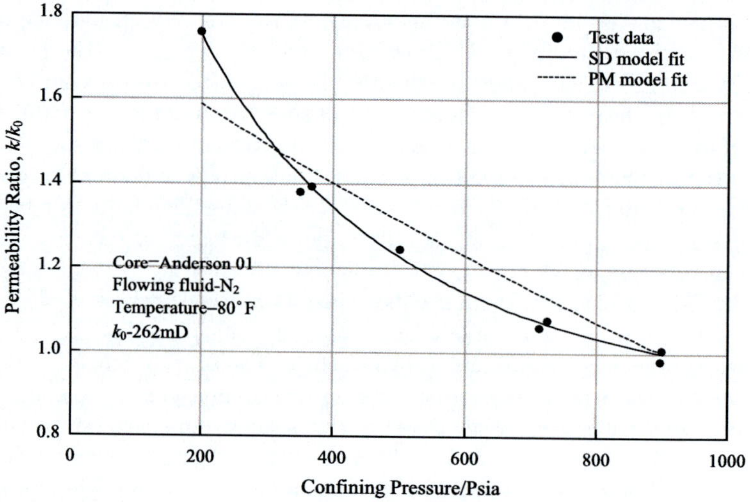

Fig. 7.10 Permeability evolution with the variation of confining pressure [14]

analysis of the RC model, for different test conditions, the PM and SD models either showed a good agreement with the experimental data, or exhibited large deviations.

Such a situation is caused by many factors. The primary factor can be attributed to the assignment of model parameters (sensitivity of each model parameter). Owing to the limitations of the current test methods, several parameters in the permeability models cannot be obtained directly, and must be derived from the theory of homogeneity. However, for actual coal seams or coal specimens, their physical and mechanical properties usually exhibit anisotropic characteristics, which significantly increase the difficulty of model parameter assignment. Thus, the accuracy and applicability of the permeability models (including classical permeability models and newly developed permeability models) are affected to some extent.

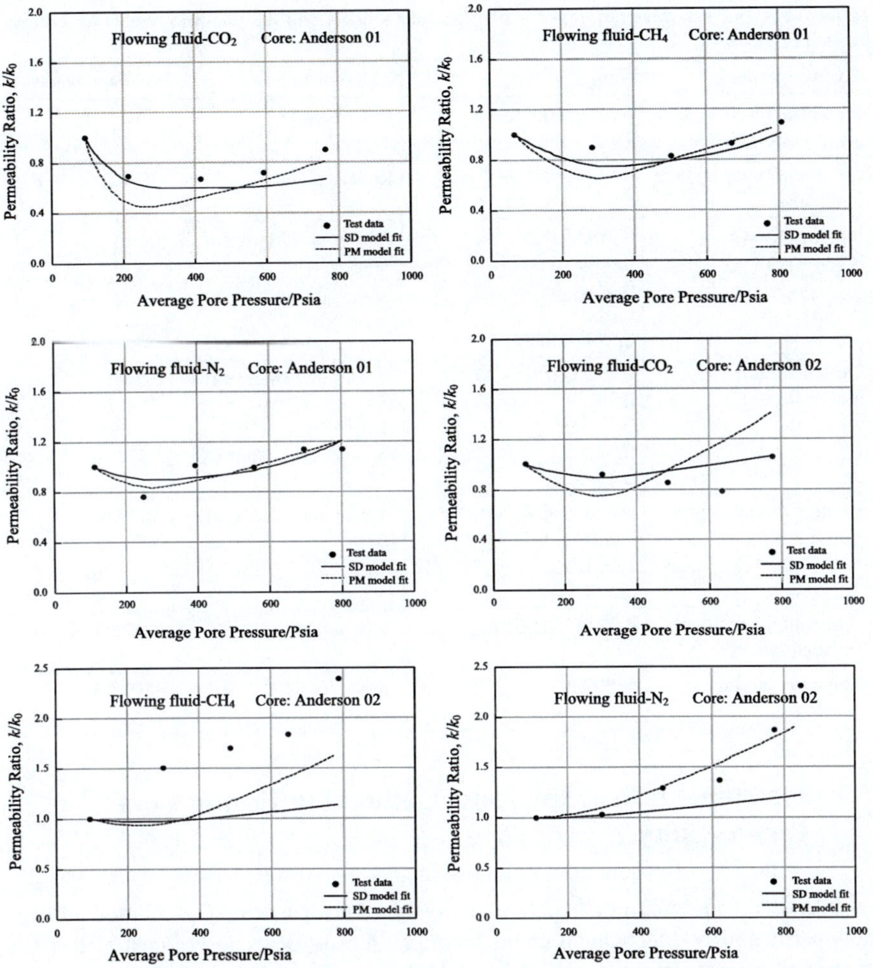

Fig. 7.11 Permeability data and the fitting results of the SD model and PM model under different test conditions [14] (confining pressure of 1000 Psia, and temperature of 80 °F)

The following is an introduction to the model parameter sensitivity of the RC model Eq. (7.86). The model parameters include fracture compressibility, initial fracture porosity, Young's modulus, Poisson's ratio, and sorption-induced deformation parameters. The ranges of these parameters are listed in Table 7.4. In the comparison analysis, the upper and lower limit values and the average value of each parameter in Table 7.4 are selected, and the permeability changes with increasing gas pressure under different parameter values are presented in Fig. 7.12.

Table 7.4 Input variables required for permeability models and the possible ranges of variable values [55]

Coal properties	Range	Reference	Average value
Initial fracture compressibility/Psia	1.69×10^{-4} to 1.88×10^{-4}	Robertson [13]	7.19×10^{-4}
Compressibility change rate/Psia	24.3×10^{-4} to 37.5×10^{-4}	Robertson [13]	30.4×10^{-4}
Initial properties (dimensionless)	0.001–0.005 0.0005–0.0075 0.03 0.016 0.024–0.028	Palmer and Mansoori [64]	0.014
Poisson's ratio (dimensionless)	0.23–0.40 0.39 0.32 0.33 0.20–0.50	Palmer and Mansoori [64] Gray [65] Shi and Durucan [61]	0.339
Young's modulus (psi)	300,000–600,000 124,000–445,000 392,000 500,000	Palmer and Mansoori [64] Gray [65]	393,500
Langmuir strain (N_2) (dimensionless)	0.00305–0.00196	Robertson [13]	0.00251
Langmuir pressure (N_2)/Psia	1193.93	Robertson [13]	1193.93

7.3 Instrument, Principle, and Method for Testing Coal Permeability

As introduced in detail in Chap. 6, a test system for measuring the coupled adsorption–permeability–mechanical characteristics of coal/rock can be used to test the permeability features of coal. Here, we only introduce the test principles of the steady-state method and transient method.

The general procedure of a coal permeability test is as follows:

(1) Before the experiment, the length and diameter of a coal specimen are measured five times, and the mean value is taken as the result.
(2) The coal specimen is installed in the test instrument, and the upstream and downstream pipelines are connected.
(3) The gas pipeline of coal sample and test system are then arranged to be evacuated.
(4) Following the experimental scheme, the pre-stress loaded on the coal sample is determined. Then, the test system and coal sample are filled with gas at a predetermined pressure, and the coal adsorbs the gas until an equilibrium state is reached.

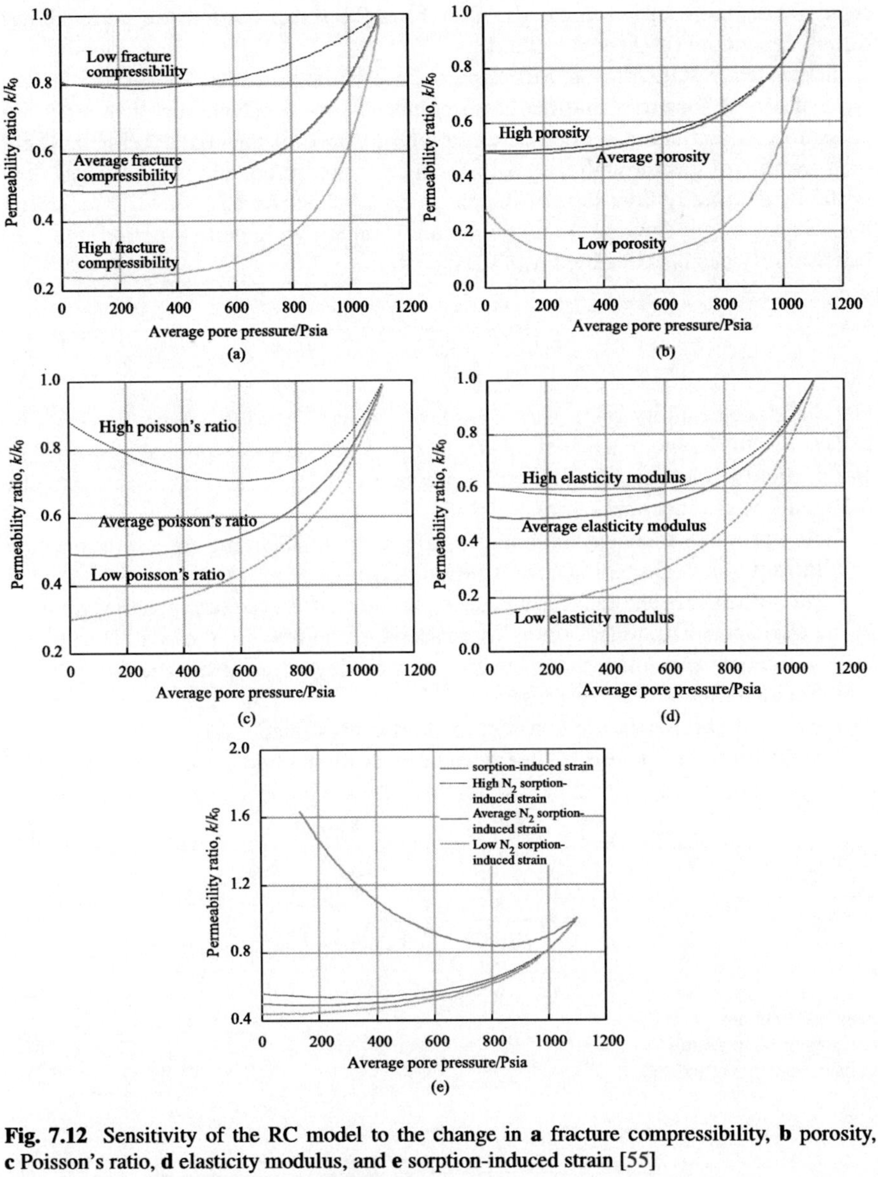

Fig. 7.12 Sensitivity of the RC model to the change in **a** fracture compressibility, **b** porosity, **c** Poisson's ratio, **d** elasticity modulus, and **e** sorption-induced strain [55]

After completing the above steps, the coal permeability test can be performed.

As we know, the laboratory determination of coal permeability is based on Darcy's law for describing gas migration in a coal fracture system. At present, the most commonly used methods for measuring coal permeability in the laboratory are the steady-state and transient methods. Therefore, the test procedure for permeability differs depending on the test method.

In the steady-state method, a certain gas differential pressure is applied to the two ends of a coal specimen and the gas pressure is kept constant, such that a certain pressure gradient is maintained between the two ends of the coal specimen. Then, the flow rate of gas through the coal specimen is measured, and when the gas flow is stable, the steady flow state of the gas is established. At this point, by recording the amount of gas flowing through the coal sample over a certain period, the coal permeability can be calculated using Eq. (7.87):

$$k = \frac{2p_0 Q L \mu}{A(p_1 - p_2)^2} \tag{7.87}$$

where k is permeability, mD; Q is the gas flow rate, cm^3/s. μ is the dynamic viscosity, MPa s; L is the length of the coal specimen, cm, A is the cross-sectional area, cm^2, p_1 and p_2 are the inlet pressure and outlet pressure of the coal specimen, respectively, MPa, and p_0 is atmospheric pressure, MPa.

In the pressure transient method [66, 67], after maintaining the gas adsorption equilibrium of a coal specimen for a period of time, the gas pressure is increased at one end of the coal sample. This creates a pressure difference between the two ends of the coal specimen, thus forming a gas seepage condition in the coal. During gas flow in the coal specimen, the upstream pressure decreases, while the downstream pressure increases until a new pressure equilibrium state of the specimen is achieved. The principle of the pressure transient method is illustrated in Fig. 7.13.

The governing equations of the transient pressure method are as follows:

$$\frac{p_u - p_d}{p_{u,0} - p_{d,0}} = e^{-\alpha t} \tag{7.88}$$

$$\alpha = \frac{k}{\mu \beta_g L^2} V_s \left(\frac{1}{V_u} + \frac{1}{V_d} \right) \tag{7.89}$$

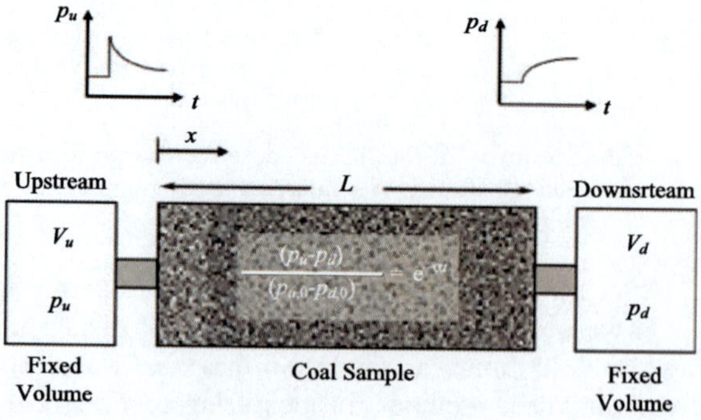

Fig. 7.13 Schematic diagram of the pressure transient pulse method [68, 69]

where p_u and p_d are the pressures of the upstream and downstream reservoirs, respectively, at time t; $p_{u,0}$ and $p_{d,0}$ are the pressures of the upstream and downstream reservoirs at time $t = 0$, respectively; μ and β_g are the dynamic viscosity and compressibility of the gas, respectively, L and V_s are the length and volume of the specimen, respectively, and V_u and V_d are the volumes of the upstream and downstream reservoirs, respectively.

When the volumes of the upstream and downstream reservoirs are equal (i.e., $V_u = V_d$), if the volume of the pipeline is ignored, Eq. (7.89) can be rewritten as follows:

$$\alpha_t = \frac{kA}{\mu L}\frac{2}{\beta_g V_u} \tag{7.90}$$

$$S_u = \beta_g V_u \tag{7.91}$$

where S_u is defined as the storage coefficient of the gas reservoirs, and indicates the change in the gas volume caused by a unit change in the gas pressure. When only the compressed gas storage is considered, it can be expressed as:

$$\alpha_t = \frac{kA}{\mu L}\frac{2}{S_u} \tag{7.92}$$

Therefore, if the volumes of the upstream and downstream reservoirs are equal, the governing equation of the pressure transient method can be rewritten as:

$$\left.\begin{aligned} p_u - p_d &= (p_u - p_d)e^{-\alpha_t t} \\ \alpha_t &= \frac{kA}{\mu L}\frac{2}{S_u} \end{aligned}\right\} \tag{7.93}$$

During the experiment, according to the obtained pressure transient curve shown in Fig. 7.14, α_t can be calculated using Eq. (7.93). Then, by calculating the storage coefficient of gas reservoirs (S_u), and by combining the length (L) and cross-sectional area (A) of the coal specimen, the coal permeability can be obtained.

Generally, the steady-state method is applied to coal specimens with high permeability, whereas the pressure transient method is usually used for samples with low permeability values. In this book, the permeability data of coal samples were determined by the pressure transient method. In the test process, the pressure transient pulse curve (see Fig. 7.14) was used to calculate the exponential fitting factor α_t based on Eq. (7.88). This was further combined with the length (L) and cross-sectional area (A) of a coal specimen to determine the coal permeability (k) under the corresponding test conditions.

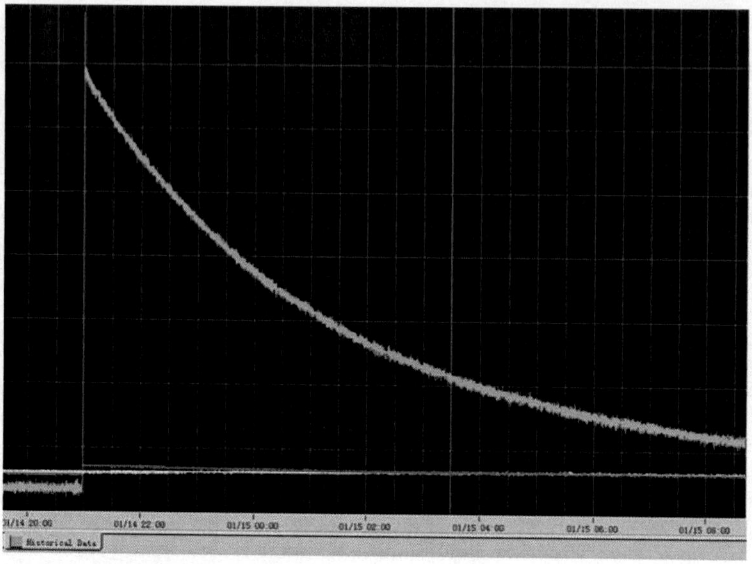

Fig. 7.14 Permeability pulse curve determined by the transient method

7.4 Impacts of Effective Stress and Sorption-Induced Deformation on Coal Permeability

The coal permeability test under hydrostatic conditions is mainly used to study the gas–solid coupling effect, that is, the effects of sorption-induced (swelling/shrinkage) deformation and effective stress on the evolution characteristics of coal permeability. At present, the commonly used test method for studying the sorption-induced deformation effect on coal permeability is that under hydrostatic conditions, whereby the gas pressure and confining pressure are changed simultaneously while maintaining a constant pressure difference. The test method for studying the impact of effective stress on coal permeability is as follows: under hydrostatic conditions, the gas pressure is kept constant while the confining pressure is changed, or the confining pressure is kept constant while changing the gas pressure.

7.4.1 Sorption-Induced Deformation of the Coal Matrix

1. Experimental device and sample preparation

The test for the sorption-induced deformation of coal is generally conducted under isothermal conditions. Figure 7.15 shows the test system, which includes a gas cylinder, sealing device of pressure vessels, thermostatic device, vacuum device, data acquisition system, and pressure transducer.

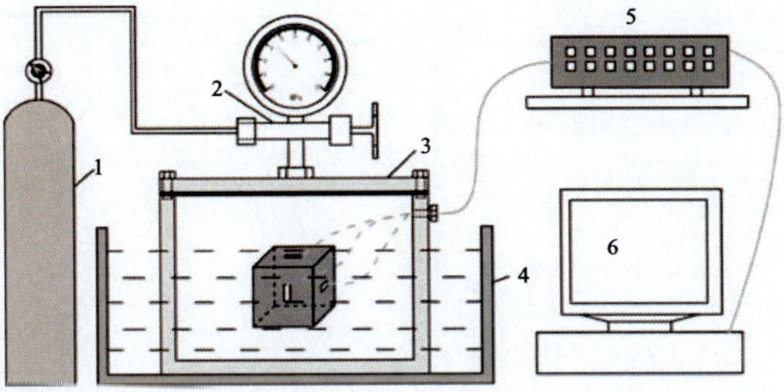

Fig. 7.15 Schematic diagram of the sorption-induced deformation test apparatus for coal. 1: High-pressure cylinder; 2: High-pressure three-way valve; 3: High-pressure vessel; 4: Thermostatic bath; 5: Data acquisition devices; 6: Computer system

During the test, a coal sample collected from an underground field is made into a cube block with a side length of 2 cm. The coal surface is then smoothed with sandpaper, and anhydrous alcohol is used to clean the coal surface. Subsequently, a strain gauge is attached to three mutually perpendicular adjacent coal surfaces, and a wire is connected to a data acquisition system. After the vacuum is pumped, gas with a predetermined pressure is filled into the test device, and the sorption-induced volumetric change in the coal matrix is measured.

2. Experimental results of sorption-induced deformation

Once coal adsorbs gas, it swells. Typical curves of sorption-induced deformation are presented in Figs. 7.16 and 7.17 for different coals. As shown in Fig. 7.16a, the sorption-induced swelling increased with an increase in the adsorbing gas pressure. The volumetric deformation increased rapidly with the increase in the adsorbing gas

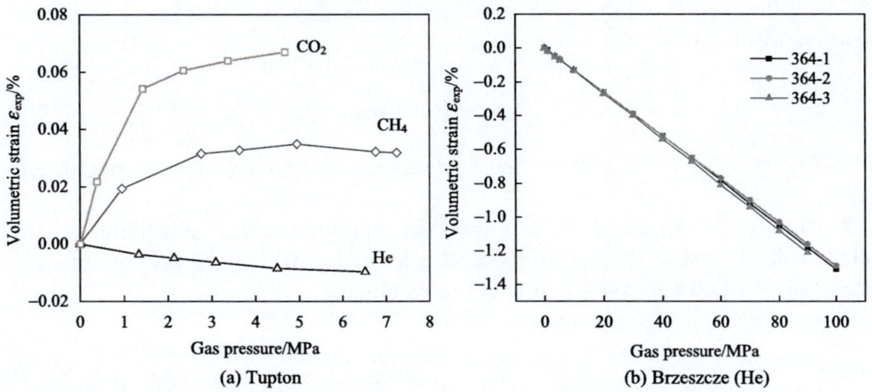

Fig. 7.16 Relationship between volumetric strain and gas pressure for high volatile bituminous coal [12, 70]

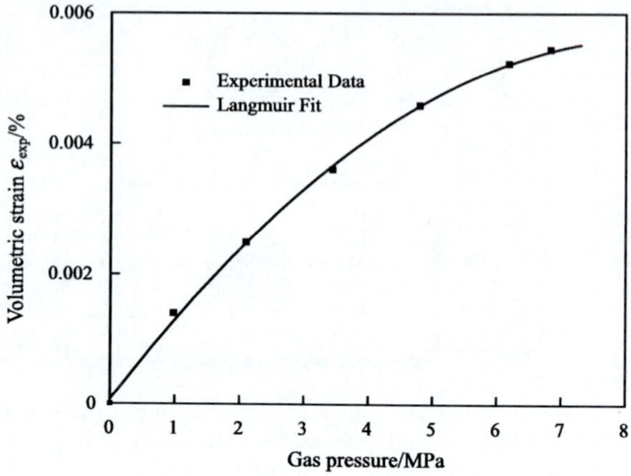

Fig. 7.17 Modeled adsorption-induced swelling deformation of coal [71]

pressure in the low-pressure area, while the increase in deformation slowed in the high-pressure area. However, with a further increase in the gas pressure, the deformation of coal decreased. Figure 7.16b reveals that, when using non-adsorptive helium gas to proceed with the test, coal was compressed and the compressive deformation exhibited a linear relationship with the gas pressure, as expressed by Eq. (7.94):

$$\varepsilon_{\exp} = \varepsilon_{cp} = -\frac{p}{K_m} \tag{7.94}$$

where p is the gas pressure, K_m is the bulk modulus of the coal matrix, $\varepsilon_{\exp}$ is the volumetric deformation measured by the test, ε_{cp} is the compression deformation due to the gas compressing coal matrix, %; The compression deformation is defined as negative.

The results also revealed that the coal matrix was compressed by free gas, and that sorption-induced swelling occurred. Hence, these experimental results correspond to the net values between sorption-induced swelling deformation and gas compression deformation, namely,

$$\varepsilon_{\exp} = \varepsilon_{sw} + \varepsilon_{cp} \tag{7.95}$$

where ε_{sw} represents the sorption-induced swelling deformation of the coal matrix, %.

Using Eqs. (7.94) and (7.95) to process the experimental data, it is found that the relationship between the sorption-induced deformation of coal and gas pressure is in accordance with the form of the Langmuir equation:

$$\varepsilon_{sw} = \frac{\varepsilon_{\max} p}{p + P_L} \tag{7.96}$$

where $\varepsilon_{\max}$ is the maximum sorption-induced volumetric strain when the gas pressure is infinite, %, and P_L is the pressure at which the volumetric strain reaches half of $\varepsilon_{\max}$, MPa.

3. Essence of adsorption deformation in the coal matrix

Coal has a large specific surface area, and its surface energy decreases when gas adsorption occurs. Research has shown that there is a fine linear relationship between the coal swelling deformation and the gas adsorption-induced surface energy reduction in the coal matrix. The sorption-induced deformation can be deduced using the following formula:

The change in surface tension (or the change in surface energy), π_s , due to gas adsorption can be written as Eq. (7.97):

$$\pi_s = \psi_0 - \psi \tag{7.97}$$

where ψ_0 is the surface free energy of coal in a vacuum state, N/m and ψ is the surface free energy of coal after gas adsorption, N/m.

According to Gibbs's equation [72]:

$$\pi_s = RT \int_0^p \Gamma d(\ln p) \tag{7.98}$$

where R is the gas constant, 8.314 J/(mol K); and Γ is the surface excess of coal, mol/m^2.

The difference between the surface concentration and bulk concentration is called the surface excess, which is defined as follows:

$$\Gamma = \frac{V}{V_{mol} \sum} \tag{7.99}$$

where $\sum$ is the specific surface area, m^2/t, V is the gas adsorption amount, m^3/t, and V_{mol} is the molar volume of gas, 22.4 L/mol.

The adsorption amount of gas in the coal matrix can be expressed by the Langmuir equation. By substituting the Langmuir equation and Eq. (7.99) into Eq. (7.98), the change in surface energy due to the gas adsorption effect can be obtained as:

$$\pi_s = \frac{aRT}{V_{mol} \sum} \ln(1 + bp) \tag{7.100}$$

According to the Bangham principle [73], the linear strain of a solid (ε_l) is proportional to the change in the surface free energy, as expressed by Eq. (7.101):

$$\varepsilon_l = \vartheta \pi = \frac{\Sigma \rho_m}{E_m} \pi_s \tag{7.101}$$

where ϑ is the deformation constant, $\vartheta = \sum \rho_m / E_m$ E_m is the elastic modulus of the coal matrix, MPa, and ρ_m is the density of the coal matrix, kg/m^3.

Substituting Eq. (7.100) into Eq. (7.101), the linear expansion strain of coal is obtained:

$$\varepsilon_l = \frac{a\rho_m RT}{E_m V_{mol}} \ln(1 + bp) \tag{7.102}$$

Assuming that coal is an isotropic homogeneous body, the volumetric strain of coal due to gas adsorption can be written as:

$$\varepsilon_{sw} = \frac{3a\rho_m RT}{E_m V_{mol}} \ln(1 + bp) \tag{7.103}$$

Based on Eq. (7.102) or Eq. (7.103), the sorption-induced swelling deformation of coal is a function of the adsorption constant, gas pressure, and physical properties (elastic modulus and density). Hence, there must be a one-to-one relationship between the adsorption swelling deformation of coal and the adsorption amount of gas.

7.4.2 Impact of Effective Stress on Coal Permeability

A coal body does not suffer damage during coalmine methane drainage or CBM exploitation. The change in effective stress and the effect of desorption shrinkage on the coal matrix mean that coal fractures will open or close as a result, thus causing a change in coal permeability. Under hydrostatic conditions, a coal body is in the elastic stage without yielding failure, which provides the conditions to study the relationship between coal permeability and gas pressure. Some test results of coal permeability are presented in Figs. 7.18 and 7.19.

As shown in Fig. 7.18, with the increase in stress, the permeability values of the coal samples decreases gradually, and the decreasing rate of coal permeability is also slowed. The test results revealed that external stress and gas pressure both had a significant impact on coal permeability. The influence of gas on coal permeability can be explained from two aspects: the effective stress and sorption-induced swelling deformation experienced by coal. On the one hand, an increased gas pressure offsets part of the stress; thus, the effective stress decreases and coal fractures open, leading to an increase in coal permeability. On the other hand, after gas adsorption in coal, the coal matrix swells and deforms, which leads to fracture compression and a decrease in coal permeability. According to Eq. (7.96), the growth trend of sorption-induced swelling slows down with increasing gas pressure. Therefore, under a constant effective stress condition, the reducing rate of coal permeability decreases with increasing gas pressure.

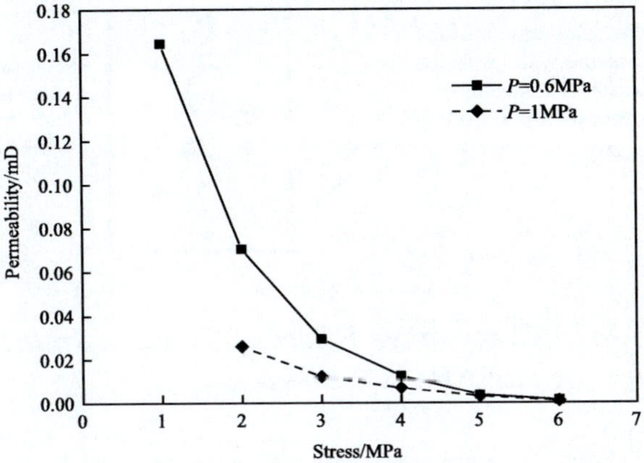

Fig. 7.18 Test results of coal permeability under constant gas pressure

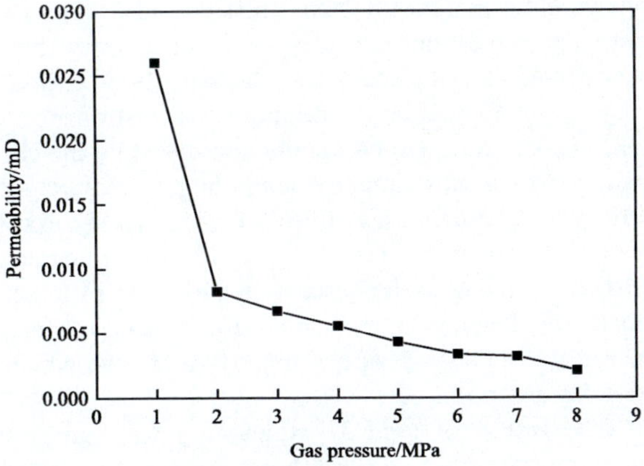

Fig. 7.19 Test results of coal permeability under constant effective confining pressure

7.4.3 Impact of Sorption-Induced Deformation on Coal Permeability

To investigate the influence of sorption-induced deformation on coal permeability, permeability tests of coal under conditions of constant effective stress and different gas pressures were carried out. The results are plotted in Fig. 7.19, which shows that the permeability values of the coal samples decreased with increasing gas pressure. According to Sect. 7.4.1, gas adsorption in coal leads to the swelling of the coal matrix, which can compress coal fractures (see Fig. 7.20) and result in a decreased coal permeability. The relationship between the swelling of the coal matrix and the gas pressure satisfies Eq. (7.96). That is, with an increase in gas pressure, the growth rate of the sorption-induced swelling deformation decreases gradually, Therefore, the reducing rate of coal permeability decreases gradually with increasing gas pressure.

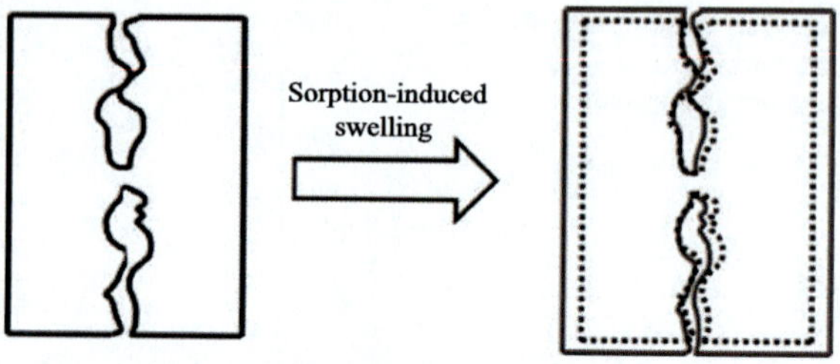

Fig. 7.20 Sketch illustrating the compression of the fracture width induced by adsorption swelling deformation in the coal matrix

7.4.4 *Competitive Effect of Sorption-Induced Deformation and Effective Stress*

Scholars have carried out numerous studies regarding the influence of effective stress and adsorption swelling on the permeability evolution in coal. The generally accepted viewpoint is that the effective stress and adsorption swelling deformation caused by the change in gas pressure play a mutually competitive role in permeability evolution. For example, in the coal seam gas extraction process, if the stress state of the coal seam is unaffected by the mining stress disturbance, then the permeability change of the coal seam will be mainly controlled by the change in gas pressure. In other words, with a substantial reduction in gas pressure during in situ gas extraction, the effective stress increases while the coal matrix shrinks. The increase in effective stress causes the compaction of coal fractures, resulting in a decrease in coal permeability. However, owing to the desorption shrinkage of the coal matrix, the fracture width increases, leading to an increased permeability. In general, the evolution of coal permeability is the result of these two effects, which agrees with the principle of dominant control.

The results of typical permeability tests under the conditions of a constant confining pressure and variable gas pressure are plotted in Figs. 7.21 and 7.22. These plots reveal that coal permeability first decreased and then increased with increasing gas pressure. As mentioned above, effective stress and sorption-induced deformation both play important roles in the permeability evolution of coal. On one hand, as the effective stress of a coal seam decreases, the fracture width of the coal body increases, thus resulting in an increased coal permeability. On the other hand, coal permeability decreases due to sorption-induced swelling deformation. Hence, the permeability evolution of coal is a superposition of these two functions.

As shown in Fig. 7.23, a further analysis of Figs. 7.21 and 7.22 indicates that when the gas pressure was low, the reduction in the fracture width due to the sorption-induced swelling deformation of the coal matrix was greater than the increase in the fracture width caused by the decreased effective stress. Hence, the fracture width and coal permeability decreased. As the gas pressure increased gradually, the effective stress decreased continuously, resulting in an increase in the fracture width, while the increase in the sorption-induced swelling deformation gradually slowed down, leading to a slow decrease in the fracture width. When the gas pressure increased to a

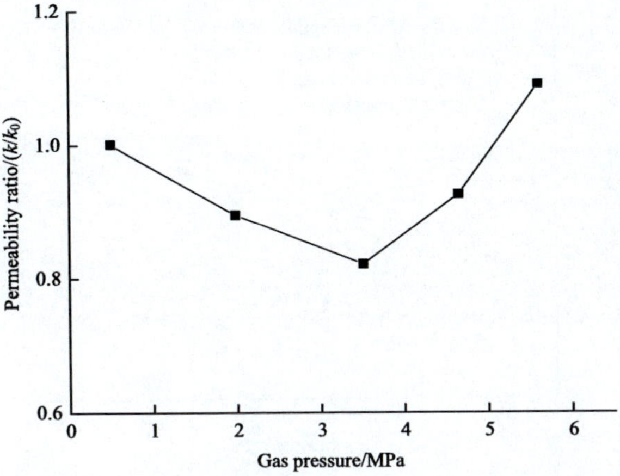

Fig. 7.21 Experimental results of permeability tests under constant effective confining pressure [74]

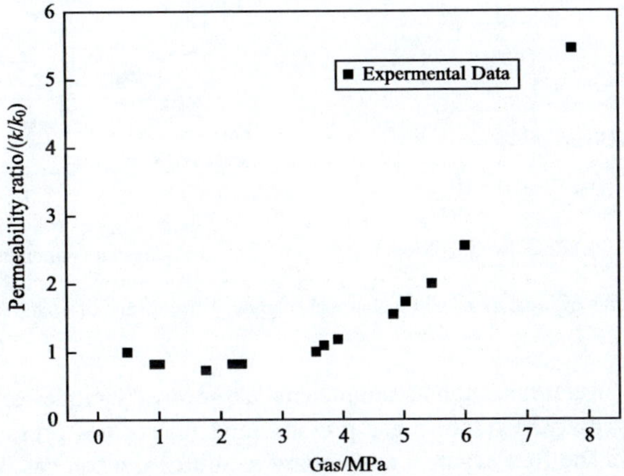

Fig. 7.22 Experimental results of permeability tests under constant effective confining pressure [14]

certain value, the fracture width of the coal ceased to decrease, and the permeability reached its minimum value. Thereafter, the fracture width began to increase with increasing gas pressure and increasing coal permeability.

Nowadays, a large number of permeability tests are carried out under hydrostatic conditions. According to the analysis by Liu et al. [57] concerning the two extreme boundary conditions when permeability models are established, the boundary condition of a coal sample belongs to the free swelling boundary under hydrostatic conditions (Fig. 7.24). However, many permeability models have been established based on

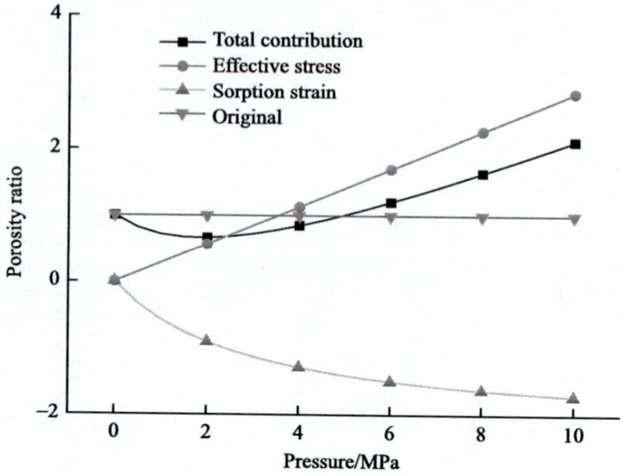

Fig. 7.23 Competitive effect between sorption-induced deformation and effective stress [78]

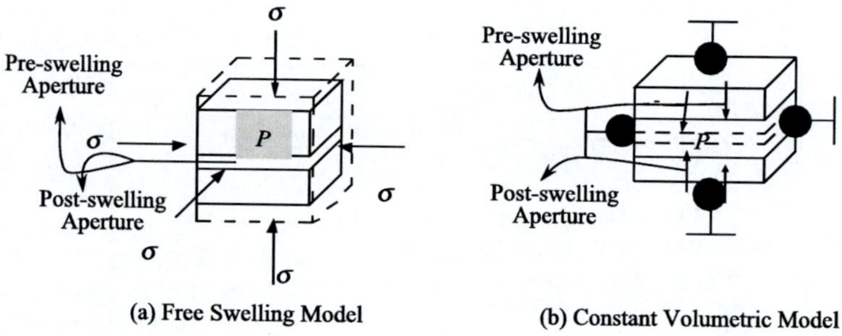

Fig. 7.24 Two kinds of ideal external mechanical boundary conditions of coal [57]

simplified coal structures, such as the matchstick geometry model or cubic model, while assuming that the fracture gas pressure and matrix gas pressure are equal. Therefore, under the free expansion boundary conditions, when gas is injected, the coal matrix is uniform and undergoing free swelling (or shrinkage), which will not cause a change in the fracture aperture, thus having no effect on fracture permeability [57, 75–77]. Using these theoretical models to analyze the relationship between coal permeability and the adsorption equilibrium time under hydrostatic pressure conditions, it can be concluded that the coal matrix swelling behavior has no effect on coal permeability, Hence, with the increase in effective stress, the permeability of coal should increase continuously until it finally stabilizes. However, this is contradictory to the observed test phenomenon.

In this book, we establish a conceptual model of permeability evolution to describe the adsorption equilibrium process (gas injection process) in order to analyze this contradiction between theory and test results. Then, we further explain why sorption-induced deformation of the coal matrix does not cause a permeability change at a certain adsorption equilibrium moment.

As illustrated in Fig. 7.25, under the initial adsorption equilibrium conditions, the fracture gas pressure and coal matrix gas pressure are equal, and the permeability is defined as k_0. According to the evolution law of radial strain, as a gas of constant pressure is injected into a coal specimen, the gas pressure in the coal fracture can quickly reach the injection pressure, while the gas pressure in the coal matrix increases slowly. Therefore, in the initial stage of gas injection, the impact of effective stress has a greater impact on permeability than the effect of sorption-induced deformation, and the permeability of coal increases rapidly under the dominant influence of effective stress. With an increase in the adsorption equilibrium time, the effect of adsorption swelling on permeability strengthens continuously, and the impact of effective stress is significantly weakened after the initial phase (after the fracture gas pressure equals the gas injection pressure, the influence of effective stress on permeability is mainly controlled by the increase in the matrix gas pressure; thus, the rate slows down). At the adsorption equilibrium moment of t_1, when the impact of effective stress can balance the effect of sorption-induced deformation, the permeability of a coal sample can be assumed to reach the maximum value of the entire evolution stage. Subsequently, the effect of sorption-induced deformation at t_1 will be greater than the impact of effective stress, and the rate of increase of the matrix adsorption swelling will exceed that of the effective stress. Hence, the permeability of the coal sample will decrease gradually.

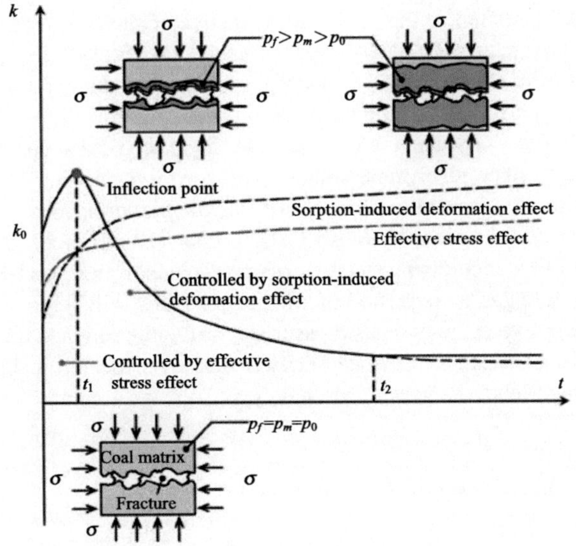

Fig. 7.25 Schematic diagram of competitive effect between sorption-induced swelling deformation and effective stress

During this period, the permeability change is inconsistent with the theoretical prediction. To address this, we introduce the "adsorption swelling region" theory proposed by Liu et al. [57, 75–77]. According to this theory, under the control effect of matrix diffusivity properties, fracture gas can only diffuse to the adjacent coal matrix during the adsorption equilibrium stage, that is, adsorption swelling only occurs in the region around the fracture. Because the diffusion rate of gas in the coal matrix is very small, the matrix gas pressure changes very little outside the adsorption swelling region. Hence, the coal matrix far away from the fracture region (with almost no adsorption swelling) prevents the free expansion of the matrix around the fractures, similar to a wall. However, matrix adsorption swelling still has an impact on the fracture aperture, which leads to a decrease in permeability.

After the moment of t_2, adsorption swelling deformation continues to increase, indicating that the complete adsorption equilibrium state has not yet been reached, therefore, the impact of effective stress is also reduced. However, the permeability values determined through tests tend to be stable and dissimilar to the dotted line in Fig. 7.25, which exhibits a continuous declining tendency. This may be attributed to two reasons. Firstly, the impacts of effective stress and sorption-induced deformation are offset at this moment. Secondly, because of the continuous adsorption equilibrium, the adsorption swelling deformation of the coal matrix around the fracture region is more balanced, Thus, the impact of the matrix wall disappears and coal permeability presents the evolution characteristics under the free expansion condition.

From the above analysis, it can be found that the inhomogeneous swelling caused by the gas diffusion characteristics in the coal matrix has a very important influence on the evolution of permeability. It can also be inferred that the degree of influence is not only closely related to the size characteristics of the coal matrix, but also to the adsorption characteristics of different gases. The greater the equivalent matrix scale of coal and the stronger the gas adsorption capacity, the greater the impact of the imbalance of matrix adsorption on the evolution of permeability. Therefore, the future application of CO_2-enhanced CBM and CO_2 geological sequestration should account for the stronger adsorption capacity and sorption swelling effect of CO_2 in the coal matrix in comparison to CH_4, as well as the higher adsorption equilibrium time for CO_2 in coal in comparison to CH_4 in coal [17]. Hence, for a large coal seam area, the imbalance characteristics of matrix adsorption, and the difference between the fracture gas pressure and matrix gas pressure, will be greater. Therefore, if the difference between the two gas pressures is still ignored when the permeability model is established, the theoretical model used to analyze the permeability evolution characteristics will lead to a great discrepancy.

7.5 Effect of Stress Perturbation on Coal Permeability

7.5.1 Evolution of Coal Permeability During Loading

1. Evolution characteristics of coal permeability in the complete stress–strain process

The complete stress–strain tests of coal can well describe the actual failure behavior of coal. Hence, studying the evolution characteristics of coal permeability in the complete stress–strain process provides good guiding significance for the prevention and control of coal and gas outbursts.

The permeability results of coal samples during the complete stress–strain test are displayed in Figs. 7.26 and 7.27, which show that the permeability of the samples first decreased and then increased with an increase in axial stress. The evolution trend of permeability agreed well with the variation trend of volumetric strain in Fig. 7.27.

Chinese and overseas Scholars [79–84] have conducted many studies on the permeability of coal and rock materials (e.g., coal, mudstone, and sandstone) during the complete stress–strain process. In summary, there are four main types of permeability evolution in the complete stress–strain process [85], which are presented in Fig. 7.28.

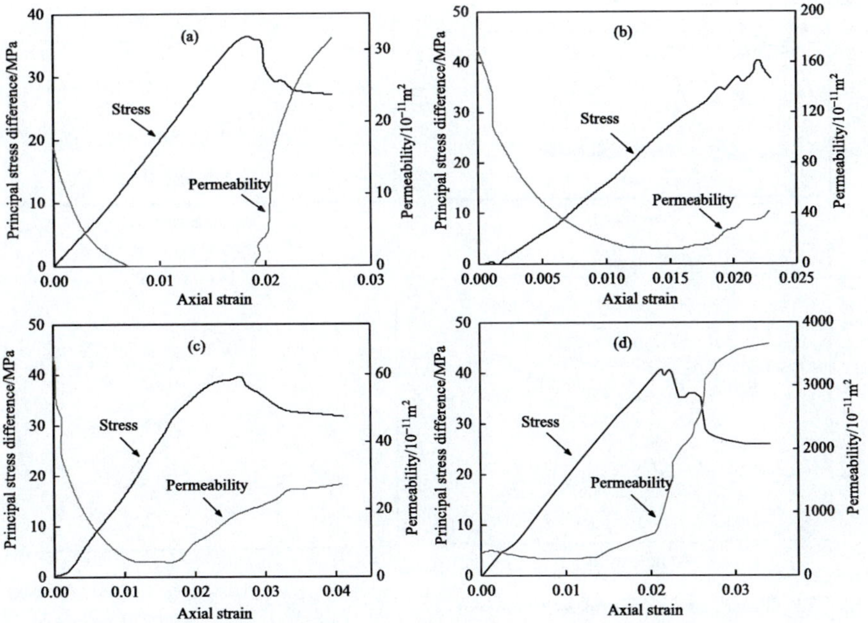

Fig. 7.26 Permeability evolution of a coal samples during the complete stress–strain process (samples from Zhangji Coalmine, Huainan Coalfield) [79]

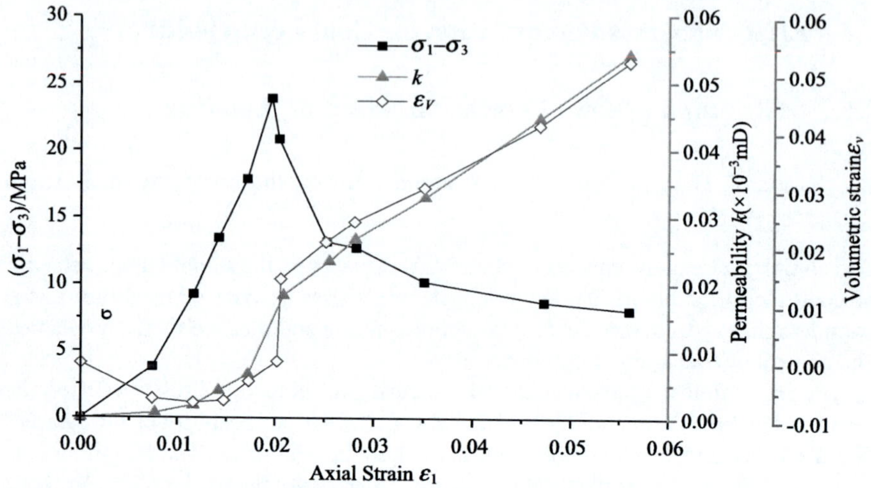

Fig. 7.27 Evolution of the permeability–strain relationship during the complete stress–strain process (coal samples from the Jingyuan Mining Bureau) [79]

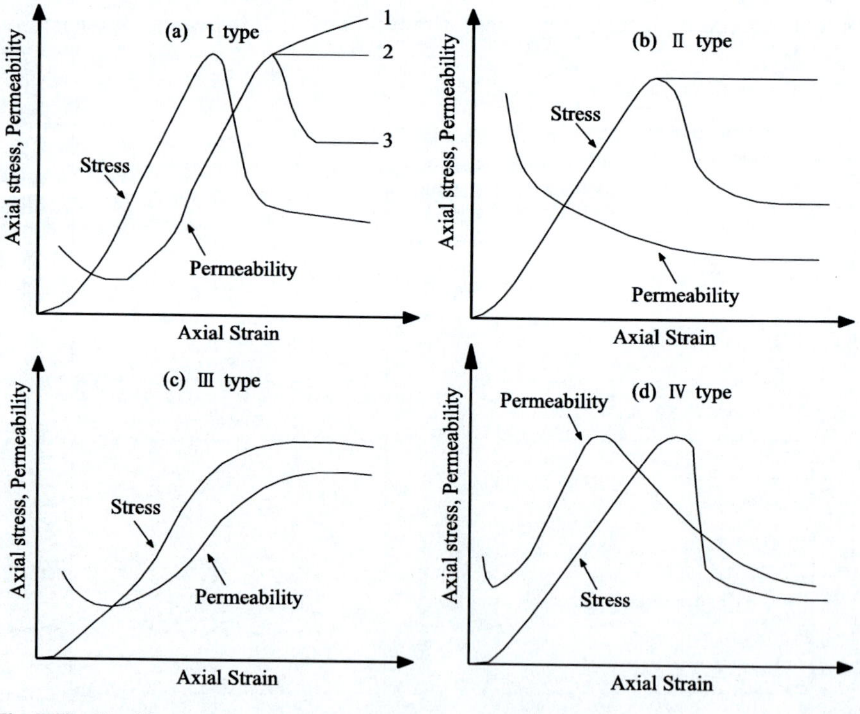

Fig. 7.28 Classification of permeability evolution in coal and rock [85]

The permeability evolution of coal in the complete stress–strain process belongs to type I. When the coal body is compressed, the seepage channels composed of primary pores and fissures are squeezed by the external force, which leads to a decrease in the fracture width and permeability. With continuous loading, the coal body enters the elastic deformation stage, the gas seepage channels continue to be compressed, and the permeability is reduced to the minimum value. In the nonlinear elastic plastic deformation stage, new fissures appear under the action of the external force, which leads to a newly created gas seepage channels, and the coal permeability recovers slowly. As the external force increases further, new fractures are produced in the coal body and continue to expand/converge, and the permeability increases continuously. When the peak strength of coal is reached, the coal body is destroyed, a large number of fractures coalesce, and the permeability continues to increase. In the residual strength stage, three different trends in coal permeability appear. Permeability first increases dramatically owing to the sustained damage experienced by the coal body. Second, for soft coal, the permeability value then remains unchanged as a result of the small change of the coal particle size in the fracture system. Third, the permeability of coal decreases rapidly, which is mainly because fractures are squeezed or coal fragments are broken into smaller particles, blocking the gas seepage channels. Under a certain extent of destruction, the coal fragment size in the fractures remains constant; thus, the permeability of coal remains stable.

Generally, coal permeability conforms to the second change trend of type I in Fig. 7.29. The permeability changes can be simplified into four stages: the elastic stage *OB,* which is the phase of decreasing permeability; the nonlinear elastic–plastic stage *BC*, which is the phase of increasing permeability; the brittle failure stage *CD*, which is the phase of rapidly increasing permeability; and the ideal plastic stage *DE*, which is the phase of constant permeability.

2. Permeability evolution of coal during loading

For underground mining, the coal body in mining affected area experiences complex loading and unloading stress paths, which control the failure and deformation law of coal under the action of stress, and influence gas seepage characteristics. Therefore, studying the controlling effect of the loading/unloading stress path is significant. However, owing to the high requirements of the experimental equipment and methods, carrying out permeability tests under the loading/unloading stress path, which corresponds completely to the on-site scenario, is almost impossible. Hence, simplified test paths are often used to perform experiments due to the limitations of equipment and methods.

In this book, to investigate the permeability evolution characteristics of coal with different bedding structures under different loading and unloading conditions, mechanical paths for a permeability test were designed, as shown in Fig. 7.30. During the test, coal samples were placed in pseudo-triaxial conditions, and the gas (methane) pressure was set to 1 MPa. The loading paths are displayed in Fig. 7.31, whereby the axial pressure and confining pressure were increased synchronously, and the differential stress constant was maintained until 80% of the peak strength was reached. Then, the differential stress constant was maintained and the confining pressure was

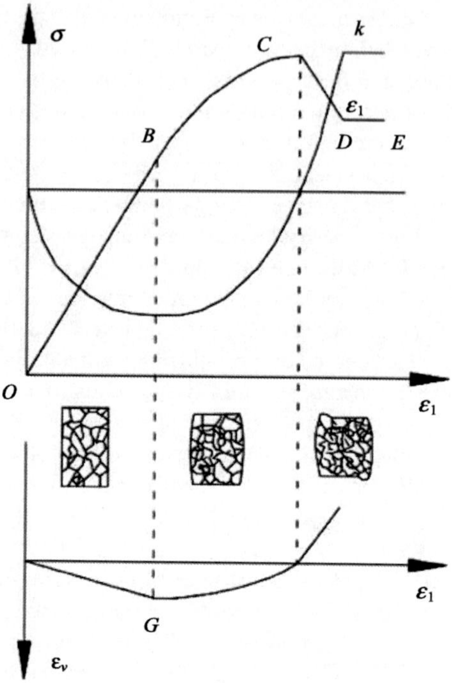

Fig. 7.29 Relationships between the change in coal permeability and volumetric strain during the complete stress–strain process

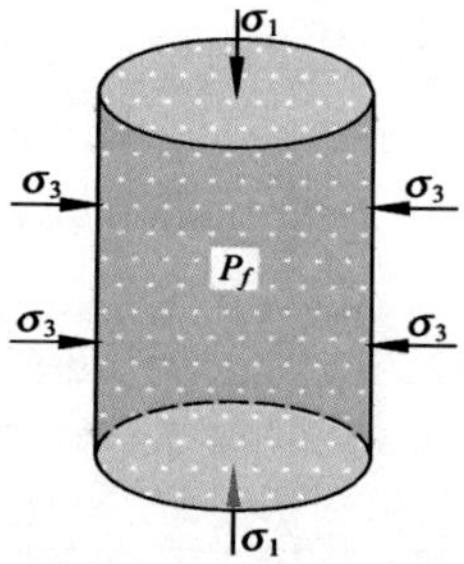

Fig. 7.30 Distribution of force on coal specimen

unloaded. The specific parameters of the loading experiment schemes are listed in Table 7.5, and the unloading scheme is detailed in Sect. 7.5.2.

The permeability results of different coal samples (with vertical bedding, parallel bedding, or oblique bedding) during the triaxial loading process are plotted in Figs. 7.32, 7.33, and 7.34.

1) Coal sample with vertical bedding

In the loading process, the results showed that the permeability of the coal sample with vertical bedding decreased with increasing effective stress (Fig. 7.32). Coal permeability decreased rapidly during the initial loading stage because the vertical

Fig. 7.31 Sketch map of the loading process

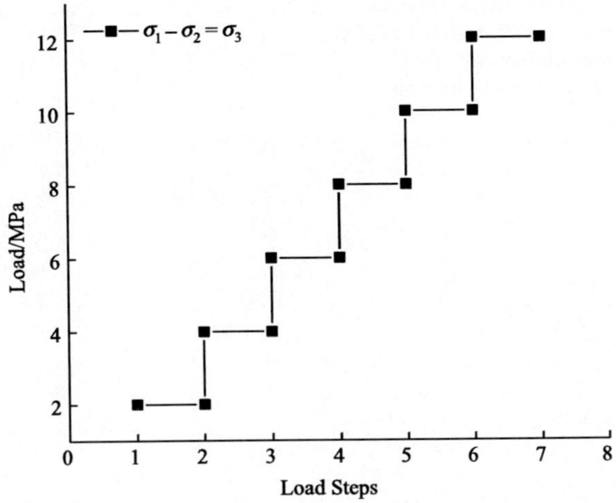

Table 7.5 Experimental scheme parameters

Test condition	No	Gas pressure P_f/MPa	Differential stress $\sigma_1-\sigma_3$/MPa	Confining pressure σ_3/MPa	Notes
Loading	1	1	2	2	Loading speed for axial and confining pressure: 0.01 MPa/s
	2	1	4	4	
	3	1	6	6	
	4	1	8	8	
	5	1	10	10	
	6	1	12	12	

Fig. 7.32 Permeability evolution characteristics of a coal sample with vertical bedding during loading

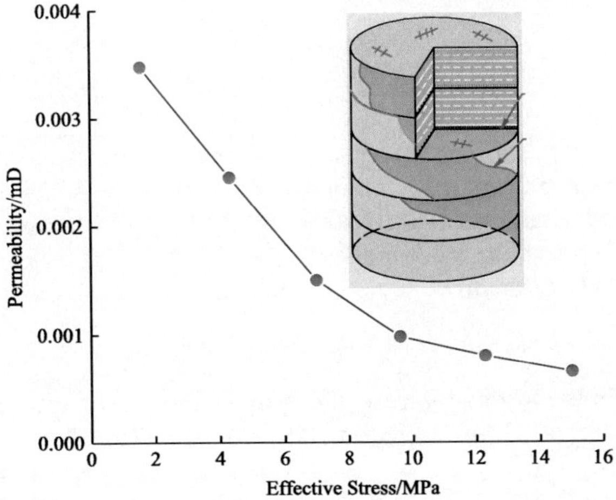

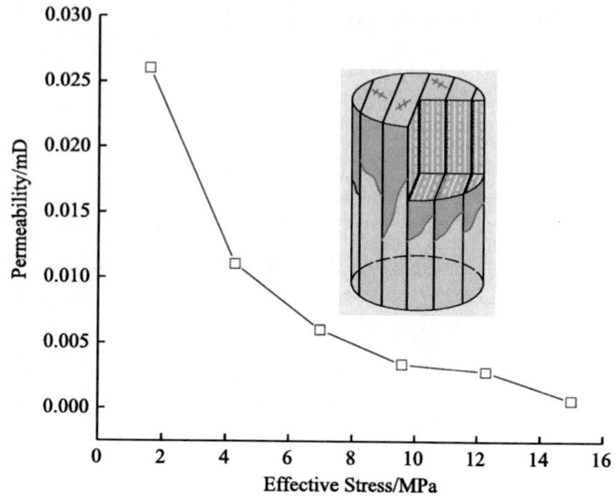

Fig. 7.33 Permeability evolution characteristics of a coal sample with parallel bedding during loading

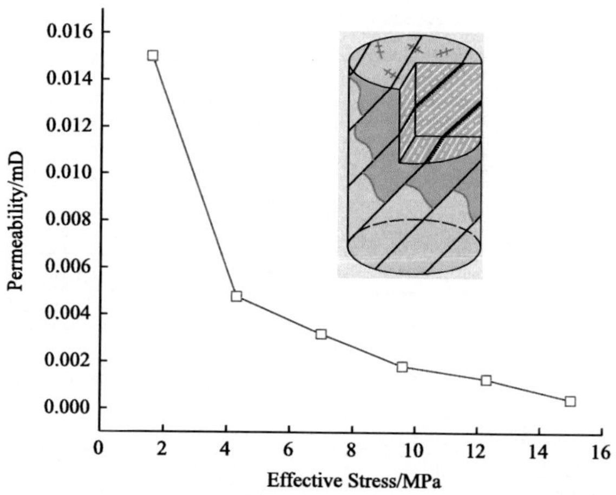

Fig. 7.34 Permeability evolution characteristics of a coal sample with oblique bedding during loading

bedding was easily compressed. Coal permeability exhibited a linear decreasing trend until an effective stress of 9 MPa, after which, it changed slowly. Overall, the permeability value reduced by 81.2% from 0.00346 mD in the initial loading stage to 0.00065 mD.

2) Coal sample with parallel bedding

In the loading process, the permeability of the coal sample with parallel bedding decreased with increasing effective stress (Fig. 7.33), and the change trend of coal permeability was consistent with that of the vertical bedding sample. In the initial loading stage, which is affected by the compression of the parallel bedding, there was a rapid reduction in permeability. When the loading reached 7 MPa, the change in permeability slowed down. When the predetermined value was reached, the permeability value was 0.00062 mD, which was nearly the same as that of the vertical bedding sample.

The permeability of the coal sample with parallel bedding was 0.026 mD in the initial loading stage, which was 7.5 higher than that of the coal sample with vertical bedding. However, when the confining pressure was loaded to 12 MPa, the permeability reduced to 0.00062 mD (i.e., a reduction of 99.7%), which indicates that the coal sample almost lost its seepage capacity.

3) Coal sample with oblique bedding

In the loading process, the permeability of the coal sample with oblique bedding decreased with increasing effective stress (Fig. 7.34), which was consistent with the change trends of the vertical and parallel bedding samples. However, the permeability of the coal sample with oblique bedding decreased significantly during the early loading stage, accounting for almost 70% of the entire change magnitude of the entire loading process, which indicates that this sample was more sensitive to stress. During the initial loading stage, the pores and fractures of the sample were much more easily compacted, and the stress effect was more obvious. When the stress was loaded at 4 MPa, the permeability value tended to be moderate. At the beginning of loading, the permeability value of the sample was 0.015 mD, which was 4.5 times higher than that of the vertical bedding coal sample. When the confining pressure was loaded to 12 MPa, the permeability value decreased to 0.00043 mD (i.e., a reduction of 97.1%), which indicates that the closure degree of pores and fractures was significant.

Of the three types of coal samples, the parallel bedding sample experienced the largest permeability drop (99.7%), which suggests that this sample was probably compacted during the loading process; thus, the decreased amplitude of permeability was the largest of the three samples.

4) Analysis and discussion

Figures 7.32, 7.33, and 7.34 reveal that the permeability values of the coal samples decreased with an increase in the loading stress, which is a key factor affecting permeability and plays a decisive role in the evolution of permeability. During the loading process, the change rate of permeability at the initial loading stage was the largest, and then tended to be gentle. The change trends of the three different coal samples were similar; however, the bedding direction of each coal sample also had an influence on the resultant permeability. Generally, the permeability values of the coal samples could be ranked as: parallel bedding > oblique bedding > vertical bedding. During the initial loading stage, the permeability of the parallel bedding sample was 0.026 mD, which was 7.5 times higher than that of the vertical bedding sample (0.00346 mD) and 1.7 times higher than that of the oblique bedding sample (0.015 mD). These findings imply that the resistance for gas seepage in coal has a close relationship with the bedding direction. This reflects some macroscopic characteristics, whereby the fracture network along the bedding direction is more developed under the same stress

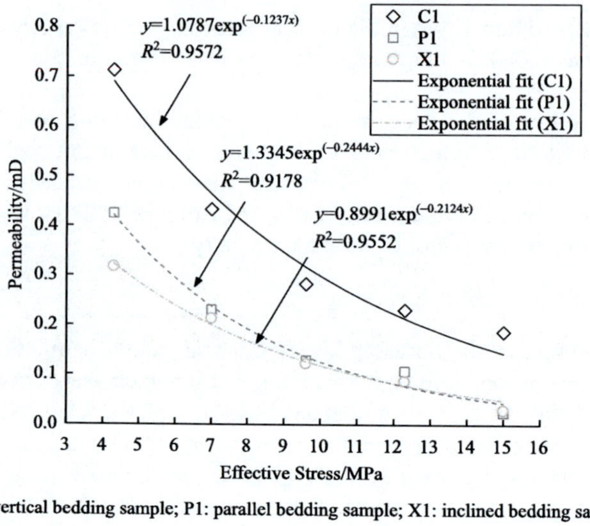

Fig. 7.35 Relationship between coal permeability and effective stress during loading

condition, thus benefiting gas seepage. However, the flow capacity of gas between the coal bedding is easily affected by the coal matrix block, such that the permeability values of the vertical bedding sample were the lowest of the three samples. The experimental results during the loading process agree with the research results of Gash [86] and some Japanese scholars [87].

Figure 7.35 shows that the permeability values of the three coal samples during the loading process decreased with an increase in effective stress. The permeability of the oblique bedding sample during the initial loading stage decreased rapidly with increasing effective stress, and the sensitivity of effective stress was the greatest, but subsequently tended to weaken. However, during the loading process, the vertical bedding and parallel bedding samples experienced almost the same degree of stress sensitivity.

Based on the analysis of the permeability evolution of these coal samples, it can be determined that the permeability and effective stress of the coal samples satisfied the following exponential relationship:

$$k = y\exp(x\sigma_e) \tag{7.104}$$

where y and x are the coupling coefficients of permeability and stress, respectively.

As shown in Table 7.6, the x values during the loading process were negative. The increase in stress caused the seepage channels to become compacted, blocked, and deformed; thus, the fracture width of each coal sample decreased rapidly, which led to a decrease in the gas seepage capacity.

Table 7.6 Exponential fitting results of coal permeability under loading/unloading conditions

Coal specimen	y	x	R^2
Vertical bedding	1.0787	−0.1237	0.9572
Parallel bedding	1.3345	−0.2444	0.9178
Inclined bedding	0.8991	−0.2124	0.9552

7.5.2 *Evolution of Coal Permeability During Unloading*

Unloading experiments were carried out based on the loading experiments described in Sect. 7.5.1. The test scheme and paths are given in Table 7.7 and Fig. 7.36. The stress state of the initial unloading point was a confining pressure of 12 MPa and a differential stress of 12 MPa. The unloaded confining pressure at each level was set to 2 MPa, and after a stable reading of the instrument, a set of permeabilities was

Table 7.7 Experimental scheme parameters

Test condition	No	Gas pressure P_f/MPa	Differential stress (σ_1–σ_3) /MPa	Confining pressure σ_3/MPa	Notes
Unloading	1	1	12	10	Unloading rate of confining pressure: 0.01 MPa/s
	2	1	12	8	
	3	1	12	6	
	4	1	12	4	
	5	1	12	2	

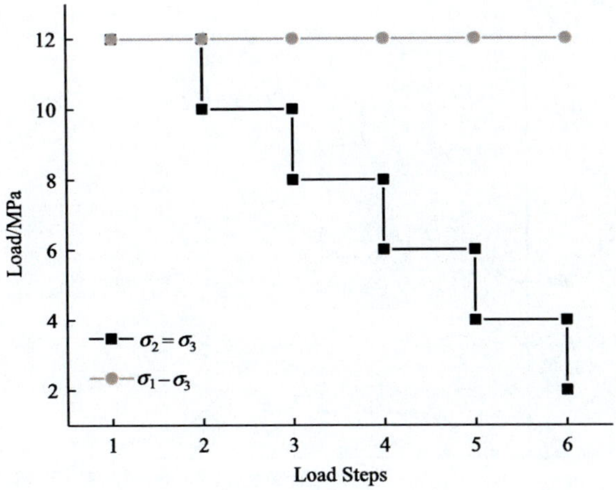

Fig. 7.36 Sketch map illustrating the unloading process

tested until the coal sample was destroyed. Once coal failure occurred, the gas pipe was rapidly shut down and the experiment was concluded.

After the conventional triaxial compression test of the three different coal samples (with vertical, parallel, and oblique beddings, respectively), axial pressure unloading tests were conducted, and the permeability of each coal sample during unloading was tested. The experimental results are shown in Figs. 7.37, 7.38, and 7.39.

1) Coal sample with vertical bedding

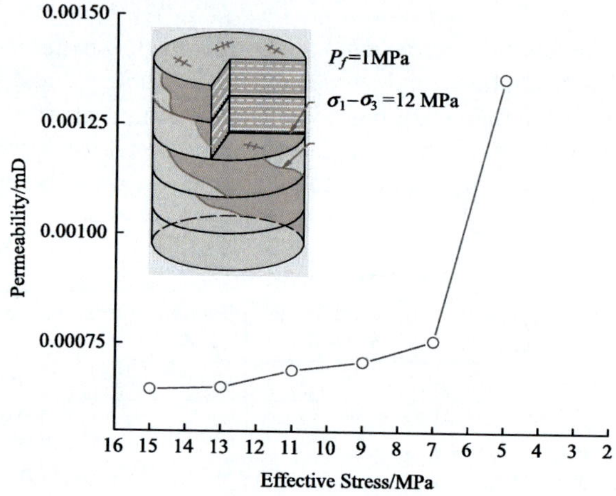

Fig. 7.37 Permeability evolution characteristics of a coal sample with vertical bedding during unloading

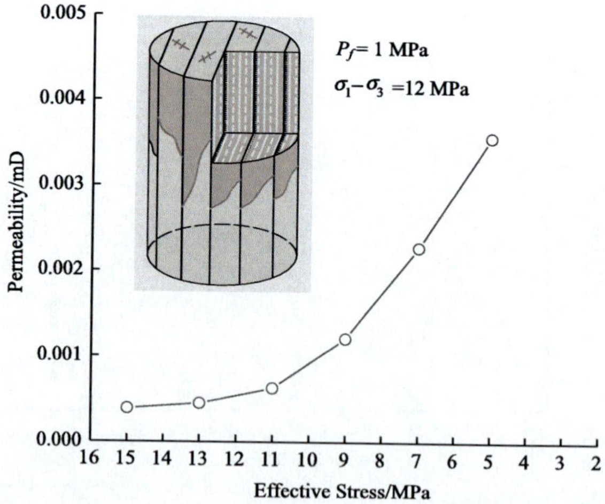

Fig. 7.38 Permeability evolution characteristics of a coal sample with parallel bedding during unloading

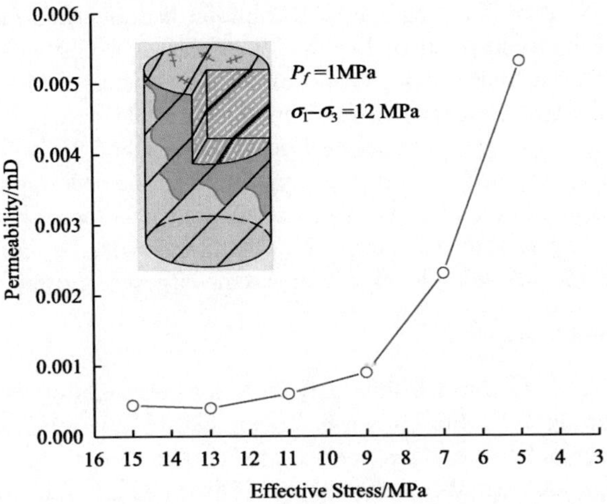

Fig. 7.39 Permeability evolution characteristics of a coal sample with oblique bedding during unloading

During unloading, the permeability of the coal sample with vertical bedding increased with decreasing effective stress. The permeability value at the unloading point was the same as that at the end point of loading (0.00065 mD). The permeability of the coal sample changed slowly during the early unloading stage. When the effective stress decreased from 15 to 7 MPa, the permeability value changed little; however, when the stress was unloaded to <7 MPa, the permeability value suddenly increased. This phenomenon is believed to be related to the recovery rate and degree of compressed fractures. During unloading, closed fractures were reopened and/or new fractures were generated. When the confining pressure was unloaded to 2 MPa, the permeability value was 0.00137 mD, which represented an increase of up to 110.8% with respect to that of the unloading point.

2) Coal sample with parallel bedding

The permeability of the coal sample with parallel bedding was the same at the unloading point and end point of loading (0.00062 mD). During unloading, the permeability value increased with decreasing effective stress. Although the change trend of permeability during the initial unloading stage was relatively small, it was larger than that of the vertical bedding sample. When the effective stress was reduced to 9 MPa, the permeability of the coal sample increased slowly to 0.00137 mD. When the confining pressure was unloaded to 2 MPa, the permeability increased to 0.00364 mD, which was 2.7 times higher than that of the vertical bedding sample under the corresponding conditions. The permeability value increased up to 487.1% with respect to that of the unloading point.

3) Coal sample with oblique bedding

The permeability of the coal sample with oblique bedding was the same at the unloading point and end point of loading (0.00062 mD). During unloading, the permeability value increased with decreasing effective stress. After the effective stress was reduced to 9 MPa, the permeability of the sample increased rapidly. When the effective stress was 5 MPa, the permeability value reached 0.0053 mD. Under the same unloading conditions, the permeability of the oblique bedding sample was 3.9 times higher than that of vertical bedding sample and 1.5 times higher than that of the parallel bedding sample. The permeability value increased up to 1132.5% with respect to that of the unloading point, which was the largest of the three coal samples.

4) Analysis and discussion

As shown in Fig. 7.40, through the analysis of the permeability evolution of the coal samples during unloading, it can be concluded that the relationship between coal permeability and effective stress also satisfied Eq. (7.104). With the decrease in effective stress, although the permeabilities of the three different coal samples all recovered and improved, the specifics varied widely. The reason for this is related to the strength characteristics of the coal samples and the degree of internal damage at unloading points. Figure 7.40 shows that the fitting correlation of each coal sample during unloading was lower than that during loading. These findings indicate that the interior of each loaded coal sample underwent plastic damage deformation, which resulted in differences in the degree of recovery during unloading as well as a large discreteness.

During the early unloading stage, the stress sensitivity of each coal sample was relatively small. When the effective stress was reduced to 9 MPa, the permeabilities of the coal samples increased quickly with decreasing effective stress. This

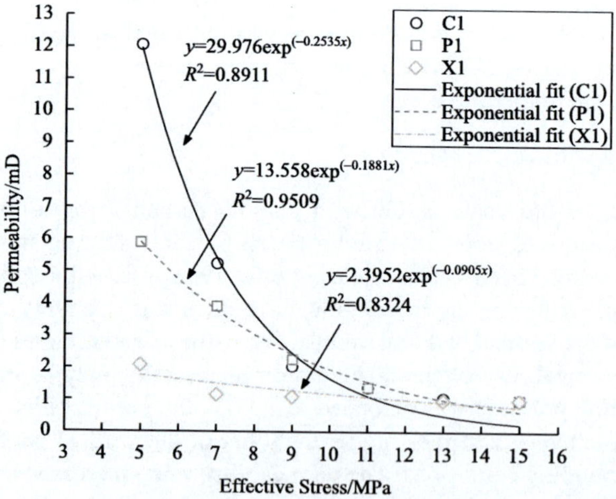

C1: vertical bedding sample; P1: parallel bedding sample; X1: oblique bedding sample

Fig. 7.40 Relationship between coal permeability and effective stress during unloading was particularly true for the oblique bedding sample, which had the greatest stress sensitivity and highest change of permeability. This was mainly because this sample experienced shear failure along the bedding plane during unloading, which generated many new fractures. Additionally, coal deformation occurring along the weak bedding face would have disconnected the cement in the bedding face, thus leading to the connection of the layer fractures and resulting in a sudden increase in coal permeability. Hence, it can be inferred that the sudden increase in coal permeability during the later unloading process was related to the dislocation of bedding faces and the generation of new fractures.

7.5.3 Analysis of the Differences in Permeability Evolution Between Loading and Unloading

Based on the experimental results of coal permeability evolution under loading and unloading conditions, the relationships between the effective stress and permeability values of the coal samples with different bedding directions under loading/unloading conditions were plotted in the same coordinate system (Figs. 7.41, 7.42, and 7.43).

Figures 7.41, 7.42, and 7.42 show that the permeability values of the three coal samples decreased with increasing effective stress. When the effective stress was reduced, the permeability of each sample could recover to some extent, but not to its original value. The permeability curves of the loading process did not coincide with

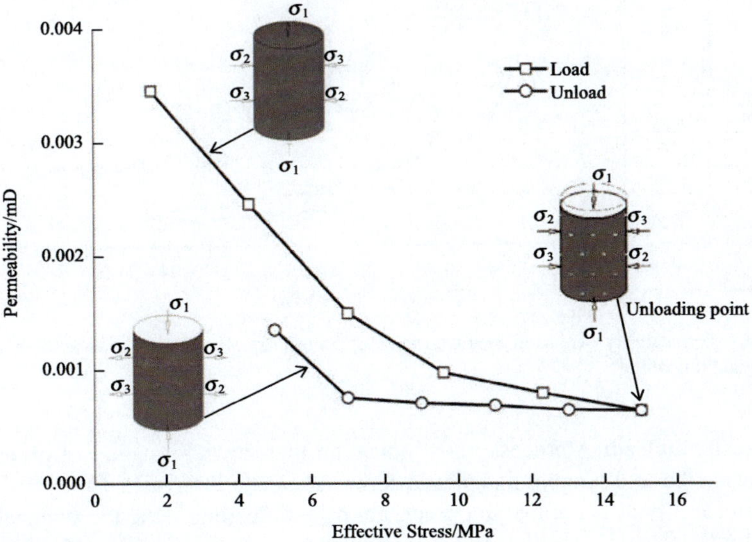

Fig. 7.41 Permeability evolution characteristics of a coal sample with vertical bedding during loading and unloading

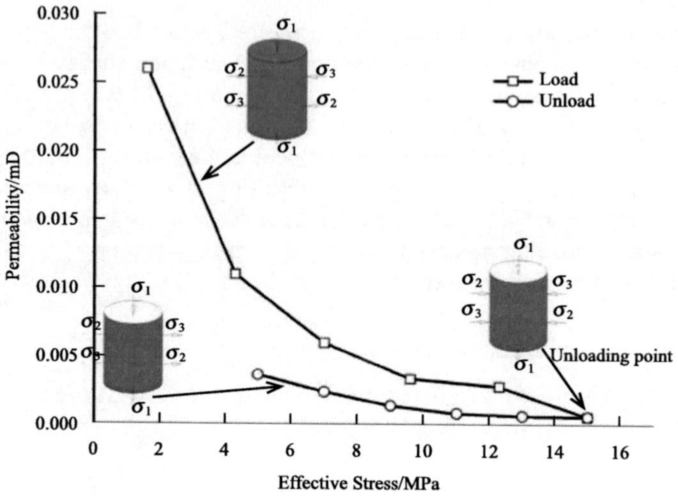

Fig. 7.42 Permeability evolution characteristics of a coal sample with parallel bedding during loading and unloading

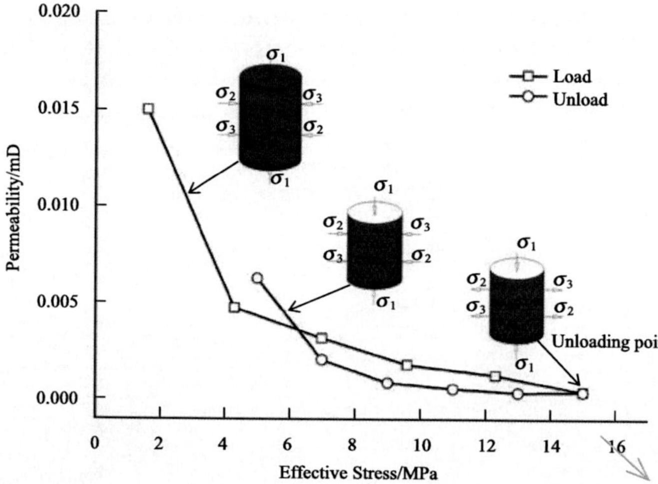

Fig. 7.43 Permeability evolution characteristics of a coal sample with oblique bedding during loading and unloading

those of the unloading process, thus indicating that the deformation of pores and fractures in the coal samples had plastic deformation characteristics. Compared with the permeability at the unloading point, after the unloading tests, the permeability values of the vertical, parallel, and inclined bedding samples increased by 110.8%, 487.1%, and 1132.5%, respectively; hence, the largest increase was for the oblique bedding sample. The deformation of pores and fractures due to the stress change in the surrounding coal and rock during coal mining will cause permanent damage to the permeability coefficient. Although the resulting deformation can be restored to a certain extent, the physical parameters of the coal seam cannot be recovered completely when the stress is restored to its original value.

The level of decrease in permeability with increasing effective stress differed between the samples, as did the recovery of permeability with decreasing effective stress. Under the corresponding stress conditions, the permeability values under the loading process were higher than those under the unloading process; hence, the unloading process was not simply the inverse of the loading process. Accordingly, it can be concluded that the complexity of the coal samples had a significant influence on their strength, stability, and seepage characteristics. From the change in permeability, we found that the oblique bedding sample was damaged when the confining pressure was unloaded to 2 MPa. This case indicates that the permeability after unloading not only recovered but was also enhanced to above that of the loading process; therefore, the original closed fracture in the coal sample was reopened and many new fractures were generated (see Fig. 7.44).

Through comparative analyses of the permeability evolution processes of three typical coal samples, it was found that the permeability recoveries of the samples during the unloading process differed significantly as a result of the strength and structural differences between the samples. When loading the samples to the same stress state of the unloading point, the coal sample with a smaller compressive strength would have experienced greater damage and deformation within the coal body. Generally, during the loading process, coal undergoes three stages: elastic deformation, elastic–plastic deformation, and destruction. If the stress at the unloading point is lower than the elastic limit of stress, the loading deformation will exhibit elastic deformation, and the unloading stress–strain curve will recover along the loading curve. If the stress of the unloading point is higher than the elastic limit of stress, the unloading stress–strain curve will deviate from the loading curve and will not be able to return to the original point. This means that both elastic deformation and plastic deformation will occur. Based on the above rationale, the corresponding deformation areas of different experimental coal samples located at the unloading point will be completely different to each other. The unloading points of the vertical, parallel, and

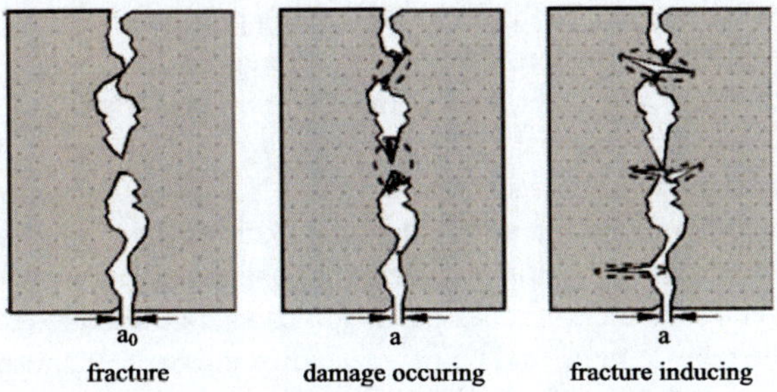

Fig. 7.44 Sketch map of coal structure and fracture damage evolution

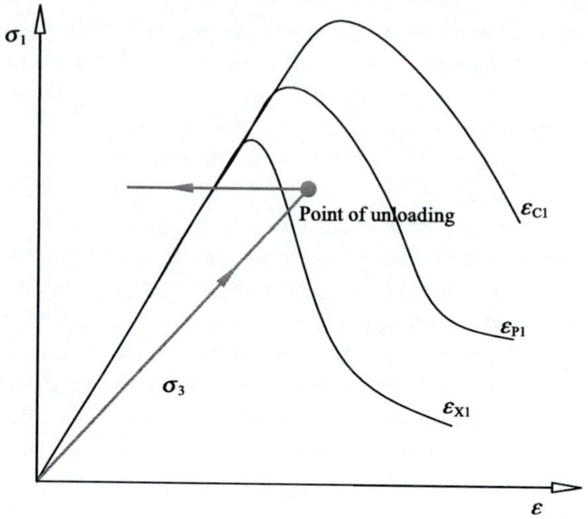

Fig. 7.45 Sketch diagram showing the relationship between the unloading point and stress–strain curve

oblique bedding samples were 81.6%, 89.8%, and 94.1% of their strengths, respectively. The stress–strain curve of the unloading point for each sample is shown in Fig. 7.45.

Under the same stress point of unloading, coal specimens with different strengths have different degrees of initial unloading damage as well as different recovery extents of damage and permeability. Hence, the permeability evolution during the unloading process exhibits three types of evolution paths: fully recovered, partially recovered, and suddenly increased. Combined with the analysis of the experimental and numerical results, we established three conceptual models of coal permeability evolution during the loading and unloading processes, as presented in Fig. 7.46. According to the experimental considerations in this book, the three types of models reflect the relationships between the permeability evolution characteristics and stress of coal samples with different strengths at the same stress state of the unloading point. However, the unloading point is variable; thus, the same sample would have different permeability evolution laws for different unloading points.

Figure 7.46 illustrates the permeability evolution path during the loading process along the path of $1 \rightarrow 2$. However, for the unloading process, owing to the differences in compressive strength, internal material, structure, and mechanical properties of samples, the damage degrees under the stress of the unloading point would vary. Thus, the permeability evolution after unloading exhibits three types of classic paths as follows. ① After unloading, the permeability changes along the path from $2 \rightarrow 1$ and shows a reversible loading process. This indicates that fractures or pores remain unchanged in the coal sample, and no new fractures appear. This result only occurs when coal undergoes elastic deformation, which is a property of a complete elastomer. This is an ideal state that is very difficult to realize experimentally, and the evolution

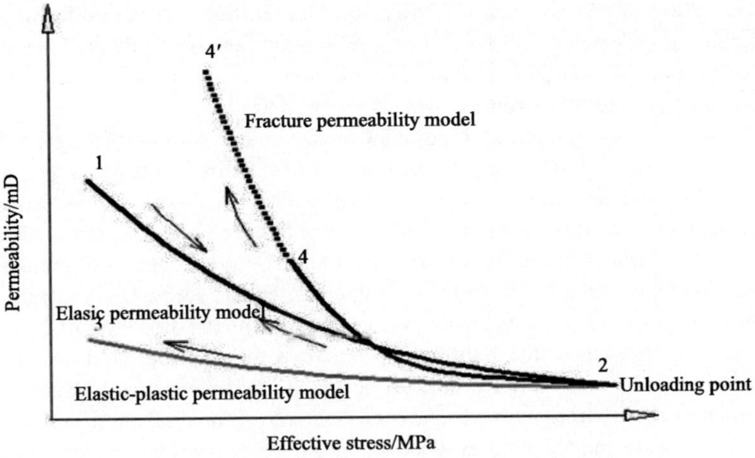

Fig. 7.46 Conceptual models of the permeability evolution for unloading coal

path of the permeability of this coal is defined as an elastic permeability model. ② After unloading, the permeability changes along the path from 2 → 3, that is, the permeability during unloading is always lower than that during loading. In this case, the sample at the unloading point not only undergoes elastic deformation but also causes plastic damage, and the fractures and pores in coal are changed. Owing to the damage caused by the loading process, the permeability in the unloading process cannot recover completely. The path of permeability evolution for coal with plastic deformation is defined as an elastic–plastic permeability model. ③ After unloading, the permeability changes along the path from 2 → 4. In this case, the permeability suddenly increases to a value that is higher than that during the loading process. To achieve this type of path, the coal sample usually has the minimum compressive strength; thus, it can have the largest internal damage extent at the same unloading point. For this path, when the stress state is unloaded to a certain value, original fractures in coal will be reopened, damaged, extended, and connected, and new fractures will also be generated. Thus, the permeability can be increased to a value that is much higher than that of the loading process. The path of permeability evolution is defined as the fracture permeability model.

7.5.4 Relationships Between Coal Body Damage Evolution and Permeability During Unloading

Real-time CT tests are an effective way to determine the internal damage development of coal samples under the effect of loading. However, in the process of real-time CT tests, the permeability of coal samples cannot be measured simultaneously because of the limitations of the experimental instrument. Moreover, because of the large dispersion of raw coal samples (distributional differences of beddings, joints, and fractures), the comparability of the experimental results of raw coal samples is poor. Until now, comprehensive laboratory research on the relationship between coal damage and permeability evolution has rarely been limited.

We used the stress paths of "fixed axial displacement with unloading confining pressure" (FADUCP) and "fixed deviatoric stress with unloading confining pressure" (FDSUCP) to measure the damage and permeability evolution characteristics of reconstituted coal samples with minimal discreteness. The effective stress principle (the same initial effective confining stress for the CT real-time test and permeability test) was used, and the effects of coal damage on methane seepage in the unloading process were analyzed according to the different experimental processes of the stress–strain curves. To analyze the relationship between the damage evolution and the permeability of the coal samples during the unloading process, real-time CT tests under the unloading stress path were carried out using FADUCP and FDSUCP. The obtained CT value and CT images were then used to analyze the damage evolution law of coal during the unloading process.

1. Damage evolution law of coal during unloading

1) Fixed axial displacement with unloading confining pressure

According to the experimental scheme, a coal sample in the initial state was subjected to a CT scan for the first time. The coal sample was then loaded to 8 MPa under hydrostatic pressure. The confining pressure was kept constant, and the sample was further loaded to a point slightly after the yield point with triaxial loading. At this time, a second CT scan was conducted on the coal sample. This mechanical state was set as the unloading point, the axial displacement was fixed, the confining pressure was gradually unloaded, and the coal sample was subjected to CT scans at different stress states during the unloading process.

The stress–strain curve under the FADUCP stress path obtained by the CT real-time tests is presented in Fig. 7.47, which shows that the unloading point was located slightly after the yield point. When loading to the unloading point, the coal sample underwent compaction, elastic deformation, and plastic deformation, thus agreeing with previous research conclusions. At the unloading point, the deviatoric stress and strain were 14.65 MPa and 3.13%, respectively. As the confining pressure was unloaded rapidly from 8 to 6 MPa, the deviatoric stress increased slightly. The peak of the deviatoric stress occurred at 15.37 MPa (with a strain of 3.21%) when the confining pressure was unloaded to 6 MPa. Subsequently, the deviatoric stress began to decrease as the confining pressure was unloaded, and the axial strain also increased slightly.

Table 7.8 and Fig. 7.48 show the changes in the CT values and CT images obtained under the FADUCP stress path. As there were many scanning layers in each scan, the 8th, 16th, and 25th layers were selected as typical representatives for analysis. According to the change of the CT values during the unloading process, which was relatively small, the deviatoric stress increased from 14.7 to 15.4 MPa when the confining pressure decreased from 8 to 6 MPa, implying that there was not extensive

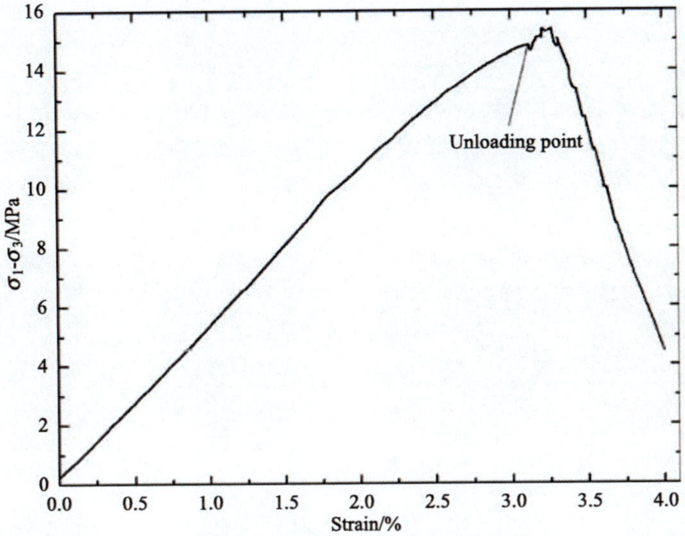

Fig. 7.47 Stress–strain curve under the FADUCP stress path

Table 7.8 Variation of CT values during FADUCP

Scan no.	Confining pressure (MPa)	$(\sigma_1 - \sigma_3)$/MPa	8th scanning layer		16th scanning layer		25th scanning layer	
			H/HU	*SD*	*H*/HU	*SD*	*H*/HU	*SD*
1	0.0	0.0	224.3	27.4	198.0	25.5	172.7	24.6
2	8.0	14.65	279.1	58.4	250.4	73.7	226.6	89.3
3	6.0	15.37	277.4	26.3	248.9	21.7	224.0	20.6
4	4.8	14.89	272.5	88.7	242.9	101.4	218.5	103.5
5	3.6	13.25	269.4	27.0	236.0	21.2	213.6	20.0
6	2.4	11.33	266.5	27.4	223.2	22.4	199.0	21.3
7	0.8	4.79	251.7	28.8	199.2	30.4	176.1	33.3

Note H represents the CT value, and *SD* represents the variance of the CT value

damage to the sample. Thus, for brevity, Fig. 7.48 only displays the CT images for a confining pressure of 0.8–4.8 MPa and a deviatoric stress of 4.79–14.89 MPa.

The damage evolution process for the FADUCP stress path was as follows:

(1) Initial scanning (1st scan) was carried out before loading. The different CT values of each scanning layer indicated differences in the initial damage.
(2) When the confining pressure reached 8 MPa and the deviatoric stress reached 14.65 MPa (2nd scan), compared with the initial scan (deviatoric stress of 0 MPa), the CT values of the 8th, 16th, and 25th scanning layers increased by 54.8, 52.4, and 53.9, respectively; hence, the sample was consolidated.

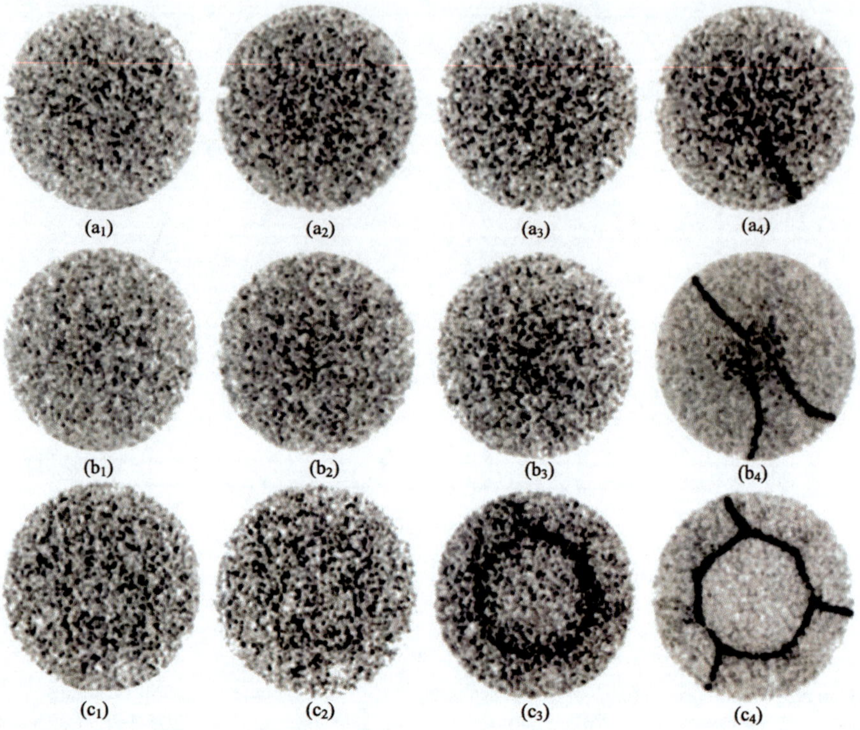

Fig. 7.48 Variation of CT images at different stress states. $\mathbf{a_1}$–$\mathbf{a_4}$, $\mathbf{b_1}$–$\mathbf{b_4}$, and $\mathbf{c_1}$–$\mathbf{c_4}$ are the CT images of the 8th, 16th, and 25th scanning layers for a confining pressure of 4.8 MPa, 3.6 MPa, 2.4 MPa, and 0.8 MPa, respectively

(3) When the confining pressure was 6 MPa and the deviatoric stress reached 15.37 MPa (3rd scan), compared with the 2nd scan, the CT values of the 8th, 16th, and 25th scanning layers were reduced by 1.7, 1.5, and 2.6, respectively; hence, the sample was damaged.

(4) When the confining pressure was 4.8 MPa and the deviatoric stress reached 14.89 MPa (4th scan; Fig. 7.48$\mathbf{a_1}$–$\mathbf{c_1}$), compared with the 3rd scan, the CT values of the 8th, 16th, and 25th scanning layers were reduced by 4.9, 6.0, and 5.5, respectively; hence, the sample experienced further damage.

(5) When the confining pressure was 3.6 MPa and the deviatoric stress reached 13.25 MPa (5th scan; Fig. 7.48$\mathbf{a_2}$–$\mathbf{c_2}$), compared with the 4th scan, the CT values of the 8th, 16th, and 25th scanning layers were reduced by 3.1, 6.9, and 4.4, respectively; hence, the damage to the sample accumulated gradually.

(6) When the confining pressure was 2.4 MPa and the deviatoric stress reached 11.33 MPa (6th scan; Fig. 7.48$\mathbf{a_3}$–$\mathbf{c_3}$), compared with the 5th scan, the CT values of the 16th and 25th scanning layers were reduced by 12.8 and 32.6, respectively, while the CT value of the 8th scanning layer was reduced by 2.9.

Areas of obvious damage were observed in the 16th and 25th scanning layers, particularly the ring-shaped damage cracks in the 25th scanning layer.

(7) When the confining pressure was 0.8 MPa and the deviatoric stress reached 4.79 MPa (7th scan; Fig. 7.48**a$_4$**–**c$_4$**), compared with the 6th scan, the CT values of the 8th, 16th, and 25th scanning layers were reduced by 14.8, 24.0, and 22.9, respectively. Damage cracks were observed in the 8th and 16th scanning layers, and the ring-shaped damage cracks of the 25th scanning layer continuously developed outward.

The CT value of the scanning layer reflects the density of the coal sample, and the heterogeneity of the CT value distribution can also indirectly reflect the damage evolution of the scanning layers. To analyze the damage evolution in different areas of the scanning layers, we define the variable Ω, which comprehensively reflects the change in the CT value and CT variance. The variable Ω is defined as the "CT damage value" and is expressed as follows:

$$\Omega = \frac{SD}{H + 1000} \tag{7.105}$$

where *SD* represents the CT variance, and *H* represents the CT value (HU).

To further analyze the damage development of the different regions in each scanning layer during unloading, the Ω values reflecting the non-uniformity of the distribution of the CT values were calculated using Eq. (7.105). As shown in Figs. 7.49, 7.50, and 7.51, where the abscissa indicates the radius of the circle to the center of the scanning layer, the Ω values changed in local areas during the FADUCP process. The values increased suddenly under the stress state when the confining pressure

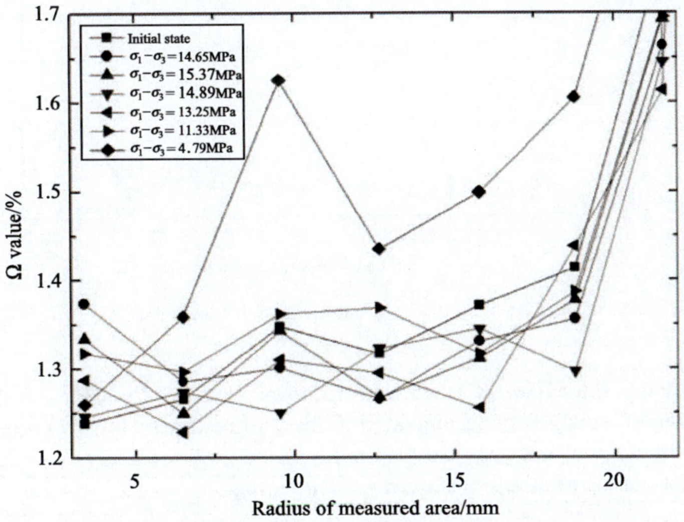

Fig. 7.49 Ω values of the 8th scanning layer during the unloading process

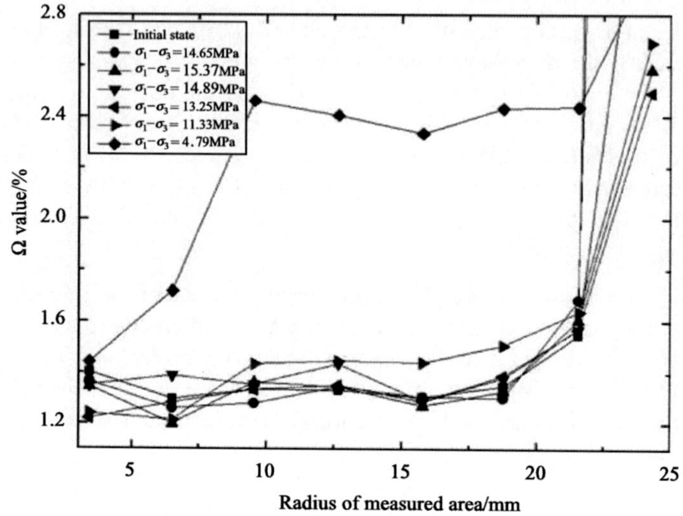

Fig. 7.50 Ω values of the 16th scanning layer during the unloading process

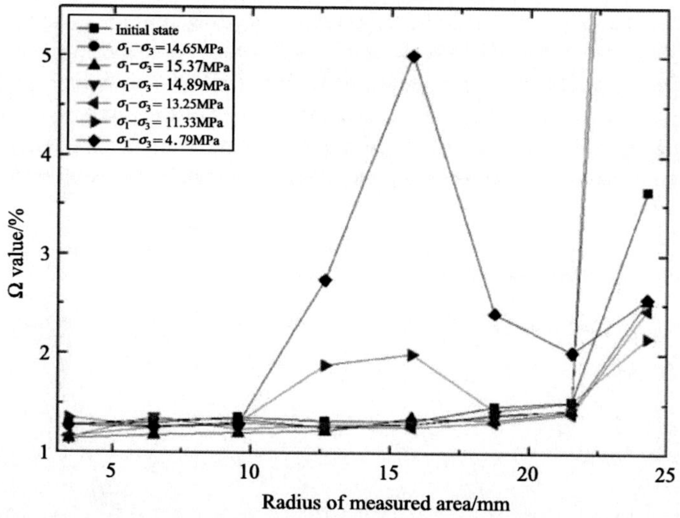

Fig. 7.51 Ω values of the 25th scanning layer during the unloading process

was unloaded to 0.8 MPa and the deviatoric stress reached 4.79 MPa. According to the CT images, damage cracks appeared in these local areas. However, the large Ω values at the edge of the coal sample could not be used for analysis owing to the influence of convolution artifacts during CT imaging.

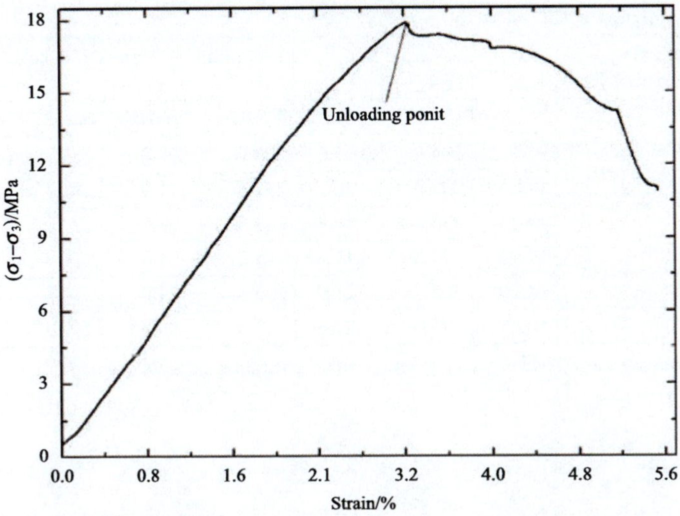

Fig. 7.52 Stress–strain curve under the FDSUCP stress path

2) Fixed deviatoric stress under an unloading confining pressure (FDSUCP)

According to the experimental scheme, a coal sample in the initial state was subjected to a CT scan for the first time. The coal sample was then loaded to 8 MPa under hydrostatic pressure. Then, the confining pressure was kept constant, and the sample was further loaded to the point slightly after the yield point with triaxial loading. At this time, a second CT scan was conducted on the coal sample. This mechanical state was set as the unloading point, the deviatoric stress was fixed, the confining pressure was gradually unloaded, and the coal sample was then subjected to CT scans at different stress states during the unloading process.

The stress–strain curve under the FDSUCP stress path obtained by the CT real-time tests is displayed in Fig. 7.52. Owing to the variations among different specimens, the stress and strain at the yield point under the FDSUCP stress path differed to those of the FADUCP stress path. The unloading point was located slightly after the yield point (Fig. 7.52). When loading to the unloading point, the coal sample underwent compaction, elastic deformation, and plastic deformation. At the unloading point, the deviatoric stress and strain were 17.84 MPa and 3.19%, respectively. Before the confining pressure was unloaded to 6 MPa, the deviatoric stress was maintained at a constant value. After that, the deviatoric stress began to decrease, while the axial strain increased rapidly.

Table 7.9 and Fig. 7.53 show the changes in the CT values and CT images obtained, respectively, under the FDSUCP stress path. The 6th, 15th, and 25th layers were selected as representatives for analysis. Additionally, for brevity, Fig. 7.53 only displays the CT images when the confining pressure was unloaded from 4.8 MPa to 2.4 MPa at a deviatoric stress of <17.13 MPa.

Table 7.9 Variation of CT values during FDSUCP

Scan no.	Confining pressure/MPa	$(\sigma_1 - \sigma_3)$ /MPa	6th scanning layer		15th scanning layer		25th scanning layer	
			H/HU	*SD*	*H*/HU	*H*/HU	*SD*	*H*/HU
1	0	0	184.0	21.6	208.6	22.3	234.3	22.1
2	8	17.84	237.3	20.1	257.8	21.8	281.4	21.3
3	6	17.73	233.1	19.8	252.3	21.5	277	21.4
4	4.8	17.13	224.3	19.2	244.2	21.4	274.4	21.0
5	3.6	14.29	206.7	21.9	226.3	21.0	270.8	22.0
6	2.4	10.89	195.6	27.5	207.3	23.8	259.9	22.4

Note *H* represents the CT value, and *SD* represents the variance of the CT value

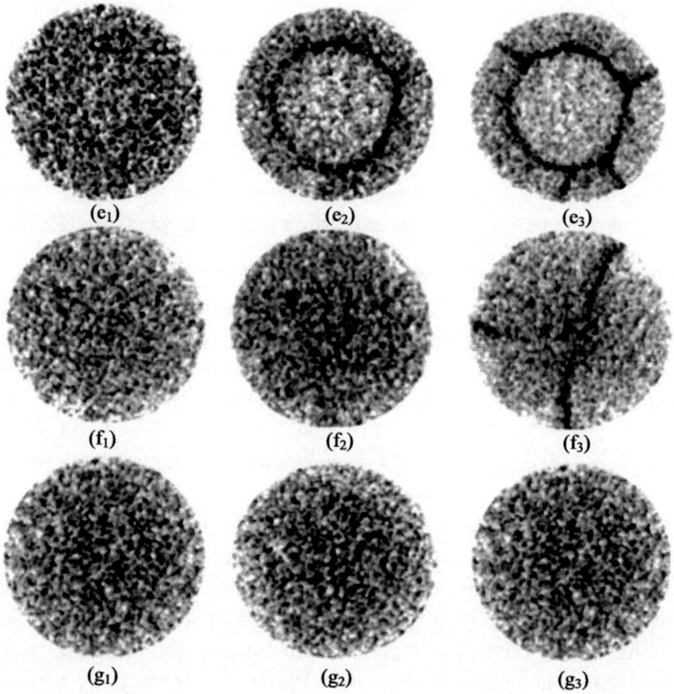

Fig. 7.53 Variation of CT images in the FDSUCP stress path. $\mathbf{e_1}$–$\mathbf{e_3}$, $\mathbf{f_1}$–$\mathbf{f_3}$, $\mathbf{g_1}$–$\mathbf{g_3}$ are the CT images of the 8th, 16th, and 25th scanning layers at a confining pressure of 4.8 MPa, 3.6 MPa, and 2.4 MPa, respectively

The damage evolution process of the FDSUCP stress path was as follows:

(1) Initial scanning (1st scan) was carried out before loading. The different CT values of each scanning layer indicated differences in the initial damage.
(2) When the confining pressure was 8 MPa and the deviatoric stress reached 17.84 MPa (2nd scan), compared with the initial scan, the CT value of the 6th, 15th, and 25th scanning layers increased by 53.3, 49.2, and 47.1, respectively; hence, the sample was consolidated.
(3) When the confining pressure was 6 MPa and the deviatoric stress reached 17.73 MPa (3rd scan), compared with the 2nd scanning, the CT values of the 6th, 15th, and 25th scanning layers were reduced by 4.2, 5.5, and 4.4, respectively; hence, damage to the sample began to appear.
(4) When the confining pressure was 4.8 MPa and the deviatoric stress reached 17.13 MPa (4th scan; Figs. 7.53e_1–g_1), compared with the 3rd scan, the CT values of the 6th, 15th, and 25th scanning layers were reduced by 8.8, 8.1, and 2.6, respectively; hence, damage to the sample continued to develop.
(5) When the confining pressure was 3.6 MPa and the deviatoric stress reached 14.29 MPa (5th scan; Figs. 7.53e_2–g_2), compared with the 4th scan, the CT values of the 6th and 15th scanning layers were reduced by 17.6 and 17.9, respectively. Obvious areas of damage were observed in the 6th and 15th scanning layers, particularly the ring-shaped damage cracks identical to Figs. 7.48c_3 in the 6th scanning layer.
(6) When the confining pressure was 2.4 MPa and the deviatoric stress reached 10.89 MPa (6th scan; Figs. 7.53e_3–g_3), compared with the 5th scan, the CT values of the 6th, 15th, and 25th scanning layers were reduced by 11.1, 19.0, and 10.9, respectively. Damage cracks were observed in the 15th scanning layer, and ring-shaped damage cracks also developed outward in the 6th scanning layer.

According to the CT values and CT variances obtained in the FDSUCP stress path, the CT damage values (Ω) were calculated using Eq. (7.105). In Figs. 7.54, 7.55, and 7.56, the abscissa indicates the radius of the circle to the center of the scanning layer. In contrast with the CT images, similar to the results of the FADUCP stress path, the Ω values of the local area increased obviously when damage cracks emerged in local areas.

2. Relationships between permeability and damage evolution of coal during unloading

1) Variation of coal permeability in the FADUCP process

According to the characteristics of the triaxial press, when the axial displacement is fixed, the axial pressure cannot be maintained and gradually decreases. According to the experimental scheme, a coal sample with an initial confining pressure of 10 MPa was loaded to a point slightly after the yield point. Then, the coal permeability of was tested using the FADUCP stress path.

Figure 7.57 presents the relationship between the deviatoric stress and confining

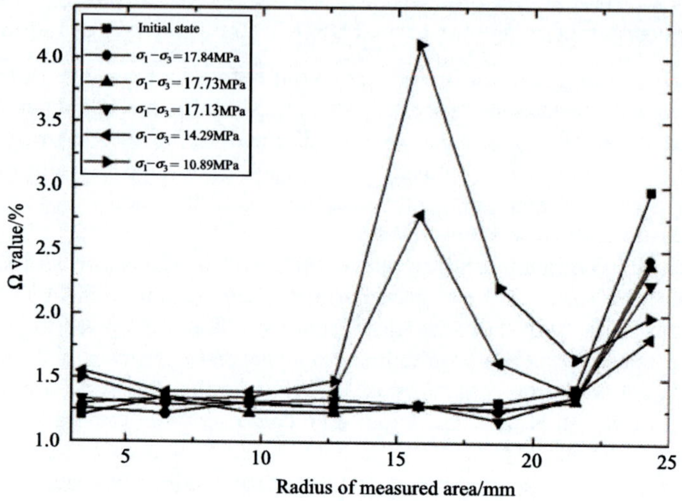

Fig. 7.54 Ω values of the 6th scanning layer in different measured areas

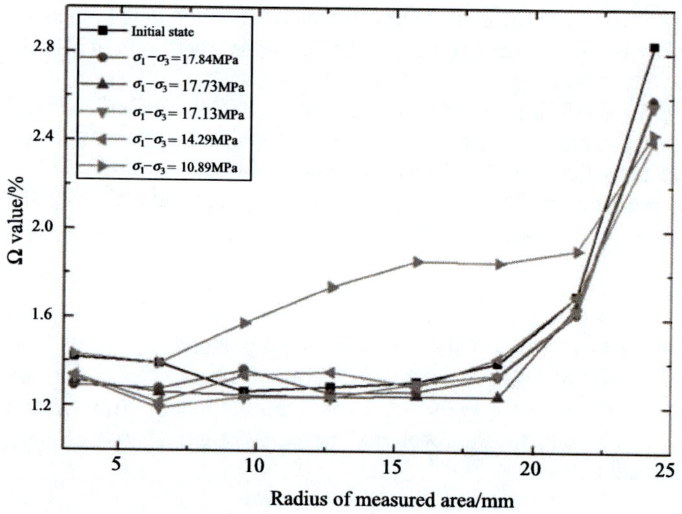

Fig. 7.55 Ω values of the 15th scanning layer in different measured areas

pressure tested by the transient pressure pulse method in the FADUCP process under a confining pressure of 10 MPa and a methane pressure of 2 MPa, while Fig. 7.58 shows the relationship between permeability and strain under the same conditions.

It can be seen from Fig. 7.57 that the deviatoric stress and confining pressure decreased at the same time under the FADUCP stress path. During the initial stage, the reduction rate of the deviatoric stress was small with the unloading of the confining

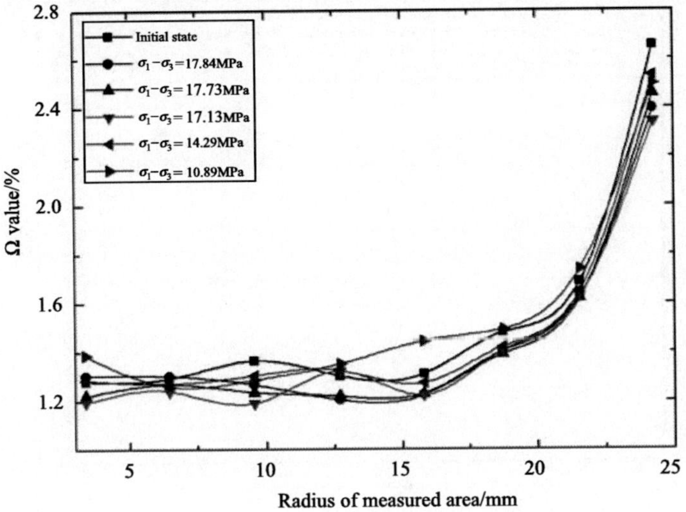

Fig. 7.56 Ω values of the 25th scanning layer in different measured areas

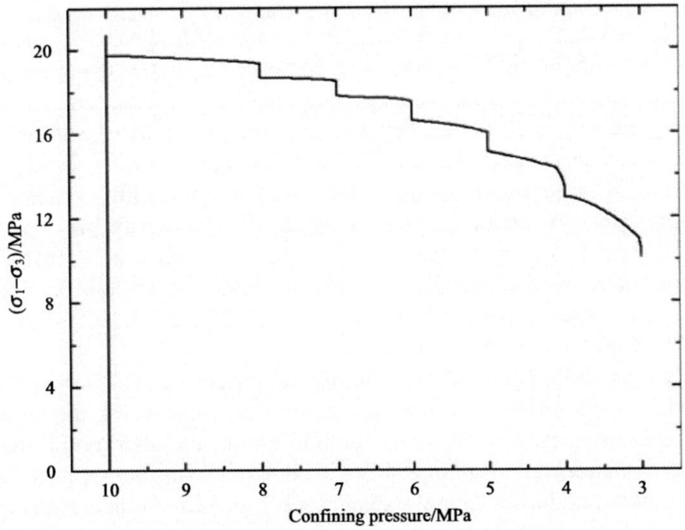

Fig. 7.57 Relationship between the deviatoric stress and confining pressure in the FADUCP process under a confining pressure of 10 MPa

pressure. When the damage accumulated to a certain extent and the confining pressure was unloaded to 7 MPa, the reduction rate of the deviatoric stress increased significantly.

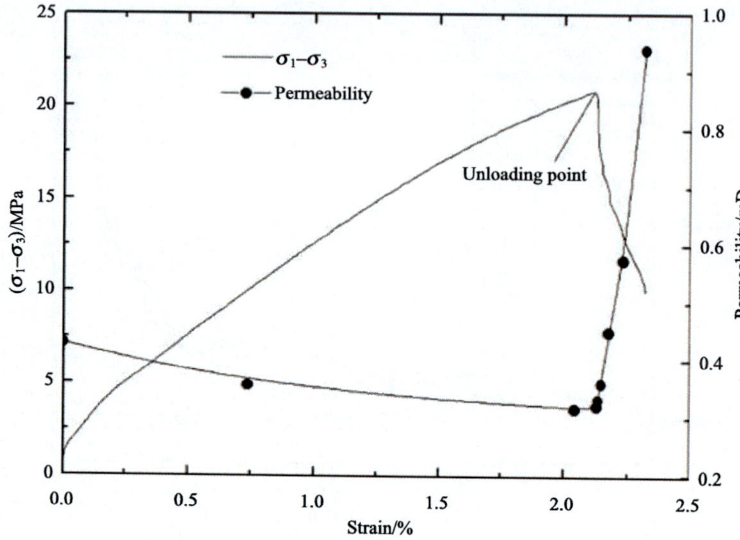

Fig. 7.58 Variation of permeability in the FADUCP process under a confining pressure of 10 MPa

Figure 7.58 plots the permeability test results under the FADUCP stress path at a confining pressure of 10 MPa. When the coal sample was in the stress state of a hydrostatic pressure of 10 MPa, the initial permeability was 0.429 mD. Maintaining the confining pressure at 10 MPa, when the sample was loaded to the unloading point (i.e., from elastic deformation to the point slightly higher than the yield point), the permeability of coal gradually reduced. Because the permeability at the unloading point is impossible to measure under the FADUCP stress path, we measured the permeability in the vicinity of the unloading point, which was 0.316 mD. When the confining pressure was unloaded to 9 MPa, 8 MPa, 7 MPa, 6 MPa, 5 MPa, and 4 MPa, the corresponding permeability values were 0.322 mD, 0.332 mD, 0.360 mD, 0.450 mD, 0.574 mD and 0.939 mD, respectively.

Before the unloading point, that is, during the process in which the coal sample was loaded from the hydrostatic pressure state to the yield point, the permeability decreased continuously. The closer the sample was to the yield point, the smaller the relative reduction rate of permeability. Compared with the permeability at the unloading point, that during the unloading process increased gradually, and the rate of increase also increased. At the same time, a permeability threshold was observed during the process of increasing permeability. When this threshold was exceeded, the increase rate of permeability increased with the continuous unloading of the coal sample. Compared with the confining pressure and deviatoric stress curves for the unloading process, the threshold point was the stress state where the deviatoric stress reduction rate increased significantly.

2) Variation of coal permeability in the FDSUCP process

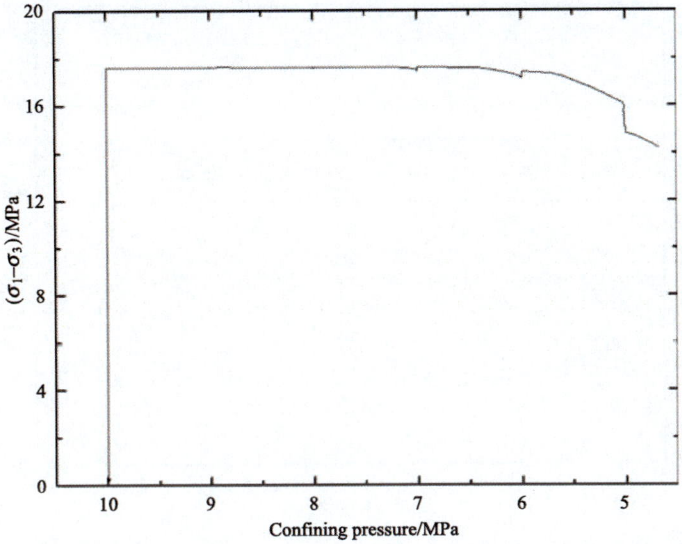

Fig. 7.59 Relationship between the deviatoric stress and confining pressure in the FDSUCP process under a confining pressure of 10 MPa

The deviatoric stress in the FDSUCP process is fixed by the permeability measurement instrument (i.e., the automatic adjustment of the axial displacement of the coal/rock stress-seepage coupling instrument). According to the experimental scheme, a coal sample with an initial confining pressure of 10 MPa was loaded to a point a little after the yield point. Then, the permeability of the coal sample was tested using the FDSUCP stress path.

Figure 7.59 shows the relationship between the deviatoric stress and confining pressure tested by the transient pressure pulse method in the process of FDSUCP under a confining pressure of 10 MPa and a methane pressure of 2 MPa, while Fig. 7.60 displays the relationship between permeability and strain under the same conditions.

As shown in Fig. 7.59, under the stress path of FDSUCP, the deviatoric stress could be kept constant with the unloading of the confining pressure before the confining pressure was unloaded to 7 MPa. After the confining pressure was unloaded to 7 MPa, the coal sample began to suffer damage, and the deviatoric stress decreased gradually with the unloading of the confining pressure.

Figure 7.60 presents the permeability test results under the FDSUCP stress path at a confining pressure of 10 MPa. When the coal sample was in the stress state of hydrostatic pressure of 10 MPa, the initial permeability was 0.31 mD. Maintaining a confining pressure of 10 MPa, when the coal sample was loaded to the unloading point, its permeability gradually reduced. The permeability of the sample at the unloading point was 0.163 mD. When the confining pressure was unloaded to 9 MPa,

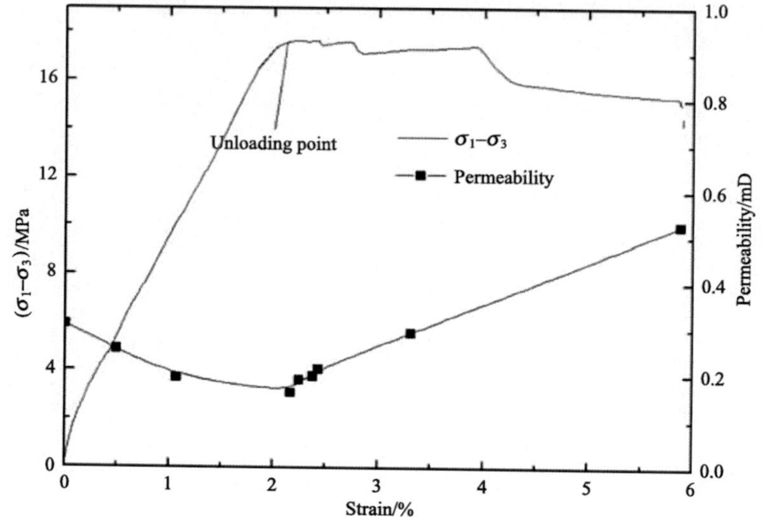

Fig. 7.60 Variation of permeability in the FDSUCP process under a confining pressure of 10 MPa

8 MPa, 7 MPa, 6 MPa, and 5 MPa, the measured permeability values were 0.190 mD, 0.199 mD, 0.213 mD, 0.295 mD, and 0.526 mD, respectively.

Compared with the results obtained from the FADUCP stress path, the axial deformation of the coal sample was large in the FDSUCP tests because the deviatoric stress was kept constant. The smaller the confining pressure, the lower the deviatoric stress at the yield point.

Before the unloading point of the coal sample, that is, during the process from the hydrostatic pressure state to the yield point, its permeability decreased continuously, and the closer to the yield point, the smaller the relative reduction rate of permeability. Compared with the permeability at the unloading point, that during the unloading process increased gradually. With the unloading of the load (i.e., axial pressure and confining pressure), the increase rate of permeability also increased. According to the variation in coal permeability during the unloading process, a permeability threshold was also observed in the process of increasing permeability, which was the same as that of the FADUCP stress path. When this threshold was exceeded, the rate of increase of permeability increased obviously. Compared with the confining pressure and deviatoric stress curves in the unloading process, the threshold point corresponded to the stress state when the deviatoric stress could no longer be maintained.

3. Effects of coal damage on permeability

In the previous section, the influence of damage on permeability was analyzed from a theoretical viewpoint. This section focuses on the effect of coal damage on permeability during the unloading process based on the experimental results of the real-time CT test and permeability experiments.

To analyze the effect of coal sample damage on the permeability during the unloading process, the FADUCP and FDSUCP stress paths were selected to conduct the permeability tests. The CT real-time scanning experiments with an initial confining pressure of 8 MPa were compared to the permeability experiments at an initial effective confining pressure of 8 MPa (confining pressure of 10 MPa and gas pressure of 2 MPa).

Owing to instrument error, sample preparation error, and the adsorption of gas in coal, the stress–strain curves obtained from the CT scan experiment and permeability experiment differed to some extent. However, the deformation laws of the coal samples obtained by the CT and permeability tests were almost the same under the same initial effective confining pressure and unloading path. This similarity verified the feasibility of the experimental design, in which the results obtained separately from the X-ray CT scan experiment and permeability experiment could be linked together to analyze the damage development and its effect on the permeability of the reconstituted coal samples.

1) Damage variable

To directly study how coal body damage affects the evolution of permeability, it is necessary to calculate the change in the damage variable. The CT values reflect the density of the material, and the damage variable can be expressed by the change in material density, as shown in Eq. (7.106) [88]. Thus, based on this equation and the CT value, we attempted to build a relationship between the CT value, material density, and damage variable.

$$D = \frac{1}{C_{\mathrm{CT}}^2}\frac{\Delta\rho}{\rho_0} \tag{7.106}$$

where D is the sample damage variable, C_{CT} is the space resolution of the CT machine, $\Delta\rho$ is the change in density, g/cm^3; and ρ_0 is the initial density (apparent density) of coal, g/cm^3.

In an experiment, the density of coal (apparent density) has the following relationship with porosity:

$$\rho = (1 - \phi)\rho_t \tag{7.107}$$

where ρ_t is the true density of the coal sample, g/cm^3; and ϕ is the porosity of coal.

The porosity of a coal sample has the following relationship with the CT value:

$$\phi = \frac{H_s - H}{H_s - H_a} \tag{7.108}$$

where H is the CT value, H_s is the initial CT value, and H_a is the CT constant.

Substituting Eq. (7.108) into Eq. (7.107) gives:

$$\rho = \frac{H - H_a}{H_s - H_a}\rho_t \quad (7.109)$$

According to the definition of the CT value, $H_a = -1000$, we obtain Eq. (7.110):

$$\rho = \frac{H + 1000}{H_s + 1000}\rho_t \quad (7.110)$$

Then, substituting Eq. (7.110) to Eq. (7.106) yields:

$$D = \frac{1}{C_{\mathrm{CT}}^2}\left(\frac{(H + 1000)\rho_t}{(H_s + 1000)\rho_0} - 1\right) \quad (7.111)$$

Equation (7.111) is the obtained expression of the damage variable expressed by the CT value according to the relationship between the CT value, coal density, and the damage variable. Hence, once the initial/subsequent CT values and initial material density are known, we can obtain the change in the damage variable.

2) Effects of damage on coal permeability

The initial density of the coal sample used in the CT real-time test with the FADUCP stress path was 1.136 g/cm^3, and that with the FDSUCP stress path was 1.1683 g/cm^3. The space resolution of the CT machine was 0.208. Based on the mean CT values of each scanning layer during unloading, the basic density of the coal samples for the FADUCP and FDSUCP stress paths were obtained according to Eq. (7.110). When the damage variable at the unloading point was set to zero, the damage variables for the FADUCP and FDSUCP stress paths were obtained according to Eqs. (7.110) and (7.111), and the CT values of different scanning layers were also determined.

Under an initial effective confining pressure of 8 MPa, the damage variable and permeability values were obtained separately from CT scan experiments and permeability tests with the FADUCP and FDSUCP stress paths. Then, the relationships between ① the damage variable and deviatoric stress, ② permeability and deviatoric stress, ③ the damage variable and the effective confining pressure, and ④ permeability and the effective confining pressure under the FADUCP (Figs. 7.61 and 7.62) and FDSUCP stress paths were obtained (Figs. 7.63 and 7.64).

The experimental error, sample preparation error, and deviatoric stress at the unloading point differed between the CT real-time test and permeability test. The trends of the damage variables and permeabilities obtained by the CT real-time test and permeability test in Figs. 7.61, 7.62, 7.63 and 7.64 were approximately consistent under the same stress path during unloading.

When the effective confining pressure was unloaded from 8 to 6 MPa, the increment of the damage variable under the FADUCP and FDSUCP stress paths increased to 0.0351 and 0.084, respectively, while the relative increment of permeability was 1.7% and 16.7%, respectively. The increases in the damage variable and permeability were relatively slight. In the CT scan tests with the FADUCP stress path, the damage variables were 0.1368, 0.2273, 0.414, and 0.7949 when the effective

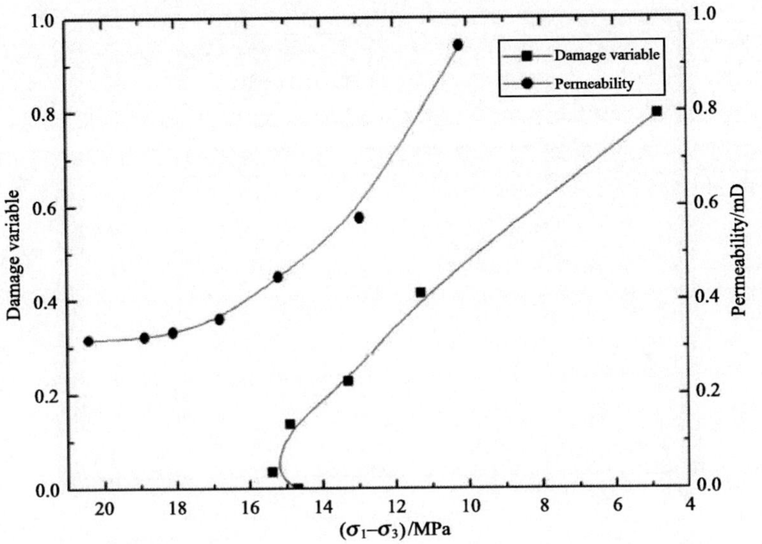

Fig. 7.61 Deviatoric stress versus the damage variable and permeability under the FADUCP stress path

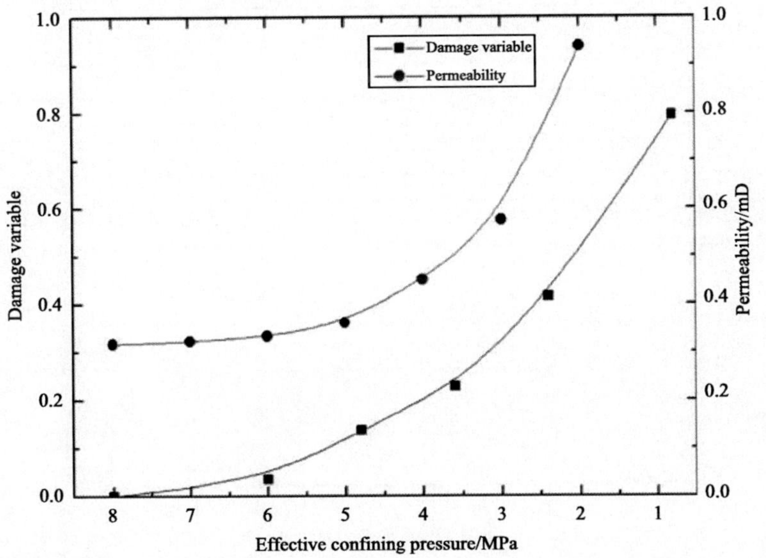

Fig. 7.62 Effective confining pressure versus the damage variable and permeability under the FADUCP stress path

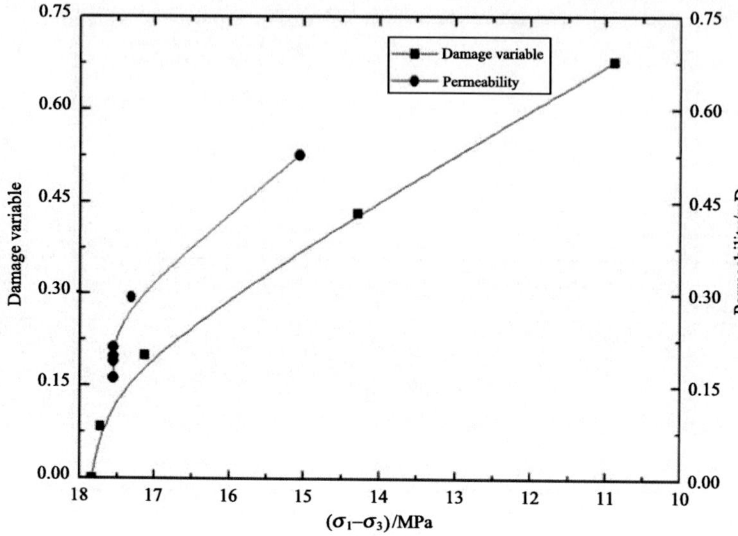

Fig. 7.63 Deviatoric stress versus the damage variable and permeability under the FDSUCP stress path

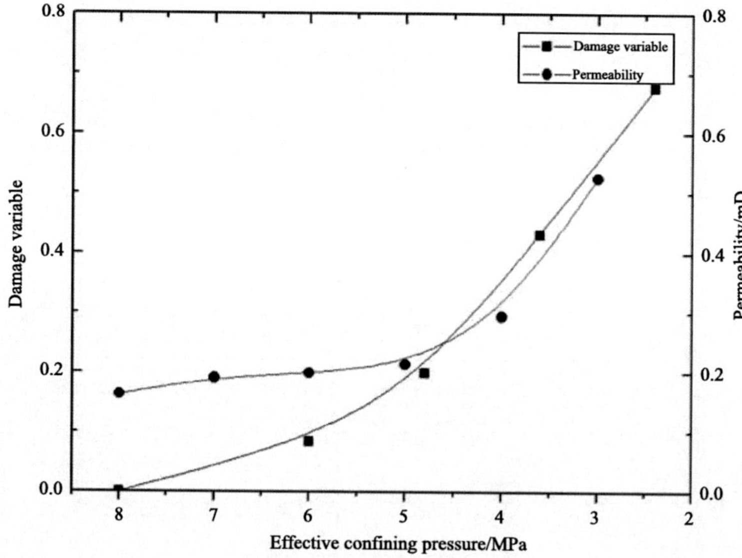

Fig. 7.64 Effective confining pressure versus the damage variable and permeability under the FDSUCP stress path

confining pressure was unloaded to 4.8 MPa, 3.6 MPa, 2.4 MPa, and 0.8 MPa, respectively, with a decreased deviatoric stress. In the permeability tests, the coal permeability values were 0.3603 mD, 0.4495 mD, 0.5744 mD, and 0.9393 mD when the effective confining pressure was unloaded to 5 MPa, 4 MPa, 3 MPa, and 2 MPa, respectively. In the CT scan tests with the FDSUCP stress path, the damage variables were 0.2001, 0.4323, and 0.6771 when the effective confining pressure was unloaded to 4.8 MPa, 3.6 MPa, and 2.4 MPa, respectively, with a decreased deviatoric stress. In the permeability tests, the coal permeability values were 0.2135 mD, 0.2946 mD, and 0.52582 mD when the effective confining pressure was unloaded to 5 MPa, 4 MPa, and 3 MPa, respectively.

Therefore, according to the variation trends of the damage variable and permeability under different unloading paths, it can be concluded that the increase rates of the damage variable and permeability increased with a decrease in the deviatoric stress when the effective confining pressure was unloaded to 5 MPa.

7.6 Permeability Evolution Model of Gas Migration in Coal in a Stress Disturbed Area

7.6.1 Basic Ideas of Model Construction

Gas extraction in underground coalmines requires a variety of excavation stress disturbances, such as those resulting from protective layer mining (regional stress disturbance), tunneling, and drilling (local stress disturbance). These situations are different from the stress environment of CBM exploitation. Thus, the permeability of coal during the process of gas migration is controlled by changes in the stress state and gas–solid coupling. These two factors have significant differences in the degree of influence and action time. That is, the stress disturbance of coal and rock mining causes a significant change in permeability, while gas–solid coupling has a weaker but more extended influence on coal permeability. Here, we assume the following two points when dealing with these factors:

(1) The change in coal permeability induced by mining stress disturbance is the change in the initial permeability of coal, whereby the existence of gas in this process only changes the mechanical strength of coal, and the gas–solid coupling effect caused by the change in gas pressure over a short period is ignored.
(2) The gas–solid coupling caused by gas migration continues to affect the evolution of permeability based on the initial permeability. In addition, the evolution of permeability in the process of gas migration is controlled only by the gas–solid coupling effect for coal without stress disturbance.

Furthermore, the effective permeability of a coal seam during the gas migration process is also affected by the Klinkenberg effect, which was originally discovered by

Muskat et al. in the laboratory [89]. For the same rock/coal sample, the permeability tested by water is less than that tested by air, and the difference is more apparent for low-permeability coal samples. Subsequently, scholars have carried out a large number of experimental and theoretical studies on the Klinkenberg effect of gas seepage in low-permeability porous media and obtained many beneficial results [20, 21, 90–92]. The Klinkenberg effect is also known as the gas slippage effect. The characteristics of gas seepage in porous media are different from those of liquids. Because of the existence of shear stress, the velocity of the liquid on the wall surface of fractures is equal to zero. However, gas molecules are not all attached to the wall of a fracture. They have a certain flow rate on the wall of a fracture, which means that the Klinkenberg permeability of gas is greater than the absolute permeability of gas. When other conditions are the same, the Klinkenberg effect is mainly affected by the gas pressure. The smaller the gas pressure, the less the collision between the gas molecules, and the more obvious the Klinkenberg effect becomes. If the gas pressure increases to infinity, the gas flow property is close to that of liquid, and the Klinkenberg effect essentially disappears. The measured permeability is, therefore, the absolute permeability, which has the following relationship with the Klinkenberg permeability [21, 93]:

$$k_e = k\left(1 + \frac{b_k}{p_f}\right) \tag{7.112}$$

where k_e is the effective permeability (Klinkenberg permeability), mD; and b_k is the Klinkenberg factor, Pa.

The value of the Klinkenberg factor depends on the pore structure of a given porous medium and the mean free path of the gas molecules. The relationship between the Klinkenberg factor and absolute permeability is expressed as $b_k = \alpha_k k^{-0.36}$, where α_k is the computing coefficient. Jones [94] conducted experiments using 100 cores ranging from 0.01 to 1000 mD, and found that the computing coefficient was 0.251.

7.6.2 Permeability Evolution Model of Coal Under the Action of Mining Stress Disturbance

In Sect. 7.5, the permeability evolution law caused by stress disturbances was described qualitatively by experiments. In this section, the permeability evolution model based on the elastic–plastic state of coal is established, which can calculate the change in the initial permeability caused by stress disturbance. Figure 7.65 displays the evolution model of permeability during the complete stress–strain process of coal.

As shown in Fig. 7.65, when the permeability evolution is located in stage I of the model, a coal body is in the elastic deformation stage. From the previous test results, it can be seen that the relationship between coal permeability and the stress

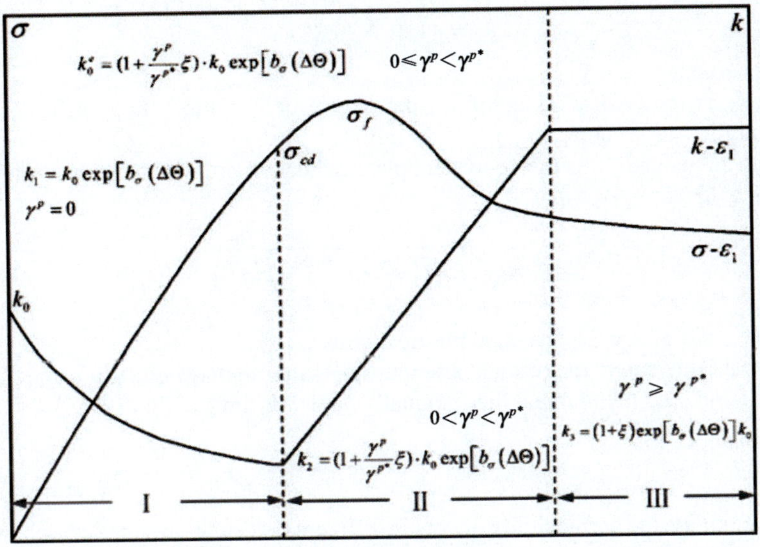

Fig. 7.65 Permeability evolution model during the complete stress–strain process of coal

difference is in good agreement with the exponential function. However, under actual engineering conditions, the stress state transformation experienced by a coal body is more complex than that under laboratory conditions, and the three principal stresses are often changed. Thus, when a coal body undergoes only elastic deformation, the disturbance of the mining stress on the permeability of coal can be expressed as the relationship between permeability and bulk stress [29, 93]:

$$k_1 = k_0 \exp[b_\sigma(\Delta\Theta)] \tag{7.113}$$

where k_1 is the coal permeability in the elastic deformation stage, mD; k_0 is the initial coal permeability, mD; b_σ is the influence coefficient of stress on permeability, MPa^{-1}; and Θ is the bulk stress ($\Theta = \sigma_1 + \sigma_2 + \sigma_3$), MPa.

When the stress state of a coal body reaches the yield condition of damage dilatation, the permeability of the coal body increases abruptly. According to the previous test results and the research results obtained by other scholars, the permeability of coal increases linearly after the yield of damage dilatation [95–97]. From the damage dilation mechanism of the coal body, it can be seen that the increase in permeability after damage dilatation is due to the increase in the fracture volumetric strain, which can be equivalent to the increase in the plastic strain. If the relationship between the permeability increase (caused by the increase in the fracture volumetric strain) and the plastic strain is linear, then the permeability surge behavior can be expressed by the plastic strain of coal [97]:

$$k_2 = (1 + \frac{\gamma^p}{\gamma^{p*}}\xi) \cdot k_0 \tag{7.114}$$

where k_2 is the coal permeability in the plastic deformation stage, mD; ξ is the permeability surge coefficient, γ^{p*} is the initial equivalent plastic strain in the residual strength stage, and γ^p is the equivalent plastic shear strain, which can be calculated using Eq. (7.115) [96, 98]:

$$\gamma^p = \sqrt{2/3(\varepsilon_1^p\varepsilon_1^p + \varepsilon_2^p\varepsilon_2^p + \varepsilon_3^p\varepsilon_3^p)} \tag{7.115}$$

where ε_1^p, ε_2^p, ε_3^p are the principal plastic strains.

A coal body enters the residual strength stage after damage dilatation and plastic deformation, and the permeability gradually tends toward stability [99].

$$k_3 = (1 + \xi)k_0 \tag{7.116}$$

In summary, the permeability of coal in different deformation states caused by the mining stress disturbance can be expressed in the form of the following piecewise function:

$$k^e = \begin{cases} (1 + \frac{\gamma^p}{\gamma^{p*}}\xi)\exp[b_\sigma(\Delta\Theta)]k_0, & 0 \le \gamma^p < \gamma^{p*} \\ (1 + \xi)\exp[b_\sigma(\Delta\Theta)]k_0, & \gamma^p \ge \gamma^{p*} \end{cases} \tag{7.117}$$

where k^e is the coal permeability after excavation stress disturbance, mD.

Furthermore, the fracture porosity of coal in different deformation states caused by the disturbance of mining stress can be obtained using the cubic law:

$$\phi_f^e = \begin{cases} (1 + \frac{\gamma^p}{\gamma^{p*}}\xi)^{1/3}\exp[\frac{b_\sigma}{3}(\Delta\Theta)]\phi_{f0}, & 0 \le \gamma^p < \gamma^{p*} \\ (1 + \xi)^{1/3}\exp[\frac{b_\sigma}{3}(\Delta\Theta)]\phi_{f0}, & \gamma^p \ge \gamma^{p*} \end{cases} \tag{7.118}$$

where ϕ_f^e is the fracture porosity of coal after the disturbance of mining stress, and ϕ_{f0} is the initial fracture porosity of coal.

7.6.3 Permeability Evolution Model of Coal Under the Action of Gas–Solid Coupling

1. Deformation equation of coal based on dual-porosity elastic theory

The presence of a freely moving fluid in a porous medium can modify its mechanical response. The change in the mechanical properties of porous media will subsequently further affect the migration characteristics of the fluid. The mechanical branch used to describe the mechanical response and volumetric response of the pore elastic medium is called the pore elasticity theory [98, 100]. This physical phenomenon is a gas–solid coupling, which is widely considered in many engineering areas. The first study of gas–solid coupling was conducted by Terzaghi, who proposed a well-known one-dimensional effective stress principle in the investigation of land subsidence in 1923 [101]. Subsequently, Rendulic generalized the Terzaghi effective stress principle to a three-dimensional case in 1936 [102]. The above studies on the principle of effective stress provide a theoretical basis for the theory of pore elasticity. In 1935 and 1941, Biot first proposed and described the linear pore elasticity theory [103, 104]. With the continuous development of the pore elasticity theory, the classical single-pore elastic theory, dual-pore elastic theory, triple-pore elastic theory, and multi-porosity elastic theory formed gradually [98].

In the content covered by the pore elasticity theory, the effective stress law is important for describing the mechanical response of pore media under pore fluid pressure [104]. Although the effective stress law is perfect for a single-pore medium (Eq. 7.119), its application for multi-porosity media, such as fractured rock and coal, is limited [105–108]. In the pore and fracture dual-porosity medium, the pore pressure is not consistent with the fluid pressure in the fractures (see Fig. 7.66), and the mechanical properties of pores or fractures are also different. Thus, the degrees of influence of the fluid on the media skeletons are also different.

$$\sigma_{ij}^{e} = \sigma_{ij} - \beta \delta_{ij} p \tag{7.119}$$

where β is the effective stress coefficient, and p is the fluid pressure, MPa.

To better describe the mechanical response of a dual-porosity medium under the action of gas pressure, we introduce the effective stress law for a dual-porosity medium when establishing a permeability model:

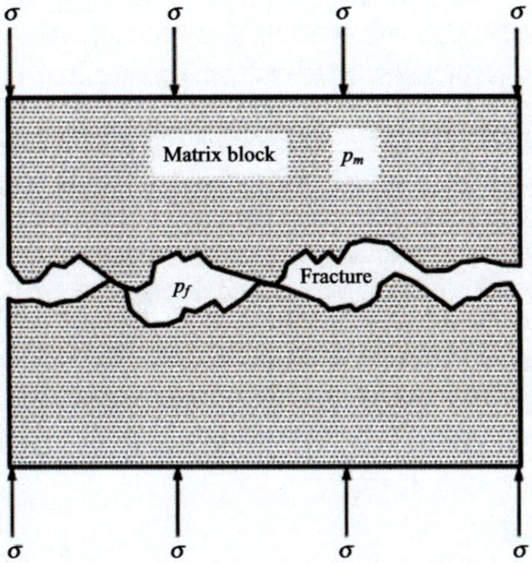

Fig. 7.66 Impact of effective stress

$$\sigma_{ij}^{e} = \sigma_{ij} - (\beta_f p_f + \beta_m p_m)\delta_{ij} \tag{7.120}$$

where β_f and β_m are effective stress coefficients for the fractures and matrix, respectively, as expressed by Eqs. (7.121) and (7.122), respectively [109]:

$$\beta_f = 1 - \frac{K}{K_m} \tag{7.121}$$

$$\beta_m = \frac{K}{K_m} - \frac{K}{K_s} \tag{7.122}$$

where K is the bulk modulus of coal, MPa; K_m is the bulk modulus of the coal matrix, MPa; and K_s is the bulk modulus of the coal skeleton, MPa.

At present, the bulk modulus of the coal skeleton cannot be directly measured, but can be obtained using a theoretical method based on its physical meaning. The above three types of bulk modulus can be calculated using the following equations [108]:

$$K = \frac{E}{3(1 - 2\upsilon)} \tag{7.123}$$

$$K_m = \frac{E_m}{3(1 - 2\upsilon)} \tag{7.124}$$

$$K_s = \frac{K_m}{1 - 3\phi_m(1 - \upsilon)/[2(1 - 2\upsilon)]} \tag{7.125}$$

where E is the elasticity modulus of the studied coal, MPa; E_m is the elasticity modulus of the coal matrix, MPa; and υ is Poisson's ratio for the studied coal.

The permeability of coal is a function of the fracture characteristics. When the adsorption equilibrium state in the coal seam is broken, the fracture characteristics of coal will change under the competitive effect of adsorption swelling and effective stress, and then the permeability will also change. In the previous sections of this book, the impact of gas pressure on the mechanical effect of coal was described. To obtain a permeability model that can describe the competitive effect of effective stress and adsorption swelling, it is necessary to deal with the relationship between the total volumetric strain of coal, skeleton volumetric strain, and fracture volumetric strain (i.e., the volumetric response of coal containing gas). In this section, the total volumetric strain of coal is described first, and the coal deformation equation is established based on the dual-porosity elastic theory.

The deformation equation of coal containing gas is composed of the stress balance equation, geometric equation, and constitutive equation. The stress balance equation can be expressed as:

$$\sigma_{ij,j} + F_i = 0 \tag{7.126}$$

where F_i is the body force, MPa.

The deformation of the coal skeleton in the elastic stage is a small deformation, and the geometric equation of coal containing gas is as follows:

$$\varepsilon_{ij} = \frac{1}{2}(u_{i,j} + u_{j,i}) \tag{7.127}$$

where ε_{ij} is the component of strain (i, j = 1, 2, 3), and u is the deformation displacement.

Based on the assumption that coal containing gas is a homogeneous and isotropic medium, the deformation in the three-dimensional stress state follows the generalized Hooke's law:

$$\sigma_{ij} = 2G\varepsilon_{ij} + \frac{2G\upsilon}{1-2\upsilon}\varepsilon_v\delta_{ij} - \beta_f p_f \delta_{ij} - \beta_m p_m \delta_{ij} \tag{7.128}$$

where G is the shearing modulus of coal elasticity ($G = E/[2(1+\upsilon)]$), MPa; and ε_v is the volumetric strain of coal, which can be expressed using Eq. (7.129):

$$\varepsilon_v = \varepsilon_{11} + \varepsilon_{22} + \varepsilon_{33} \tag{7.129}$$

Combining Eqs. (7.126)–(7.128) yields the Navier-form coal deformation equation in terms of displacement as well as in consideration of the pore pressure:

$$Gu_{i,jj} + \frac{G}{1-2\upsilon}u_{j,ji} - \beta_f p_{f,i} - \beta_m p_{m,i} + F_i = 0 \tag{7.130}$$

Equation (7.130) is the governing equation for the deformation of coal containing gas (elastic deformation).

2. Permeability evolution model of coal under the action of gas–solid coupling

When establishing a permeability model, many scholars often match the stress state of the coal seam and the simplified assumption of the physical coal model. For example, based on the uniaxial strain hypothesis and matchstick model for fractures, Gray [65] and Palmer and Mansoori [64] established a theoretical model of coal permeability (the PM model). Shi and Durucan [110], Cui and Bustin [111], and Zhang and Liu [112] improved these models (the SD and CB models). Based on the assumption of triaxial strain and the cubic model for fractures, Warren and Root [113] established a theoretical model of coal permeability. Robertson and Christiansen [55], Connell and Liu [114] improved these models (the RC model). Because the construction of roadways and boreholes only generates local stress disturbance in a coal seam, it can be considered that the pre-extraction area of the coal seam is unaffected by coal mining when the coal seam is unaffected by other mining actions. For coal seams unaffected by mining, it is considered that the top load is constant. When extracting gas from coal seams, the coal seams exhibit lateral deformation and radial deformation owing to the decrease in pore pressure. However, the vertical scale (thickness) of a coal seam is generally smaller than its lateral scale (length and width); thus, the lateral strain is much smaller than the vertical strain. For convenience, the lateral strain can be assumed to be zero, which is the so-called uniaxial strain assumption. In this book, we establish a permeability model using the matchstick model for fractures.

Under the assumption of uniaxial strain, we deduce a permeability evolution model of the dual-porosity medium based on the matchstick model for fractures. As discussed earlier, in this book, coal is treated as a dual-porosity and single-permeability system, considering only the permeability of the fractures and ignoring the permeability of the coal matrix. Thus, the permeability below refers specifically to the fracture permeability, which can be characterized as a function of the fracture properties. According to the theory of pore elasticity, it can be seen that the relationship among the changes in coal volumetric strain, coal skeleton volumetric strain, and fracture volumetric strain in the process of gas migration (the coal body is in the elastic deformation stage) can be expressed using Eq. (7.131) [64, 115]:

$$\mathrm{d}\varepsilon_p = \frac{d\varepsilon_v}{\phi_f} - \left(\frac{1-\phi_f}{\phi_f}\right)\mathrm{d}\varepsilon_g \tag{7.131}$$

where $\mathrm{d}\varepsilon_p$ is the variation in the fracture volumetric strain, $\mathrm{d}\varepsilon_v$ is the variation in the coal volumetric strain, and $\mathrm{d}\varepsilon_g$ is the variation in the coal skeleton volumetric strain.

Under the assumption of uniaxial strain, the change in the fracture porosity caused by a change in the fracture volumetric strain can be expressed as follows [116]:

$$\begin{aligned}-\mathrm{d}\phi_f = &\left[\frac{1}{M} - (1-\phi_f)f\gamma\right](\mathrm{d}\sigma - \mathrm{d}p) + \left[\frac{K}{M} - (1-\phi_f)\right]\gamma \mathrm{d}p\\ &- \left[\frac{K}{M} - (1-\phi_f)\right]\alpha \mathrm{d}T\end{aligned} \tag{7.132}$$

where M is the constrained axial modulus ($M = E(1-\upsilon)/[(1+\upsilon)(1-2\upsilon)]$), MPa; f is the simplified coefficient and ranges from 0 to 1; γ is the compressibility coefficient of the coal skeleton, MPa^{-1}; and α is the thermal diffusion coefficient of coal skeleton, °F^{-1}.

It can be observed that Eq. (7.133) was built based on the effective stress law of a single porosity medium. When introducing the effective stress law of a dual-porosity medium (Eq. 7.120), the change in the fracture porosity caused by the change in the fracture volumetric strain can be expressed as Eq. (7.133):

$$\begin{aligned}-\mathrm{d}\phi_f = &\left[\frac{1}{M} - (1-\phi_f)f\gamma\right](\mathrm{d}\sigma - \beta_f \mathrm{d}p_f - \beta_m \mathrm{d}p_m)\\ &+ \left[\frac{K}{M} - (1-\phi_f)\right]\gamma(\beta_f \mathrm{d}p_f + \beta_m \mathrm{d}p_m) - \left[\frac{K}{M} - (1-\phi_f)\right]\alpha \mathrm{d}T\end{aligned} \tag{7.133}$$

Coal is a type of porous media with a strong adsorption capacity. In this book, gas migration in coal seams is treated as an isothermal process. The relationship between the deformation caused by gas adsorption/desorption and the deformation caused by the temperature change is given by Eq. (7.134) [64]:

$$\alpha \mathrm{d}T \equiv \frac{\mathrm{d}}{\mathrm{d}p_m}\left(\frac{\varepsilon_l p_m}{p_m + P_L}\right)\mathrm{d}p_m \tag{7.134}$$

The fracture porosity of coal is considerably less than 1 ($\phi_f \ll 1$), and the top load is constant under the uniaxial strain condition. Substituting Eq. (7.134) into Eq. (7.133), we obtain the following:

$$-\mathrm{d}\phi_f = -\frac{1}{M}(\beta_f \mathrm{d}p_f + \beta_m \mathrm{d}p_m) - \left(\frac{K}{M} - 1\right)\frac{\mathrm{d}}{\mathrm{d}p_m}\left(\frac{\varepsilon_l p_m}{p_m + P_L}\right)\mathrm{d}p_m \tag{7.135}$$

Recombining the parameters in Eq. (7.135) and transforming the equation into a fully differential form yield:

$$\mathrm{d}\phi_f(p_f, p_m) = \frac{\beta_f}{M}\mathrm{d}p_f + \left[\frac{\beta_m}{M} + \left(\frac{K}{M} - 1\right)\frac{\mathrm{d}}{\mathrm{d}p_m}\left(\frac{\varepsilon_l p_m}{p_m + P_L}\right)\right]\mathrm{d}p_m \tag{7.136}$$

Then, by solving the multivariable differential equation (Eq. 7.136) of the fully differential form, we obtain Eq. (7.137):

$$\phi_f - \phi_{f_0} = \frac{\beta_f}{M}(p_f - p_0) + \frac{\beta_m}{M}(p_m - p_0) + \left(\frac{K}{M} - 1\right)\left[\frac{\varepsilon_l p_m}{p_m + P_L} - \frac{\varepsilon_l p_0}{p_0 + P_L}\right] \tag{7.137}$$

where p_0 is the initial gas pressure ($p_{f_0} = p_{m0} = p_0$) at the initial equilibrium state (MPa).

Dividing by ϕ_{f_0} on both sides of Eq. (7.137) gives:

$$\begin{aligned}\frac{\phi_f}{\phi_{f_0}} &= 1 + \frac{1}{M\phi_{f_0}}[\beta_f(p_f - p_0) + \beta_m(p_m - p_0)] \\ &\quad + \frac{\varepsilon_l}{\phi_{f_0}}\left(\frac{K}{M} - 1\right)\left(\frac{p_m}{P_L + p_m} - \frac{p_0}{P_L + p_0}\right)\end{aligned} \tag{7.138}$$

According to the research results of Chilingar [117], the relationships among the fracture porosity, matrix scale, and permeability can be expressed as follows:

$$k = \frac{\mathrm{d}_e^2\phi_f^3}{72(1 - \phi_f)^2} \tag{7.139}$$

where d_e is the equivalent matrix scale.

Then, we can obtain the common cubic law for describing the relationship between the permeability and fracture porosity as follows:

$$\frac{k}{k_0} = \left(\frac{\phi_f}{\phi_{f0}}\right)^3 \left(\frac{1-\phi_{f0}}{1-\phi_f}\right)^2 \tag{7.140}$$

Because the fracture porosity is considerably less than 1 ($\phi_f \ll 1$), Eq. (7.140) can be simplified as Eq. (7.141):

$$\begin{aligned}\frac{k}{k_0} &= \left(\frac{\phi_f}{\phi_{f0}}\right)^3 \\ &= \left\{1 + \frac{1}{M\phi_{f0}}[\beta_f(p_f - p_0) + \beta_m(p_m - p_0)]\right. \\ &\left. + \frac{\varepsilon_l}{\phi_{f0}}\left(\frac{K}{M} - 1\right)\left(\frac{p_m}{P_L + p_m} - \frac{p_0}{P_L + p_0}\right)\right\}^3\end{aligned} \tag{7.141}$$

The above equation is the permeability evolution model for describing the change in coal permeability with ① the gas pressure in the fractures, and ② the gas pressure in the matrix during the process of gas migration.

7.6.4 *Permeability Evolution Model of Coal Considering the Actions of Mining Stress Disturbance and Gas–Solid Coupling*

In the previous sections, a permeability evolution model considering the mining stress disturbance, and a gas–solid coupling model based on the dual-porosity elastic theory were established. To analyze the effects of mining disturbance and gas–solid coupling (i.e., coupling effects of the diffusion field, seepage field, and coal deformation field), the above established permeability equations are combined.

Considering the degree of influence and action time, the mining stress disturbance is believed to change the initial stage of a coal seam (initial permeability and initial fracture porosity) as follows:

$$k_0^{e} = \begin{cases} (1 + \frac{\gamma^p}{\gamma^{p*}}\xi)\exp[b_\sigma(\Delta\Theta)]k_0, & 0 \le \gamma^p < \gamma^{p*} \\ (1 + \xi)\exp[b_\sigma(\Delta\Theta)]k_0, & \gamma^p \ge \gamma^{p*} \end{cases} \tag{7.142}$$

$$\phi_{f0}^{e} = \begin{cases} (1 + \frac{\gamma^p}{\gamma^{p*}}\xi)^{1/3}\exp[\frac{b_\sigma}{3}(\Delta\Theta)]\phi_{f0}, & 0 \le \gamma^p < \gamma^{p*} \\ (1 + \xi)^{1/3}\exp[\frac{b_\sigma}{3}(\Delta\Theta)]\phi_{f0}, & \gamma^p \ge \gamma^{p*} \end{cases} \tag{7.143}$$

If gas extraction is carried out after the influence of mining stress disturbance on a coal seam, the gas–solid coupling during the gas migration process will continue to affect the coal seam after excavation.

$$\begin{aligned} k_{\mathrm{e}} = k_0^{\mathrm{e}}\left(1+\frac{b_k}{p_f}\right) \\ \times \left\{1+\frac{1}{M\phi_{f0}^{\mathrm{e}}}[\beta_f(p_f-p_0)+\beta_m(p_m-p_0)] \right. \\ \left. +\frac{\varepsilon_l}{\phi_{f0}^{\mathrm{e}}}\left(\frac{K}{M}-1\right)\left(\frac{p_m}{P_L+p_m}-\frac{p_0}{P_L+p_0}\right)\right\}^3 \end{aligned} \tag{7.144}$$

The coal permeability evolution model considering the mining stress disturbance and gas–solid coupling can be obtained by combining Eqs. (7.142) and (7.144). It is worth mentioning that Eq. (7.142) or Eq. (7.144) can be used separately to describe the coal permeability evolution under the corresponding conditions.

References

1. Zhou, S., & Sun, J. (1965). Theory and application of gas flow in coal seams. *Journal of China Coal Society, 2*(1), 24–36 (in Chinese).
2. Zhou, S., & Lin, B. (1999). *The theory of gas flow and storage in coal seams*. Beijing: China Coal Industry Publishing House (in Chinese).
3. Zhou, S. (1980). Measurement and calculation of gas permeation coefficients of coal seams. *Journal of China University of Mining and Technology, 1*, 1–5 (in Chinese).
4. Ma, Y. (1998). The determination and analysis of permeability coefficient in coal seam. *Journal of Liaoning Technical University (Natural Science), 17*(3), 240–243 (in Chinese).
5. Mitra, A., Harpalani, S., & Liu, S. (2012). Laboratory measurement and modeling of coal permeability with continued methane production: Part 1—Laboratory results. *Fuel, 94*, 110–116.
6. Harpalani, S., & Schraufnagel, R. A. (1990). Shrinkage of coal matrix with release of gas and its impact on permeability of coal. *Fuel, 69*(5), 551–556.
7. Mazumder, S., Scott, M., & Jiang, J. (2012). Permeability increase in Bowen Basin coal as a result of matrix shrinkage during primary depletion. *International Journal of Coal Geology, 96*, 109–119.
8. Chen, Z. W., Pan, Z. J., Liu, J. S., et al. (2011). Effect of the effective stress coefficient and sorption-induced strain on the evolution of coal permeability: Experimental observations. *International Journal of Greenhouse Gas Control, 5*(5), 1284–1293.
9. Chen, Z., Liu, J., Pan, Z., et al. (2012). Influence of the effective stress coefficient and sorption-induced strain on the evolution of coal permeability: Model development and analysis. *International Journal of Greenhouse Gas Control, 8*, 101–110.
10. Pan, Z. J., Connell, L. D., & Camilleri, M. (2010). Laboratory characterisation of coal reservoir permeability for primary and enhanced coalbed methane recovery. *International Journal of Coal Geology, 82*(3), 252–261.
11. Seidle, J., & Huitt, L. (1995). Experimental measurement of coal matrix shrinkage due to gas desorption and implications for cleat permeability increases. In *International Meeting on Petroleum Engineering*. Society of Petroleum Engineers.
12. Durucan, S., Ahsanb, M., & Shia, J. Q. (2009). Matrix shrinkage and swelling characteristics of European coals. *Energy Procedia, 1*(1), 3055–3062.
13. Robertson, E. P. (2005). *Measurement and modeling of sorption-induced strain and permeability changes in coal*. United States: Department of Energy.
14. Robertson, E. P., & Christiansen, R. L. (2005). Modeling permeability in coal using sorption-induced strain data. In *SPE Annual Technical Conference and Exhibition*. Society of Petroleum Engineers.
15. Van Bergen, F., Spiers, C., Floor, G., et al. (2009). Strain development in unconfined coals exposed to CO_2, CH_4 and Ar: Effect of moisture. *International Journal of Coal Geology, 77*(1), 43–53.
16. Jassinge, D., Ranjith, P., Choi, X., et al. (2012). Investigation of the influence of coal swelling on permeability characteristics using natural brown coal and reconstituted brown coal specimens. *Energy, 39*(1), 303–309.
17. Battistutta, E., Van Hemert, P., Lutynski, M., et al. (2010). Swelling and sorption experiments on methane, nitrogen and carbon dioxide on dry Selar Cornish coal. *International Journal of Coal Geology, 84*(1), 39–48.
18. Liu, S., Harpalani, S., & Pillalamarry, M. (2012). Laboratory measurement and modeling of coal permeability with continued methane production: Part 2—Modeling results. *Fuel, 94*, 117–124.
19. Perera, M., Ranjith, P., & Choi, S. (2013). Coal cleat permeability for gas movement under triaxial, non-zero lateral strain condition: A theoretical and experimental study. *Fuel, 109*, 389–399.
20. Klinkenberg, L. J. (1941). The permeability of porous media to liquids and gases [M]. American Petroleum Institute.

21. Wu, Y. S., & Karsten, P. (1998). Gas flow in porous media with Klinkenberg effects. *Transport in Porous Media, 32*(1), 117–137.
22. Dawson, G. K. W., & Esterle, J. S. (2010). Controls on coal cleat spacing. *International Journal of Coal Geology, 82*(34), 213–218.
23. Perera, M. S. A., Ranjith, P. G., Choi, S. K., et al. (2012). Investigation of temperature effect on permeability of naturally fractured black coal for carbon dioxide movement: An experimental and numerical study. *Fuel, 94*, 596–605.
24. Tan, X., & Xian, X. (1994). Research on the permeability of coal. *Journal of Xian Mining Institute, 14*(1), 22–25 (in Chinese).
25. Li, Z., Xian, X., & Long, Q. (2009). Experiment study of coal permeability under different temperature and stress. *Journal of China University of Mining and Technology, 38*(4), 523–527 (in Chinese).
26. Jiang, D., Yang, X., Xian, X., et al. (2010). The infiltration equation of coalbed under the cooperation of stress field, temperature field and sound field. *Journal of China Coal Society, 35*(3), 434–438 (in Chinese).
27. Zhou, J., Xian, X., Jiang, Y., et al. (2009). A permeability model considering the effective stress and coal matrix shrinking effect. *Journal of Southwest Petroleum University (Science and Technology Edition), 31*(1), 4–8 (in Chinese).
28. Lin, B., & Zhou, S. (1987). Experimental investigation on the permeability of the coal samples containing methane. *Journal of China University of Mining and Technology, 16*(1), 21–28 (in Chinese).
29. Zhao, Y., Hu, Y., Yang, D., et al. (1999). The experimental study on the gas seepage law of rock related to adsorption under 3-D stresses. *Chinese Journal of Rock Mechanics and Engineering, 18*(6), 651–653 (in Chinese).
30. Hu, Y., Zhao, Y., Wei, J., et al. (1996). Experimental study of permeating law of coal mass gas under action of 3-dimension stress. *Journal of Xian Mining Institute, 16*(4), 308–311 (in Chinese).
31. Fu, X., Qin, Y., Jiang, B., et al. (2003). Physical and numerical simulations of permeability of coal reservoirs in central and southern part of the Qinshui basin, Shanxi. *Chinese Journal of Geology, 38*(2), 221–229 (in Chinese).
32. Chen, J., Qin, Y., & Fu, X. (2006). Numerical simulation on dynamic variation of the permeability of high rank coal reservoirs during gas recovery. *Journal of China University of Mining and Technology, 35*(1), 49–53 (in Chinese).
33. Sun, P., & Ling, Z. (2000). Experimental study of the law for permeability of coal under action of 3-triaxia compression. *Journal of Chongqing University (Natural Science Edition)* (z1), 28–31 (in Chinese).
34. Wu, S. (2006). *Research of methane-coalbed coupling movement theory and its application.* Dongbei University (in Chinese).
35. Li, X., Guo, Y., & Wu, S. (2005). Analysis of the relation of porosity, permeability and swelling deformation of coal. *Journal of Taiyuan University of Technology, 36*(3), 264–246 (in Chinese).
36. Yin, G., Huang, Q., Zhang, D., et al. (2010). Test study of gas seepage characteristics of gas-bearing coal specimen during process of deformation and failure in geostress field. *Chinese Journal of Rock Mechanics and Engineering, 29*(2), 336–343 (in Chinese).
37. Yin, G., Jiang, C., Li, X., et al. (2011). An experimental study of gas permeabilities of outburst and non outburst coals under complete stress-strain process. *Rock and Solid Mechanics, 32*(6), 1613–1619 (in Chinese).
38. Yin, G., Jiang, C., Wang, W., et al. (2011). Experimental study of influence of confining pressure unloading speed on mechanical properties and gas permeability of containing-gas coal rock. *Chinese Journal of Rock Mechanics and Engineering, 30*(1), 68–77 (in Chinese).
39. Yin, G., Li, M., Li, W., et al. (2012). Influence of gas pressure on mechanical and seepage characteristics of coal under unloading condition. *Journal of China Coal Society, 37*(9), 1499–1504 (in Chinese).

40. Yin, G., Li, X., Zhao, H., et al. (2008). Experimental research on effect of geostress on outburst coal's gas seepage. *Chinese Journal of Rock Mechanics and Engineering, 27*(12), 2557–2561 (in Chinese).
41. Yin, G., Jiang, C., Xu, J., et al. (2011). Experimental study of thermo-fluid-solid coupling seepage of coal containing gas. *Journal of China Coal Society, 36*(9), 1495–1500 (in Chinese).
42. Xu, J., Li, B., Zhou, T., et al. (2012). Experimental study of coal deformation and permeability characteristics under loading-unloading conditions. *Journal of China Coal Society, 37*(9), 1493–1498 (in Chinese).
43. Xu, J., Peng, S., Tao, Y., et al. (2009). Experimental analysis of influence of creep on permeability of gas-bering coal. *Chinese Journal of Rock Mechanics and Engineering, 28*(11), 2273–2279 (in Chinese).
44. Xu, J., Peng, S., Yin, G., et al. (2010). Development and application of triaxial servocontrolled seepage equipment for thermo-fluid-solid coupling of coal containing methane. *Chinese Journal of Rock Mechanics and Engineering, 29*(5), 907–914 (in Chinese).
45. Peng, Y., Qi, Q., Deng, Z., et al. (2008). Experimental research on sensibility of permeability of coal samples under confining pressure status based on scale effect. *Journal of China Coal Society, 33*(5), 509–513 (in Chinese).
46. Xie, H., Gao, F., Zhou, H., et al. (2013). On theoretical and modeling approach to mining-enhanced permeability for simultaneous exploitation of coal and gas. *Journal of China Coal Society, 38*(7), 1101–1108 (in Chinese).
47. Chen, H. (2013). *Damage and permeability development of unloading coal body during mining the protective coal seam.* China University of Mining and Technology (in Chinese).
48. Chen, H., Cheng, Y., Ren, T., et al. (2014). Permeability distribution characteristics of protected coal seams during unloading of the coal body. *International Journal of Rock Mechanics and Mining Sciences, 71*, 105–116.
49. Pan, R. (2014). *The permeability evolution characteristics of loaded coal and its application in the drainage of pressure-relief gas.* Xuzhou: China University of Mining and Technology (in Chinese).
50. Seelheim, F. (1880). Methoden zur Bestimmung der Durchlässigkeit des Bodens. *Analytical and Bioanalytical Chemistry, 19*(1), 387–418.
51. Kozeny, J. (1927). Über kapillare Leitung des Wassers im Boden. *Sitzungsber Wien, Aksd Wiss, 136*(2a), 271–306.
52. Carman, P. (1997). Fluid flow through granular beds. *Chemical Engineering Research and Design, 75*, S32–S48.
53. Pan, Z. (2012). Modelling permeability for coal reservoirs: A review of analytical models and testing data. *International Journal of Coal Geology, 92*, 1–44.
54. Reiss, L. H. (1980). *The reservoir engineering aspects of fractured formations.* Editions Technip.
55. Robertson, E. P., & Christiansen, R. L. (2008). A permeability model for coal and other fractured sorptive-elastic media. *Spe Journal, 13*(3), 314–324.
56. Fairhurst, C. (2003). Stress estimation in rock: A brief history and review. *International Journal of Rock Mechanics and Mining Sciences, 40*(7), 957–973.
57. Liu, J., Chen, Z., Elsworth, D., et al. (2011). Interactions of multiple processes during CBM extraction: A critical review. *International Journal of Coal Geology, 87*(3), 175–189.
58. Palmer, I., & Mansoori, J. (1996). How permeability depends on stress and pore pressure in coalbeds: A new model [J]. *Spe Reservoir Evaluation and Engineering, 1*(6), 539–544.
59. Palmer, I. (2009). Permeability changes in coal: Analytical modeling. *International Journal of Coal Geology, 77*(1), 119–126.
60. Durucan, S., Daltaban, T., Shi, J., et al. (1993). Permeability characterisation for modelling methane flow in coal seams. In *Proceedings of the 1993 International Coalbed Methane Symposium* (pp. 453–460).
61. Shi, J., & Durucan, S. (2003). Changes in permeability of coalbeds during primary recovery—Part 1: Model formulation and analysis. In *Proceedings of the 2003 International Coalbed Methane Symposium* (p. 341). University of Alabama, Tuscaloosa, Alabama.

62. Mckee, C., Bumb, A., & Koenig, R. (1987). Stress-dependent permeability and porosity of coal and other geologic formations. In *International Coalbed Methane Symposium* (pp. 183–193). University of Alabama, Tuscaloosa, Alabama.
63. Clarkson, C., & Mcgovern, J. (2003). A new tool for unconventional reservoir exploration and development applications. In *International Coalbed Methane Symposium* (pp. 5–9), Tuscaloosa, Alabama.
64. Palmer, I., & Mansoori, J. (1998). How permeability depends on stress and pore pressure in coalbeds: A new model. *SPE Reservoir Evaluation and Engineering, 1*(6), 539–544.
65. Gray, I. (1987). Reservoir engineering in coal seams: Part 1—The physical process of gas storage and movement in coal seams. *SPE Reservoir Engineering, 2*(1), 28–34.
66. Brace, W., Walsh, J., & Frangos, W. (1968). Permeability of granite under high pressure. *Journal of Geophysical Research, 73*(6), 2225–2236.
67. Wang, S., Elsworth, D., & Liu, J. (2011). Permeability evolution in fractured coal: The roles of fracture geometry and water-content. *International Journal of Coal Geology, 87*(1), 13–25.
68. Siriwardane, H., Haljasmaa, I., Mclendon, R., et al. (2009). Influence of carbon dioxide on coal permeability determined by pressure transient methods. *International Journal of Coal Geology, 77*(1), 109–118.
69. Li, X., Gao, Q., Wu, Z., et al. (2001). Transient pulse technique and its application to conventional triaxial compressive tests. *Chinese Journal of Rock Mechanics and Engineering, 20*(z1), 1725–1733 (in Chinese).
70. Hol, S., & Spiers, C. J. (2012). Competition between adsorption-induced swelling and elastic compression of coal at CO_2 pressures up to 100 MPa. *Journal of the Mechanics and Physics of Solids, 60*(11), 1862–1882.
71. Liu, S., & Harpalani, S. (2013). A new theoretical approach to model sorption induced coal shrinkage or swelling. *AAPG Bulletin, 97*(7), 1033–1049.
72. Adamson, A. W., & Gast, A. P. (1990). *Physical chemistry of surfaces*. New York: Wiley.
73. Bangham, D. H., & Fakhoury, N. (1931). The translation motion of molecules in the adsorbed phase on solids. *Journal of the Chemical Society* 1324–1333.
74. Pini, R., Ottiger, S., Burlini, L., et al. (2009). Role of adsorption and swelling on the dynamics of gas injection in coal. *Journal of Geophysical Research: Solid Earth, 114*(B4), 2415–2440.
75. Qu, H., Liu, J., Chen, Z., et al. (2012). Complex evolution of coal permeability during CO_2 injection under variable temperatures. *International Journal of Greenhouse Gas Control, 9*, 281–293.
76. Liu, J., Wang, J., Chen, Z., et al. (2011). Impact of transition from local swelling to macro swelling on the evolution of coal permeability. *International Journal of Coal Geology, 88*(1), 31–40.
77. Qu, H., Liu, J., Pan, Z., et al. (2014). Impact of matrix swelling area propagation on the evolution of coal permeability under coupled multiple processes. *Journal of Natural Gas Science and Engineering, 18*, 51–66.
78. Xie, H., Zhang, Z., Gao, F., et al. (2016). Stress-fracture-seepage field behavior of coal under different mining layouts. *Journal of China Coal Society, 41*(10), 2405–2417 (in Chinese).
79. Wang, G., Xue, D., Hao, H., et al. (2012). Study on permeability characteristics of coal rock in complete stress strain process. *Journal of China Coal Society, 37*(1), 107–112 (in Chinese).
80. Peng, S., Qu, H., Luo, L., et al. (2000). An experimental study on the penetrability of sedimentary rock during the complete stress-strain path. *Journal of China Coal Society, 25*(2), 113–116 (in Chinese).
81. Li, S., Qian, M., & Shi, P. (2001). Permeability-strain equation relation to complete stress-strain path of coal sample. *Coal Geology and Exploration, 29*(1), 22–24 (in Chinese).
82. Laubach, S. E., Marrett, R. A., Olson, J. E., et al. (1998). Characteristics and origins of coal cleat: A review. *International Journal of Coal Geology, 35*(1), 175–207.
83. Huang, Q. (2010). Effect of gas pressure on gas seepage in complete stress-strain process of coal material. *Materials Review, 24*(16), 80–83 (in Chinese).
84. Wang, C., Xian, X., Zhou, J., et al. (2013). Experimental study on permeability of coal during the complete stress-strain process with different gases. *Chinese Journal of Underground Space and Engineering, 9*(3), 492–496 (in Chinese).

85. Han, G., Wang, E., & Liu, X. (2011). Seepage characteristics of rock during damage process. *Journal of Civil, Architectural and Environmental Engineering, 33*(5), 41–50 (in Chinese).
86. Gash, B. W., Volz, R. F., Potter, G., et al. (1992). The effects of cleat orientation and confining pressure on cleat porosity, permeability and relative permeability in coal. *93*(21), 17–21.
87. Li, H., Shimada, S., & Zhang, M. (2004). Anisotropy of gas permeability associated with cleat pattern in a coal seam of the Kushiro coalfield in Japan [J]. *Environmental Geology, 47*(1), 45–50.
88. Qin, H., Huang, G., & Wang, W. (2012). Experimental study of acoustic emission characteristics of coal samples with different moisture contents in process of compression deformation and failure. *Chinese Journal of Rock Mechanics and Engineering, 31*(6), 1115–1120 (in Chinese).
89. Muskat, M., & Wyckoff, R. D. (1937). *The flow of homogeneous fluids through porous media.* New York: McGraw-Hill.
90. Zhu, W., Liu, J., Sheng, J., et al. (2007). Analysis of coupled gas flow and deformation process with desorption and Klinkenberg effects in coal seams. *International Journal of Rock Mechanics and Mining Sciences, 44*(7), 971–980.
91. Hu, G., Wang, H., Fan, X., et al. (2009). Mathematical model of coalbed gas flow with Klinkenberg effects in multi-physical fields and its analytic solution. *Transport in Porous Media, 76*(3), 407–420.
92. Zhu, Y., Tao, G., Fang, W., et al. (2007). Research progress of the Klinkenberg effect in tight gas reservoir. *Progress in Geophysics, 22*(5), 1591–1596 (in Chinese).
93. Klinkenberg, L. (1941). The permeability of porous media to liquids and gases. In *Drilling and production practice* (pp. 200–213). Tulsa, Oklahoma: American Petroleum Institute.
94. Jones, F. O., & Owens, W. (1980). A laboratory study of low-permeability gas sands. *Journal of Petroleum Technology, 32*(9), 1631–1640.
95. Chen, J., Lyv, J., Guo, D., et al. (2011). Factors and development technology of coalbed methane production capability. *Resources and Industries, 13*(1), 108–113 (in Chinese).
96. Viete, D., & Ranjith, P. (2006). The effect of CO_2 on the geomechanical and permeability behaviour of brown coal: Implications for coal seam CO_2 sequestration. *International Journal of Coal Geology, 66*(3), 204–216.
97. Wang, J. A., & Park, H. (2002). Fluid permeability of sedimentary rocks in a complete stress-strain process. *Engineering Geology, 63*(3), 291–300.
98. Bai, M. (1999). *Introduction theory of pore fracture elasticity and application.* Petroleum Industry Press (in Chinese).
99. Zhao, Y., Hu, Y., Zhao, B., et al. (2004). Nonlinear coupled mathematical model for solid deformation and gas seepage in fractured media. *Transport in Porous Media, 55*(2), 119–136.
100. Hajiabdolmajid, V., & Kaiser, P. (2003). Brittleness of rock and stability assessment in hard rock tunneling. *Tunnelling and Underground Space Technology, 18*(1), 35–48.
101. Terzaghi, K. V. (1923). Die berechnung der durchlassigkeitsziffer des tones aus dem verlauf der hydrodynamischen spannungserscheinungen. *Akad Wissensch Wien Sitzungsber Mathnatur-wissensch Klasse IIa, 132*, 125–138.
102. Rendulic, L. (1936). *Porenziffer und porenwasserdruck in Tonen.* Berlin: Springer.
103. Biot, M. A. (1941). General theory of three-dimensional consolidation. *Journal of Applied Physics, 12*(2), 155–164.
104. Biot, M. A. (1935). Le problème de la consolidation des matières argileuses sous une charge. *Annales de la Société Scientifique de Bruxelles. Serie B, 55*, 110–113.
105. Rice, J. R., & Cleary, M. P. (1976). Some basic stress diffusion solutions for fluid-saturated elastic porous media with compressible constituents. *Reviews of Geophysics, 14*(2), 227–241.
106. Rudnicki, J. (1985). Effect of pore fluid diffusion on deformation and failure of rock. *Mechanics of Geomaterials* 315–347.
107. Zhang, J. (2002). *Dual-porosity approach to wellbore stability in naturally fractured reservoirs.* University of Oklahoma.
108. Zhang, J., Bai, M., & Roegiers, J. C. (2006). On drilling directions for optimizing horizontal well stability using a dual-porosity poroelastic approach. *Journal of Petroleum Science and Engineering, 53*(1), 61–76.

109. Chen, M., & Chen, Z. (1999). Effective stress laws for multi-porosity media. *Appliled Mathematics and Mechanics, 20*(11), 1121–1127 (in Chinese).
110. Shi, J., & Durucan, S. (2004). Drawdown induced changes in permeability of coalbeds: A new interpretation of the reservoir response to primary recovery. *Transport in Porous Media, 56*(1), 1–16.
111. Cui, X., & Bustin, R. M. (2005). Volumetric strain associated with methane desorption and its impact on coalbed gas production from deep coal seams. *Aapg Bulletin, 89*(9), 1181–1202.
112. Zhang, H., Liu, J., & Elsworth, D. (2008). How sorption-induced matrix deformation affects gas flow in coal seams: A new FE model. *International Journal of Rock Mechanics and Mining Sciences, 45*(8), 1226–1236.
113. Warren, J., & Root, P. J. (1963). The behavior of naturally fractured reservoirs. *SPE Journal.*
114. Connell, L. D., Lu, M., & Pan, Z. (2010). An analytical coal permeability model for tri-axial strain and stress conditions. *International Journal of Coal Geology, 84*(2), 103–114.
115. Detournay, E. (1993). Fundamentals of poroelasticity. In *Comprehensive rock engineering: Principles, practice & projects* (p. 2).
116. Detoumay, E., & Cheng, A. H. D. (1993). Fundamentals of poroelasticity. *Analysis & Design Methods, 2*(1), 113–171.
117. Bradley, J. S., & Powley, D. E. (1994). Pressure compartments in sedimentary basins: A review. *Basin Compartments and Seals, 61*, 3–26.

Chapter 8
Geostress and Coal Seam Gas Occurrence

Learning Outcomes

- Basic concept of geostress and its simplified model;
- Basic principle and method for measuring geostress;
- Influence of tectonic stress on the occurrence of coal seam and the dynamic metamorphism of coal;
- Vertical zoning characteristics of coal seam gas and the variation laws of gas pressure/content with burial depth.

8.1 Basic Concept of Geostress

8.1.1 Concept of Geostress

Geostress is the natural stress that has not been disturbed by engineering actions affecting strata. The geostress is also known as the initial stress of the rock mass, absolute stress, or the original rock stress [1].

In contrast, due to the effects of roadway excavation, mineral resource mining, and/or other engineering, the equilibrium state of the original stress within the influence scope of engineering is damaged, and the stress after damage is called the secondary stress or induced stress, as shown in Fig. 8.1. From the viewpoint of mining engineering, the stress of surrounding rock underground is composed of the original rock stress as well as the secondary stress, and the characteristic of ground pressure manifestation is determined by the mechanical properties of the surrounding rock and the geostress field in which the surrounding rock is located [2].

Geostress is the fundamental force that causes the deformation and failure of mining and other underground/open-pit rock excavation works. It is also necessary to determine the mechanical properties of the engineering rock mass, carry out a stability analysis of the surrounding rock, realize the rock excavation design, and make scientific decisions.

Y. Cheng et al., *Coal Mechanics*,
https://doi.org/10.1007/978-981-16-3895-4_8

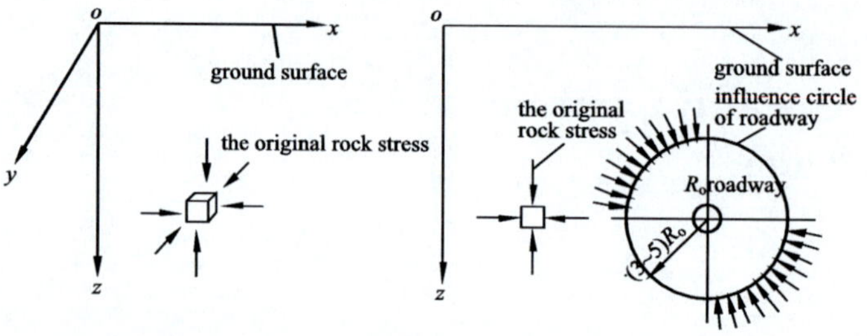

Fig. 8.1 Original rock stress and secondary stress

Geostress is mainly composed of the coupling of gravity stress, tectonic stress, pore pressure, and thermal stress [3]. Gravity stress is the geostress component induced by the gravity of the overlying strata, which is the stress caused by the gravity of the overlying strata. Tectonic stress is the stress that can lead to tectonic movement, resulting in tectonic deformation and forming a variety of tectonic traces. Tectonic stress mainly affects the stresses in two horizontal directions. Thermal stress refers to the internal stress increment caused by changes in the temperature of the stratum, which is related to the change in temperature and the thermal properties of rocks. Finally, the geostress within pore–fracture coal/rock layers is composed of two parts: one is borne by the fluid in the pores (i.e., the pore pressure), and the other is borne by the coal/rock skeleton (i.e., the effective stress).

8.1.2 *Simplified Model of Geostress*

The stress state of any point in an underground rock mass can be determined by three principal stresses both in magnitude and direction. Because the shear stress on the free surface is zero, the positive stress vector of any unloaded free surface must be a principal stress vector. Therefore, it can be understood that one of the principal stresses in the flat or near horizontal region should be vertical in direction [4]. It can be assumed that at the burial depth H, the positive compressive stress must be able to support the weight of the overlying strata, that is,

$$\sigma_{zz} = \int_0^H \rho(H) g(H) \mathrm{d}H \tag{8.1}$$

where ρ is rock density and g is acceleration due to gravity, which both generally change with burial depth. However, for a shallow-burial area (i.e., <1000 m), which engineering usually focuses on, g can be regarded as a constant. Thus, Eq. (8.1) can be written as Eq. (8.2):

$$\sigma_{zz} = \bar{\rho} g H \tag{8.2}$$

where $\bar{\rho}$ is the average density of the rock whose burial depth is shallower than H.

Although Eq. (8.1) is generally accepted for the calculation of vertical stress, it should be noted that it is only application when the shear stress (τ_{xz} and τ_{yz}) is equal to zero. τ_{xz} and τ_{yz} can support the gravity of overburden and achieve balance. In Sect. 8.3, the results of the geostress measurement are in good agreement with the calculated results from Eq. (8.1), except for areas where the depth is very shallow (e.g., 10–100 m in burial depth) and/or the surface has irregular landforms.

The simplest assumption is that the complete stress state refers to the other two principal stresses being equal to the overburden gravity, that is, $\sigma_{xx} = \sigma_{yy} = \bar{\rho} g H$. The situation where three principal stresses are equal to each other generally occurs in a static water column; thus, in the field of mechanics, any stress state in which the three principal stresses are equal is called the hydrostatic pressure state. In the field of rock mechanics, in order to avoid confusion with the pressure caused by the pore fluid, the stress state of $\sigma_{xx} = \sigma_{yy} = \sigma_{zz} = \bar{\rho} g H$ is called the rock static pressure state. This stress state is characterized by the absence of deviatoric stress, that is, there is no shear stress in the stress tensor under any coordinate system.

This assumption of the rock static pressure state is called the Heim hypothesis. The initial basis for this hypothesis is that if a rock is viscoelastic and conforms to the Maxwell model or the Burgess model, its stress state will be gradually close to the static rock pressure conditions over a sufficiently long period of time. However, this model ignores the role of tectonic stress. In fact, the tectonic stress often changes, and its change time is less than the rock relaxation time. Moreover, most of the rocks near the surface show the properties of brittleness rather than viscoelasticity.

Therefore, another model is often used to predict the magnitudes of the other two principal stresses. This model assumes that when the lateral deformation of a rock is limited, the gravity caused by self-weight can take effect instantaneously. Then, assuming that the rock is homogeneously isotropic, the vertical stress ($\sigma_{zz} = \rho g H$) and stress state are as follows:

$$\sigma_{zz} = \rho g H, \sigma_{xx} = \sigma_{yy} = \frac{\upsilon}{1-\upsilon}\sigma_{zz}, \tau_{xy} = \tau_{yz} = \tau_{zx} = 0 \tag{8.3}$$

where υ is Poisson's ratio.

For a typical Poisson's ratio (e.g., $\upsilon = 0.25$), this model determines that the vertical stress (rock static pressure) is the maximum principal stress, while the other two principal stresses are equal to 1/3 of the rock static pressure.

The assumption of no lateral strain obviously only occurs in very special cases. Another extreme assumption is that there is no lateral stress when gravity starts to act. At that time, a single stress state will appear, namely,

$$\sigma_{zz} = \rho g H, \sigma_{xx} = \sigma_{yy} = \tau_{xy} = \tau_{yz} = \tau_{zx} = 0 \tag{8.4}$$

The above three models can be written in the following uniform manner:

$$\sigma_{zz} = \rho g H, \sigma_{xx} = \sigma_{yy} = \lambda \sigma_{zz}, \tau_{xy} = \tau_{yz} = \tau_{zx} = 0 \tag{8.5}$$

where λ is the side pressure coefficient.

In the law of Heim $\lambda = 1$; in the single-strain model $\lambda = \frac{\upsilon}{1-\upsilon}$; and in the single stress state $\lambda = 0$.

By solving the above models, the obtained lateral stress from each model does not exceed the vertical stress. However, the model proposed by McCutchen can be drawn for $\lambda > 1$. This model assumes that the crust is a thin, elastic, isotropic, and uniform shell that rests at the top of the rigid mantle and deforms under gravity. The difference between McCutchen's model and the single strain model is that the former takes the spherical geometry of Earth into consideration. The boundary condition is that there is no radial normal stress at the surface and no tangential displacement at the junction between the crust and mantle. Setting Earth's radius to R and the radius of the junction of the crust and the mantle to R_1, then the elastic equilibrium equation of this model can yield Eqs. (8.6) and (8.7) [4]:

$$\sigma_{rr} = \frac{\rho g R}{4(1-\upsilon)} \left\{ \left[2 - (1+\upsilon)\frac{R_1}{R} \right] \frac{(1+\upsilon) + 2(1-2\upsilon)(R_1/r)^3}{(1+\upsilon) + 2(1-2\upsilon)(R_1/R)^3} + (1+\upsilon)\frac{R_1}{R} - 2\frac{r}{R} \right\} \tag{8.6}$$

$$\tau_{\theta\theta} = \tau_{\phi\phi} = \frac{\rho g R}{4(1-\upsilon)} \left\{ \begin{array}{l} \left[2 - (1+\upsilon)\frac{R_1}{R} \right] \frac{(1+\upsilon) + 2(1-2\upsilon)(R_1/r)^3}{(1+\upsilon) + 2(1-2\upsilon)(R_1/R)^3} \\ +(1+\upsilon)\frac{R_1}{R} - (1+2\upsilon)\frac{r}{R} \end{array} \right\} \tag{8.7}$$

where r is the distance from the center of the Earth, the burial depth is $H = R - r$, and $\tau_{\theta\theta}$ and $\tau_{\phi\phi}$ are tangential stresses.

The thickness of the crust $(R - R_1)$ is only tens of kilometers, which is much smaller than Earth's radius (approximately 6.4×10^3 km). By substituting $(R - R_1)$ into Eqs. (8.6) and (8.7), and expanding the series of H/R, we ignore $[(R - R_1)/R]^2 \approx 0.0001$ and obtain:

$$\sigma_{rr} = \sigma_v = \rho g H \tag{8.8}$$

$$\tau_{\theta\theta} = \tau_{\phi\phi} = \sigma_h = \frac{R - R_1}{4R}\rho g(R - R_1) + \frac{\upsilon}{1-\upsilon}\rho g H \tag{8.9}$$

where σ_{rr}, σ_v, and σ_h are the radial, vertical, and horizontal stresses, respectively.

Assuming that the average thickness of the crust is 40 km, the average density is 2700 kg/m^3, and Poisson's ratio is 0.25, these can be substituted into Eqs. (8.8)

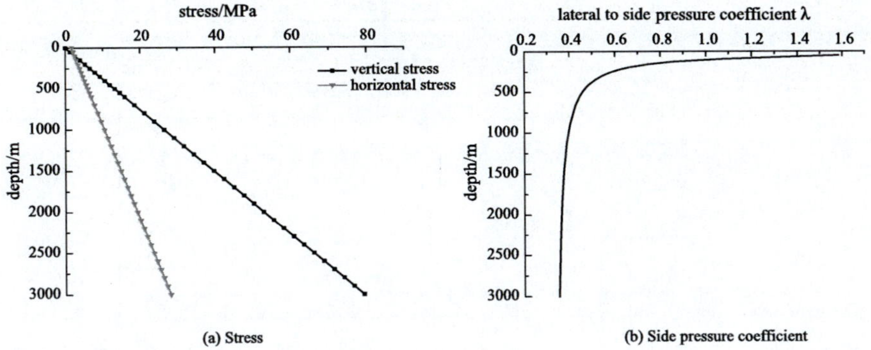

Fig. 8.2 Computed results from the spherical shell model

and (8.9) to obtain the variation of vertical/horizontal stresses with the change in burial depth (Fig. 8.2a). When using the spherical shell model to predict geostress, the results show that there exists shear stress at the surface (i.e., the horizontal stress of the ground is not zero).

Dividing Eq. (8.9) by Eq. (8.8) yields the geostress side pressure coefficient λ as follows:

$$\lambda = \frac{\sigma_h}{\sigma_v} = \frac{\upsilon}{1-\upsilon} + \frac{(R-R_1)^2}{4RH} \tag{8.10}$$

In Fig. 8.2b, the relationship between the lateral pressure coefficient λ and the burial depth exhibits a non-linear decrease, which is in good agreement with the measured results.

8.2 In Situ Measurement Methods for Geostress

8.2.1 Fundamental Principles

Geostress is the most fundamental original data in various subject areas, such as those pertaining to Earth's crust and strata. In terms of mining engineering, the magnitude and orientation of geostress are the primary basis for optimizing the form of roadway sections, selecting roadway orientation in a reasonable manner, and supporting the scheme and parameter of roadways. The measurement of the original geostress ascertains the three-dimensional stress state of the planned excavating rock and the unaffected surrounding area. This type of measurement is completed point by point. The three-dimensional stress state of any point in the rock can be described by six components of stress (σ_x, σ_y, σ_z, τ_{xy}, τ_{yz}, τ_{zx}) in the selected coordinate system, as shown in Fig. 8.3. This coordinate system can be selected based on need and

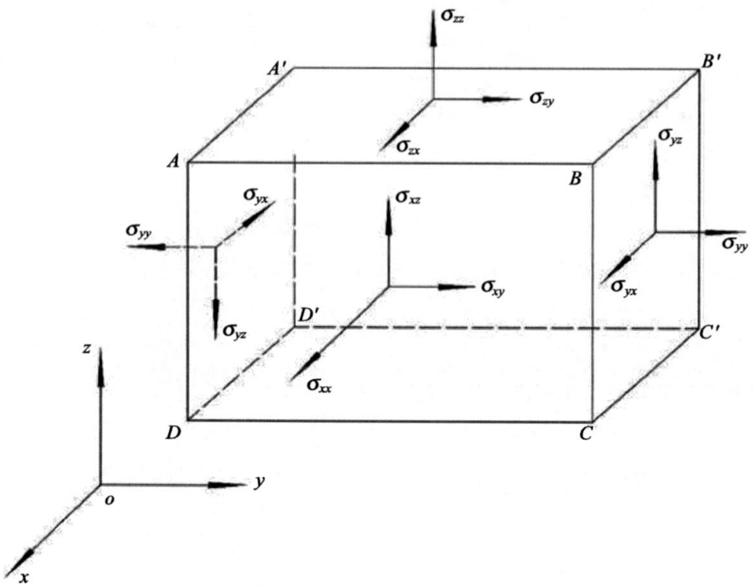

Fig. 8.3 Schematic diagram showing the three-dimensional stress state of any point in rock mass

convenience; however, it usually takes Earth's coordinates as a surveying coordinate system. The magnitude and orientation of the three principal stresses at the selected point can be calculated from these six stress components, which are unique. In practical measurements, each measurement point can range from several cubic centimeters to thousands of cubic meters, and depends on the adopted measuring method. Nevertheless, regardless of the scale, it can still be treated as a point for the whole rock mass. Although there are also some methods for determining the average stress in a wide range of rock mass (e.g., the ultrasonic wave method and other geophysical methods), these methods are still very immature and not as popular as the "point" measuring method. Because of the complexity and variability of the geostress state, in order to accurately measure the stress of a region, a sufficient number of "point" measurements are required. On this basis, the geostress state field of a region can be finally delineated through a fitting analysis using methods such as numerical analysis, mathematical statistics, gray modeling, and artificial intelligence.

Since the 1950s–1960s, the measurement of geostress has developed vigorously, with 10–20 measurement methods and >100 measuring apparatus having been proposed. Of these, the five most representative and relatively mature methods were standardized as the ***Proposed Method for Determination of Rock Stress*** by the test committee of the International Society of Rock Mechanics (ISRM) in 1987. These include the flat jacking method, aperture deforming method, hydraulic fracturing method, hole wall strain method, and hollow package strain method. The last two methods can determine six stress components at a time in drilling one borehole, which belong to the three-dimensional stress measurement method.

Because there is no uniform standard for the classification of geostress measurement methods, according to differences in measurement methods, some people classify the methods used in actual measurements into five categories: the construction method, deformation method, electromagnetic method, seismic method, and radioactivity method. In addition to this, according to different measurement principles, other people classify the measurement methods into stress recovery, stress relief, strain recovery, strain relief, hydraulic fracturing, acoustic emission, X-ray, and gravity methods.

However, according to the viewpoint of most people in this field, the measurement method can be classified into direct and indirect measurement methods based on differences in basic principles. The direct measurement method is used to directly determine and record various stresses by measuring apparatus, such as compensation stress, recovery stress, and equilibrium stress. The original stress of a rock is obtained by calculating the relationship between the recorded stress and the stress of the original rock. Using this method, the conversion of different physical quantities is not involved in the calculation; thus, there is no need to know the physical mechanical properties of rocks or the relationship between stress and strain. The flat jacking method, hydraulic fracturing method, rigid package stress meter method, and acoustic emission method are all direct measurement methods, among which, the hydraulic fracturing method has been most widely used worldwide, followed by the acoustic emission method.

For the indirect measurement method, stresses are not directly measured; instead, various indirect physical quantities are measured and recorded through certain sensing elements or certain media. Examples of these indirect physical quantities include the deformation or strain in rock mass, density of rock mass, permeability, water-absorbing quality, change of resistance, capacitance, and change of elastic wave propagation velocity. As these indirect physical quantities are related to the stress experienced by a rock mass, then, from the measured changes of indirect physical quantity, the corresponding stress value can be calculated by the known formulae. In the indirect measurement method, to calculate the stress value, the physical and mechanical properties of a given rock mass and the relationship between the physical quantity and measured stress must be determined first. The hole stress relief method and other stress or strain relief methods as well as geophysical methods are commonly used indirect methods. Among these, the hole stress relief method is one of the most widely used and relatively mature geostress measurement methods worldwide.

In the following subsection, the hydraulic fracturing and stress relief methods, which are widely used in coalmines, will be briefly introduced.

8.2.2 Hydraulic Fracturing Method

1. Principle of measurement

Hydraulic fracturing was widely used in oilfields in the 1950s to increase oil production by creating artificial fractures in drilling oil reservoirs [1]. Hubbert and Willis determined the relationship between hydrocracking fractures and original rock stress in practice. This finding was adopted by Fairhurst and Haimson to measure geostress.

From the theory of elastic mechanics, we can see that when a drill in an infinite body is subjected to an infinite two-dimensional stress field (σ_1, σ_2), the part of the distance from the end of the borehole is in a plane strain state. At these locations, the stress around the borehole can be expressed as follows:

$$\sigma_\theta = \sigma_1 + \sigma_2 - 2(\sigma_1 - \sigma_2)\cos 2\theta \tag{8.11}$$

$$\sigma_r = 0 \tag{8.12}$$

where σ_θ is the tangential stress around the borehole, σ_r is the radial stress around the borehole, and θ is the angle of any point around the borehole with the axis σ_1.

From Eq. (8.11) we know that when $\theta = 0°$, σ_θ obtains a minimum value, then

$$\sigma_\theta = 3\sigma_2 - \sigma_1 \tag{8.13}$$

If the hydraulic fracturing system (Fig. 8.4) is used to enclose a section of the borehole and inject high-pressure water into this enclosed section, when the water pressure exceeds the sum of ($3\sigma_2 - \sigma_1$) and the tensile strength of the rock (σ_c), at the point of $\theta = 0°$, fractures will occur at the borehole wall in the position of σ_1. Assuming the water pressure of the initial cracking of the borehole wall is P_i, we obtain Eq. (8.14):

$$P_i = 3\sigma_2 - \sigma_1 + \sigma_c \tag{8.14}$$

If the enclosed borehole continues to be injected with high-pressure water, the fracture will expand further. When the fracture depth reaches three times the borehole diameter, the stress state at the distal end of the borehole is close to the stress state of the original rock. The injection is stopped and the pressure constant is maintained, and the constant pressure at that moment is recorded as P_s. As can be seen from Fig. 8.6, P_s should be equal to the original rock stress σ_2, that is,

$$P_s = \sigma_2 \tag{8.15}$$

As long as the tensile strength of the rock is measured, the stresses σ_1 and σ_2 can be obtained by using Eqs. 8.14) and (8.15) based on the pressures P_i and P_s; thus, the magnitudes and directions of σ_1 and σ_2 are both determined.

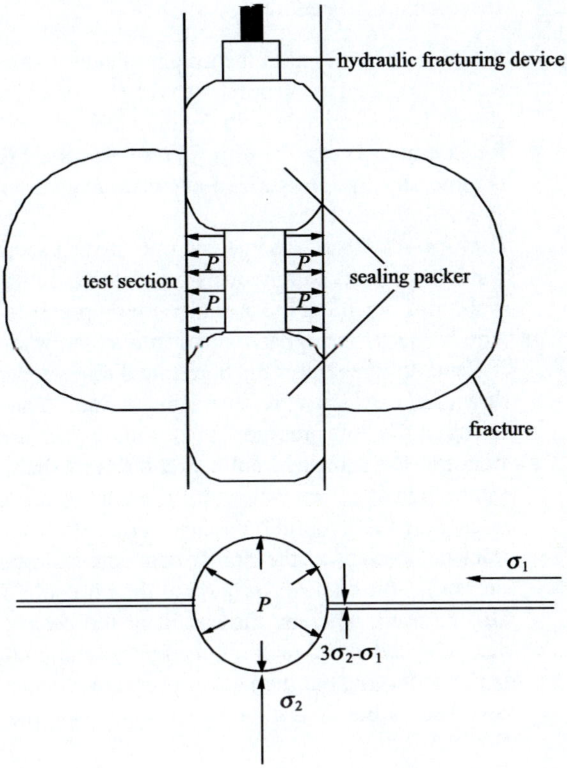

Fig. 8.4 Schematic of the stress measurement method by hydraulic fracturing

If water exists in the fractures of the borehole, then by assuming the fracture water pressure in the enclosed borehole as P_0, Eq. (8.14) can be rewritten as Eq. (8.16):

$$P_i = 3\sigma_2 - \sigma_1 + \sigma_c - P_0 \tag{8.16}$$

By using Eqs. (8.15) and (8.16) to calculate σ_1 and σ_2, it is necessary to know the tensile strength of the rock first, which is very difficult. In order to overcome this difficulty, an extra step is added to the hydraulic fracturing test, whereby the hydraulic pressure is removed when the initial fracture occurs close to the new fracture. Then, by repressurizing the enclosed section to make the fractures open again, the pressure when the fractures are reopened is recorded as P_r, then it can be determined that:

$$P_r = 3\sigma_2 - \sigma_1 - P_0 \tag{8.17}$$

Therefore, it is not necessary to know the rock tensile strength when using Eqs. (8.15) and (8.16) to solve σ_1 and σ_2. The measurement of the original rock stress by the hydraulic fracturing method will no longer involve the mechanical properties of the rock, but will be entirely determined by the measured/recorded values of pressure.

2. Measurement procedure

(1) A borehole is drilled at the prepared site of stress measurement, then the test section is sealed in the borehole with the sealing packer. The borehole diameter should be coordinated with the diameter of the selected packer, which can be, for example, 38 mm, 51 mm, 76 mm, 91 mm, 110 mm, or 130 mm. The packer is generally the pressurized expansion type, with both liquid and gas being available for pressurization.

(2) High-pressure water is injected into the test section between two packers. The water pressure is continuously increased until fractures appear on the borehole wall, thus obtaining the initial cracking pressure (P_i). Then, a continued pressure is exerted on expanding fractures, and when the fractures have expanded in depth to three times the borehole diameter, the high-pressure system is shut down and a constant pressure is maintained. The water pressure at this moment is called the "off pressure" (P_s). Finally, the pressure is relieved to close the fracture. The same hydraulic circuit can be shared to pressurize the packer and inject high-pressure water into the enclosed section. In general, a drill pipe is used as the hydraulic passage. First, the packer is pressurized, the packer inlet is closed, and the line is connected through the change-over switch to pressurize the enclosed section of the borehole. However, a double circuit can also be used, whereby the circuit of the packer and the hydraulic fracturing are independent of each other; the hydraulic channel of hydraulic fracturing is the drill pipe, but the packer pressure channel is the high-pressure hose. A Swedish hydraulic fracturing stress measurement system [5] is illustrated in Fig. 8.5.

During the entire injection process, the pressure–time and flow–time curves should be recorded simultaneously, as shown in Fig. 8.6. By using the appro-

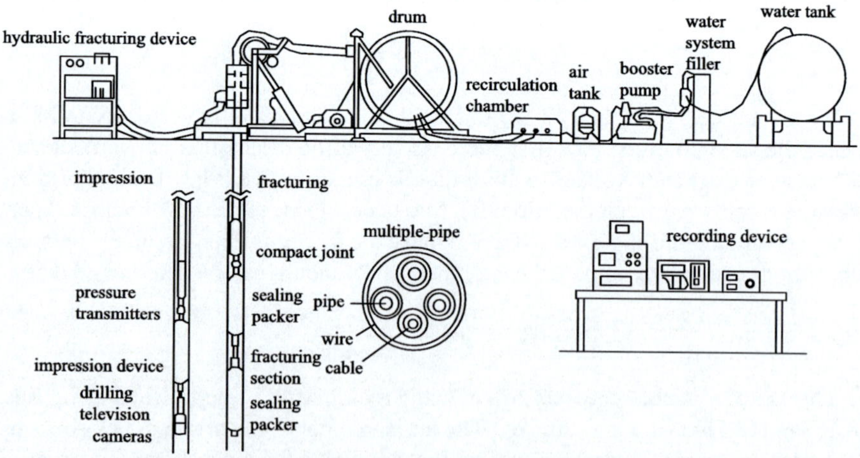

Fig. 8.5 Sketch map of the hydraulic fracturing stress measurement system

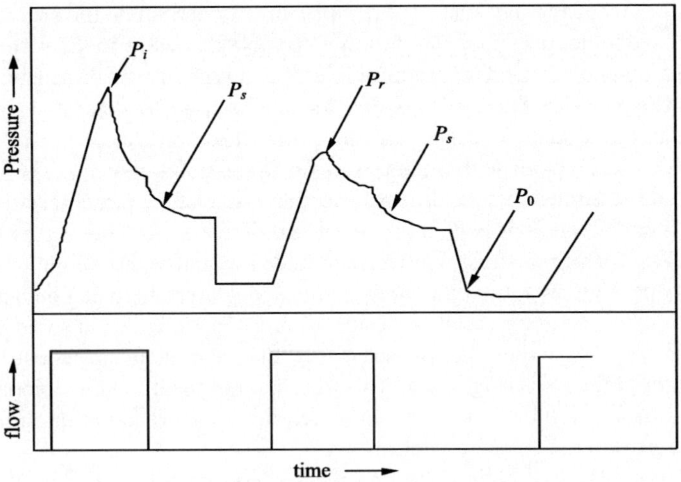

Fig. 8.6 Pressure–time and flow–time curves of the hydraulic fracturing method

priate method, the P_i and P_s values can be determined from the pressure–time curve, while the depth of fracture propagation can be determined from the flow–time curve.

(3) High-pressure water is reinjected into the test section and the fracture is reopened. The pressure P_r and subsequent constant closing pressure P_s are recorded. This pressure relief/repressure process is repeated 2–3 times to improve the accuracy of the test data. The values of P_i and P_s are also determined by the pressure–time and flow–time curves.

(4) The packer is completely decompressed and removed from the borehole along with all the equipment (e.g., the pressure pipe).

(5) The location, direction, and size of the hydraulic fractured fractures and natural joints of the drilling test section are measured. The measurement devices can select the down-hole cameras, down-hole television, down-hole optical telescopes, or impression devices. Of these, the cost of the first three methods is expensive and complicated to operate, while using the impression device is relatively simple and practical. The structure and shape of the impression device is similar to that of the packer. It is covered with a layer of plastic rubber or similar material outside. The impression device is sent with the pressure pipeline into the underground fracture site, where the impression device is pressurized to expansion in order to imprint all the joints and fissures of the borehole on it. This impression can be kept for the time required to record the fractures on the impression device after getting back to the surface, and can determine the fracture orientation. In general, hydraulic fracturing fractures are very easy to recognize and appear as a group of longitudinal cracks that are radially facing each other.

Hydraulic fracturing measurement results can only determine the size and direction of the maximum principal and minimum principal stresses in the plane perpendicular to a borehole; therefore, in principle, it is a type of two-dimensional stress measurement method. Therefore, if this method is used to determine the three-dimensional stress state of the measuring point, three boreholes must be drilled non-parallel to each other with the intersection at one point, which is very difficult. In general, it is assumed that the drilling direction is one of the principal stresses, For example, if one borehole is drilled in the vertical direction, and the vertical stress is considered as a principal stress, whose magnitude is equal to the weight of the overlying strata per unit area, then the three-dimensional stress field can be determined by the hydraulic fracturing results of a single borehole. However, in some cases, the vertical direction is not the principal stress direction, and the magnitude is not equal to the weight of the overlying strata. If the direction of the borehole compared with the actual stress has a deviation of >15°, then the above assumption will cause a significant error in the measurement results.

Furthermore, the hydraulic fracturing method considers that the initial cracking occurs in the parts of the borehole wall with the minimum tangential stress, which is parallel to the direction of the maximum principal stress. This is based on the hypothesis that the rock is continuous, homogeneous, and isotropic. If natural joints on the hole wall exist, the initial cracks are likely to occur in these areas, rather than at the site with the minimum tangential stress. Therefore, the hydraulic fracturing method is suitable for brittle rock with good integrity.

8.2.3 *Stress Relief Method*

1. Basic principles

The stress relief method is a widely used method for measuring the original rock stress [1]. The basic principle of this method is that when it is necessary to determine the stress state of a certain point in a rock mass, the rock mass unit at this location should be artificially separated from the surrounding rock mass. Because the stress on the rock unit is released, the geometry of the unit body will produce elastic recovery. Thus, by using some instruments to determine the strain or deformation of this elastic recovery, we can use the elastic theory to calculate the original rock stress based on the assumption that the rock mass is a continuous, homogeneous, and isotropic elastomer. This process can be summarized as follows: break the bond and release the stress; measure the deformation of the rock mass at the point of elastic recovery; obtain the stress according to the deformation.

There are many specific approaches for the stress relief method, which can be divided according to the test depth as: surface stress relief, shallow-hole stress relief, and deep-hole stress relief. Based on the difference in testing deformation or strain method, the radial deformation test, hole-wall strain test, and drilling stress relief method can be applied. The following subsection introduces the commonly used complete stress relief method.

2. Complete stress relief method

The complete stress relief method is a way to ensure that a rock mass is measured without the influence of geostress. Usually, drilling is used to achieve the complete stress relief of the sleeve core borehole, also known as the stress relief method by over-coring. The complete stress relief method is the longest developed and more mature method for geostress measurement. At present, there is no other method that can be compared with the complete stress relief method in determining the applicability and reliability of rock stress.

At present, the complete stress relief method forms a set of standard measurement procedures, and the specific steps are as follows (also see Fig. 8.7).

(1) First, a large hole is drilled from the rock surface to the interior of the rock mass until it reaches the point where the stress of the rock mass needs to be measured (Fig. 8.7a). Drilling is usually performed from an underground roadway, tunnel, or other excavation chamber. The diameter of the large borehole should be >3 times the diameter of the small borehole that is used to install the probe for the next step. The diameters of the small and large boreholes are 36–38 mm and 130–150 mm, respectively. The depth of the large borehole should be >2.5 times the span of the roadway, tunnel, or excavated chamber, such that the measurement point can be located at the original stress location unaffected by rock mass excavation disturbance. Thus, the larger the chamber span, the deeper the borehole depth should be. The measurement should be done within the smallest span of the excavation space as possible to save manpower and material resources, ensure the success of the test, and avoid arranging the test points at a fork or other locations where the excavation disturbance is large. In order to facilitate the installation of the test probe in the next step, the large borehole should maintain a certain degree of concentricity; hence, the drilling process requires a guide device. After the large borehole is drilled, the bottom

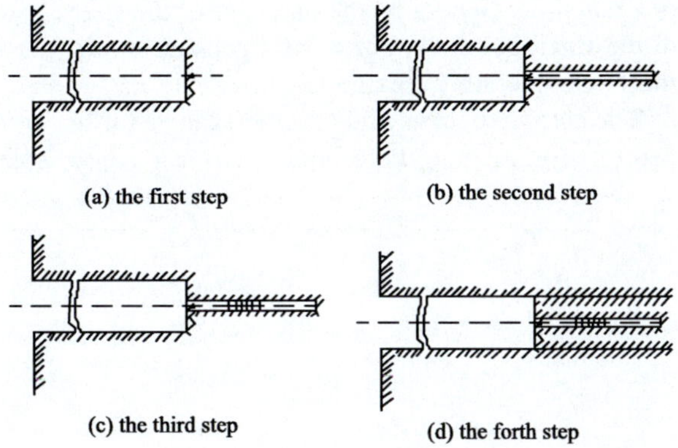

Fig. 8.7 Sketch map of the measuring steps of the complete stress relief method

of the hole should be ground flat, and a small conical hole should be made to facilitate the next step of drilling the concentric hole, cleaning the borehole and making the probe enter smoothly.

(2) Second, a concentric small borehole is drilled from the bottom of the large borehole for the installation of the probe (Fig. 8.7b). The diameter of the small borehole is determined by the probe diameter, which usually ranges from 36 to 38 mm. The depth of the small borehole is generally approximately 10 times its diameter to ensure that the central part of the small borehole is in a plane strain state. When the small borehole is drilled, it should be flushed with water to clean up all the cuttings and other debris within it. In addition, the borehole needs to be tilted up to 1°–3°.

(3) Third, the measuring probe is installed (fixed or glued) using a set of special devices, for example, a radial deformation gauge and hole-wall strain gauges in the central part of the borehole (Fig. 8.7c).

(4) Fourth, the thin-walled drill that was used in the first step is used again to further extend the depth of the large borehole in order to relieve the stress around the small borehole (Fig. 8.7d). The deformation or strain of the small borehole due to stress relief is measured by a measurement system that includes the test probe, and is then recorded by a recording instrument. According to the measured deformation or strain of the small borehole, the original stress state around it can be obtained using the relevant equations.

Theoretically, the stress within the sleeve core will be completely lifted regardless of the shape and size of the sleeve. However, if the measuring probes have limitations or constraints on the deformation of a borehole during stress relief, they will influence the stress release in the casing core; thus, the shape and size of the sleeve must be considered in this situation. Generally, the greater the stiffness of the probe, the greater the constraint on the deformation of the borehole and the larger the diameter of the sleeve borehole. For probes with absolute stiffness, the size of the sleeve should be infinite to achieve complete stress relief. Therefore, a rigid probe cannot be used for stress relief measurements. For the radial deformation gauge, hole-wall strain gauge, and hollow inclusion cell among other instruments, the size and shape of the sleeve are not very strict because they almost have no constraint on the deformation of a borehole. Generally, the diameter of the sleeve should be >3 times the diameter of the small borehole. For a solid package strain gauge, the diameter of the sleeve must be moderately enlarged.

The complete stress relief method can be further divided into five types: the radial deformation method, hole bottom strain method, hole-wall strain method, hollow inclusion strain method, and solid inclusion strain method. The following subsection mainly deals with the hole-wall strain method.

3. Hole-wall strain method

1) Measurement technology

Under the three-dimensional stress field, the state of stress distribution on the borehole surface and its surroundings in an infinite body can be solved using the modern elasticity theory. Thus, the stress of the borehole surface can be calculated by measuring the strain on the borehole surface according to the stress relief method, and the state of the original rock stress can be calculated accurately. The CSIR triaxial hole-wall strain gauge from South Africa was developed based on this principle.

The main part of the CSIR triaxial hole-wall strain gauge includes three measuring pistons, which are made of rubber and have a diameter of approximately 1.5 cm each. They have arc-shaped ends so that they can be consistent with the borehole radian in order to maintain close contact with the borehole. On the surface of the end, four resistance strain gauges are pasted with an interval of 45° to each other, thus composing a circular strain gauge array. Three pistons (i.e., three sets of rosette gauges) are located in the same circumference and are distributed equally in a spacing of 120°. The shell consists of two parts: the front shell and the back shell. There is a circular slot in the front shell and a strain gauge is attached, while the end of the back shell has a plug of 14 resistance strain wires (see Fig. 8.8). During the use of the gauge, the first step is to stick a piece of rock with a diameter of approximately 1.2 cm and a thickness of approximately 0.8 cm to the end of the front shell for temperature compensation. The second step is to apply a cementing agent to the three piston heads, and then use a special tool to send the strain gauge into the borehole to perform the measurement. The third step is to push the piston out using wind pressure, with the end of the gauge in close contact with the borehole wall until cement curing is achieved. Finally, the borehole stress is relieved. The stress readings of 12 gauges are measured before and after stress relief, and the difference between each gauge reading is used to calculate the stress. A single-hole stress measurement

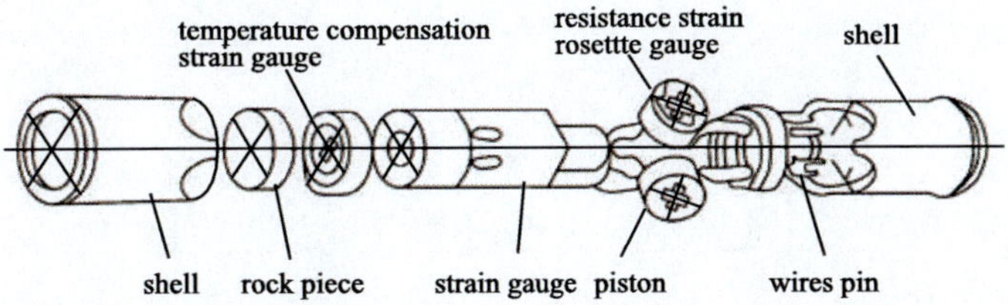

Fig. 8.8 CSIR triaxial hole-wall strain gauge

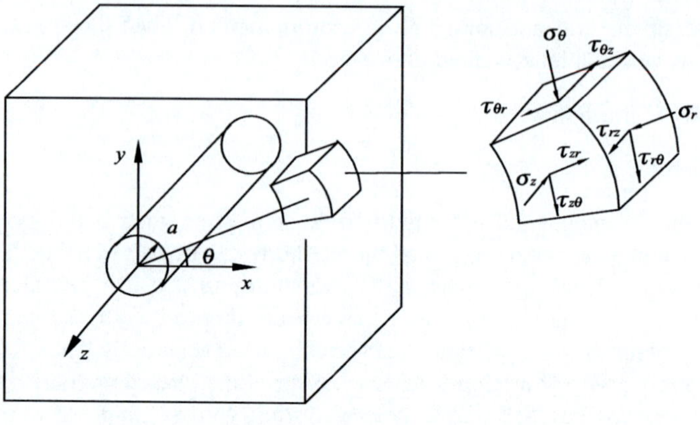

Fig. 8.9 Stress distribution in the surrounding rock of a three-dimensional borehole

can determine the magnitude and direction of the three-dimensional stress at the measuring point. The applicable borehole diameter for the CSIR hole-wall strain gauge is 36–38 mm.

2) Principles of the stress calculation

(1) Stress distribution in the surrounding rock of a borehole
The borehole in an infinite body is subjected to a three-dimensional stress field (σ_x, σ_y, σ_z, τ_{xy}, τ_{yz}, and τ_{zx}) at infinity (as shown in Fig. 8.9), and the stress distribution in the surrounding rock can be expressed as follows:

$$\sigma_r = \frac{\sigma_x + \sigma_y}{2}\left(1 - \frac{a^2}{r^2}\right) + \frac{\sigma_x - \sigma_y}{2}\left(1 - 4\frac{a^2}{r^2} + 3\frac{a^4}{r^4}\right)\cos 2\theta + \tau_{xy}\left(1 - 4\frac{a^2}{r^2} + 3\frac{a^4}{r^4}\right)\sin 2\theta \quad (8.18)$$

$$\sigma_\theta = \frac{\sigma_x + \sigma_y}{2}\left(1 + \frac{a^2}{r^2}\right) - \frac{\sigma_x - \sigma_y}{2}\left(1 + 3\frac{a^4}{r^4}\right)\cos 2\theta - \tau_{xy}\left(1 + 3\frac{a^4}{r^4}\right)\sin 2\theta \quad (8.19)$$

$$\sigma_z' = -\upsilon\left[2(\sigma_x - \sigma_y)\frac{a^2}{r^2}\cos 2\theta + 4\tau_{xy}\frac{a^2}{r^2}\sin 2\theta\right] + \sigma_z \quad (8.20)$$

$$\tau_{r\theta} = \frac{\sigma_x - \sigma_y}{2}\left(1 + 2\frac{a^2}{r^2} - 3\frac{a^4}{r^4}\right)\sin 2\theta - \tau_{xy}\left(1 + 2\frac{a^2}{r^2} - 3\frac{a^4}{r^4}\right)\cos 2\theta \quad (8.21)$$

$$\tau_{\theta z} = (-\tau_{zx}\sin\theta + \tau_{yz}\cos\theta)\left(1 + \frac{a^2}{r^2}\right) \quad (8.22)$$

$$\tau_{rz} = \left(\tau_{zx}\cos\theta + \tau_{yz}\sin\theta\right)\left(1 - \frac{a^2}{r^2}\right) \quad (8.23)$$

where a is the borehole radius, and σ_z' is the stress component of the Z axial direction at any point of the surrounding rock affected by excavation.
In the above equations, the rectangular coordinate system is adopted to describe the original rock stress state, and the stress state of the surrounding rock around the borehole is described by the cylindrical coordinate system. The Z-axis of a cylindrical coordinate system is the same as that of the rectangular coordinate system, and the angle θ of the cylindrical coordinate system is defined as positive when rotated counterclockwise from the X-axis. We note that in Eq. (8.20), σ_z is the stress component of the original rock, and σ_z' is the stress component of the Z-axial direction at any point of the surrounding rock of a borehole affected by excavation. When $r \to \infty$, we obtain $\sigma_z' = \sigma_z$.

(2) Relationship between the hole-wall strain and the three-dimensional stress component
The hole wall is in the state of plane stress, and there are only three stress components: σ_θ, σ_z', $\tau_{\theta z}$ (Fig. 8.10). The relationships between the measured strain values of the four strain gauges of each rosette gauge (i.e., ε_θ, ε_z, ε_{45}, ε_{-45}) are as follows:

$$\varepsilon_\theta = \frac{1}{E}\left(\sigma_\theta - \upsilon\sigma_z'\right)$$

$$\varepsilon_z = \frac{1}{E}\left(\sigma_z' - \upsilon\sigma_\theta\right)$$

$$\gamma_{\theta z} = 2\varepsilon_{45} - (\varepsilon_\theta + \varepsilon_z) = (\varepsilon_\theta + \varepsilon_z) - 2\varepsilon_{-45} = \frac{\tau_{\theta z}}{G} \quad (8.24)$$

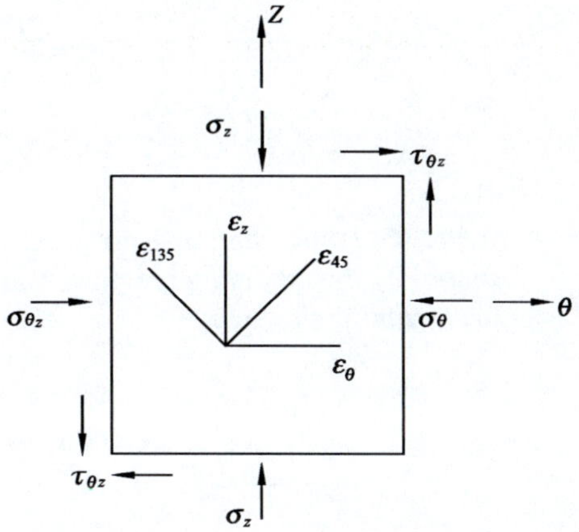

Fig. 8.10 Stress state of a resistance rosette gauge

where ε_θ, ε_z, and $\varepsilon_{\pm 45}$ are the hole-wall strain values in the circumferential direction, axial direction, and ±45° from the drilling axis direction, respectively, and $\gamma_{\theta z}$ is the shear strain value.

Using Eqs. (8.18)–(8.23) to transform σ_θ, σ_z', and $\tau_{\theta z}$ to the expression of the stress component σ_x, σ_y, $-\sigma_z$, τ_{xy}, τ_{yz}, and τ_{zx} of the original rock, we obtain Eqs. (8.25)–(8.28):

$$\varepsilon_\theta = \frac{1}{E}\left\{(\sigma_x + \sigma_y) + 2(1 - \upsilon^2)\left[(\sigma_x - \sigma_y)\cos 2\theta - 2\tau_{xy}\sin 2\theta\right] - \upsilon\sigma_z\right\} \tag{8.25}$$

$$\varepsilon_z = \frac{1}{E}\left[\sigma_x - \upsilon(\sigma_x + \sigma_y)\right] \tag{8.26}$$

$$\gamma_{\theta z} = \frac{4}{E}(1 + \upsilon)\left(\tau_{yz}\cos\theta - \tau_{zx}\sin\theta\right) \tag{8.27}$$

$$\varepsilon_{\pm 45} = \frac{1}{2}(\varepsilon_\theta + \varepsilon_z \pm \gamma_{\theta z}) \tag{8.28}$$

The measured results from each rosette gauge can yield 4 equations; hence, 3 sets of rosette gauges can provide a total of 12 equations, of which there are at least 6 independent equations. Therefore, the six components of the original rock stress can be solved.

(3) Equations for calculating the elastic modulus and Poisson's ratio of the measured rock

After the stress is completely relieved, the sleeve core is taken out and the hole-wall strain gauge is still cemented in the borehole. According to the results of the confining pressure test on the sleeve core, the elastic modulus and Poisson's ratio of the measured rock can be obtained from Eqs. (8.29) and (8.30), respectively:

$$E = \frac{P_0}{\varepsilon_\theta}\frac{2R^2}{R^2 - r^2} \tag{8.29}$$

$$\upsilon = \frac{\varepsilon_z}{\varepsilon_\theta} \tag{8.30}$$

where P_0 is the confining pressure, ε_θ is the average circumferential strain caused by the confining pressure, and ε_z is the average axial strain caused by the confining pressure.

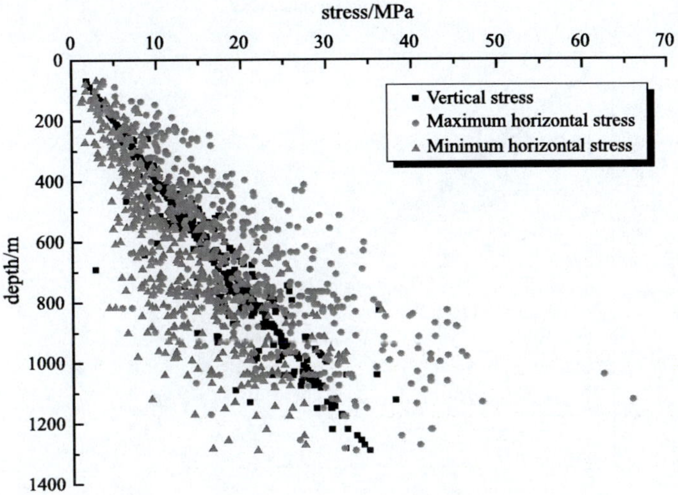

Fig. 8.11 Geostress measurement results from Chinese coalmines

8.3 Evolution Rule of Geostress with Burial Depth

8.3.1 Measurement Results of Geostress

During the exploration and development phase of coal resources, coal miners and researchers have carried out numerous measurements on the geostress distribution in coal-bearing strata. As shown in Fig. 8.11, 585 groups of geostress have been measured in the major coal-producing areas of China [6–74]. In general, the geostress of coal-bearing strata increases with increasing burial depth. The detailed rules will be presented in the next section. The horizontal stress in each direction is mostly presented as follows: one greater than the vertical stress and the other less than the vertical stress.

8.3.2 Evolution Rules of Geostress with Burial Depth

1. Evolution rule of vertical stress with burial depth

The statistical analysis results of vertical stress in the main coal mining areas of China are shown in Fig. 8.12. From this, the relationship between the vertical stress of coal-bearing strata and the burial depth in China can be described by Eq. (8.31):

$$\sigma_V = 0.0257H \tag{8.31}$$

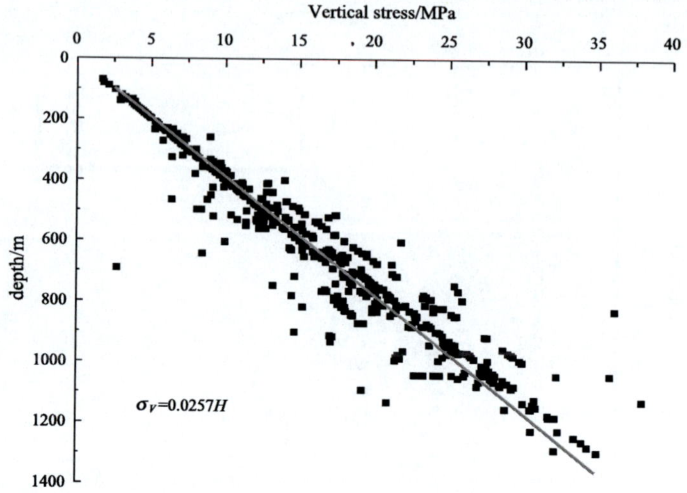

Fig. 8.12 Evolution rule of vertical stress with burial depth in China

where σ_V is the vertical stress of coal-bearing strata, MPa; and H is the burial depth, m.

Hoek and Brown statistically analyzed the vertical stress in different countries around the world (Fig. 8.13), and found that the relationship between vertical stress and burial depth could be expressed using Eq. (8.32) [1]:

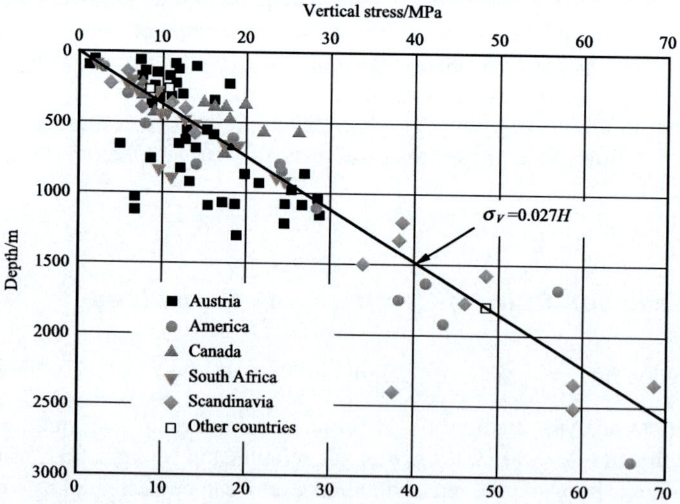

Fig. 8.13 Relationship between vertical stress and burial depth in different countries worldwide

$$\sigma_V = 0.027H \tag{8.32}$$

Comparing Eq. 8.31 with Eq. 8.32, it can be noticed that the relationship between the vertical stress and burial depth in China is similar to that of other countries, and that the vertical stress of the strata is basically equal to the self-weight of the overlying strata. The difference between the coefficients in the two equations is mainly caused by the variation in the rock density of the overlying strata in different areas. For coal-bearing strata, the rock is mostly sedimentary, whose lithology is mainly mudstone, shale, and sandstone; thus, the density is relatively small.

2. Evolution rule of horizontal stress with burial depth

The statistical analysis results of the horizontal stress in the main coal mining areas of China are shown in Figs. 8.14 and 8.15. From this, the relationships between the maximum and minimum horizontal stresses (σ_H and σ_h, respectively) of coal-bearing strata and the burial depth (H) in China are as follows:

$$\sigma_H = 0.0272H + 2.90 \tag{8.33}$$

$$\sigma_h = 0.0179H + 1.312 \tag{8.34}$$

In Figs. 8.14 and 8.15, the horizontal stress increases linearly with increasing burial depth. Equations (8.33) and (8.34) further indicate that the horizontal stress is not zero at Earth's surface, which is the same as the prediction results from Eq. (8.9) in Sect. 8.1.2.

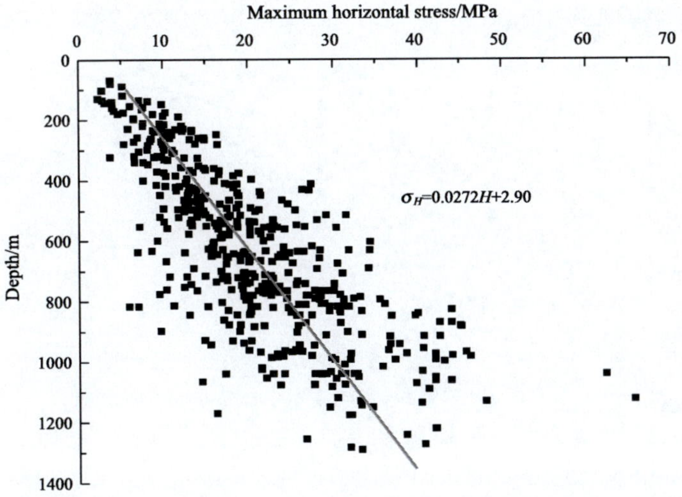

Fig. 8.14 Evolution rule of the maximum horizontal stress with burial depth

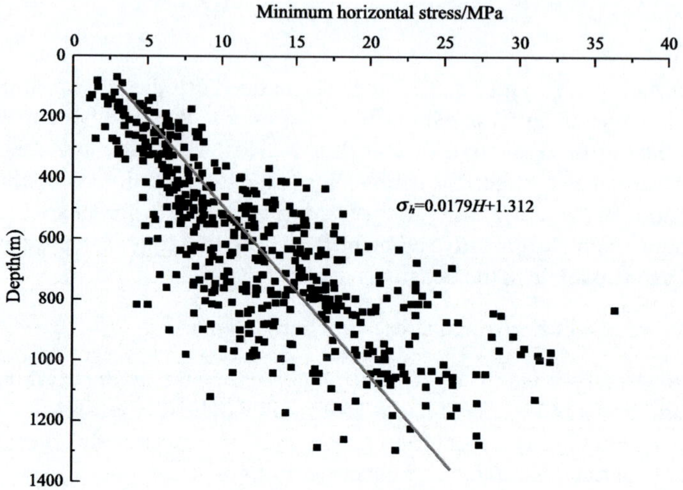

Fig. 8.15 Evolution rule of the minimum horizontal stress with burial depth

The average lateral stress ratio of coal-bearing strata in China was calculated using $k = \frac{\sigma_H + \sigma_h}{2\sigma_V}$. The results are plotted in Fig. 8.16, which shows that the average lateral stress ratio of the coal-bearing strata in China decreases nonlinearly with increasing burial depth. The function relationship of this trend is consistent with the form of

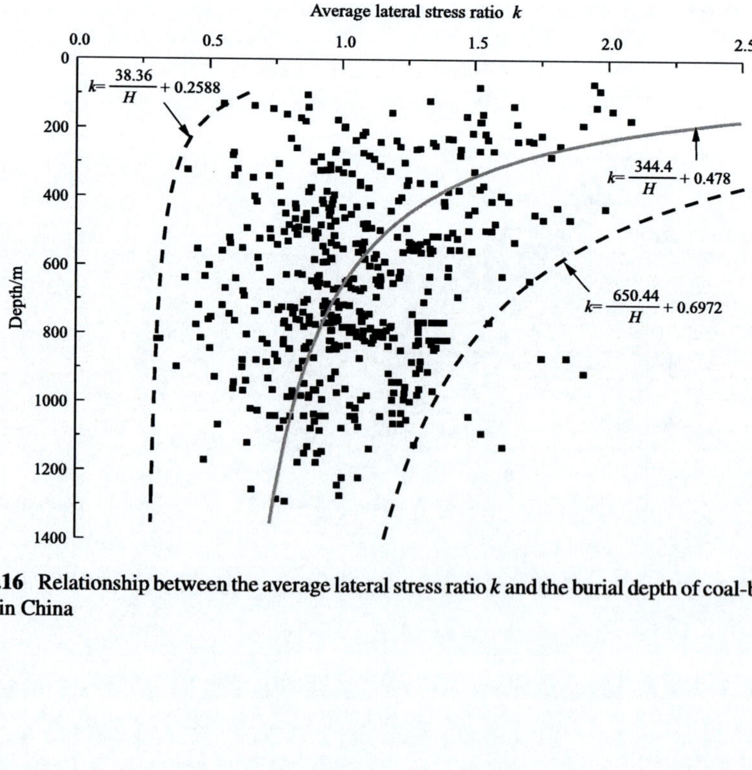

Fig. 8.16 Relationship between the average lateral stress ratio k and the burial depth of coal-bearing strata in China

Eq. (8.10). The internal and external envelope lines are expressed by Eqs. (8.35) and (8.36), respectively:

$$k = \frac{38.36}{H} + 0.2588 \tag{8.35}$$

$$k = \frac{650.44}{H} + 0.6972 \tag{8.36}$$

The average value of which is given by Eq. (8.37):

$$k = \frac{344.4}{H} + 0.478 \tag{8.37}$$

The relationship between the average lateral stress ratio and the burial depth of coal-bearing strata is similar to that of the statistical analysis results of Hoek and Brown [1] (see Fig. 8.17). However, it can be found that their specific parameters are different, which is related to the difference in the stress field of various regions, and to the lithologies of the stress test sites. Zhao et al. [75] summarized the geostress measurement results from mines, oilfields, transportation, and other industries in China, and analyzed the relationship between the average lateral stress ratio and burial depth in different lithology strata, as shown in Table 8.1.

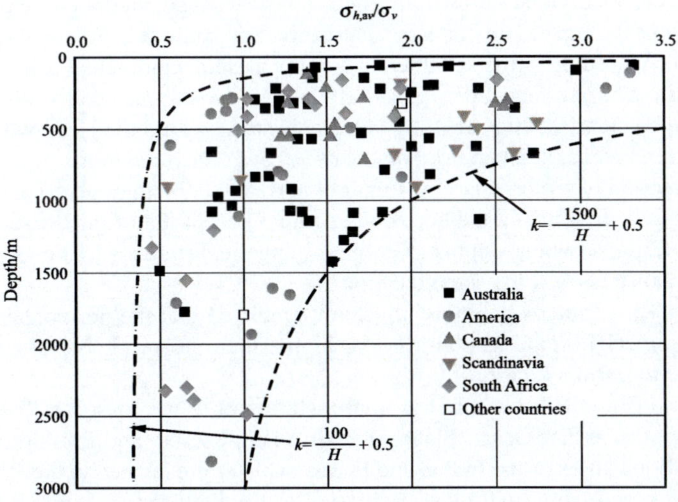

Fig. 8.17 Regularity of the average horizontal stress/vertical stress ratio versus burial depth worldwide

Table 8.1 Relationship between the average lateral stress ratio and burial depth in different lithology strata [75]

Lithology	Average value	External envelope line (small)	Internal envelope line (large)
Magmatic rock	$k = \frac{200}{H} + 0.7$	$k = \frac{76}{H} + 0.3$	$k = \frac{309}{H} + 1.2$
Sedimentary rock	$k = \frac{104}{H} + 0.9$	$k = \frac{41}{H} + 0.4$	$k = \frac{673}{H} + 1.3$
Metamorphic rock	$k = \frac{82}{H} + 1.0$	$k = \frac{45}{H} + 0.4$	$k = \frac{172}{H} + 1.3$

8.4 Tectonic Stress of Coal-Bearing Strata

8.4.1 Tectonic Stress

From the previous section, the vertical stress in coal-bearing strata is basically the intermediate stress, while the maximum stress is the horizontal stress. Thus, a gap exists between the measured results and theoretical analysis in Sect. 8.1. This shows that, besides the lateral stress caused by the self-gravity of strata, there are other origins that can induce additional geostress. This additional geostress is the stress caused by the tectonic movement of various structures within the crust, which is often called tectonic stress.

According to the theory of Continental Drift, Earth's crust and the upper part of the mantle can be divided into several rigid spherical shape plates. These plates can have a thickness of up to 100 km in the depth direction, and are in a state of tangential movement. According to the satellite positioning system, continental drift can reach a few centimeters per year. Continental drift is bound to produce collision and shear at the plate boundaries, thus resulting in tectonic stress within the plate, which tends to exceed the vertical stress caused by the effect of gravity on strata.

The direction of tectonic stress is mainly horizontal. Tectonic stress is the main reason for the difference in the horizontal components of the original rock stress at different locations (whose depths are similar), and the difference in the levels of the two principal stresses at the same location.

Siguang Li, a famous geologist in China, predicted that the horizontal stress in many geological formations could be far higher than the vertical stress, which is largely due to Earth's rotation.

Mainland China is located in the southeastern part of the Eurasian Plate, and is clamped by the Indian Ocean Plate, Siberian Plate, Pacific, and Philippine Ocean Plate. The boundaries of the Indian and Eurasian plates are located in the Himalayan tectonic belt, and Chinese Taiwan is located at the boundary tectonic belt of the Eurasian and Philippine Sea plates. The interactions between the plates, which constitute the dynamic boundary of mainland China, and joint actions of deep dynamics inside the Eurasian Plate, create the complexity of the internal active structure and modern tectonics of mainland China.

The activity of the block in the interior of mainland China is closely related to the active fractures, active folds, active basins, active blocks, and seismic activities. The stress field caused by the block activity reflects the regional characteristics which constitute the overall background of continental dynamics in China. Mainland China suffers from the push effect from the Indian Ocean and Pacific plates, with a speed of several centimeters per year, while it is also bound by the Siberian and Philippine Sea plates. Under such boundary conditions, plates are deformed to produce a horizontal compressive stress field. The modern stress field reflects the general dynamic background of mainland China: extrusion in the western region and tension in eastern region [76]. Generally, the western region of China is affected by the north–south push effect; thus, the stress level is higher, while the stress field of the eastern region is generally east–west directed, and the stress level is relatively low.

8.4.2 Influence of Tectonics on Coal Seam Occurrence

The formation periods of coal-bearing basins in China were synchronous with those worldwide, mainly occurring after the Late Paleocene-Carboniferous. The main coal-forming periods were concentrated in the Carboniferous-Permian, Triassic (Late Triassic), Jurassic (Early and Middle Jurassic), Cretaceous (Early Cretaceous), Paleogene, and Neogene. Most of the coal-bearing basins in China have experienced complex tectonic movements and destruction; thus, they are now fragmented [76].

After their formation, the Carboniferous-Permian coal-bearing basins in China mainly experienced the Indo-China, Yanshanian, and Himalayan movements. The magnitude, scope, and tectonic stress field induced by each tectonic movement differed. The generation and storage conditions of coal seam gas control the occurrence and distribution of this gas. As a type of gas stored in coal seams and surrounding rock, coal seam gas is easily dissipated. After the formation of a coal seam, the tension and depression of tectonic movement would have caused a considerable dissipation of the coal seam gas. Different levels of tectonic activities and tectonic stress fields can control ① the scope and strength of tectonism, ② the coal seam gas occurrence/distribution in different areas and over different ranges, and ③ the coal seam occurrence conditions and the destruction conditions/scopes of the coal body structure. Generally, plate tectonics control the scope and strength of the regional structure, which itself controls the scope and strength of a mining area (coalfield). The mining area structure then controls the scope and strength of a coalmine/mine working face; thus, tectonics control the gas formation and distribution of a coal seam with different ranges.

Here, we take the Carboniferous-Permian coal-bearing series, which is generally found on the North China Plate, as an example. Due to the impact of subduction collision from the Pacific Kula Plate, Eastern China started to uplift during the Indo-China period. The Luxi fault-uplift zone was the first to uplift, resulting in a wide denudation of Permian sediments in the Luxi area. The Permian coal seam in this area suffered serious erosion and the coal seam gas was largely dissipated. The coal seam at a depth of 600–800 m is still located in the gas weathering belt. At present, more than 90% of the coalmines in Shandong Province and its adjacent area, which is controlled by the Luxi fault-uplift zone, are low gas mines. In the western Henan coalfields (e.g., Pingdingshan, Xinmi, and Dengfeng) and the northern Henan coalfields (e.g., Jiaozuo, Hebi, and Anyang), thousands of thick Triassic caprocks are deposited above the Carboniferous-Permian coal seam and the gas storage condition for coal seam gas is good; thus, the high gas mines and coal/gas outburst mines (mining area) are clustered together.

8.4.3 Dynamic Metamorphism of Tectonic Stress on Coal

Coal is an organic rock that is sensitive to temperature, pressure, and other geological environmental factors. The various tectonic–thermal events in the geological evolution process inevitably lead to a series of physical and chemical changes in coal, thus forming different types of tectonic coal. Under tectonic stress, coal rock not only exhibits brittle and ductile deformation, but also produces different degrees of dynamic metamorphism.

Based on the existing research results of scholars worldwide, the mode and mechanism of the evolution of organic dynamics that are affected by tectonic stress (tectonism) can be summarized from three viewpoints: friction heat, strain energy, and mechanical chemistry [77].

According to the theory of friction heat, the tectonic stress applied to a coal body can cause mechanical displacement and produce frictional heat, which provides thermal energy to promote coalification. When explaining the localized elevation of the coal rank in the Luhr–Sutan thrust fault zone, TeichmÜLler and TeichmÜLler [78] determined that this is caused by the frictional heat generated during the movement of the fault. Meanwhile, because the fault generally moves at a very slow speed, it is difficult to accumulate the generated frictional heat. Therefore, the influence scope of this type of metamorphism is small, and the coal rank change due to this effect is also not significant. In explaining the anthracitization effect on the reverse fault plane of the Rocky Mountains and the coal graphitization induced by structural shear in some areas of the Alps, Bustin [79] and Suchy et al. [80] indicated that frictional heat is the reason for the above results. Bustin [79] considered that stick–slip faulting can lead to an instantaneous heating effect, whose maximum temperature can reach 350–650 °C. The dynamic metamorphism hypothesis put forward by Chinese scholars also postulates that coal metamorphism is promoted as a result of the fragmentation/flow of coal seams under the effect of force and the conversion of mechanical energy into heat [81]. The frictional heat viewpoint is still essentially a temperature factor, although thermal energy is derived from the displacement friction of tectonic stress.

According to the theory of strain energy, tectonic stress–strain energy and/or kinetic energy provide the energy for the evolution of organic matter, while mechanical energy without heat promotes the physical activation of material and directly causes a chemical reaction of organic matter. Ross and Bustin [82, 83] proposed the strain energy conversion mechanism of kerogen graphitization. Under shear stress, the strain energy produced by the shear creep of a coal seam can reduce the thermal activation energy required by graphitization by nearly two orders of magnitude. Their experimental results showed that under the condition of simple shear, the graphitizing effect of high volatile bituminous coal could occur at 800 °C. However, in a hydrostatic pressure test, the graphitization requires an overall increase in temperature and strain. Therefore, the simple shear condition is the best stress condition for graphitization, and the thermal activation energy demand is the lowest. Under the simple shear condition, the strain energy can cause the collapse of the pore wall in coal as well as the steering and orientation of the basic structure unit of coal to form graphite. In addition, the optical graphite reported by Suchy et al. [80] is actually a typical strain of graphite. Cao et al. [84] demonstrated that the energy of the dynamic metamorphism of tectonic coal was mainly the strain energy. Strain energy, especially the strain energy under the simple shear condition, can cause highly volatile bituminous coal to generate a graphitizing effect at 800 °C. Under the normal geothermal field, the strain energy is also likely to cause the dynamic metamorphism of the so-called dealkylation side chain and aromatic nucleation rising.

Барамбой and Амрскийthe of former Soviet Union proposed the mechanical chemistry view. Based on the chemical conversion of organic macromolecular polymers under mechanical action, they considered that the micro-shearing action is concentrated in individual segments, the critical stress is exceeded to break the chemical bonds break, and the organic macromolecules are cracked to produce small molecular hydrocarbons. Liu et al. [85] conducted a hydrocarbon generation simulation under low temperature and shear stress conditions, found that the humic organic matter involved the transformation of large molecular hydrocarbons into small molecular hydrocarbons. Zhang et al. [86, 87] applied mechanical chemistry theory and coal structure analysis to discuss the mechanism of the mechanical chemical action of coal. The authors considered that the extrusion stress of a geological structure is an important energy source for the coal mechanical chemistry effect, whose extruding shear function not only deforms and damages coal, but also causes a series of chemical effects within the coal structure. The coal structure is polymerized, heterogeneous, rearranged, and aromatic, forming a larger network structure. The molecules break and decompose into liquid hydrocarbons (low molecular weight compounds) and gaseous hydrocarbons (gas) in a short time. This type of tectonic coal is a product of mechanical chemistry effects, which not only accelerate the process of coal metamorphism, but also accelerate the conversion of hydrocarbons. These accelerated processes are advantageous for producing low molecular compounds and enriching gas, which are the main causes of dynamic metamorphism and hydrocarbon formation in coal and rock.

8.5 Variation of Gas Parameters with Burial Depth

8.5.1 Variation of Gas Pressure with Burial Depth

The gas pressure of a coal seam depends on the gas emission conditions. Over geological time, the gas emission conditions of a coal seam represent a very complicated issue. This is not only related to the thickness/permeability of caprocks, permeability of the coal seam, and coal geological structure conditions, but also to the water content of the covering layer. When the coal seam gas pressure is less than the hydrostatic pressure, the coal seam gas emission stops and the gas pressure can be preserved. The most unfavorable condition for the gas emission of a coal seam is that the pores in the covering layer are filled with water. At that time, the gas pressure of the coal seam is the highest and equal to the hydrostatic pressure at the same level. In China and abroad, only a small number of measured gas pressure values have been reported to exceed the hydrostatic pressure at the same level, which can be explained by the following reasons. The first reason is that the measured point is affected by the concentrated stress from the mining roadway. When the gas adsorption of coal is close to saturation, the stress concentration causes the pore volume to shrink in order to increase the gas pressure. The second reason is that fractures are connected with deep pressurized gas. Third, when the terrain is appropriate, fountain type conditions are formed.

The general path of coal seam gas migration involves escaping from the deep part of the stratum to the surface, which determines that the gas pressure and gas content of the coal seam both increase with depth. Figure 8.18 displays the variation in the measured coal seam gas pressure with depth in the main mining areas of China.

The results of the coal seam gas occurrence in China and abroad indicate that the gas pressure increases with depth in the methane zone, while the gas pressure gradient varies under different geological conditions. In geological blocks with similar geological conditions, the same coal seam at the same depth has essentially the same gas pressure, which mostly exhibits a linear increase with increasing burial depth.

Owing to the linear relationship between gas pressure and burial depth, in general, the deep gas pressure can be predicted by Eq. (8.38):

$$p = \omega H + C = p_c + \omega(H - H_c) \tag{8.38}$$

where p and p_c are the coal seam gas pressure at burial depths of H and H_c, respectively (MPa); H and H_c are the burial depths of the coal seam (m); ω is the gas pressure gradient (i.e., the increment of gas pressure with increasing burial depth per 1 m), the value of which is general 0.01 ± 0.005 MPa/m; and C is a constant.

Based on the statistical analysis of measured data from outbursts in the mining area of Chongqing, Beipiao, and Hunan [90], a study found that the gas pressure distribution variation line with burial depth was usually close to the hydrostatic pressure line (line OC in Fig. 8.19), and that the coal seam gas pressure had a limiting value: the hydrostatic pressure value. The gas pressure variation with burial

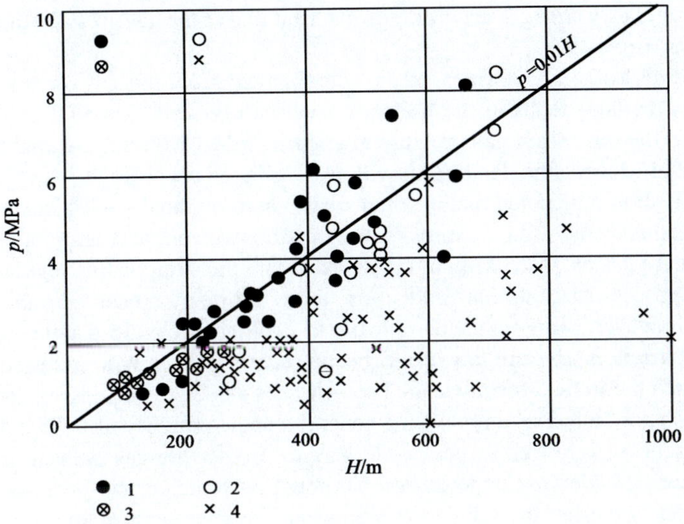

1: Chongqing Mining Bureau; 2: Beipiao Mining Bureau; 3: Hunan Mining Bureau; 4: Other mining bureaus

Fig. 8.18 Relationship between the coal seam gas pressure (p) and depth (H) [88, 89]

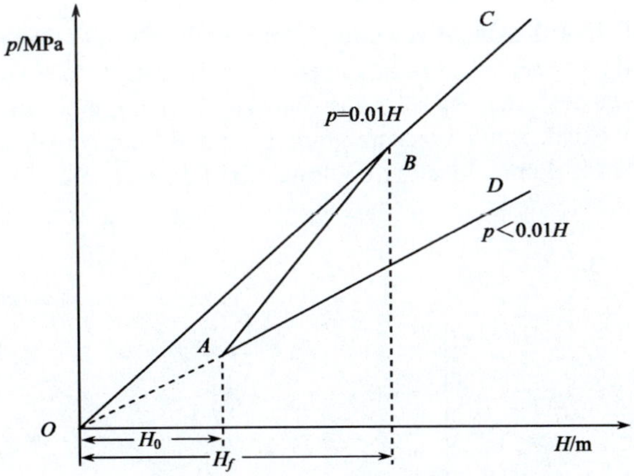

Fig. 8.19 Two types of gas pressure variation with burial depth [90]

depth can be classified into two types. For type I, the gas pressure gradient ω is <0.01 MPa/m (line AD in Fig. 8.19), and the gap between the gas pressure and the limit value becomes larger with increasing depth. For type II, the gas pressure gradient ω is >0.01 MPa/m (line AB in Fig. 8.19), and the coal seam gas pressure becomes closer to the gas pressure limit of the coal seam with increasing depth. The depth (H_f) of point B is the ultimate gas pressure depth, beyond which the gas pressure regularity changes according to the limit line of the gas pressure distribution (line *BC* in Fig. 8.19).

According to the above rules, in the actual calculation, the gas pressure growth line over a certain depth in the methane zone can be approximated as a straight line. When the coal seam gas pressure gradient is >0.01 MPa/m, the coal seam gas pressure that exceeds the depth limit will be distributed according to the hydrostatic pressure. It should be noted that in some areas where regional geological conditions and properties change (e.g., caprocks, magmatic intrusion, and large open faults), the coal seam gas pressure will largely deviate from the straight-line regularity [91].

In the process of coalmine production, there are many measured values of gas pressure; however, owing to the complexity of geological conditions, it is very difficult to determine a uniform law of the gas pressure variation with increasing burial depth. When theoretical formulae are used, there is a need to consider many factors; hence, this approach does not have the general maneuverability of field technicians, and it is necessary to find a reasonable way to predict the gas pressure of a coal seam. After considerable experience of production practice, we propose a safety line method to predict the relationship between the gas pressure and burial depth. We statistically analyzed a large number of gas pressure values in the process of production, and then excluded abnormal measured points due to confined water. We subsequently selected two true symbol points to construct the safety line [92, 93] by allowing all points excluding the abnormal points to stay below the line (as shown in Fig. 8.20). The selection of the symbol points requires the consideration of the lower limit critical values of the gas weathering, in which the pressure values are 0.15–0.2 MPa. A selection is correct if the safety line passes near the critical values; otherwise, other points need to be chosen. If two suitable symbol points are not selected, then by selecting one symbol point, the safety line is constructed by using the hydrostatic pressure gradient (0.01 MPa/m) as the gas pressure gradient. All the work should consider the lower limit critical values of the gas weathering zone and

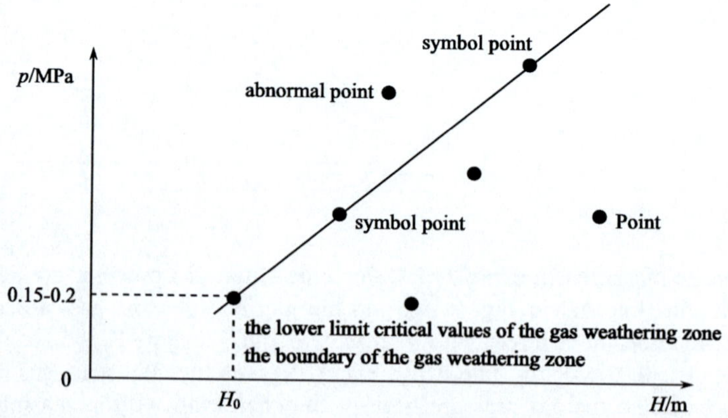

Fig. 8.20 Schematic diagram showing the gas pressure prediction method of the safety line

ensure that the measured pressure values are below the line (except for the abnormal points). It has been noted that some measured pressure values at deep levels are lower than expected because of the pressure measurement environment and the quality of the borehole sealing, which are both inconsistent with the basic variation law of deep gas pressure and cannot be adopted as symbol points.

In addition, the gradient of the predicted gas pressure with buried depth should generally vary in the range of 0.01 ± 0.005. In the methane zone, the gas pressure is usually less than or equal to the hydrostatic pressure owing to the small geostress at a shallow level, namely $p = 0.01H$. However, at the deeper level of a coalmine, the geostress (including self-weight stress, tectonic stress, and temperature stress) linearly increases with increasing vertical depth, and the gas pressure can exceed the hydrostatic pressure. Thus, the pressure value p can be 0.013–0.015 times greater than the value of H [94]. Moreover, abnormal gas pressure regions exist in individual zones with high concentration stress and mining concentration stress.

According to the predicted curve obtained from the safety line method, it is possible to avoid situations where the predicted gas pressure of the coal seam is smaller than the actual gas pressure, thus providing a favorable basis for the formulation of a mine gas control scheme. Meanwhile, when the gas pressure in deep areas of a coalmine is higher than the safe line predicts, then the symbol points need to be redefined and the safe line revised.

The measured gas pressure data (in the same geological unit) of the No. 8_2 coal seam at Taoyuan Coalmine, Huaibei Coalfield, are shown in Fig. 8.21. When adopting the one-variable linear regression and polynomial regression methods, the predicted values of the deep gas pressure were smaller than the measured data. The gas control countermeasure based on these values would not guarantee the safety of mining work. In addition, the lower limit elevation of the gas weathering zone of the No. 8_2 coal seam at Taoyuan Coalmine obtained through these two methods was around −420 m, which is 100 m deeper than the real elevation of the gas weathering zone

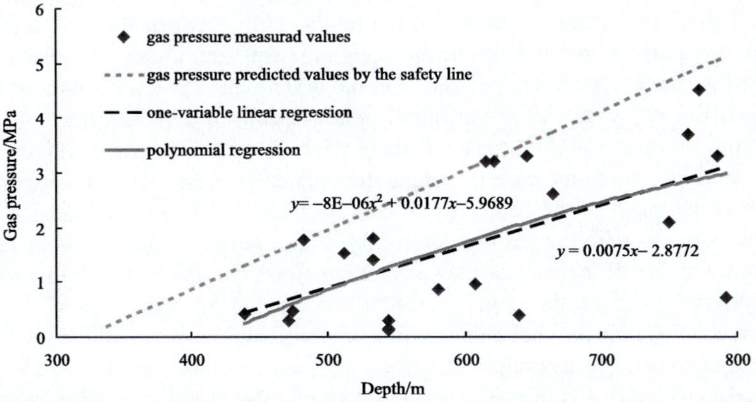

Fig. 8.21 Gas pressure prediction curves using different methods for the No. 8_2 coal seam of Taoyuan coalmine

(−310 m) in this mine. However, when the safety line method was adopted, all the actual gas pressure values measured in the field were below the line. The obtained gas pressure curve equation of the safety line diagrams was $P = 0.01082 \times H - 3.45473$, and the gas weathering zone depth calculated using the equation was −330 m (corresponding elevation of −306 m), which is almost the same as that of the actual gas weathering zone. The gas pressure gradient was 0.01082 MPa/m within the range of 0.01 ± 0.005 MPa/m, which agrees with the variation law of the deep coal seam. Moreover, all gas pressure values measured in the later production process were near the safety line, verifying the safety and rationality of the safety line method.

8.5.2 Variation of Gas Content with Burial Depth

The original gas content of a coal seam is the amount of geological residual gas retained within it. In addition to the relationship with the amount of gas generated, the gas content of a coal seam mainly depends on the emission and migration conditions of gas after coal seam formation, as well as the ability of the coal seam and its surrounding environment for gas preservation.

The burial depth of a coal seam is the main factor determining its gas content. As stated in Sect. 4.1.4, the gas content is the sum of the free and adsorbed gas content, both of which vary with increasing burial depth. On one hand, the permeabilities of a coal seam and surrounding rock reduce due to the increased geostress, and the distance of gas migrating to the surface is also increased, both of which are beneficial to gas preservation. The gas pressure in the methane zone increases linearly with increasing depth. When omitting the influence of the coal seam temperature, the free gas content increases linearly with increasing burial depth according to Eq. (4.14), while the adsorption gas increases with increasing burial depth according to Eq. (4.15) and approximately satisfies the Langmuir equation. Because coal seam gas is mainly in a state of adsorption, the gas content of coal is basically determined by the Langmuir equation. Therefore, the relationship between the coal seam gas content and burial depth is close to the Langmuir equation under the precondition of ignoring the coal seam temperature. On the other hand, the temperature of strata increases linearly with increasing burial depth, and according to Eqs. (4.14) and (4.15), the free and adsorbed gas contents of the coal seam decrease with increasing burial depth due to an increase in temperature. Therefore, the actual coal seam gas content is the result of the competition between these two factors. Because the gas content increases with the gas pressure of the coal seam according to the Langmuir equation, the rate of increase of the gas content slows continuously with increasing gas pressure. Therefore, the magnitude of the influence of the former factor is increasingly weakening, and the coal seam gas content may exhibit a trend of first increasing and then decreasing as a result.

Based on Eqs. (4.13) to (4.15) and the data in Table 4.3, Fig. 8.22 presents the relationship between the estimated gas content and burial depth of the No. 10 coal seam of Wolonghu Coalmine under the condition of either considering or ignoring

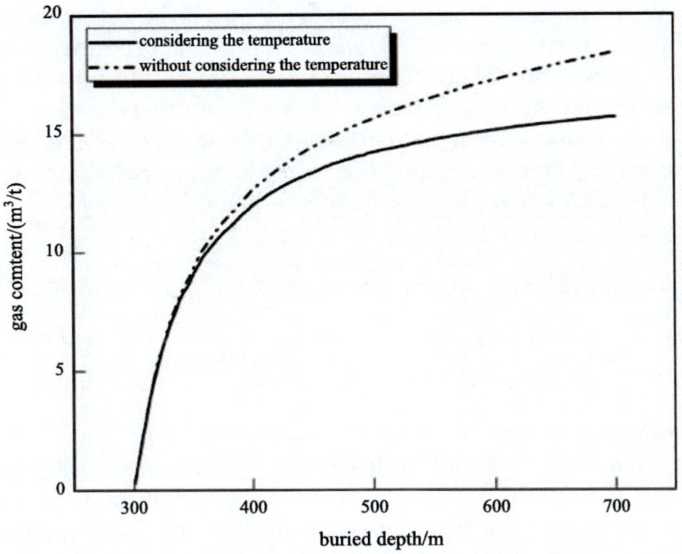

Fig. 8.22 Relationship between the estimated gas content and burial depth of the No. 10 coal seam at Wolonghu coalmine

the influence of temperature. The temperature was defined as 30 °C at a depth of 300 m, and the temperature gradient was set to 0.03 °C/m. The ground temperature and depth met the following relationship:

$$T = 0.03 \times (H - 300) + 30 \tag{8.39}$$

where T is the coal seam temperature, °C; and H is the burial depth, m.

As shown in Fig. 8.22, although the calculated gas content increased with increasing burial depth of the coal seam, this was not a simple linear relationship. At a relatively shallow burial depth, the gas content increased rapidly with depth, whereas it increased relatively slowly at a deep burial depth. In addition, with the increase in burial depth, the influence of temperature gradually increased, and the difference between the estimated gas content (both considering and ignoring the influence of temperature) gradually became significant.

8.5.3 Variation of Permeability with Burial Depth

Coal is a type of pore-fissure structural body, and there are great variations in the pore/fracture size, structure form, and development degree of different types of coal. The flow of fluid under a certain pressure gradient indicates that coal has a permeable performance. Coal seam permeability is an important parameter that reflects the difficulty of gas flowing in a coal seam, and is also one of the indices for evaluating the feasibility of gas drainage, which is used to judge the risk of coal and gas outburst.

Many factors influence the permeability of coal seam gas, although it is generally believed that the geostress, gas pressure, and coal seam temperature are the main factors [95, 96]. Guo and Cheng [95] proposed a single-pore model to predict coal seam permeability. This model considers the effective stress, gas adsorption-induced swelling, and thermal expansion, which can be expressed as follows:

$$\frac{k}{k_0} = \left\{1 - \frac{\alpha}{\phi_{f0} K}[(\overline{\sigma} - \overline{\sigma}_0) - (p - p_0)] - \frac{f_m}{\phi_{f0}}\left[\chi\left(\frac{abp}{1+bp} - \frac{ab_0 p_0}{1+b_0 p_0}\right) + \eta(T - T_0)\right]\right\}^3 \tag{8.40}$$

where k is the permeability of the coal seam, mD; k_0 is the permeability of the coal seam at the initial depth, mD; α is the Biot coefficient; ϕ_{f0} is the porosity of the coal seam at the initial depth, %; $\overline{\sigma}$ is the average geostress of the coal seam, MPa; $\overline{\sigma}_0$ is the average geostress of the coal seam at the initial depth, MPa; p is the coal seam gas pressure, MPa; p_0 is the coal seam gas pressure at the initial depth, MPa; f_m is the effective matrix deformation coefficient of the studied coal; χ is the adsorption deformation coefficient of the studied coal, t/m^3; a is the ultimate gas adsorption capacity of coal, m^3/t; b is the adsorption constant of coal, MPa^{-1}; K is the bulk modulus ($\frac{K}{M} = \frac{1}{3}\left[\frac{1+V}{1-V}\right]$), MPa; M is the constrained axial modulus ($M = \frac{(1-V)E}{(1+V)(1-2V)}$), MPa; b_0 is the adsorption constant of coal under the temperature of the initial depth, MPa^{-1}; η is the thermal deformation coefficient of coal, K^{-1}; T is the temperature of the coal seam, K; and T_0 is the temperature of the coal seam at the initial depth, K.

The relationship between the adsorption constant b and temperature can be expressed using Eq. (8.41) [97, 98]:

$$b = \frac{b_c}{\sqrt{T}} \exp\left(\frac{\Delta H}{RT}\right) \tag{8.41}$$

where b_c is the adsorption constant coefficient of coal, MPa^{-1} K$^{0.5}$; ΔH is the adsorption heat of coal, J/mol; and R is the perfect gas constant, 8.314 J/(mol K).

The relationship between the permeability and depth of a coal seam in the southern Qinshui Basin was studied using Eq. (8.40). The coal seam gas pressure, average geostress, and coal seam temperature satisfied the following linear relationships with increasing burial depth, respectively:

$$P = 0.0136 \times H - 2.1003 \tag{8.42}$$

$$\overline{\sigma} = 0.027 \times H - 2.2367 \tag{8.43}$$

$$T = G \times (H - 20) + 282.15 \tag{8.44}$$

where H is the burial depth of the coal seam, m; G is the geothermal temperature gradient, °C/m.

The parameters for calculating the variation of the permeability of the No. 3 coal seam with burial depth are shown in Table 8.2. The parameters of the gas weathering zone at a depth of 160 m were chosen as the initial values, and the calculated relationships of the average geostress, gas pressure, and coal seam permeability with increasing burial depth are displayed in Fig. 8.23.

Table 8.2 Predicted parameters of the permeability of the No. 3 coal seam in the southern Qinshui basin

Parameter	Value
Initial permeability k_0	3 mD
Biot coefficient α	0.80
Initial fissure porosity ϕ_{f0}	0.009
Effective matrix deformation coefficient f_m	0.25
Adsorption deformation coefficient χ	2.23×10^{-4} t/m^3
Langmuir volume a	31.43 m^3/t
Adsorption constant b_c	8.95×10^{-3} MPa^{-1} K$^{0.5}$
Thermal deformation coefficient η	3.813×10^{-5} K^{-1}
Adsorption heat ΔH	1.70×10^4 J/mol
Geothermal gradient G	0.0353 °C/m

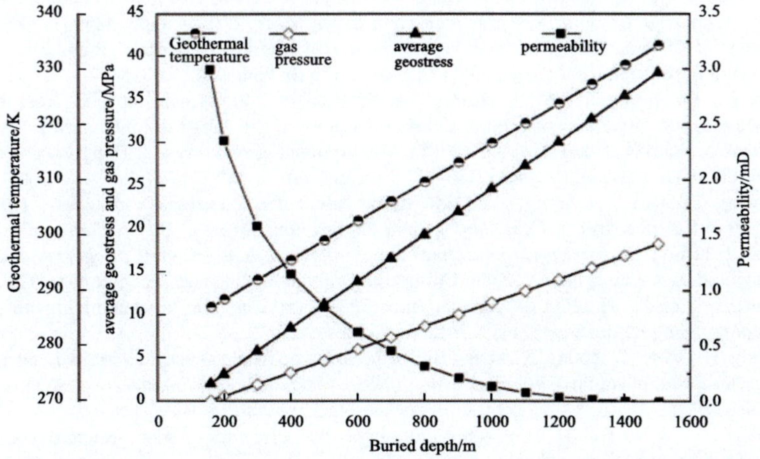

Fig. 8.23 Relationships between the coal seam geothermal temperature, average geostress, gas pressure, and permeability with the burial depth of the No. 3 coal seam in the Qinshui basin

Figure 8.23 shows that the average geostress, gas pressure, and geothermal temperature of the No. 3 coal seam in the southern Qinshui Basin increased linearly with increasing burial depth. The rate of increase in the average geostress was greater than that of the gas pressure; thus, the effective stress increased with the depth of the coal seam, which compressed the coal fractures and reduced the permeability of the coal. At the same time, owing to the increased gas pressure and geothermal temperature, the effects of adsorption swelling and thermal expansion on the coal seam gradually became significant, which further compressed the fractures to reduce the permeability of the coal. Under the combined effect of effective stress, gas pressure, and geothermal temperature, the permeability of the coal seam decreased with increasing burial depth, and the rate of decrease gradually slowed down.

References

1. Cai, M., He, M., & Liu, D. (2013). *Rock mechanics and engineering*. Beijing: Science Publishing House (in Chinese).
2. Yang, Y., Zhou, G., Li, Y., et al. (2010). *In situ stress measurement and the technical of the gob-side entry supporting in the fully-mechanized coal winning*[M]. Beijing: China Coal Industry Publishing House (in Chinese).
3. Li, Z., & Zhang, J. (1997). *In-situ stress and oil and gas exploration and development*. Beijing: Petroleum Industry Press (in Chinese).
4. Jaeger, J. C., Cook, N. G. W., & Zimmerman, R. W. (2007). *Fundamentals of rock mechanics*. Oxford: Blackwell Publishers.
5. Yang, J. (2008). *Rock mechanics*. Beijing: Metallurgical Industry Press (in Chinese).
6. Wei, F., Hu, G., Wang, H., et al. (2007). Measurement and variation characteristics of strata stress in Yongchuan coal mine. *Mining Safety and Environmental Protection, 34*(4), 1–3 (in Chinese).
7. Han, J., Zhang, H., Song, W., et al. (2008). In situ stress field of coal and gas outburst mining area. *Chinese Journal of Rock Mechanics and Engineering, 27*(S2), 3852–3859 (in Chinese).
8. Kang, H. (2008). Analysis on types and interaction of stress fields in underground coal mines. *Journal of Chinese Coal Society, 33*(12), 1329–1335 (in Chinese).
9. Yu, L., Wang, A., & Jang, G. (2008). Application of hollow inclusion geo-stress survey technology in Hongmiao coal mine. *Coal Technology, 27*(7), 128–129 (in Chinese).
10. Zhao, W., Sun, D., Wang, H., et al. (2008). Measurement of rock stress of Hongling coal mine. *Journal of Geomechanics, 14*(3), 286–233 (in Chinese).
11. Zheng, Z., Zhang, J., & Wang, W. (2008). Measurement of in situ stresses with acoustic emission effect in Haizi coal mine. *Coal Geology and Exploration, 36*(2), 68–71 (in Chinese).
12. Gao, F. (2009). *Research on ground stress distribution characteristics and its effect on roadway surrounding rock stability*. Xuzhou: China University of Mining and Technology (in Chinese).
13. Han, J., & Zhang, H. (2009). Characteristic of in situ stress field in Huainan mining area. *Coal Geology and Exploration, 37*(1), 17–21 (in Chinese).
14. Kang, H., Jiang, T., Zhang, X., et al. (2009). Research on in situ stress field in Jincheng mining area and its application. *Chinese Journal of Rock Mechanics and Engineering, 28*(1), 1–8 (in Chinese).
15. Kang, H., Lin, J., Yan, L., et al. (2009). Study on characteristics of underground in situ stress distribution in Shanxi coal mining fields. *Chinese Journal of Geophysics, 52*(7), 1782–1792 (in Chinese).
16. Kang, H., Yan, L., & Jian, Z. (2009). Measurements and analysis of ground stresses in Fenxi coal mining area. *Journal of Mining and Safety Engineering, 26*(3), 263–268 (in Chinese).

17. Li, Y. (2009). *The theory and method of prediction for in situ stress based on tectonic control—A case study of Zhaolou coal mine*. Xuzhou: China University of Mining and Technology (in Chinese).
18. Wang, L., Sun, D., Zhang, L., et al. (2009). Application of in situ stress measurement on bursts disasters of rock and CO_2 in coal mine. *Journal of China Coal Society, 34*(1), 28–32 (in Chinese).
19. Kang, H., Zhang, X., Si, L., et al. (2010). In situ stress measurements and stress distribution characteristics in underground coal mines in China. *Engineering Geology, 116*(3–4), 333–345.
20. Pan, Y., Tang, Z., Yan, H., et al. (2010). Measurement and numerical simulation analysis of in situ stresses in Daanshan coal mine. *Journal of China Coal Society, 35*(S1), 49–53 (in Chinese).
21. Wang, L., Lu, Y., Yang, X., et al. (2010). Study of distribution characteristics of in situ stresses for Huozhou mining area. *Chinese Journal of Rock Mechanics and Engineering, 29*, 2768–2774 (in Chinese).
22. Jiang, Y., Zhou, W., & Mei, S. (2011). Test and distribution rule of strata stress field in Bide coal mine. *Mining Safety and Environmental Protection, 38*(1), 1–3 (in Chinese).
23. Luo, C., Li, H., & Liu, Y. (2011). Characteristics of in situ stress and variation law of plastic zone of surrounding rocks around deep tunnels in a coal mine. *Chinese Journal of Rock Mechanics and Engineering, 8*(8), 1613–1618 (in Chinese).
24. Meng, Y., & Wang, R. (2011). Characteristics of in situ stress field in Kailuan coal mine field and its research significance. *Coal Geology of China, 23*(10), 29–34 (in Chinese).
25. Han, J., Zhang, H. W., Li, S., et al. (2012). The characteristic of in situ stress in outburst area of China. *Safety Science, 50*(4), 878–884.
26. Kang, H., Wu, Z., Gao, F., et al. (2012). Effect of geological structures on in situ stress distribution in underground coal mines. *Chinese Journal of Rock Mechanics and Engineering, 31*(1), 2674–2680 (in Chinese).
27. Liu, Q., & Liu, K. (2012). Characteristics of in situ stress field for deep levels in Huainan coal mine. *Rock and Soil Mechanics, 7*(7), 2089–2096 (in Chinese).
28. Zhang, G., Wei, Z., & Pei, Z. (2012). In situ stress measurements and analysis of action of geological structures of deep coal mines in Xuzhou. *Chinese Journal of Geotechnical Engineering, 34*(12), 2318–2324 (in Chinese).
29. Zhang, Y. (2012). *Panxie mining area: A coal mining group of geostress measurement and research on mine roadway support technology*. Huainan: AnHui University of Science and Technology.
30. Hua, J., Zheng, X., Zhang, N., et al. (2013). In situ stress measurement and control of roadway surrounding rock in 1015m mine floor of Kongzhuang coal mine. In *Mining technology of Kilometer deep coal mine in China* (pp. 57–63), China National Coal Association. Xuzhou: China University of Mining and Technology Press (in Chinese).
31. Meng, S., Lan, J., Liu, C., et al. (2013). In situ stress and coal reservoir pressure in Southeast margin of Ordos basin and their coupling relations. *Journal of China Coal Society, 38*(1), 122–128 (in Chinese).
32. Wang, Y., Cai, C., Zhao, S., et al. (2013). Measurement of in situ stress and relationship between the stress and geologic structure in Sikeshu mining area. *Safety in Coal Mines, 44*(8), 213–215 (in Chinese).
33. Zhao, S., Cai, C., Liu, Z., et al. (2013). Relationship between regional geological tectonic feature and in situ stress field distribution in Northern Tianshan mountain mining areas. *Chinese Journal of Underground Space and Engineering, 9*(S1), 1599–1603 (in Chinese).
34. Zheng, G., Lian, H., & Ling, B. (2013). Measurement of geostress in deep coal seams in Guqiao coalmine of Huainan coal field. *Journal of Liaoning Technical University (Natural Science), 32*(10), 1324–1328 (in Chinese).
35. Zhou, X., & Zheng, X. (2013). Analysis of working area stress distribution law and its impact on stability of the surrounding rock of roadway. *Journal of Anhui Institute of Architecture and Industry, 21*(4), 9–13 (in Chinese).

36. Zhu, X. (2013). Application of the measurement method of empty core package and analysis of the characteristics of the in situ stress. *Coal Science and Technology Magazine, 4*, 35–36 (in Chinese).
37. Liu, Q., Cheng, Y., Yuan, L., et al. (2014). CMM capture engineering challenges and characteristics of in situ stress distribution in deep level of Huainan coalfield. *Journal of Natural Gas Science and Engineering, 20*, 328–336.
38. Chai, H., & Li, Y. (2014). Study on in situ stress distribution law of coal and gas outburst mine. *Journal of Henan Science and Technology, 6*, 163–165 (in Chinese).
39. Chai, S., Wang, Z., & Zhang, C. (2014). Study of stress measurement practice in west Zizhuang coal mine. *Science and Technology Information, 4*, 144–145 (in Chinese).
40. Dong, H., & Sun, S. (2014). Research on the in situ stress field in thick coal seam of Liuhuanggou coal mine. *Shandong Coal Science and Technology, 10*, 12–16 (in Chinese).
41. Han, Z. (2014). *Research on pre-stressed anchor support technology in dynamic pressure and high geo-stress roadway of Zaoquan coal mine*. Xi'an: Xi'an University of Science and Technology (in Chinese).
42. Hao, P. (2014). *3D geostress measurement by sleeve fracturing technique and engineering application*. Beijing: China University of Mining and Technology (in Chinese).
43. He, C., Yao, Z., & Chen, Z. (2014). Stress detection of a large mine in western area. *Coal Technology, 33*(9), 91–93 (in Chinese).
44. He, S., Lin, Q., Zhang, C., et al. (2014). Ground stress field characteristics and their influence on coal and gas outburst in Qidong well field. *Coal Geology and Exploration, 42*(2), 9–13 (in Chinese).
45. Li, D. (2014). *The mine of Taoyi those ground stress test and its research of changes in the distribution*. Handan: Hebei University of Engineering (in Chinese).
46. Li, J., Jin, S., Jiao, X., et al. (2014). Study on the stress measurement of large mining face in Wangzhuang coal mine. *Coal, 6*, 91–92 (in Chinese).
47. Li, Y., Tang, D., Xu, H., et al. (2014). Characteristic of in situ stress field in Liulin area, Ordos basin and its control on coal fractures. *Journal of China Coal Society, 39*(S1), 164–168 (in Chinese).
48. Liu, H. (2014). *Study on the relationship of the ground stress and roadway surrounding rock stability in Gucheng coal mine*. Xuzhou: China University of Mining and Technology (in Chinese).
49. Luo, K., He, L., & Sun, Y. (2014). Underground ground stress field measurements and the analysis of stress fields distribution laws in Qianyingzi coal mine. *Coal Mine Safety, 45*(1), 32–35 (in Chinese).
50. Ouyang, D., An, T., & Zhang, H. (2014). Analysis and study on ground stress test in Qianjiaying mine. *Coal Science and Technology, 42*(S), 230–232 (in Chinese).
51. Shi, X., Ju, Y., & Meng, Z. (2014). Characteristics of in situ stress field in Xinji coal mine. *Coal Geology and Exploration, 42*(6), 68–72 (in Chinese).
52. Sun, M. (2014). *Study on in situ stress distribution law and its application in Pingdingshan mining area*. Xuzhou, Beijing: China University of Mining and Technology (in Chinese).
53. Wang, Q. (2014). *Study on the relationship between Kou Zidong mine deep crustal stress distribution and roadway layout*. Huainan: AnHui University of Science and Technology (in Chinese).
54. Wang, Y., Jing, H., Chen, K., et al. (2014). Study of distribution regularities and regional division of in situ stresses for Pingdingshan mining area. *Chinese Journal of Rock Mechanics and Engineering, 33*(S1), 2620–2627 (in Chinese).
55. Wu, B., Zhang, X., Wang, Y., et al. (2014). The stress measurement technique of hydraulic fracturing is used in the Ho family tower coal mine. *Shanxi Coal, 4*, 93–94 (in Chinese).
56. Xu, H., Sang, S., Yi, T., et al. (2014). Characteristics of in situ stress field and its tectonic origin in Western Guizhou. *Journal of Central South University (Science and Technology), 45*(6), 1960–1966 (in Chinese).
57. Yang, W. (2014). The characteristics of the stress field and the rationality analysis of roadway layout in Changping mining area. *Support in Coal Mines* 1, 45–47 (in Chinese).

58. Yin, S. (2014). *Research on supporting technology and roadway stability of −720m auxiliary air return roadway of Chensilou mine*. Huainan: AnHui University of Science and Technology (in Chinese).
59. Dong, P., & Cai, H. (2015). Statistical analysis of stress field characteristics in Huainan mining area. *Coal Science and Technology Magazine, 3*, 98–101 (in Chinese).
60. Gao, Z., Rong, H., Bao, X., et al. (2015). Study on in situ stress field characteristic and regional tectonic force in Linxi coal mine. *Chinese Journal of Underground Space and Engineering, 11*(S1), 129–133 (in Chinese).
61. Gu, L. (2015). Deep in situ stress measurement and its effect on rock burst in Donghuantuo coal mine. *Journal of North China Institute of Science and Technology, 12*(4), 36–38 (in Chinese).
62. Hu, B. (2015). Measurements and applications of ground stresses at Hejiata colliery with shallow buried depth coal seam. *Shandong Coal Science and Technology, 6*, 152–155 (in Chinese).
63. Jia, N. (2015). Study on the stress measurement of coal seam in Hengxing coal mine and its influence on its roadway stability. *Inner Mongolia Coal Economy, 1*, 142–144 (in Chinese).
64. Jia, X., Li, Y., Lu, L., et al. (2015). Analysis of ground stress distribution and its influencing factors in Yangying coal mine. *Coal, 3*, 4–6 (in Chinese).
65. Liang, Y., & Feng, R. (2015). Study on distribution regularity of large mining depth stress in Liliangdian colliery. *Jiang Xi Coal Science and Technology, 2*, 146–148 (in Chinese).
66. Lv, W., Yang, Y., Ma, J., et al. (2015). Study on geo-stress measurement and distribution features in Jibei mining area. *Coal Science and Technology, 43*(11), 32–37 (in Chinese).
67. Ma, J., Gui, H., Sun, L., et al. (2015). Distribution characteristics of in situ stress field and tectonic evolution in Huaibei coalfield. *Coal Engineering, 47*(10), 97–100 (in Chinese).
68. Meng, N. (2015). *The measuring method and research of in situ stress*. Baotou: Inner Mongolia University of Science and Technology (in Chinese).
69. Ren, Q., Ai, D., & Li, J. (2015). Application of hydraulic fracturing in situ stress measurement through CBM wells in No.13 mine of Pingdingshan. *Coal Technology, 34*(5), 193–195 (in Chinese).
70. Su, Y., & Liu, Y. (2015). Study on field test and regularities of distribution of ground stress in Xinxing mine. *Coal Technology, 34*(11), 125–127 (in Chinese).
71. Wang, B., Gao, Y., & Zhu, W. (2015). Geostress measurement at deep position of Qishan mine and distribution law of geostress field. *Journal of North China Institute of Science and Technology, 12*(2), 37–42 (in Chinese).
72. Yang, R., Li, C., & Han, J. (2015). Liangbaoshi coal mine in situ stress measurement and mining and excavating design optimization. *Shandong Coal Technology, 4*, 160–165 (in Chinese).
73. Zhang, Z. (2015). The measurement of geostress and its application in Longquan coalmine. *Shanxi Science and Technology, 30*(4), 77–79 (in Chinese).
74. Pu, W., & Zhang, G. (2006). The ore crustal stress measure of ChanggouYu mine and analyse the stability of country rock and controlled. *Coal Engineering, 10*, 68–70 (in Chinese).
75. Zhao, D., Cheng, Z., Cai, X., et al. (2007). Analysis of distribution rule of geostress in China. *Chinese Journal of Mechanics and Engineering, 26*(6), 1265–1271 (in Chinese).
76. Zhang, Z. (2010). *Gas geology*. Xuzhou: China University of Mining and Technology Press (in Chinese).
77. Ju, Y., Lin, H., Li, X., et al. (2009). Tectonic deformation and dynamic metamorphism of coal. *Earth Science Frontiers, 16*(1), 158–166 (in Chinese).
78. TeichmÜLler, M., & TeichmÜLler, R. (1966). *Geological causes of coalification* (pp. 133–155). Coal Science. American Chemical Society.
79. Bustin, R. M. (1983). Heating during thrust faulting in the rocky mountains: Friction or fiction? *Tectonophysics, 95*(3–4), 309–328.
80. Suchy, V., Frey, M., & Wolf, M. (1997). Vitrinite reflectance and shear-induced graphitization in orogenic belts: A case study from the Kandersteg area, Helvetic Alps, Switzerland. *International Journal of Coal Geology, 34*(1–2), 1–20.
81. Han, D. (1996). *Coal petrology of China*. Xuhzou: China University of Mining and Technology Press (in Chinese).

82. Ross, J. V., & Bustin, R. M. (1990). The role of strain energy in creep graphitization of anthracite. *Nature, 343*(6253), 58–60.
83. Bustin, R. M., Ross, J. V., & Rouzaud, J. N. (1995). Mechanisms of graphite formation from kerogen: Experimental evidence. *International Journal of Coal Geology, 28*(1), 1–36.
84. Cao, Y., Mitchell Gareth, D., Davis, A., et al. (2000). Deformation metamorphism of bituminous and anthracite coals from China. *International Journal of Coal Geology, 43*(1–4), 227–242.
85. Liu, W., Xu, Y., Zhang, S., et al. (2000). A new kind of mechanism for hydrocarbon generated-mechanochemistral function and experimental prove. *Acta Sedimentologica Sinica, 18*(2), 314–318 (in Chinese).
86. Zhang, Y., Zhang, Z., & Xie, K. (2005). Coal mechanochemistry action in coal evolution process and tectonic coal structure. *Journal of Henan Polytechnic University, 24*(2), 95–99 (in Chinese).
87. Zhang, Y., Zhang, Z., Zhang, X., et al. (2008). Mechanochemical action mechanism of tectonically deformed coal evolvement. *Coal Geology of China, 20*(10), 11–21 (in Chinese).
88. Yu, Q. (1992). *Mine gas prevention and control*. Xuzhou: China University of Mining and Technology Press (in Chinese).
89. Yu, B. (2000). *Technical manual for coal mine gas disaster prevention and utilization*. Beijing: China Coal Industry Publishing House (in Chinese).
90. The First Research office of Liaoning Coal Research Institute. (1974). Relationship between gas pressure and depth in coal seam. *Safety in Coal Mines* (3), 24–27 (in Chinese).
91. Cheng, Y., Zhang, X., & Wang, L. (2013). Controlling effect of ground stress on gas pressure and outburst disaster. *Journal of Mining and Safety Engineering, 30*(3), 408–414 (in Chinese).
92. Tian, J., Wang, L., Cheng, Y., et al. (2008). Research on distribution rule and forecast method of gas pressure in coal seam. *Journal of Mining and Safety Engineering, 25*(4), 481–485 (in Chinese).
93. Wang, L., Cheng, Y. P., Wang, L., et al. (2012). Safety line method for the prediction of deep coal-seam gas pressure and its application in coal mines. *Safety Science, 50*(3), 523–529.
94. Zhou, S., & Lin, B. (1999). *Theory of gas occurrence and flow in coal seam*. Beijing: China Coal Industry Publishing House (in Chinese).
95. Guo, P., & Cheng, Y. (2013). Permeability prediction in deep coal seam: A case study on the No. 3 coal seam of the Southern Qinshui basin in China. *Scientific World Journal,* 161457.
96. Guo, P., Cheng, Y., Jin, K., et al. (2014). Impact of effective stress and matrix deformation on the coal fracture permeability. *Transport in Porous Media, 103*(1), 99–115.
97. Gregg, S. J., & Sing, K. S. W. (1982). *Adsorption surface area and porosity* (2nd ed.). London: Academic Press.
98. Do, D. D. (1998). *Adsorption analysis: Equilibria and kinetics*. London: Imperial College Press.

Chapter 9
Flow Theory of Gas in Coal Seams

Learning Outcomes

- Fundamental physical process of gas migration;
- Basic assumptions and physical significance of the gas–solid coupling model;
- Derivation of the governing equations for gas diffusion in the coal matrix;
- Derivation of the governing equations for gas seepage in coal fractures;
- Mass exchange between the diffusion and seepage fields, and a description of the model concept.

9.1 Fundamental Laws of Gas Migration

9.1.1 Gas Migration Process and in Coal Seams and its Basic Assumptions

The form of gas migration in coal seams is either diffusion or seepage, and is closely related to the fracture scale and pore size distribution. However, the characteristic of inhomogeneous pore structure exists in natural coal seams, and has been simplified by some researchers on the basis of the priority of seepage and diffusion in coal seams. In general, the pore structure can be simplified into three models: the homogeneous porous media model that only considers diffusion, the homogeneous fracture media model that only considers seepage, and the dual-porosity model that considers both seepage in fractures and diffusion in the coal matrix [1].

In the homogeneous porous media model that only considers diffusion, the essence of gas diffusion movement is that gas molecules move from places with high gas concentration to places with low gas concentration in the diffusion system, where the diffusion rate is linearly proportional to the concentration gradient [2]. Since the 1950s, Chinese and foreign scholars have studied the gas diffusion theory of coal particles, and have established desorption diffusion models under the first boundary

Y. Cheng et al., *Coal Mechanics*,
https://doi.org/10.1007/978-981-16-3895-4_9

condition. Combining these models with adsorption and desorption experiments of coal particles, some empirical and semi-empirical function expressions for the relationship between diffusion and adsorption time have been obtained. In addition, some related research achievements have been widely applied for predicting the coal seam gas content and the risk of coal and gas outburst, including some representative research results. For instance, Wang et al. [3–5] analyzed the relationship between the coal pore structure and gas desorption rate with respect to the risk of coal and gas outbursts, and proposed a method to determine the gas content in coal seams based on the coal desorption rate. Yang and Wang [6] systematically studied the gas diffusion theory experimentally, and discussed the relationship between gas emissions and coal and gas outbursts with respect to gas diffusion. The authors then proposed a hypothesis for critical size of coal grains and established a diffusion model. Wu [7] firstly applied the third boundary condition to set up a model of gas desorption and diffusion in coal grains. They also studied the measurement mechanism of the diffusion coefficient and convective mass transfer coefficient, and discussed the mutual transformation between the first and third boundary conditions. Nie et al. [8] established gas diffusion model for coal grains under the condition of a third boundary, and then determined some diffusion parameters and the theoretical solution of the model. In addition, Cheng et al. [9, 10] used a self-developed test device for high-pressure adsorption-water injection–desorption to improve the even distribution of moisture content after water injection, which is helpful for studying the comprehensive effect of adding moisture to different types of coal on gas desorption.

In the homogeneous fracture media model that only considers seepage, the theory of gas seepage provides the theoretical basis to guide the prevention of coalmine gas. Zhou et al. [11–13] conducted a series of fundamental and pioneering works regarding the occurrence and flow theory of the coal seam gas. Moreover, the author established some gas seepage models of the unidirectional, radial, and ball flow fields, which can yield the approximate theoretical solution. These models can further improve the field test method for coal seam permeability. Xian et al. [14, 15] performed a lot of work on the theory of gas seepage and obtained some useful results. For example, the ideal gas state equation was replaced with the real gas state equation, and a new permeability model, which considers that the gas involved in seepage is only desorption gas, was established. In addition, Zhou studied the effects of the geoelectric field on gas seepage.

In recent decades, the pore-fracture dual-porosity model has been recognized and used by an increasing number of researchers [16–20]. The model assumes that a coal sample is typically a dual-porosity system consisting of a coal matrix surrounded by intersecting fractures. Most of the gas exists in the coal matrix, which meets Fick's law of diffusion, while a small portion of free gas occurs in the fractures, which satisfies Darcy's law [19, 21, 22]. Under undisturbed conditions, the adsorbed gas and free gas are in a state of dynamic equilibrium. If the adsorption equilibrium state is broken, the gas migration process can be regarded as a continuous two-step process because the seepage velocity in fractures is much greater than the diffusion velocity in the coal matrix. Gas migration in coal seams can be commonly divided into two processes: first, the gas diffuses from the coal matrix into the fractures following

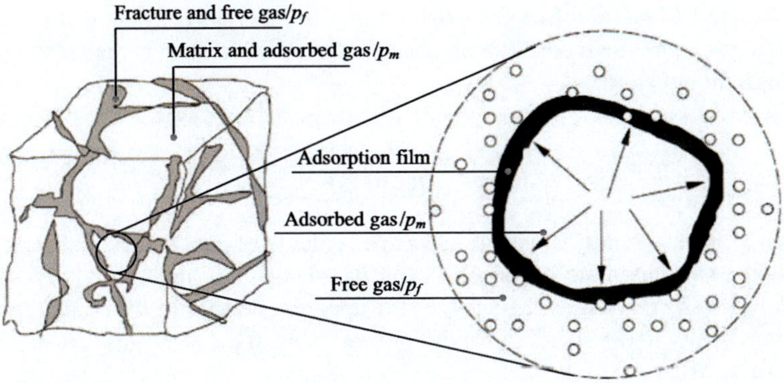

Fig. 9.1 Illustration of the typical dual-porosity system of coal [7]

Fick's law, then it flows into boreholes or roadways from the fractures following Darcy's law.

To study the migration form of coal seam gas, the pore structure of coal is simplified as a pore-fracture double porosity model, as shown in Fig. 9.1. Coal is a typical dual-porosity system that consists of a coal matrix surrounded by intersecting fractures. The fracture network provides the occurrence site and flow channel for free gas, and the migration law of free gas complies with Darcy's law. The pore network in the coal matrix offers an occurrence site for adsorbed gas and a small amount of free gas, and the migration law of adsorbed gas complies with Fick's law.

Free gas exists in the fracture system in the gas phase state, and the quantity of free gas per volume of coal can be expressed as:

$$m_f = \phi_f \rho_f \tag{9.1}$$

where m_f is the quantity of free gas per volume of coal, kg; ϕ_f is the fracture porosity, %; and ρ_f is the free gas density, kg/m^3.

By applying the ideal gas law, the relationship between the free gas density and pressure can be described by Eq. (9.2):

$$\rho_f = \frac{M_c}{RT} p_f \tag{9.2}$$

where M_c is the molar mass of methane, kg/mol; R is the universal gas constant, J/(mol K); T is temperature, K; and p_f is the gas pressure in the fracture. Therefore, the quantity of free gas per volume of coal can be expressed as:

$$m_f = \phi_f \frac{M_c}{RT} p_f \tag{9.3}$$

Both adsorbed gas and free gas exist within the coal matrix system, although the adsorbed gas is the main constituent, the quantity of which can be calculated using the Langmuir equation:

$$m_{\mathrm{m1}} = \frac{V_{\mathrm{L}} p_{\mathrm{m}}}{p_{\mathrm{m}} + P_{\mathrm{L}}} \rho_a \rho_{\mathrm{s}} \tag{9.4}$$

where m_{m1} is the quantity of adsorbed gas per volume of coal matrix blocks, kg/m^3; V_L denotes the maximum adsorption capacity of the coal, m^3/kg; P_L denotes the Langmuir pressure constant, MPa; ρ_a is the apparent density of the coal, kg/m^3; ρ_s is the gas density under standard conditions, kg/m^3; and p_m is the gas pressure in the coal matrix, MPa.

The gas density under standard conditions can be described using Eq. (9.5):

$$\rho_s = \frac{M_c}{V_M} \tag{9.5}$$

where V_M is the molar volume of methane under standard conditions (m^3/mol).

In current technology, the gas pressure (p_{m}) cannot be measured directly, and is defined as the "virtual" pressure that would be in equilibrium with the current concentration of an adsorbate in the coal matrix [22]. In addition, V_L and P_L have the following relationships with the common adsorption constants in the engineering of coal mine gas control:

$$V_L = a \tag{9.6}$$

$$P_L = \frac{1}{b} \tag{9.7}$$

where a is the gas limit adsorption capacity, m^3/kg; and b is the adsorption constant, MPa^{-1}.

The free gas in the pore system can also be calculated using the ideal gas law:

$$m_{m2} = \phi_m \frac{M_c}{RT} p_m \tag{9.8}$$

where m_{m2} is the quantity of free gas per volume of coal matrix, kg/m^3; and ϕ_m is the coal matrix porosity, %.

The total quantity of gas per volume of coal matrix blocks can be calculated using Eq. (9.9):

$$m_m = \frac{V_L p_m}{p_m + P_L} \frac{M_c}{V_M} \rho_a + \phi_m \frac{M_c}{RT} p_m \tag{9.9}$$

where m_m is the total quantity of gas per volume of coal matrix, kg/m^3.

The total quantity of gas per volume of coal can be described as follows:

$$m = \frac{V_L p_{\mathrm{m}}}{p_{\mathrm{m}} + P_L} \frac{M_c}{V_M} \rho_{\mathrm{a}} + \phi_{\mathrm{m}} \frac{M_c}{RT} p_{\mathrm{m}} + \phi_f \frac{M_c}{RT} p_f \tag{9.10}$$

where m is the total quantity of gas per volume of coal, kg/m^3.

The adsorbed gas and free gas are in a state of dynamic equilibrium when a coal seam is unaffected by mining or extraction. Gas mass exchange between the coal matrix and fractures always exists; however, there is no mass transfer at the macroscale. Once a coal seam is mined or affected by gas extraction, the free gas and adsorbed gas will simultaneously flow toward the coal wall driven by pressure and concentration gradient, which are based on Darcy's law for seepage and Fick's law for diffusion respectively. Meanwhile, because the seepage velocity in a fracture is much larger than that in the coal matrix, the gas pressure in the coal matrix exceeds that in the fractures after the equilibrium state is broken. Under these conditions, the gas in the pore system will transport from matrixes to fractures in the form of diffusion. This process can be regarded as a mass exchange between the coal fractures and coal matrix. For pore systems, the mass is an inflow, whereas it is an outflow in fracture systems. The pore system is equivalent to an even mass source for the fracture system [19, 23]. In fact, gas migration of seepage and diffusion in a real coal seam is a parallel and series coexistence process; however, the process can be simplified as a continuous two-step process because the amount of gas diffusion is insignificant from the coal matrix directly to a borehole or roadway. First, the gas diffuses from the coal matrix into fractures based on Fick's law, and then it flows into a borehole or roadway from the fractures based on Darcy's law (see Fig. 8.1) [24].

Because the gas–solid coupling effect of coal and gas is an extremely complex physical process, many researchers have made reasonable assumptions to simplify the physical processes when establishing the coupled motion equation. When combining the previous research results with our new insights into the coupling effect [7, 13], we made the following assumptions [25, 26]:

(1) The coal seam is dry; hence, the influence of water may be ignored.
(2) The CBM reservoir is a homogeneous isotropic medium, and the effect of anisotropy on the directionality of permeability is ignored.
(3) The system can be regarded as isothermal during gas migration.
(4) The coal fracture is saturated by free gas, and the gas in the coal matrix blocks is in an adsorbed or free state.
(5) Methane behaves as an ideal gas, and its viscosity is constant under isothermal conditions.
(6) The gas–solid coupling effect (mechanical and volume response) between coal and gas can be treated using the double-pore elasticity theory.
(7) The pore and fracture systems are continuous media systems. In these continuous media systems, every physical parameter is continuously changed in space. For example, the seepage velocity of gas is not the true velocity in

the fracture, but rather the ratio of the volume flow of gas to the fracture cross-section, that is, the seepage velocity is a continuous function in the spatial coordinate. In addition, the mass concentration of the adsorbed gas in the matrix pore system is not the mass in the unit pore volume, but the mass in the unit space volume.

9.1.2 Classification of the Gas Flow Field in Coal Seams

When the gas pressure distribution in a coal seam is uneven, it often forms a certain gas flow area, which can be called the flow field. To facilitate the study of gas flow, the gas flow fields can be classified into different types under engineering conditions based on its spatial and temporal properties.

1. Classification based on flow field geometry

Gas flow conditions can be divided into three types (unidirectional, radial, and spherical) based on the space geometry shape of the flow field.

(1) Unidirectional flow occurs when the flow velocity of gas exists only in one direction in a three-dimensional space, and the flow velocity is zero in the other two directions. When the thickness of a coal seam is less than the height of the roadway and digging along the coal seam, the gas flow field on both sides of the roadway will flow along the direction perpendicular to the roadway, which is a unidirectional flow. The streamlines are parallel to each other and flow in the same direction, as shown in Fig. 9.2.
(2) Radial flow occurs when the flow velocity of gas exists in two directions in a three-dimensional space, and the flow velocity is zero in the other direction. The gas flow in a coal wall follows radial flow when the crosscut, vertical shaft, and borehole are vertical through the coal seam. Under normal circumstances,

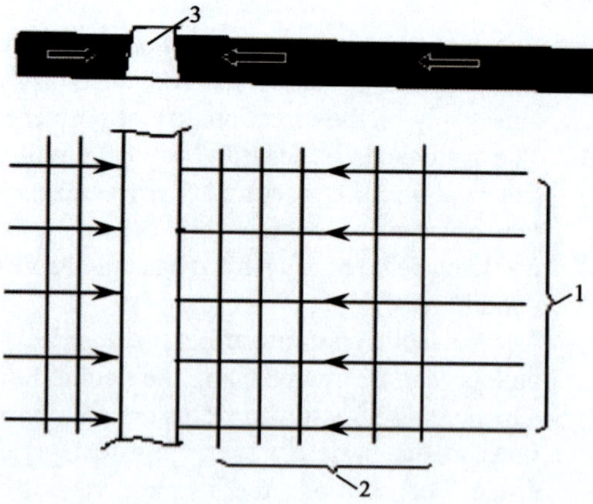

Fig. 9.2 Schematic diagram of unidirectional flow

isobars parallel to the coal wall are approximately concentric circles, as shown in Fig. 9.3. Because the radial flow is generally a planar flow, it can be expressed in Cartesian rectangular coordinates in actual research work, although it can also be expressed by polar coordinates.

(3) Spherical flow occurs when the flow velocity of gas exists in every direction in a three-dimensional space. The gas flow ① of an excavation working face of a thick coal seam, ② in a coal wall when the borehole and crosscut enter the coal seam, and ③ from the falling coal blocks, all belong to spherical flow (Fig. 9.4). Spherical flow is characterized by the formation of nearly concentric spherical isobars in a coal body, while the streamlines are generally radiated. Because spherical flow is a three-dimensional space flow, it can be represented by Cartesian rectangular coordinates in actual research work, although it can also be expressed by spherical polar coordinates.

Fig. 9.3 Schematic diagram of radial flow

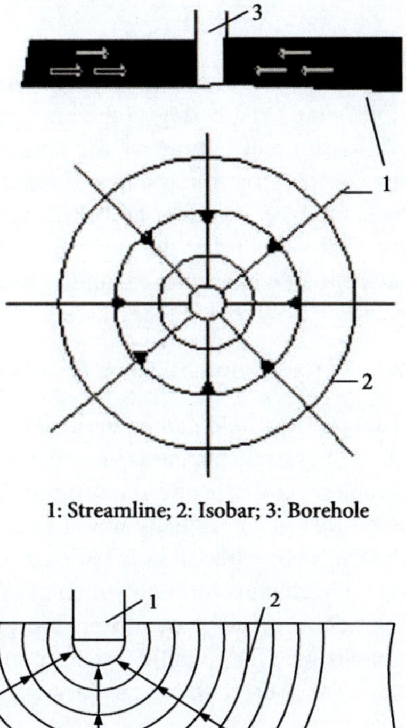

1: Streamline; 2: Isobar; 3: Borehole

Fig. 9.4 Schematic diagram of spherical flow

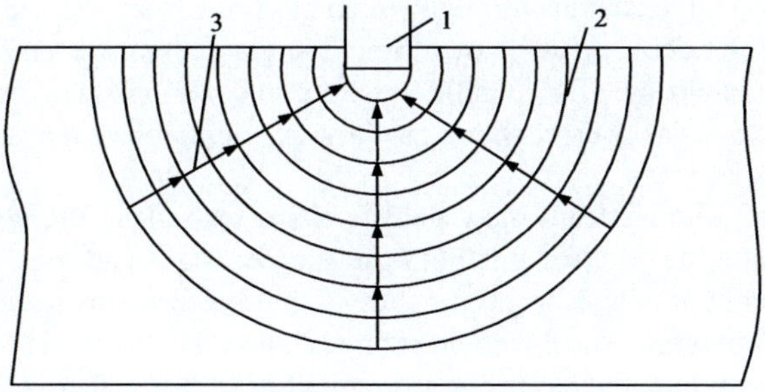

1: Driving working face; 2: Isobar; 3: Streamline

Table 9.1 Common types of flow field in coalmines

Engineering classification	Condition	Flow field type	Note
Roadway	Height of roadway < thickness of the coal seam	Unidirectional flow	Starts with radial flow, and after a period of extraction, converts to radial flow
	Height of roadway ≥ thickness of the coal seam	Unidirectional flow	
Borehole	Crossing borehole (including surface well)	Radial flow	
	Bedding borehole	Unidirectional flow	Starts with radial flow, and after a period of extraction, converts to radial flow

The above three flow types are typical and basic. In actual coalmines several basic flows may exist owing to the heterogeneity of the coal seam, the variable lithology of the top and bottom of the coal seam, the gas flow in the actual roadway, and the complexity of boreholes. For example, the gas flow in the head coal wall of the working face is similar to that of radial flow, and is a single flow field at the back of the coal wall during the excavation of the coal seam roadway. Therefore, a specific analysis should be carried out in actual research work in order to establish a relatively accurate flow model (see Table 9.1).

2. Classification based on flow field stability

The gas flow field can be divided into a stable flow field and an unsteady flow field according to the change over time. In a stable flow field, the gas flow velocity, flow direction, and pressure at any point do not change over time, whereas they do change over time in an unsteady flow field.

The stable flow field is characterized by a fixed gas source, and the coal seam is only the channel through which gas flows. Because the gas pressure at the outlet and inlet do not change over time, the gas pressure at each point of the flow field does not change. This condition is rare in coalmines, and only occurs in the vicinity of a large gas source, where gas flows through the coal seam or fractures to the roadway space.

The unsteady flow field is characterized by the absence of a fixed gas source, and the coal seam is both the source of gas and the gas channel. As gas flows out continuously, the gas pressure in the coal seam decreases, and the flow field expands. Therefore, the flow field changes, and both gas pressure and its gradient change. For example, the gas pressure gradient near the coal surface decreases with an increased outflow of gas when the exposed surface of the coal seam is fixed, which means that the amount of gas emission decreases over time. In addition to the explored coal

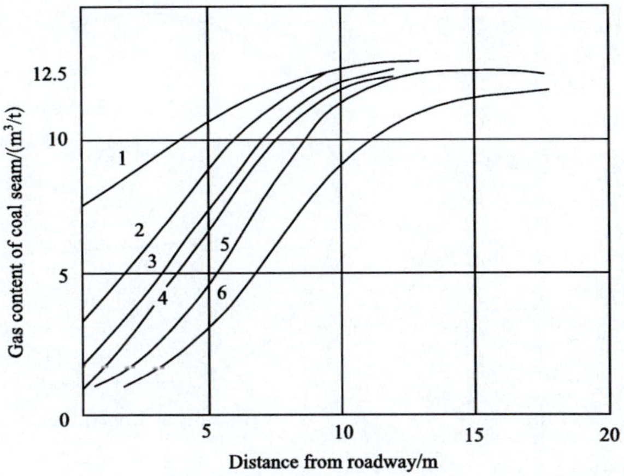

Fig. 9.5 Variation curve of the gas content of coal seam in an unsteady flow field. Curves 1–6 represent the variation curves of the coal seam gas content after being driven for (1) several hours, (2) 4 d, (3) 10 d, (4) 15 d, (5) 55 d, and (6) 150 d

seam itself, the gas flows of the adjacent layer of the surrounding rock and the falling coal also belong to unsteady flow.

The change in the gas content of coal seam in the unstable flow field is illustrated in Fig. 9.5. This shows that, when the roadway was excavated, the gas flow and change were severe in the initial unsteady flow field, and then gradually stabilized. After excavating for 150 d, the gas flow in both sides of the roadway was stabilized.

3. Classification based on flow-field scale

The gas flow field can be divided into an infinite or finite flow field according to its scale. In the finite flow field domain, the gas flow velocity, flow direction, and pressure vary with the entire flow field. In the infinite flow field domain, the dimensions of each direction of the outer boundary can be regarded as infinite compared to the inner boundary of the research object, and the outer boundary has no effect on the gas flow field. The gas pressure and flow velocity of the boundaries are p and u, as shown in Fig. 9.6.

9.2 Theory of Gas Diffusion

Coal is a typical naturally porous medium, and gas migration in its pore channels follows diffusion. The diffusion mechanisms of different-sized channels in porous media vary. According to the relative size of the pore radius and the mean free path of molecular motion, the gas-phase diffusion of the gas in different pores can be divided into three forms: fast diffusion, Knudsen diffusion, and transition diffusion [27, 28]. In addition, the adsorbed gas undergoes surface diffusion, while the inhaled

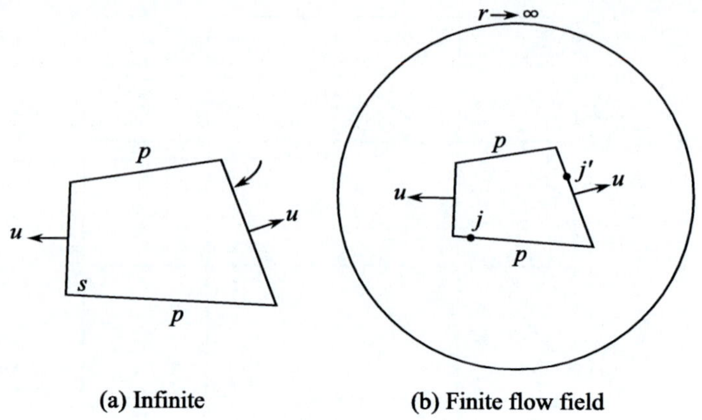

Fig. 9.6 Schematic diagram of infinite and finite flow fields

gas undergoes crystal diffusion. The quantity of surface diffusion is smaller than that of gas-phase diffusion [29], and the flux of crystal diffusion in coal is relatively small owing to a large resistance [30]. Therefore, the gas diffusion in a coal seam is mainly dominated by gas-phase diffusion [31], which is driven by the concentration difference between the coal matrix and fractures diffusion power is the difference between the gas-phase gas concentration in the pores of the coal matrix and the gas-phase gas concentration in coal fractures [32]. After the gas concentration in the pores reduces, the adsorbed gas rapidly desorbs as the gas source to ensure that diffusion continues.

The gas migrates from high to low concentrations along the concentration gradient, and the diffusion flux is proportional to the concentration gradient and diffusion coefficient, which can be expressed by Fick's diffusion law:

$$q = -D\frac{\mathrm{d}c}{\mathrm{d}l} \tag{9.11}$$

where q is the diffusion flux, kg/s, D is the diffusion coefficient, m^2/s; and $\frac{\mathrm{d}c}{\mathrm{d}l}$ is the concentration gradient.

The diffusion coefficient of the coal matrix is not only affected by coal/rock components, coal rank, particle size, and other "internal" factors, but also by the coal temperature, diffusion gas type, and other "external" factors. Zhang studied the diffusion coefficient of different types of coal and showed that the effective diffusion coefficient of CH_4 and CO_2 were in "U" relation to the coal rank. The diffusion coefficient was found to range from 10^{-8} to 10^{-9} cm^2/s, and the effective diffusion coefficient of CO_2 was higher [33]. A study of CO_2 adsorption and diffusion showed that the diffusion coefficient of CO_2 in coal decreased with the decrease in mass fraction [34], which was related to desorption shrinkage, and the measured range was approximately 10^{-6}–10^{-4} cm^2/s. The above diffusion coefficients are the overall diffusion coefficients of the coal matrix, and are based on a single-pore diffusion model.

Coal matrix diffusion is the basic form of gas migration and is closely related to coal and gas outbursts. According to the study on gas release of tectonic coal, it was shown that the release velocity of tectonic coal was faster than that of raw coal [35]. For the same coal sample, the gas desorption characteristics were found to be related to the gas pressure and gas content of the coal [36]. The initial methane release rate and desorption index of drilling cuttings are based on the results of coal grain desorption, which can be used to analyze coal and gas outbursts. The gas diffusion process also affects the exploitation of gas resources, and it has been shown that the capacity of coal matrix diffusion will have an impact on the gas extraction rate during coalbed methane recovery [37, 38].

Figure 9.7 shows that the coal matrix can be idealized as cubes. The adsorbed gas in the coal matrix expands as a mass source to facilitate diffusion and seepage during gas extraction, and the gas exchange rate between the coal matrix and fracture system can be expressed as [39, 40]:

$$Q_s = D\sigma_s(c_m - c_f) \tag{9.12}$$

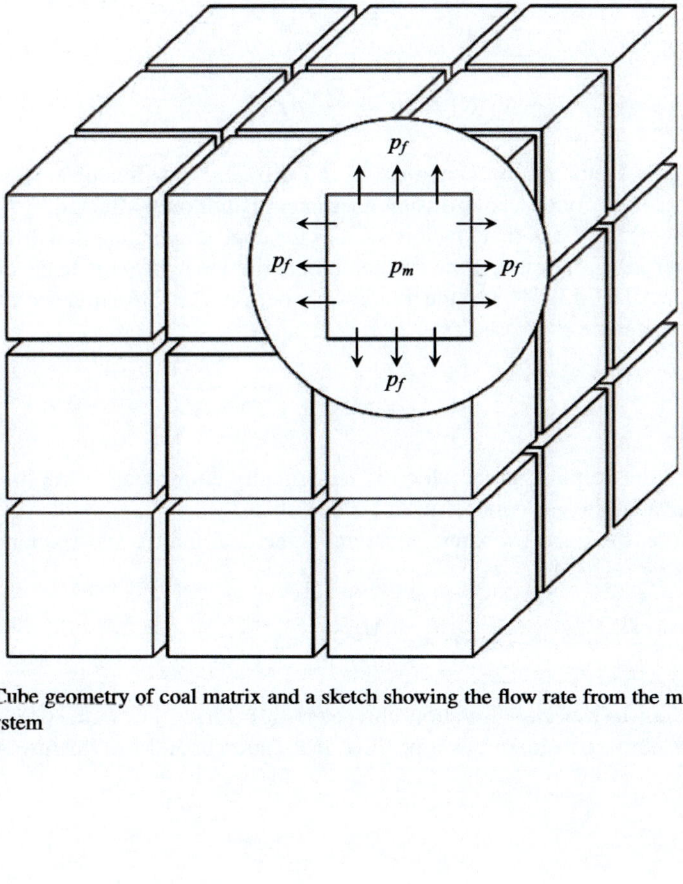

Fig. 9.7 Cube geometry of coal matrix and a sketch showing the flow rate from the matrix to the fracture system

where Q_s is the gas exchange rate per volume of coal matrix, kg/(m^3 s); D is the gas diffusion coefficient, m^2/s; σ_s is the coal matrix shape factor, m^{-2}; c_m is the concentration of gas in the coal matrix blocks, kg/m^3; and c_f is the concentration of gas in the fractures, kg/m^3.

The coal matrix shape factor σ_s is a very important parameter in Eq. (9.12). It is a special factor associated with the geometry of the coal matrix and controls the mass exchange of the coal matrix and fracture systems [41]. The representative geometries include plate, rectangular, and cubic coal matrix, corresponding to different shape factors and calculation formulae. For the cubic type, the shape factor is calculated using Eq. (9.13) [42]:

$$\sigma_s = \frac{3\pi^2}{L^2} \tag{9.13}$$

where L is the fracture spacing, m.

According to gas migration characteristics and the simplified model of the physical structure of coal, the gas concentration in Eq. (9.12) can be regarded as the gas phase concentration. The ideal gas law can be used to calculate the gas concentration in the matrix and fractures:

$$c_m = \frac{M_c}{RT} p_m \tag{9.14}$$

$$c_f = \frac{M_c}{RT} p_f \tag{9.15}$$

The shape factor of the coal matrix and diffusion coefficient are difficult to measure. As mentioned, the diffusion coefficient is not only affected by "internal" factors, but also by "external" factors such as the coal temperature and diffusion gas type. Therefore, the sorption time τ is commonly used to approximate the diffusivity of coal matrix [39, 43, 44], and the relationship between the diffusion coefficient and shape factor can be expressed as:

$$\tau = \frac{1}{\sigma_s D} \tag{9.16}$$

where τ is the sorption time, which is numerically equivalent to the time during which 63.2% of the gas content in coal is desorbed.

Therefore, the mass exchange equation of the coal matrix and fractures can be defined using Eq. (9.17):

$$Q_s = \frac{M_c}{\tau RT}(p_m - p_f) \tag{9.17}$$

According to the gas migration characteristics during gas extraction, the coal matrix system is considered as a positive mass source, and the fracture system is considered as a negative mass source. By applying the mass conservation law to the coal matrix, we obtain:

$$\frac{\partial m_m}{\partial t} = -\frac{M_c}{\tau RT}(p_m - p_f) \tag{9.18}$$

By substituting Eq. (9.9) into Eq. (9.18), we obtain the governing equation for the gas pressure change in the coal matrix:

$$\frac{\partial p_m}{\partial t} = -\frac{V_M(p_m - p_f)(p_m + P_L)^2}{\tau V_L RT P_L \rho_a + \tau \phi_m V_M (p_m + P_L)^2} \tag{9.19}$$

9.3 Theory of Gas Seepage

The coal fracture system is the location of free gas as well as a channel for seepage. When a coal seam is unaffected by mining or extraction, the gas in the coal matrix and fractures are in a state of dynamic equilibrium. There is a mass exchange between the coal fractures and the coal matrix, but no material transmission from the macroscale point of view. Once a coal seam is affected by mining or gas extraction, the adsorbed gas and free gas will flow to the coal wall in different forms under pressure and concentration differences, respectively. In Sect. 9.2, it was pointed out that the migration form of the adsorbed gas meets Fick's law, and that the migration form of free gas is closely related to a fluid's viscous force and inertial force.

These forces are the main forms of energy consumption in the gas flow process. When the viscous force dominates, the fluid behaves in a state of motion as laminar flow, while it behaves in a turbulent flow when the inertial force dominates. Zhou pointed out some laws regarding gas seepage when studying the basic law of gas migration in coal seams [13]. As the form and scale of coal pores and fractures are not uniform, the corresponding gas flow belongs to turbulent flow in large fractures, and to laminar flow in microfractures. According to the gas flow laws based on laboratory and field measurements, the gas of the coal fracture system can be considered as laminar flow and mainly follows Darcy's seepage law:

$$V = -\frac{k_e}{\mu}\nabla p_f \tag{9.20}$$

where k_e is the effective gas permeability, m_2; and μ is the viscosity of methane 1.08 $\times$ 10^{-5} Pa s.

It is important to note that the velocity of gas seepage in fractures is not the true velocity but average velocity. The average velocity is the ratio of the volume flow through a spatial cross-section to the area of the spatial cross-section, while the true velocity is the ratio of the volumetric flow through the fracture cross-section to the

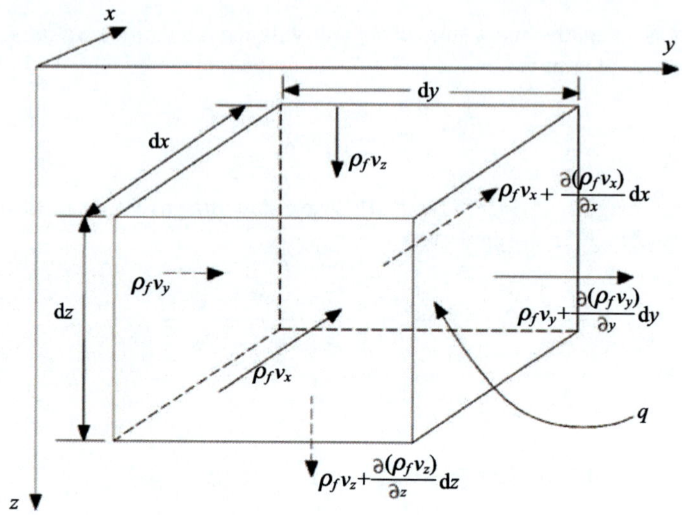

Fig. 9.8 Conservation of mass in a fracture system

area of the fracture cross-section. Darcy's law is used to avoid the description of the real gas flow characteristics of fractures, which can greatly simplify the governing equation for gas seepage (Fig. 9.8).

On this basis, we further deduced the governing equation for gas seepage in the fracture system. A microbody in the fracture is selected, where the length, width, and height are represented by dx, dy, and dz, respectively, and the seepage velocities in these different directions are set as v_x, v_y, and v_z, respectively. The free gas in fracture system with mass equal to $\phi_f \rho_f$ participates in seepage process, and it is known that the pore system is the mass source for the fracture system. According to the mass conservation law, the mass change of gas in per unit time and each direction can be expressed as follows:

$$\begin{aligned} \frac{\partial(\phi_f \rho_f)}{\partial t} \mathrm{d}x\mathrm{d}y\mathrm{d}z = & \left\{\rho_f v_x - \left[\rho_f v_x + \frac{\partial(\rho_f v_x)}{\partial x}\mathrm{d}x\right]\right\}\mathrm{d}y\mathrm{d}z \\ & + \left\{\rho_f v_y - \left[\rho_f v_y + \frac{\partial(\rho_f v_y)}{\partial y}\mathrm{d}y\right]\right\}\mathrm{d}x\mathrm{d}z \\ & + \left\{\rho_f v_z - \left[\rho_f v_z + \frac{\partial(\rho_f v_z)}{\partial z}\mathrm{d}z\right]\right\}\mathrm{d}x\mathrm{d}y \\ & + Q_s(1-\phi_f)\mathrm{d}x\mathrm{d}y\mathrm{d}z \end{aligned} \tag{9.21}$$

Equation (9.21) can be simplified as Eq. (9.22):

$$\frac{\partial(\phi_f \rho_f)}{\partial t} = -\left[\frac{\partial(\rho_f v_x)}{\partial x} + \frac{\partial(\rho_f v_y)}{\partial y} + \frac{\partial(\rho_f v_z)}{\partial z}\right] + Q_s(1-\phi_f) \tag{9.22}$$

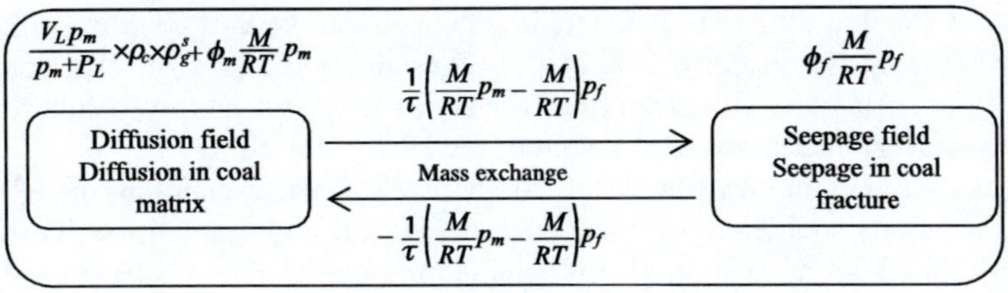

Fig. 9.9 Mass exchange between the coal matrix and fracture systems

Equation (9.22) can be rewritten as:

$$\frac{\partial}{\partial t}(\phi_f \rho_f) = -\nabla(\rho_f V) + Q_s(1 - \phi_f) \tag{9.23}$$

As already stated in the previous section, the coal matrix system is considered as a positive mass source, and the fracture system is considered as a negative mass source. Figure 9.9 and Eq. (9.22) describe the influence of mass exchange on the coal matrix and fracture systems.

Combining Eqs. (9.2), (9.16), (9.20), and (9.23) yields the governing equation for the change in gas pressure over time:

$$\phi_f \frac{\partial p_f}{\partial t} + p_f \frac{\partial \phi_f}{\partial t} = \nabla\left(\frac{k_e}{\mu} p_f \nabla p_f\right) + \frac{1}{\tau}(1 - \phi_f)(p_m - p_f) \tag{9.24}$$

Equations (9.18) and (9.24) constitute the governing equations for gas migration in coal seams. The change in fracture porosity over time is related to the results of gas–solid coupling, which will further influence gas migration. Therefore, the change in fracture porosity over time can be expressed as:

$$\frac{\partial \phi_f}{\partial t} = \frac{1}{M}\left(\beta_f \frac{\partial p_f}{\partial t} + \beta_m \frac{\partial p_m}{\partial t}\right) + \frac{\varepsilon_L P_L}{(P_L + p_m)^2}\left(\frac{K}{M} - 1\right)\frac{\partial p_m}{\partial t} \tag{9.25}$$

9.4 Gas–Coal Coupling Model

The presence of fluid in a porous medium affects its mechanical response. The change in the mechanical properties of the porous medium will subsequently further affect the migration characteristics of the fluid [45], which is the fluid–solid coupling effect in many areas. The fluid–solid coupling effect was first seen in Terzaghi's study of land subsidence in 1923, in which he proposed a well-known formula of effective stress [46]. It was then developed by Biot and a new mechanical branch, i.e., pore elastic mechanics, was formed gradually [47, 48]. Because of the importance of fluid-solid coupling, many researchers investigated that kind of coupling effect on gas-coal system. For instance, Zhu et al. [49] established a new gas-coal coupling model considering the Klinkenberg effect under low permeability condition. Xu et al. [50] developed a gas-coal coupling model for unloading coal seams under different stress conditions. Sun et al. [51] built a gas–water coupling model for a low-permeability coal seam and conducted a numerical study on multi-well methane recovery. In addition, Zhao et al. [52] studied the percolation mechanism of coal under a gas-coal coupling model.

9.4.1 Model Establishment of Gas-Coal Coupling

Chapter 7 discussed the differences in the stress environment between coal seam gas extraction during mining and CBM recovery. For example, protective seam mining (regional stress disturbance), roadway excavation (local stress disturbance), and borehole drilling (local stress disturbance). Therefore, the permeability of coal is controlled by the change in stress and the effect of gas–solid coupling during gas migration. These two factors have obvious differences in the degree of influence and action time on coal permeability. The stress change in coal/rock has a significant effect on coal permeability, but the effect of gas–solid coupling is weak and prolonged. The effect of the degree of gas–solid coupling is much smaller than that of the stress change, but its action time is much larger. We also used the same consideration when establishing a gas–solid coupling permeability model. The change in coal permeability is affected by excavation, which is equivalent to the change in the initial permeability; that is, the existence of gas only changes the mechanical strength of the coal body, and the gas–solid coupling effect can be ignored for a short period. Then, the effect of gas–solid coupling on coal permeability can be evaluated based on the new initial permeability.

In the previous section, permeability evolution models were established, which are affected by mining stress disturbance and the gas–solid coupling effect. The governing equations of gas diffusion and seepage were also obtained. To clarify the coupling effect between coal and gas (the coupling effect between the diffusion field, seepage field, and coal deformation field), the main governing equations of the above-mentioned physical fields are summarized and analyzed.

Considering the degree of influence and action time on coal permeability, the effect of stress change on the initial coal permeability and fracture porosity can be expressed as follows:

$$k_0^e = \begin{cases} \left(1 + \dfrac{\gamma^p}{\gamma^{p*}}\xi\right) \exp[b_\sigma(\Delta\Theta)]k_0, & 0 \le \gamma^p < \gamma^{p*} \\ (1+\xi)\exp[b_\sigma(\Delta\Theta)]k_0, & \gamma^p \ge \gamma^{p*} \end{cases} \tag{9.26}$$

$$\phi_{f_0}^{e} = \begin{cases} \left(1 + \frac{\gamma^{p}}{\gamma^{p*}}\xi\right)^{1/3} \exp\left[\frac{b_\sigma}{3}(\Delta\Theta)\right]\phi_{f_0}, & 0 \le \gamma^{p} < \gamma^{p*} \\ (1+\xi)^{1/3} \exp\left[\frac{b_\sigma}{3}(\Delta\Theta)\right]\phi_{f_0}, & \gamma^{p} \ge \gamma^{p*} \end{cases} \tag{9.27}$$

The governing equations of the gas diffusion field, gas seepage field, and coal deformation field during gas migration can be expressed using Eq. (9.28):

$$\begin{cases} \frac{\partial p_m}{\partial t} = -\frac{V_M(p_m - p_f)(p_m + P_L)^2}{\tau V_L RT P_L \rho_c + \tau\phi_m V_M (p_m + P_L)^2} \\ \phi_f \frac{\partial p_f}{\partial t} + p_f\left[\frac{1}{M}\left(\beta_f \frac{\partial p_f}{\partial t} + \beta_m \frac{\partial p_m}{\partial t}\right) + \frac{\varepsilon_L P_L}{(P_L + p_m)^2}\left(\frac{K}{M} - 1\right)\frac{\partial p_m}{\partial t}\right] \\ \quad = \nabla\left(\frac{k_e}{\mu} p_f \nabla p_f\right) + \frac{1}{\tau}(1 - \phi_f)(p_m - p_f) \\ G u_{i,jj} + \frac{G}{1-2\nu} u_{j,ji} - \beta_f p_{f,i} - \beta_m p_{m,i} + F_i = 0 \\ \frac{\partial \phi_f}{\partial t} = \frac{1}{M}\left(\beta_f \frac{\partial p_f}{\partial t} + \beta_m \frac{\partial p_m}{\partial t}\right) + \frac{\varepsilon_L P_L}{(P_L + p_m)^2}\left(\frac{K}{M} - 1\right)\frac{\partial p_m}{\partial t} \\ k_e = k_0^e\left(1 + \frac{b_k}{p_f}\right) \\ \quad \times \left\{1 + \frac{1}{M\phi_{f0}^{e}}[\beta_f(p_f - p_0) + \beta_m(p_m - p_0)] \right. \\ \quad \left. + \frac{\varepsilon_L}{\phi_{f0}^{e}}\left(\frac{K}{M} - 1\right)\left(\frac{p_m}{P_L + p_m} - \frac{p_0}{P_L + p_0}\right)\right\}^3 \end{cases} \tag{9.28}$$

Equation (9.28) includes the governing equation for gas diffusion in coal matrix, governing equation for gas seepage in coal fractures, the governing equation of coal deformation, the governing equation of fracture porosity over time, and the coal permeability evolution equation under the assumption of a dual-porosity model. The dynamic change in the effective permeability k_e depends on the dynamic change in the fracture porosity ϕ_f, and the change in ϕ_f is the result of competition between effective stress and desorptive shrink; with a decrease in gas pressure, ① the effective stress increases, so that the coal body is compressed (causing the decrease of ϕf); ② the coal matrixes shrink due to the gas desorption (causing the increase of ϕf). The effective permeability k_e is affected by the gas migration in Eqs. (9.19) and (9.24). The process of gas migration in a coal body involves gas diffusion and seepage, and the desorbed gas diffuses to participate in gas seepage, which becomes a positive mass source. In turn, this seepage also affects the diffusion rate, both of which are boundary conditions [(as shown in Eqs. (9.19) and (9.24)]. The influence of gas migration on the coal deformation field can be obtained using the effective stress principle of porous media. After the gas pressure balance in a coal body is broken,

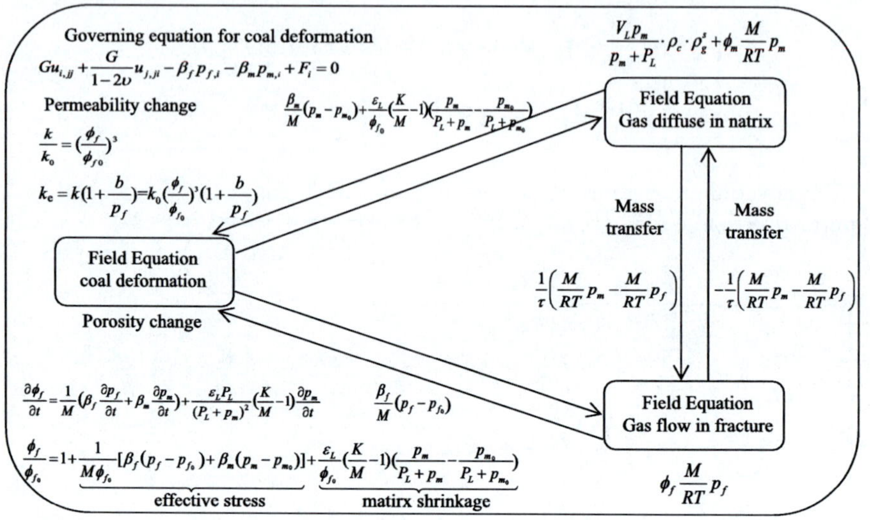

Fig. 9.10 Cross-coupling relationships between the coal deformation equation, gas diffusion equation in the coal matrix system, and gas seepage equation in the fracture system

the pressure gradient can be equivalent to the volume force. The above governing equations constitute the gas–coal coupling model, as shown in Fig. 9.10.

9.4.2 Boundary and Initial Conditions

For the above governing equations for every physical field, only combining them with the initial and boundary conditions can constitute the dual-porosity gas-coal coupling model in the process of gas migration.

The boundary conditions can be divided into Dirichlet, Neuman, and Cauchy boundary conditions. The Dirichlet boundary condition is also called the necessary condition, which gives the value on the boundary (u). The Neumann boundary condition is also called the natural condition, which gives the values of the derivative on the boundary (u_x and u_y).

1) Initial conditions

For the diffusion, seepage, and deformation fields, the initial values are the initial gas pressure of the coal matrix/fracture, stress, and displacement, which are defined as follows:

$$p_m(0) = p_f(0) = p_0 \tag{9.29}$$

$$u_i(0) = u_0 \tag{9.30}$$

$$\sigma_{ij}(0) = \sigma_o \tag{9.31}$$

where p_0, u_0, and σ_0 are the initial values of the gas pressure, displacement, and stress in the domain, respectively.

2) Boundary conditions

For the deformation field of coal, the definition of the displacement and stress conditions should be based on the actual stress conditions:

$$u_i = \tilde{u}_i(t) \tag{9.32}$$

$$\sigma_{ij}\vec{n} = \tilde{f}_i(t) \tag{9.33}$$

where $\tilde{u}_i(t)$ and $\tilde{f}_i(t)$ are the known displacement and stress at time t on the boundary, respectively.

For the boundary conditions of the gas diffusion and seepage fields, the pressure boundary (the common boundary is a constant pressure boundary) can be defined using Eq. (9.34):

$$p_m = p_f = \tilde{p}(t) \tag{9.34}$$

where $\tilde{p}(t)$ is the pressure at time t on the boundary, MPa.

The flux boundary (the common boundary is a zero-flux boundary) can be expressed as follows:

$$\frac{k_m}{u}\nabla p_m \cdot \vec{n} = \tilde{Q}_m(t) \tag{9.35}$$

$$\frac{k_f}{u}\nabla p_f \cdot \vec{n} = \tilde{Q}_f(t) \tag{9.36}$$

where $\tilde{Q}_m(t)$ and $\tilde{Q}_f(t)$ are the gas flux of the coal matrix and fracture at time t through the boundaries, respectively, kg/(m^2 s).

9.5 Evolution Characteristics of Pore Pressure and Coal Permeability During Adsorption Equilibration

As is well known, for mathematical and physical problems, if we want to obtain its quantitative solution, the mathematical model should first be built based on the physical nature of the problem. Then, the initial conditions and boundary conditions are extracted according to the actual situation, and the analytical or numerical method is selected to solve the model according to the complexity. Because the coal–gas coupling model is composed of multiple complex partial differential equations, it is difficult to obtain the exact solution using an analytical method.

The numerical solution of the gas–solid coupling model usually includes the finite difference method (FDM) and finite element method (FEM). Comparing the FDM and FEM from the perspective of mathematics, the FDM is more accurate than the FEM, while the FEM is more suitable than the FDM. The FEM can be used to deal with numerical calculations with complex geometric models and boundary condition so that the FEM is used more extensively than the FDM.

In the numerical calculation, the realization of gas–solid coupling mainly includes full coupling, explicit coupling, iterative coupling, and decoupling coupling [53]. Full coupling and iterative coupling have been widely used to solve multi-physics field problems in this book. The full-coupling solution is the simultaneous calculation of variables in a multi-physics field. The exact solution strictly depends on the real-time coupling results of the physical fields. For example, the solution of the diffusion and seepage fields must be solved using the full coupling model. However, it should also be noted that the calculation is difficult to converge because it needs to solve multiple partial differential equations simultaneously. The solution process is close to a real physical process. We can choose the iterative coupling method when the interdependence between the physical fields is not strong. This involves solving the physical field control equation in a specific order, and then using the solution results of the pre-physical field control equation as the initial or boundary conditions for the post-physical field control equations. The iterative coupling method is relatively efficient and easy, and has been widely used in other disciplines.

A numerical experiment was performed using COMSOL Multiphysics, which is a reliable numerical simulation software based on the FEM and partial differential equations and can solve some engineering problems. The experiment aimed to examine the accuracy of using COMSOL Multiphysics to simulate gas migration in porous media. Because it is generally not possible to obtain the analytical solution of the coupling model, some scholars have used the results of a single physical field to verify the accuracy of the coupling model [49]. The steady and transient flow fields calculated by Wu et al. [54] are regarded as reference to analyze the solution feasibility of COMSOL Multiphysics.

9.5.1 Introduction to the Physical Model

Many researchers have pointed out that the adsorption equilibrium time for coal samples is very important for the accurate measurement of adsorption swelling deformation (Fig. 9.11). Van Bergen et al. [55] measured the sorption-induced linear strain of coal samples exposed to CO_2 and CH_4 and found that although the gas pressures were constant during the measurements, the linear strain did not stabilize after 35 h

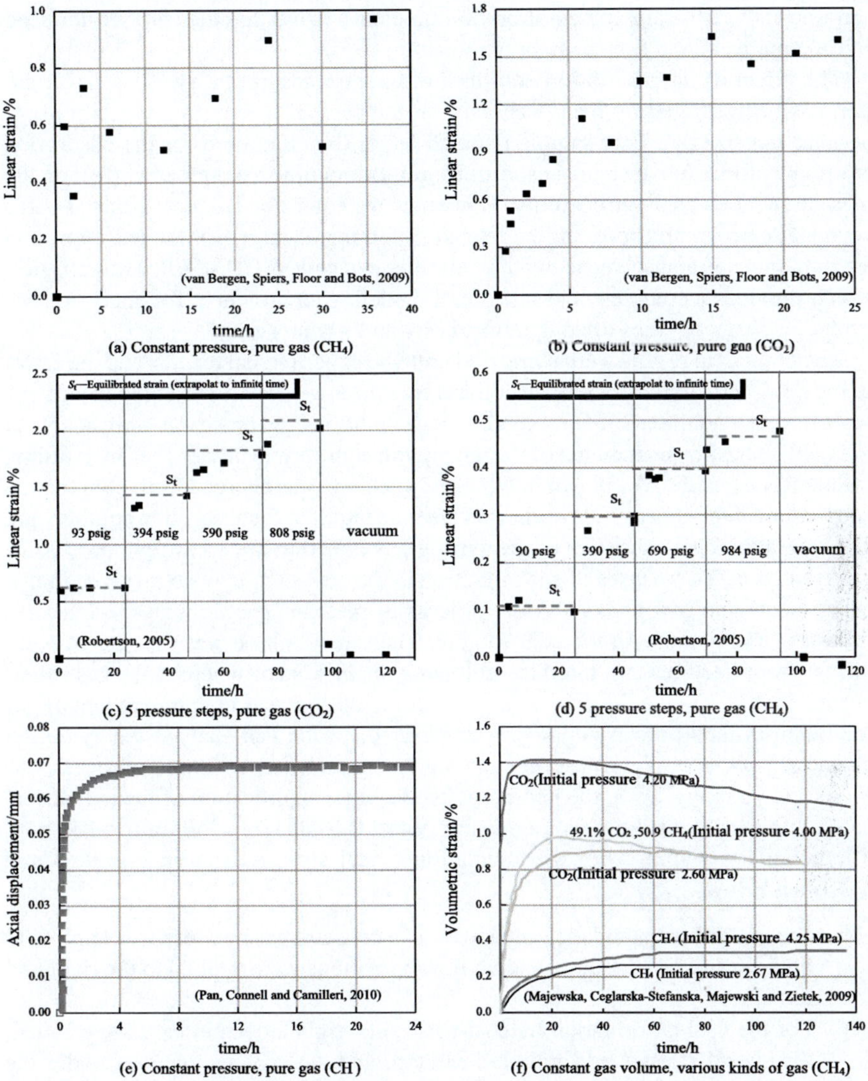

Fig. 9.11 Sorption-induced strain plotted against equilibration time

for CH_4 and 25 h for CO_2 (see Fig. 9.11 (a) and (b)). Robertson [56] measured and modeled sorption-induced strain and permeability changes in coal samples, and used the extended Langmuir theory to simulate the variation of the sorption-induced strain to avoid the long time required for his experiments. As shown in Fig. 9.11 (c)and (d), the equilibrated strain (S_t, extrapolated to infinite time) was obtained by fitting experimental data measured of the first 24 h of equilibration. As shown in Fig. 9.11, Pan et al. [57] and Majewska et al. [58] determined the relationship between the sorption-induced strain and equilibration time, and found that the sorption-induced strain requires a long time to stabilize.

The effect of the adsorption equilibrium time on adsorption swelling deformation has been studied by many scholars. In the case of the coal permeability test, because the size of a coal sample is much larger than that used for the adsorption swelling deformation test, the adsorption equilibrium time has a greater effect on the measurement of coal permeability. However, this condition has not attracted sufficient attention. In this book, all the related governing equations of the gas migration are implemented and solved using the finite element code in COMSOL Multiphysics, which provides a complete and integrated modeling environment for analyzing the evolution characteristics of coal permeability and gas pressure.

The numerical results were compared with the experimental results, and the effect of the dual-porosity characteristics on gas migration was analyzed at the laboratory scale (i.e., at a small scale). The experimental conditions and processes were analyzed to establish the physical model of the adsorption equilibrium process before building the numerical model. As shown in Fig. 9.12, methane was charged from both the top and bottom of the coal sample to shorten the time required for adsorption equilibrium. The migration characteristics of gas during gas injection are an inverse process of gas extraction. Two phases of gas migration in the coal sample were involved in this experiment. The first phase was the process of methane gas through the fractures following Darcy's law in the coal sample. The second phase was that the free gas diffuse from fractures into matrixes following the Fick's law whereby the gas would eventually adsorb onto the pore wall. Therefore, the coal–gas coupling model of gas migration in dual-porosity coal was established, which can analyze the evolution characteristics of coal permeability and gas pressure during adsorption process.

As shown in Fig. 9.13, the geometric model is a simplification of the cylindrical coal sample used for the usual permeability test, and its size is 50 mm × 100 mm. The boundary conditions of the simulation model are presented in Fig. 9.13 and described below.

(1) For gas diffusion and the seepage field, a constant gas pressure was applied to both the top and the base, and no flow conditions were applied to the right and left sides.
(2) For the coal deformation field, the left side, right side, and base were rolled, a constant stress was applied to the top, and the coal sample was under the condition of uniaxial strain.
(3) For the initial condition, the gas pressures p_m and p_f were both equal to 0.1 MPa.

These numerical simulations used two modules of COMSOL Multiphysics: the diffusion of methane was implemented through the "PDE module", and the seepage of methane was implemented through "Darcy's Law module". The input parameters used in these simulations are listed in Table 9.2. Most were chosen from appropriate data obtained from recently published studies to ensure that the parameters were within a reasonable range [59].

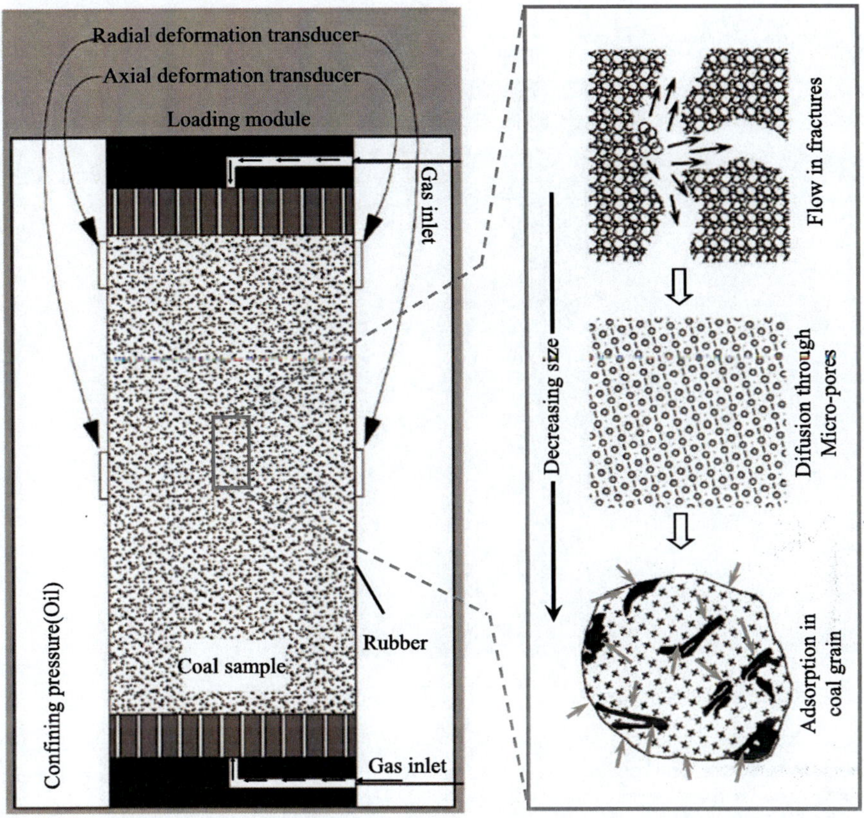

Fig. 9.12 Migration process of methane gas in a coal sample under laboratory conditions

9.5.2 *Variation of Gas Pressures with Equilibration Time*

It is clear that coal permeability varies constantly with the effective stress and sorption-induced strain during the gas adsorption process of a coal permeability experiment. These changes are closely related to the two gas pressures p_f and p_m. The variation of these gas pressures with equilibration time is discussed first.

The gas pressures p_f and p_m obtained at equilibration times of 1 h, 6 h, 12 h, 24 h, and 48 h are displayed in Fig. 9.14 using the same color range. The minimum and maximum values of the color range are 0.1 MPa and 2.0 MPa, respectively, and p_f corresponds to the upper cloud, while p_m corresponds to the lower cloud.

As shown in Fig. 9.14, p_f and p_m exhibited the following spatiotemporal evolution characteristics:

(1) During the initial stage of adsorption, the maximum values of p_f and p_m were located along the top and bottom boundaries, and the gas pressures subsequently decreased gradually toward the middle of the solution domain. Both

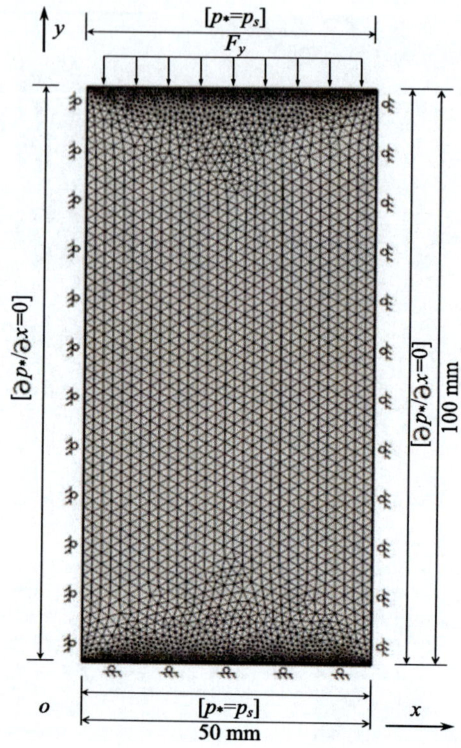

Fig. 9.13 Geometry and boundary conditions for coupled gas migration–coal deformation in a coal sample

p_f and p_m increased with the equilibration time, and simultaneously became more equally distributed over the entire model. However, there were still many differences in the speed to even distribution between p_f and p_m, that is, the speed of p_f was far greater than that of p_m.

(2) For the same equilibration time, there still exist a large difference between p_f and p_m in the domain. Figure 9.14 reveals that p_f was much larger than p_m when the equilibration time was 1 h and 6 h. The p_m value was still lower than the p_f value when the equilibration time was 12 h, and there was a small difference between them when the equilibration time was 24 h. The two gas pressures presented a substantially equal distribution when the equilibration time was 48 h. The evolution of the two gas pressures obtained by the numerical simulation agreed with the experimental results.

For the purpose of comparison, the two gas pressures along the *AB* detection line are presented in Fig. 9.15, where the coordinates of the two endpoints are *A* (2.5 mm, 0 mm) and *B* (2.5 mm, 10 mm).

Table 9.2 Parameters used in the numerical simulation

Parameter	Value
Young's modulus of coal, E	2713 MPa
Young's modulus of coal matrix, E_m	8139 MPa
Bulk modulus of the coal skeleton, K_s	10,335 MPa
Passion's ratio of coal, υ	0.339
Initial porosity of the coal matrix, ϕ_{m0}	0.06
Initial porosity of fractures, ϕ_{f0}	0.012
Initial permeability, k_0	5×10^{-4} mD
Sorption time, τ	8 h
Langmuir pressure, P_L	1 MPa
Langmuir volume, V_L	0.02 m^3/kg
Langmuir volumetric strain, ε_L	0.012
Density of coal, ρ_a	1250 kg/m^3
Stress, F_y	−6 MPa
Molar mass of methane, M_c	0.016 kg/mol
Temperature, T	293 K

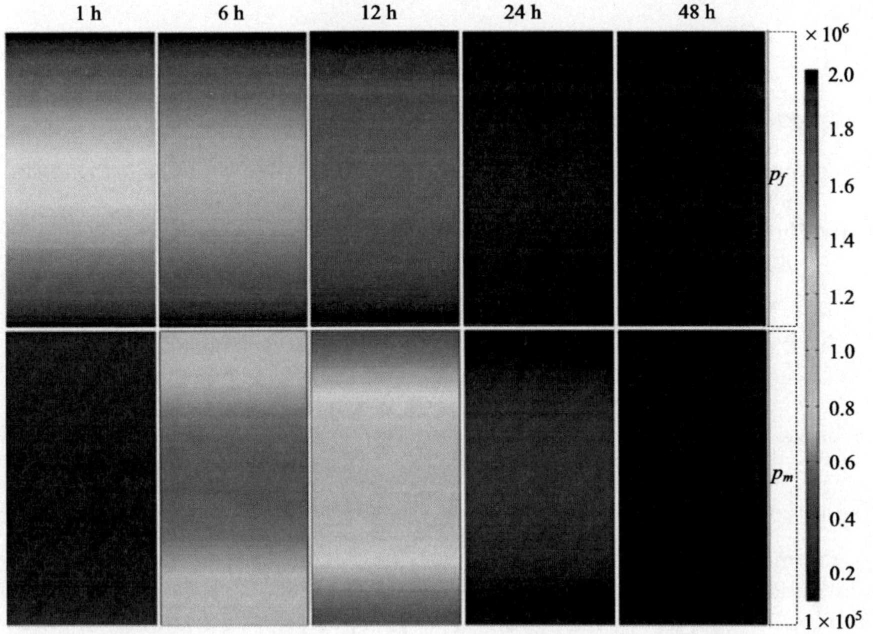

Fig. 9.14 Gas pressure distribution in a coal sample at different equilibration times

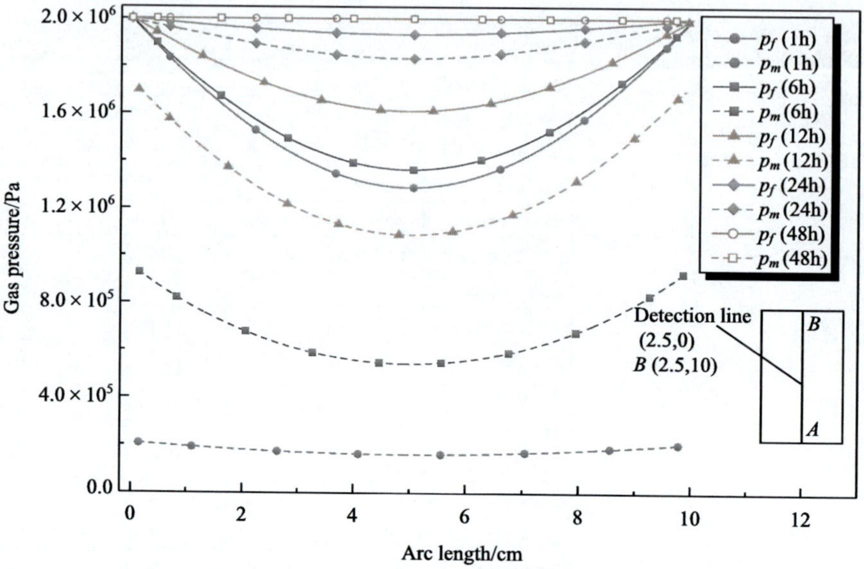

Fig. 9.15 Gas pressure distribution along the *AB* detection line [*A* (2.5 cm, 0 cm), *B* (2.5 cm, 10 cm)]

The characteristics of gas pressures p_f and p_m at equilibration times of 1 h, 6 h, 12 h, 24 h, and 48 h (Fig. 9.15) were follows:

(1) The minimum pressures in fractures were 1.28 MPa, 1.36 MPa, 1.60 MPa, 1.93 MPa, and 2.00 MPa.
(2) The minimum pressures in the coal matrix were 0.16 MPa, 0.54 MPa, 1.08 MPa, 1.83 MPa, and 2.00 MPa, respectively.

In the initial stage of adsorption, p_f was clearly at a high level, while p_m was at a relatively low level. It can be concluded that the pressure p_f could quickly reach the target through seepage, and the equilibration time was determined by the pressure p_m, which was mainly influenced by the diffusion characteristics of the coal sample.

In Eq. (9.37), the variable χ represents the ratio between p_f and p_m:

$$\chi(t) = p_f(t)/p_m(t) \tag{9.37}$$

The minimum values of the two pressures were used in the calculation, and χ was found to equal 800%, 252%, 148%, and 105% at equilibration times of 1 h, 6 h, 12 h, and 24 h, respectively. Large differences in the χ value were observed for most of the equilibration times.

The adsorption process is a dynamic equilibrium process, which means that the equilibrium state of the gas pressure in the coal fractures and matrix converts to another equilibrium state. The above analysis of the evolution characteristics demonstrates that the variation of gas diffusion and gas seepage with the change in gas pressure differed considerably when the equilibrium state of the gas pressure was broken. Moreover, it took a long time to achieve a new state of equilibrium again. At the same time, the difference between p_f and p_m was considerable during adsorption equilibrium if the external pressure conditions changed significantly. The phenomenon of this equilibrium state not only exists in laboratory tests, but also exists widely in gas extraction, coalbed methane mining, and geological sequestration of CO_2. In some permeability models, only one gas pressure was used, thus ignoring the difference between the gas pressure in the coal fractures and matrix. As seen from the above evolution characteristics of the gas pressures in coal fractures and matrix, if these permeability models are used to analyze the dynamic variation of coal permeability, there will be a large deviation from the actual situation. This further demonstrates that the permeability model based on the dual-porosity theory in this book has a better applicability than other models.

9.5.3 *Permeability Variation with Equilibration Time*

The evolution characteristics of coal permeability with respect to the equilibration time were studied based on the evolution characteristics of p_f and p_m. The relative permeability values along the A–B detection line at different equilibration times (1/12 h, 1/2 h, 1 h, 6 h, 12 h, 24 h, and 48 h) are displayed in Fig. 9.16, and comply with the following laws:

(1) The relative permeabilities were >1.0 at equilibration times of 1/12 h and 1/2 h, and presented a concave form. It is well known that coal permeability is controlled by the competing influences of effective stress and sorption-induced strain. At the same equilibration time, it can be concluded that the impact of effective stress was greater than that of the sorption-induced strain because the change rate of p_f was much larger than p_m at the beginning of adsorption. The impact of the effective stress at the top and bottom of the coal sample was greater than that at the center of the coal sample. This was because the p_f was greater at the top and bottom of the coal sample, and because the coal permeability at the top and bottom of the coal sample was greater than that at the center of the coal sample.

(2) The relative permeability was approximately 1.0 at an equilibration time of 1 h, and presented a slightly convex form. These findings indicate that the effect of the sorption-induced strain on relative permeability was greater than that of the effective stress at the top/bottom of the coal sample and vice versa at the center of the coal sample. The effects of sorption-induced strain and effective stress were dominant in different areas, although their respective advantages were not obvious. Therefore, the permeability of the coal sample was essentially equal to the original permeability at that time.

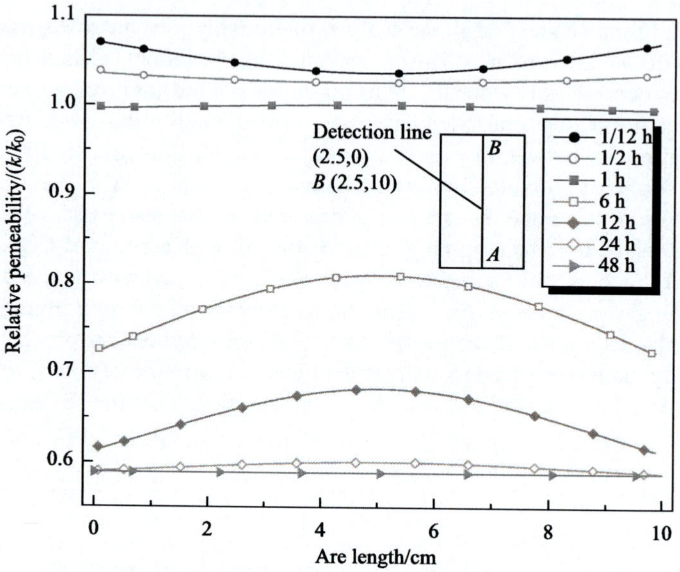

Fig. 9.16 Relative permeability along the *AB* detection line [*A* (2.5 cm, 0 cm), *B* (2.5 cm, 10 cm)]

(3) The relative permeabilities were <1.0 at equilibration times of 6 h, 12 h, and 24 h, and presented a convex form. These findings indicate that the effect of sorption-induced strain on relative permeability was greater than that of effective stress. In addition, the effect of sorption-induced strain on relative permeability at the top and bottom of the coal sample was greater than that at the center of the coal sample because the p_m was greater at the top and bottom of the coal sample. The permeabilities at these time points were lower than the initial permeability, and were also lower at the top and bottom of the coal sample in comparison to the center of the coal sample.

(4) The distribution of permeability was uniform and equal inside the coal sample at an equilibration time of 48 h because adsorption equilibrium was achieved. The effects of sorption-induced strain and effective stress tended to be consistent when the equilibration time was 48 h. This was because new equilibrium states of p_f and p_m were roughly formed.

To consider the overall relationships between the equilibration time and coal permeability evolution characteristics, the permeability evolution characteristics of three different points inside a coal sample during the adsorption stage of the coal permeability test were also studied. As shown in Fig. 9.17, the coordinates of the three detection points were *A* (2.5 mm, 9.0 mm), *B* (2.5 mm, 8.0 mm), and *C* (2.5 mm, 5.0 mm). The patterns of permeability variation over time at the three detection points were consistent: permeability first increased and then decreased with increasing equilibration time before adsorption equilibrium was achieved. Both the effects of sorption-induced volume change and effective stress increased with an increase in

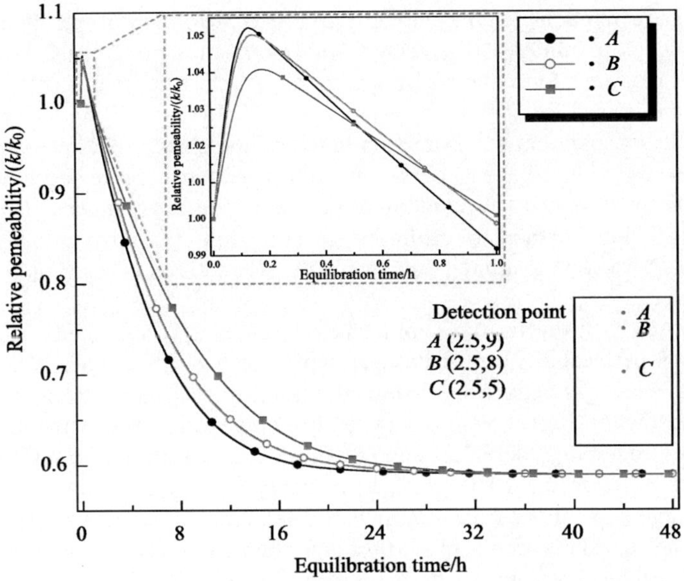

Fig. 9.17 Gas pressure distribution of three detection points inside a coal sample

the two pressures before adsorption equilibrium was achieved. The impact of effective stress was greater than that of the sorption-induced volume change when the permeability was greater than the initial value; however, the reverse was true when the permeability was less than the initial value. Some differences were also observed in the permeability evolution characteristics of the three detection points at the same equilibration time because the distances between the three detection points and the top of the coal sample differed. Thus, the permeabilities at the three detection points were only equal when a new adsorption equilibrium was formed.

The above-mentioned simulation results reveal that the relationships of the gas pressure and permeability with equilibration time agreed well with the laws obtained in the experiment. The simulation results could be a good response to the competing effect between effective stress and sorption-based volume at different equilibration times, which indicates that the coal–gas coupling model in this study has a good applicability for a small-scale flow field.

9.6 Characteristics of the Gas Flow Field During Extraction

Through years of research and development, the concept of coalmine gas drainage, mining methods and techniques, sealing methods and materials, and drilling equipment have made considerable progress in China. The development of the concept of coalmine gas extraction has gone through four stages: ① local measures for outbursts, ② pumping before mining, ③ assessing whether extraction meets the standards, and ④ regional measures for outbursts [60]. Currently, there is no uniform classification of the methods used for coalmine gas extraction. Yu [61] and Wang [62] classified the methods of coalmine gas extraction based on the spatial relationship between mined coal seams and adjacent coal seams (rock layers). In order to adapt the assessment requirement of the "basic index of gas extraction in coalmines" Cheng [60, 63] reclassified the methods of coalmine gas extraction. As shown in Fig. 9.18, the first level is based on the mining time, while the second level is based on the spatial relationship.

There have been several stages of coalmine gas extraction in China: gas extraction for high-permeability coal seams, pressure-relief gas extraction in adjacent coal seams, enhanced gas extraction for low-permeability coal seams, and comprehensive gas extraction [62]. For low-permeability coal seams, gas extraction is very difficult before mining; therefore, some enhanced gas extraction methods have been used in China since the 1960s (e.g., hydrofracture, hydraulic slotting, loosening blasting, large dimeter boreholes, dense boreholes, presplit-controlled, blasting, and cross boreholes). Many scholars have postulated numerous basic theories and a large amount of technical research on these methods [64–72].

The average permeability of coal in China is generally low (10^{-4}–10^{-3} mD) because of the deep buried depth, complex deposition environment and geological

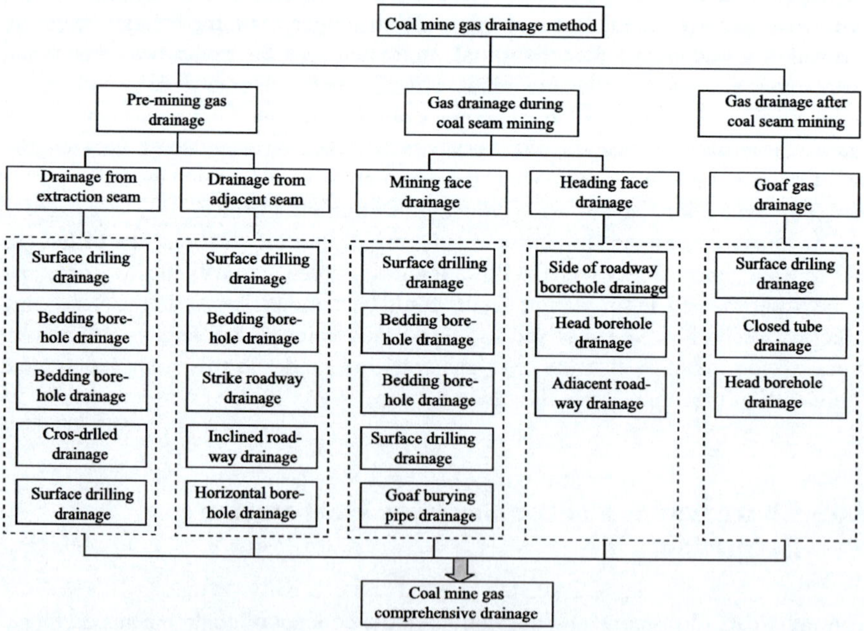

Fig. 9.18 Classification of gas drainage methods in a coalmine [60, 63]

evolution of the coal seams [73]. More than 65% of coal resources are below a depth of 1000 m, which is not conducive to gas extraction. Numerical simulations of the gas extraction of original low-permeability coal seams are discussed in this section, and the characteristics of the gas flow field in the extraction process are explained in detail.

9.6.1 Introduction to the Physical Model

It is appropriate to use the Panyi coal mine, a state owned coal mine located in the northeastern of the Huainan coal area, as a case study. A numerical simulation model was created to study the gas extraction evolution of the original coal seam. The simplified geometric model and its boundary conditions are shown in Fig. 9.19. The two-dimensional geometric model is a simplification of a real three-dimensional coal seam, and was set to a rectangle of 100 m in length and 50 m in width. The diameter of the extraction borehole was set to 10 cm.

The initial and boundary conditions were as follows:

(1) The initial gas pressure of the coal seam was set to 4.4 MPa.
(2) No flow conditions were applied to all boundaries.
(3) The boundaries of the extraction borehole were applied at a constant gas pressure of 25 kPa.

The numerical simulation was implemented using COMSOL Multiphysics, which was described in Sect. 9.5. The input parameters used in the original coal seam model are listed in Table 9.3.

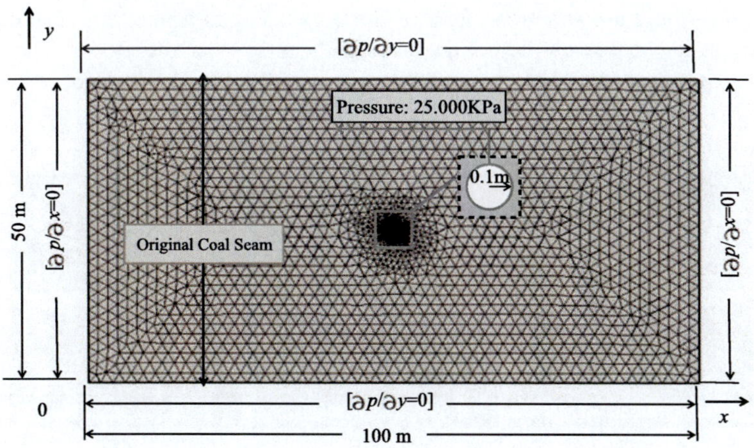

Fig. 9.19 Geometry and boundary conditions of an original coal seam

Table 9.3 Parameters used in the numerical simulation model

Parameter	Value
Young's modulus of coal, E	2713 MPa
Young's modulus of coal matrix, E_m	8139 MPa
Bulk modulus of coal skeleton, K_s	10,335 MPa
Passion's ratio of coal, υ	0.339
Initial porosity of coal matrix, ϕ_{m0}	0.06
Molar mass of methane, M_c	0.016 kg/mol
Thermodynamic temperature, T	293 K
Initial porosity of fractures, ϕ_{f0}	0.012
Initial permeability, $k_{\infty 0}$	2.8375×10^{-4} mD
Langmuir pressure, P_L	1.41 MPa
Langmuir volume, V_L	0.0228 m^3/kg
Langmuir volumetric strain, ε_L	0.004
Initial pressure of original coal seam, p	4.4 MPa
Density of coal, ρ_a	1250 kg/m^3

9.6.2 Gas Pressure Distribution Characteristics During Gas Extraction

The gas pressure levels in the original coal seam at different times (1 d, 10 d, 50 d, 100 d, 200 d, and 365 d) are presented in Fig. 9.20, which shows that the extraction effect of the original coal seam was limited. A gas pressure of 4.4 MPa did not change in almost all areas except for a small area around the borehole after extraction for 1 d. Although the effect of drilling extraction increased gradually with increasing extraction time, the extraction effect of the borehole was limited because the gas pressure in the coal seam did not decline significantly even after 365 d. Therefore, the effect of single borehole gas extraction was limited, and could not reduce the

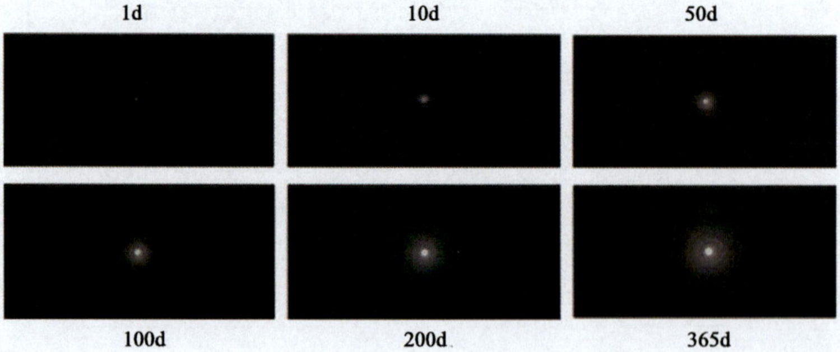

Fig. 9.20 Gas pressure distribution in an original coal seam over different extraction times

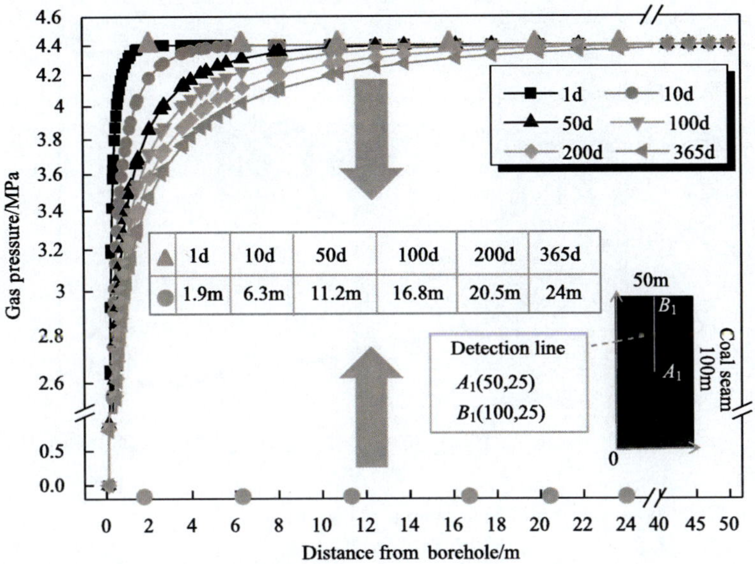

Fig. 9.21 Gas pressure distribution along the detection line of an original coal seam

gas pressure of the coal seam over a wide area or eliminate the risk of coal and gas outbursts.

To quantify the extraction effect on the original coal seam, detection line *AB* was set to monitor the gas pressure. The coordinates of the endpoints were as follows: *A* (50 m, 25 m) and *B* (100 m, 25 m). With an increase in the gas extraction time, the affected area in the original coal seam gradually increased. Figure 9.21 displays the gas pressure in the coal seam fractures when the extraction time was 1 d, 10 d, 50 d, 100 d, 200 d, and 365 d.

(1) With an increase in the gas extraction time, the affected area in the original coal seam increased gradually. The affected radii were approximately 1.9 m, 6.3 m, 11.2 m, 16.8 m, 20.5 m, and 24 m at 1 d, 10 d, 50 d, 100 d, 200 d, and 365 d, respectively.

(2) Although the affected area in the original coal seam increased gradually with increasing gas extraction time, most regions in the original coal seam were not influenced by gas extraction, even when the gas extraction time was 365 d.

From the distribution characteristics of gas pressure in the process of gas extraction, we can infer that it would take a long time to eliminate the risk of coal and gas outbursts for low-permeability coal seams if only gas extraction boreholes are used. If we arrange many extraction boreholes to reduce the distance between them (even <1 m), the extraction cycle may be shortened. However, an increase in the number of boreholes would also increase the cost of the project.

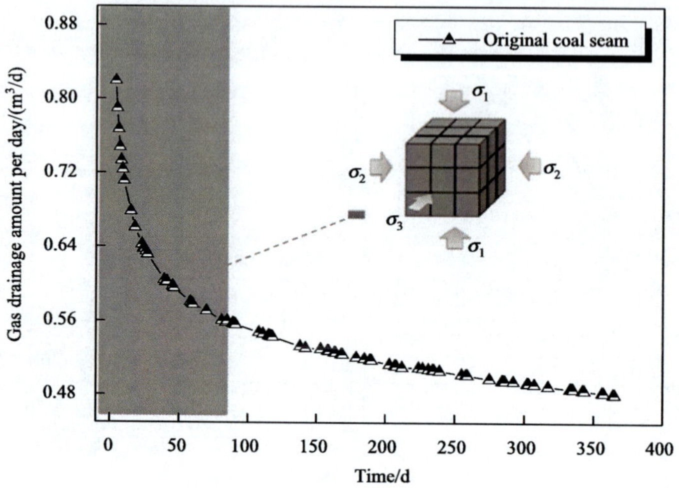

Fig. 9.22 Daily gas drainage volume in an original coal seam

9.6.3 Daily Gas Drainage Volume of an Original Coal Seam

The daily gas drainage volume is another significant parameter for gas extraction, and can be regarded as an objective criterion to judge the effect of gas extraction. The greater the daily gas drainage volume and the smaller the attenuation rate, the better the effect of extraction. For the numerical simulation results, the daily gas drainage volume can be calculated based on the difference between the residual gas content of the adjacent 2 d. The residual gas content can be expressed by the gas pressure. Because the gas pressure is not always the same in the solving domain, the residual gas content of every point should be calculated first, and then the surface integral should be calculated to obtain the residual gas content of the entire coal seam.

Figure 9.22 shows that the daily gas drainage volume of an original coal seam declined rapidly with extraction time. The rate of decrease in gas extraction slowed when the extraction time was 75–80 d. As seen from extraction volume of a single borehole, the daily maximum gas drainage volume in the original coal seam only reached 0.82 m^3/d, which eventually stabilized to approximately 0.48 m^3/d.

References

1. Wu, S. (1994). Preliminary discussion on gas diffusion and seepage in coal seam. *Shanxi Mining Institute Learned Journal, 12*(3), 259–263 (in Chinese).
2. Xia, Y. (1985). *Engineering mass transfer theory*. China: China Machine Press (in Chinese).
3. Wang, Y., & Pu, C. (1982). Gas desorption index of underground coal and its relationship with the outburst risk in coal seam area. *Safety in Coal Mines, 7*, 16–22 (in Chinese).

4. Wang, Y., & Pu, C. (1981). A preliminary study on gas content of coal seam was determined by coal desorption rate method. *Safety in Coal Mines, 11*, 9–17 (in Chinese).
5. Wang, Y., & Yang, S. (1980). Some characteristics of coal and gas outburst in coal seam. *Journal of China Coal Society, 5*(1), 47–53 (in Chinese).
6. Yang, Q., & Wang, Y. (1986). The theory and application of coal dust gas diffusion. *Journal of China Coal Society, 3*, 87–94 (in Chinese).
7. Wu, S. (2009). *Theory of coupling motion in coal seam and its application: Gas-solid coupled motion theory with adsorption gas*. China: Science Press (in Chinese).
8. Nie, B., Guo, Y., Wu, S., et al. (2001). Theoretical model of gas diffusion through coal particles and its analytical solution. *Journal China University of Minging and Technology, 30*(1), 19–22 (in Chinese).
9. Cheng, X., Cheng, Y., He, T., et al. (2013). Water injection impact on gas diffusion characteristic of coal. *Journal of Mining and Safety Engineering, 02*, 296–301 (in Chinese).
10. Mou, J., Cheng, Y., & Liu, H. (2012). Research on gas diffusion characteristic of injection water coal. *Journal of Mining and Safety Engineering, 05*, 746–749 (in Chinese).
11. Zhou, S. (1980). Measurement and calculation of gas permeation coefficients of coal seams. *Journal China University of Minging and Technology, 1*, 1–5 (in Chinese).
12. Zhou, S., & Sun, J. (1965). Theory of coal seam gas flow and its application. *Journal of China Coal Society, 2*(1), 24–36 (in Chinese).
13. Zhou, S., & Lin, B. (1999). *Coal seam gas occurrence and flow theory*. China: China Coal Industry Publishing House (in Chinese).
14. Zhou, J., Xian, X., Jiang, Y., et al. (2010). A permeability model including effective stress and coal matrix shrinking effect. *Rock and Soil Mechanics, 31*(7), 2317–2323 (in Chinese).
15. Jiang, Y., Yang, X., Xian, X., et al. (2010). The infiltration equation of coalbed under the cooperation of stress field temperature field and sound field. *Journal of China Coal Society, 35*(3), 434–438 (in Chinese).
16. Pan, Z. (2012). Modelling permeability for coal reservoirs: A review of analytical models and testing data. *International Journal of Coal Geology, 92*, 1–44.
17. Reiss, L. H. (1980). *The reservoir engineering aspects of fractured formations*. Editions Technip.
18. Robertson, E. P., & Christiansen, R. L. (2008). A permeability model for coal and other fractured sorptive-elastic media. *Spe Journal, 13*(3), 314–324.
19. Valliappan, S., & Zhang, W. (1996). Numerical modelling of methane gas migration in dry coal seams. *International Journal for Numerical and Analytical Methods in Geomechanics, 20*(8), 571–593.
20. Thararoop, P., Karpyn, Z. T., & Ertekin, T. (2012). Development of a multi-mechanistic, dual-porosity, dual-permeability, numerical flow model for coalbed methane reservoirs. *Journal of Natural Gas Science and Engineering, 8*, 121–131.
21. King, G., Ertekin, T., & Schwerer, F. (1986). Numerical simulation of the transient behavior of coal-seam degasification wells. *SPE Formation Evaluation, 1*(2), 165–183.
22. Gilman, A., & Beckie, R. (2000). Flow of coal-bed methane to a gallery. *Transport in Porous Media, 41*(1), 1–16.
23. Ge, J., Ning, Z., Liu, Y., et al. (2001). *The modern mechanics of fluids flow in oil reservoir*. China: Petroleum Industry Press (in Chinese).
24. Liu, Q., Cheng, Y., Haifeng, W., et al. (2015). Numerical assessment of the effect of equilibration time on coal permeability evolution characteristics. *Fuel, 140*, 81–89.
25. Liu, Q. (2015). *Damage and permeability evolution mechanism of dual-porosity coal under multiple stress paths and its application*. China University of Mining and Technology (in Chinese).
26. Liu, Q., Cheng, Y., Zhou, H., et al. (2015). A mathematical model of coupled gas flow and coal deformation with gas diffusion and Klinkenberg effects. *Rock Mechanics and Rock Engineering, 48*(3), 1163–1180.
27. Zhao, H., Wang, Z., & Hu, G. (2016). Effect of dynamic loads on internal microstructure of coal by nuclear magnetic resonance (NMR). *Chinese Journal of Rock Mechanics and Engineering, 35*(8), 1569–1577 (in Chinese).

28. Yin, G., Li, W., Jiang, C., et al. (2013). Mechanical property and permeability of raw coal containing methane under unloading confining pressure. *International Journal of Mining Science and Technology, 23*(6), 789–893.
29. Su, G., Hu, L., Feng, X., et al. (2016). Ture triaxial experimental study of rockburst process under low frequency cyclic disturbance load combined with static load. *Chinese Journal of Rock Mechanics and Engineering, 35*(7), 1309–1322 (in Chinese).
30. Meng, W., Zhu, Z., & Xiong, W. (2016). The growth of mixed-model I/II crack under impacting loads. *Chinese Journal of Rock Mechanics and Engineering, 35*(7), 1323–1332 (in Chinese).
31. Zhou, X., Men, J., Song, D., et al. (2016). Research on optim borehole parameters of antireflection in coal seam by liquid CO_2 blasting. *Chinese Journal of Rock Mechanics and Engineering, 35*(3), 524–529 (in Chinese).
32. Li, S., Wang, H., Qian, Q., et al. (2008). In-situ monitoring research on zonal disintegration of surrounding rock mass in deep mine roadways. *Chinese Journal of Rock Mechanics and Engineering, 27*(8), 1545–1553 (in Chinese).
33. Zhao, H., Yang, X., Gao, F., et al. (2014). Discussion on preparing methods of rock samples with different water contents. *Journal of Luoyang Institute of Science and Technology (Natural Science Edition), 24*(1), 4–7 (in Chinese).
34. Qian, Q., & Li, S. (2008). A review of research on zonal disintegration phenomenon in deep rock mass engineering. *Chinese Journal of Rock Mechanics and Engineering, 27*(6), 1278–1284 (in Chinese).
35. Zhou, X., & Qina, Q. (2007). Zonal fracturing mechanism in deep tunnel. *Chinese Journal of Rock Mechanics and Engineering, 26*(5), 877–885.
36. Wang, H., Zhang, Q., Zhang, Y., et al. (2010). Numerical simulation on disintegration of surrounding rock mass in deep mine roadways. *Journal of China Coal Society, 4*, 21–25.
37. Zhang, Q., Chen, X., Lin, B., et al. (2009). Study of 3D geomechanical model test of zonal disintegration of surrounding rock of deep tunnel. *Chinese Journal of Rock Mechanics and Engineering, 28*(9), 1757–1766 (in Chinese).
38. Zhang, X., Zhang, Q., Xiang, W., et al. (2016). zonal disintegration mechanism based on strain gradient theory. *Chinese Journal of Rock Mechanics and Engineering, 4*, 724–734 (in Chinese).
39. Mora, C., & Wattenbarger, R. (2009). Analysis and verification of dual porosity and CBM shape factors. *Journal of Canadian Petroleum Technology, 48*(2), 17–21.
40. Wang, J., Kabir, A., Liu, J., et al. (2012). Effects of non-Darcy flow on the performance of coal seam gas wells. *International Journal of Coal Geology, 93*, 62–74.
41. Fang, Z., & Xia, Z. (2009). The theory solution of matrix factor in coalbed reservoir. *Coal Geology and Exploration, 37*(3), 15–18 (in Chinese).
42. Lim, K., & Aziz, K. (1995). Matrix-fracture transfer shape factors for dual-porosity simulators. *Journal of Petroleum Science and Engineering, 13*(3), 169–178.
43. An, F. H., Cheng, Y. P., Wu, D. M., et al. (2013). The effect of small micropores on methane adsorption of coals from Northern China. *Adsorption, 19*(1), 83–90.
44. Zuber, M., Sawyer, W., Schraufnagel, R., et al. (1987). The use of simulation and history matching to determine critical coalbed methane reservoir properties. In *Low Permeability Reservoirs Symposium*. Society of Petroleum Engineers.
45. Cheng, A. H. D. (2014). Fundamentals of poroelasticity. *Analysis and Design Methods: Comprehensive Rock Engineering: Principles, Practice and Projects, 113*.
46. Terzaghi, K. V. (1923). Die berechnung der durchlassigkeitsziffer des tones aus dem verlauf der hydrodyn- amischen: Spannungserscheinungen. *Akad Wissensch Wien Sitzungsber Mathnaturwissensch Klasse IIa, 142*, 125–138.
47. Biot, M. A. (1941). General theory of three-dimensional consolidation. *Journal of Applied Physics, 12*(2), 155–164.
48. Le, B. M. (1935). Problème de la consolidation des matières argileuses sous une charge. *Annales de la Société Scientifique de Bruxelles. Serie B, 55*, 110–113.
49. Zhu, W., Liu, J., Sheng, J., et al. (2007). Analysis of coupled gas flow and deformation process with desorption and Klinkenberg effects in coal seams. *International Journal of Rock Mechanics and Mining Sciences, 44*(7), 971–980.

50. Xu, T., Tang, C., Song, L., et al. (2005). Numerical simulation of coupled gas flow in failure process of gassy coal-rock. *Chinese Journal of Rock Mechanics and Engineering, 24*(10), 1667–1673 (in Chinese).
51. Sun, K., Pan, Y., & Liang, B. (2007). Numerical simulation of deep coal-bed methane multi-well exploitation under fluid-solid coupling. *Chinese Journal of Rock Mechanics and Engineering, 26*(5), 994–1001 (in Chinese).
52. Zhao, Y., Hu, Y., Chen, D., et al. (1999). The experimental study on the gas seepage law of rock related to adsorption under 3-D stresses. *Chinese Journal of Rock Mechanics and Engineering, 18*(6), 651–653 (in Chinese).
53. Zhou, Z. (2003). *The theory and application research of fluid-solid coupling seepage flow in low permeability reservoir*. Daqing: Northeast Petroleum University (in Chinese).
54. Wu, Y. S., & Karsten, P. (1998). Gas flow in porous media with Klinkenberg effects. *Transport in Porous Media, 32*(1), 117–137.
55. Van Bergen, F., Spiers, C., Floor, G., et al. (2009). Strain development in unconfined coals exposed to CO_2, CH_4 and Ar: Effect of moisture. *International Journal of Coal Geology, 77*(1), 43–53.
56. Robertson, E. P. (2005). *Measurement and modeling of sorption-induced strain and permeability changes in coal.* United States: Department of Energy.
57. Pan, Z. J., Connell, L. D., & Camilleri, M. (2010). Laboratory characterisation of coal reservoir permeability for primary and enhanced coalbed methane recovery. *International Journal of Coal Geology, 82*(3–4), 252–261.
58. Majewska, Z., Ceglarska-stefańska, G., Majewski, S., et al. (2009). Binary gas sorption/desorption experiments on a bituminous coal: Simultaneous measurements on sorption kinetics, volumetric strain and acoustic emission. *International Journal of Coal Geology, 77*(1), 90–102.
59. Chen, J., Lu, J., Guo, D., et al. (2011). Factors and development technology of coalbed methane production capability. *Resources and Industries, 13*(1), 108–113 (in Chinese).
60. Cheng, Y., Fu, J., & Yu, Q. (2009). Development of gas extraction technology in coal mines of China. *Journal of Mining and Safety Engineering, 26*(2), 127–139 (in Chinese).
61. Yu, Q. (1992). *Coal seam gas prevention and control.* China: China University of Mining and Technology Press (in Chinese).
62. Yu, B., & Wang, Y. (2000). *Technical manual of mine gas disaster control and utilization.* Beijing: China Coal Industry Publishing House (in Chinese).
63. Cheng, Y. (2010). *Theories and engineering applications on coal mine gas control.* China: China University of Mining and Technology Press (in Chinese).
64. Xiao, Z., & Wang, Z. (2009). Status and progress of studies on mechanism of preventing coal and gas outburst by coal seam infusion. *China Safety Science Journal, 10*, 150–158 (in Chinese).
65. Zhai, C., Li, X., & Li, Q. (2012). Research and application of coal seam pulse hydraulic fracturing technology. *Journal of China Coal Society, 36*(12), 1996–2001 (in Chinese).
66. Zhao, L., & Feng, Z. (2001). Testing study of improving seepage flow laws of low seepage coalbed by hydratic-cutting seams. *Journal of Taiyuan University of Technology, 32*(2), 109–111 (in Chinese).
67. Feng, Z., & Kang, J. (2001). Experimental and mechanism study of coal and gas burst in the process of hydraulic cutting seam. *Journal of Liaoning Technical University (Natural Science), 20*(4), 443–445 (in Chinese).
68. Shi, B., & Yu, Q. (2002). Deep hole precracking control looseness blasting anti-process analysis in the low permeability coal seam. *Mine Construction Technology, 23*(5), 27–30 (in Chinese).
69. Li, C., & Fan, S. (2007). Application of vibration blasting technology with deep boreholes to fully mechanized longwall mining face under complicated mining technology. *Coal Science and Technology, 35*(5), 33–35 (in Chinese).
70. Zhao, Y., Xie, S., Wen, B., et al. (2009). Gas drainage technique by 100m long and large diameter roof boreholes in high gas coal seam group. *Journal of China Coal Society, 34*(6), 797–801 (in Chinese).

71. Yi, L. (2008). *Laboratory and numerical test studies of gas pre-drainage with dense boreholes in outburst coal seam*. Xuzhou: China University of Mining and Technology (in Chinese).
72. Wang, H. (2012). *Study on the mechanism of enhancing permeability of high stressed and how permeable coal seam in deep mining by pre-splitting controlled blasting technology*. Harbin: Institute of Engineering Mechanics, China Earthquake Administration (in Chinese).
73. Cheng, Y., Liu, H., Guo, P., et al. (2014). A theoretical model and evolution characteristic of mining-enhanced permeability in deeper gassy coal seam. *Journal of China Coal Society, 39*(8), 1650–1658 (in Chinese).

Chapter 10
The Application of Coal Mechanics in the Stress Relief Gas Extraction

Learning Outcomes

- The definition of stress relief gas extraction;
- Temporal evolution characteristics of the stress and strain in the stress-relief coal seam;
- Temporal evolution regulation of the permeability in the protected coal seam;
- Stress relief gas extraction and its gas flow characteristics in the protected coal seam;
- The extraction method of the stress relief gas and its field application.

10.1 Technology Synopsis of the Stress Relief Gas Extraction

Most of the coal seams in China are characterized by high in-situ stress, high gas pressure, high gas content and low permeability, which makes the gas extraction rather difficult. This infrastructure not only limits CBM production but also presents considerable challenges to the safety and efficiency of coal mining in China.

Most of the coal seams in China exist in the coal seam group conditions. As a result, the coal seam with lower coal and gas outburst risk could be chosen as the initial mining coal seam. Upon the mining of this coal seam, the in-situ stress unloads and the movement and deformation occur in the adjacent coal seams, as a result the fractures develop and the permeability improves. Therefore, if the main minable coal seam is of high outburst risk with great gas content and gas pressure, the adjacent coal seam with low gas content and low outburst risk could be mined firstly to unload the in-situ stress in the main mineable coal seam. At that moment, combining with a well-designed stress-relief gas extraction measure, the gas pressure/content in the main mineable coal seam could be significantly reduced and the outburst prone could be completely eliminated as well, which provides good mining conditions for the

Y. Cheng et al., *Coal Mechanics*,
https://doi.org/10.1007/978-981-16-3895-4_10

main mineable coal seam. This technology is called the protective seam mining technology.

The coal seam that is mined firstly to reduce the gas content and to eliminate the outburst risk in the adjacent coal seam is called as the protective coal seam, while the adjacent coal seam with high outburst risk is called as the protected coal seam [1, 2]. If the protective coal seam is located above the protected coal seam, this protective coal seam is usually called as the upper protective coal seam. On the contrary, the protective coal seam that located below the protected coal seam is usually called as the lower protective coal seam.

The extensive theoretical researches and the mining practice of the outburst-prone coal seams show that the protective seam mining technology is the most safe and economic technology in preventing the coal and gas outburst. Since firstly adopted in 1933 in France, this technology has been widely used in many countries [3, 4]. During 1958–1985, this technology has been widely used in China in the Beipiao, Tianfu, Nantong, Zhongliangshan and Songzao mine areas. Among these applications, the stress unloading effect of the remote lower protective seam in the Tianfu coal mine was systematically evaluated with the help of the China University of Mining and Technology (CUMT) and Chongqing branch of Coal Science Research Institute (CBCSRI) in the middle of 70s of last century [5–7].

Since the middle of 1980s, the top coal caving mining method has been widely used in mining of the extremely thick coal seam in China, with the improving of the mechanization level in the coal mine mining. As a result, the mining speed is greatly accelerated, leading to serious gas accidents, especially the coal and gas outburst accidents. Given this, the research on the remote lower protective seam mining and the stress relief gas extraction technology was carried out by the Huainan Mining Group, the CUMT and the CBCSRI, which achieved remarkable economic, social and safety benefits, and provides a solid theoretical and practical foundation for proposing the idea that the protective coal seam mining technology should be used and the pressure relief gas should be extracted as much as possible [8–14]. Afterwards, the protective coal seam mining and stress relief gas extraction technology has been widely tested under different conditions, such as different dip angle of coal seam (the horizontal, inclined and steeply inclined coal seams), different protection pattern (the upper and the lower protective coal seams), different ventilation type (the U and the Y shape types), different layer spacing (the remote and the short distance protective coal seams) and different gas extraction method (such as the underground net-like upward cross-borehole and the ground well gas extraction methods). The above related researches greatly promote the development and the field application of the protective coal seam mining and pressure relief gas extraction technology [15–19].

According to the coal and gas outburst prevention detailed rules and regulations promulgated on August 1, 2009, the regional outburst prevention measures must be adopted firstly and the local outburst prevention measures should be adopted to enhance the regional measures. Moreover, the protective coal seam mining technology is the preferential technology in the regional outburst prevention measures, which indicating the importance of the protective coal seam mining technology in the outburst prevention works [20–22].

In the past few years, protective seam mining and the stress relief gas extraction technology has been widely used in China. According to the their own coal-bearing stratum occurrence characteristics, gas accumulation characteristics and mining method, many new improvements have been made and many new technologies have been proposed in different mine areas, such as the short distance upper [23], the near lower [19], and extremely thin protective coal seam mining technology [24], and so on. With the continuous improvement of the protective coal seam mining technology, this technology has been widely used at more than 60 coal mines in more than 19 mine areas, such as the Huainan, Huaibei, Wanbei, Tiefa, Yaojie, Zhengzhou mine areas, and so on [25].

With the mining of the protective coal seam, the in situ stress would be redistributed nearby. After the mining of the coal seam, the overlying coal and rock masses break and collapse. The overlying coal and rock masses could be divided into three zones, i.e. caving, a fracture and bending zones [26, 27]. In the caving zone, the rock masses cave and fill the gob. Moreover, in the fracture zone, the fractures develop and form gas migration channel. However, the rock masses in the bending zone subside as a whole. The formation mechanism and division basis of the three zones has been recognized by many researches. In the caving zone, the broken coal and rock masses filled the gob loosely, which provides good conditions for the gas accumulation, as a result the gob burying pipes and the strike high level borehole technologies are widely used to extract the gas in the gob. In the fracture zone, plenty of new fractures develop. Moreover, considering that the gob has been filled in by the broken coal and rock masses in the caving zone, the coal and rock masses would not be seriously destroyed, which provides good construction conditions for the gas extraction. In the field gas extraction process, the protected coal seam should be located in the fracture zone or the bending zone under the mining of the lower protective coal seam. This is because the mining condition of the protected coal seam would not be destroyed under this circumstance. At the same time, due to the fact that plenty of new fractures develop in these two zones, notable stress unloading and permeability increasing effect could be achieved in the protected coal seams.

In spatial, the fracture zone is usually located between the protective and the protected coal seams. Therefore, the gas both in the protective and in the protected coal seams could flow into this zone, making this zone the gas accumulation zone. Consequently, enhancing the gas extraction in the fracture zone is of significant importance. Plenty of engineering practice has shown that the gas extraction effect of the strike high level gas extraction rock roadway technology is rather significant both in extracting the accumulated gas in the gob and in extracting the stress relief gas in the protected coal seam, as a result the gas accumulation in the working face of the protective coal seam could be successfully prevented. This technology is specially suitable under the mining conditions of short distance coal seam group and the fully mechanized mining technologies, such as the fully mechanized mining with large mining height, the fully mechanized caving mining, and so on (Fig. 10.1).

Along the strike direction, the coal and rock masses above or below the gob move and deform, as a result the in-situ stress would be redistributed with the advance of the working face [28]. According to the stress redistribution characteristics, four different

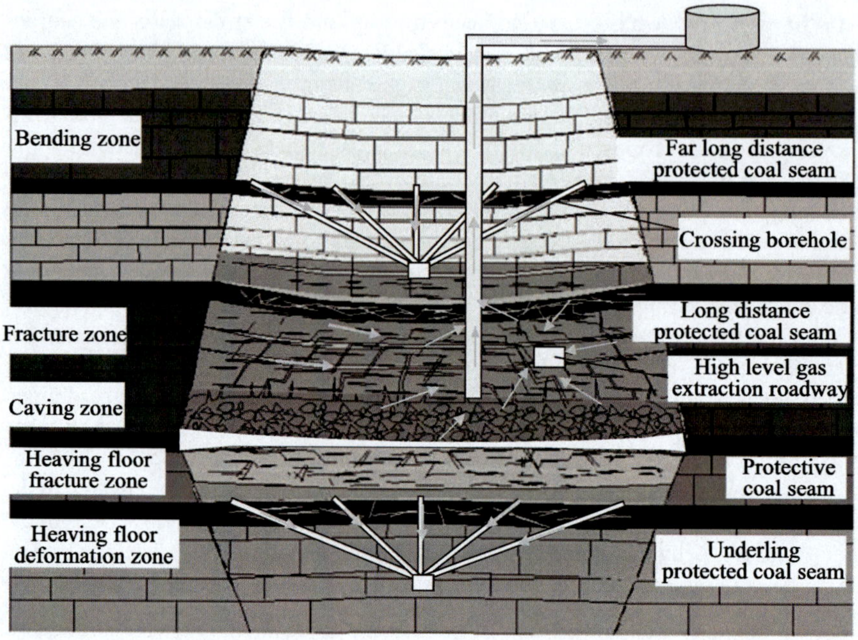

Fig. 10.1 The location of the fracture zone and the gas extraction method

zones could be formed along the strike direction, i.e., the original stress zone, the stress concentration zone, the stress unloading zone and the stress recovering zone, as shown in Fig. 10.2. The stress and gas migration characteristics in each zone are illustrated as follows:

1) Original stress zone

In the original stress zone, the in-situ stress is not affected by the mining of the protective coal seam, as a result the vertical stress is directly proportional to the buried depth. Generally speaking, the original stress zone is usually located at 50–100 m in front of the working face. Moreover, the gas parameters also doesn't change, and thus the gas extraction is rather difficult and the gas extraction volume decreases rapidly in negative exponential form.

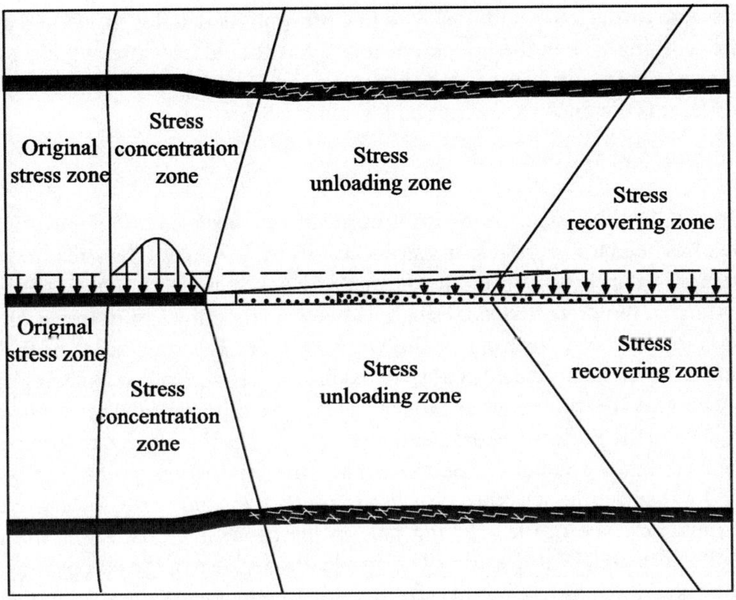

Fig. 10.2 Division of four zones in the rock mass along the strike direction of the working face

2) Stress concentration zone

The stress concentration zone is formed by the mining induced stress transfer during the mining of the protective coal seam, and is usually located at 50 m in front of the working face to 20 m behind the working face. The length of the stress concentration zone is usually affected by the buried depth of the protective coal seam, the length of the working face, the mining height, the dip angle of the coal seam, and the layer spacing between the protective coal seam and the protected coal seam. The maximum concentration stress usually occur at 2–30 m in front of the working face, especially in the 0–10 m in front of the working face. In this zone, the protected coal seam suffers greater in-situ stress. According to the field test results at the Panyi coal mine in the Huainan mine area, the compression deformation in the protected coal seam (the C13 coal seam) could reach up to 3.37‰ during the mining process of the protective coal seam (the B11 coal seam). Higher in-situ stress would shrink the fractures, as a result the permeability decreases in this zone and the gas extraction is much more difficult.

The stress concentration coefficient is related to the layer spacing between the protective and the protected coal seams. With the increase of the layer spacing, the stress concentration coefficient decreases while the length of the stress concentration zone increases. In the field application, coal and gas outburst accident may occur in the protected coal seam if the working face is located in the stress concentration zone. Therefore, the working face in the protected coal seam should keep a certain distance behind the working face in the protective coal seam.

According to the coal and gas outburst prevention detailed rules and regulations, the working face in the protective coal seam should leave the working face in the protected coal seam behind at a distance of at least 3 times the layer spacing. Meanwhile, this distance should not be less than 100 m.

3) Stress unloading zone

In the stress unloading zone, the in-situ stress decreases because part of the original in-situ stress has been transferred to the stress concentration zone. The stress unloading phenomenon usually occurs in the protected coal seam at 0–20 m behind the working face in the protective coal seam. However, sometimes, the stress unloading phenomenon also occur in front of the working face. The start point of the stress unloading zone in the protected coal seam is directly related to layer spacing and the lithology between the protective and the protected coal seams. With the advance of the working face in the protective coal seam, the stress unloading degree in the protected coal seam increases gradually. The maximum stress unloading point usually occur at 20–130 m behind the working face in the protective coal seam. Afterward, with the continual advance of the working face in the protective coal seam, the in-situ stress in the protected coal seam recovers gradually because of the compacting of the overlying rock mass, as a result the stress unloading degree decreases. In the stress unloading zone, the in-situ stress is significantly lower than its initial value.

With the decreasing of the in-situ stress in the stress unloading zone, expansion deformation occurs in the protected coal seam, as a result the primary fractures open and the secondary fractures develop, which leading to significant increase in the permeability of the coal seam. Therefore, the gas extraction condition could be notably improved. With the advance of the working face in the protected coal seam, the gas extraction volume of the borehole also increase gradually. According to field experimental results in the Panyi coal mine, the expansion deformation could reach up to 26.33‰, and the permeability increases by 2880 times, and the gas extraction volume of one single borehole could reach up to 1 m^3/min in the protected coal seam, with the mining of the remote lower protective coal seam. After the gas extraction, the gas content in the protected coal seam decreases significantly.

In the protected coal seam, the stress unloading changes over time and space. During the mining of the protective coal seam, the coal mass in the protected coal seam, originally in the stress unloading zone, would become a coal mass in stress recovering zone. In the stress recovering zone, the in-situ stress recovers, as a result the permeability decreases and the gas extraction condition is relatively difficult compared with the stress unloading zone. Therefore, measures much be taken to enhance the gas extraction in the stress unloading zone before the recovery of the in-situ stress. Considering this, the gas extraction boreholes in the protected coal seam must be constructed and put into use before the mining of the protective coal seam.

4) Stress recovering zone

The stress recovering zone, formed by the movement of the overlying rock mass, is usually located in the gob far behind the working face. Although the permeability in the stress recovering zone is lower than that in the stress unloading zone, the permeability in this zone is also significantly greater than the original permeability due to the fact that the in-situ stress is smaller than the original stress. After the efficient gas extraction in the stress unloading zone, the gas extraction volume decreases significantly in the stress recovering zone.

Under the mining of the protective coal seam, stress unloading and the expansion deformation occur, as a result fractures widely develop and the permeability could increase by 3–4 orders of magnitude. According to the field application results in the Panyi coal mine, the permeability in the protected coal seam (C13 coal seam) increases by 2880 times during the mining of the remote lower protective coal seam. The permeability in the protected coal seam (2_1 coal seam) also increases by 403 times during the mining of the extremely thin lower protective coal seam in the Cuimiao coal mine in the Zhengzhou mine area. The fracture development characteristics, expansion deformation, and permeability increasing magnitude at different zones in the overlying and underlying strata of the protective coal seam are shown in Table 10.1.

The in-situ permeability evolution in the lower protected coal seam is shown is shown in Fig. 10.3 during the mining of the upper protective coal seam. The layer spacing between the protective and the protected coal seams is 70–80 m, and the dip angle of the coal seam is 60–65°. The permeability evolution regulation is as follows: original permeability → decrease slightly → increase significantly → stable. From Fig. 10.3, we can see that the permeability evolution characteristics are consistent with those of the stress, which proves that the permeability evolution is related to stress and vertical expansion deformation. In the stress concentration zone, compression deformation occurs in the protected coal seam, as a result the permeability decreases. In the stress unloading zone, expansion deformation occurs in the protected coal seam, which leads to permeability increasing. The original permeability in the lower protected coal seam is 0.0008 mD. When the working face advance the protected coal seam by 3 times the layer spacing, the permeability in the protected coal seam increases up to 0.2825 mD, which is 560 times the original value. Therefore, the stable permeability increasing in the protected coal seam after the mining of the protective coal seam could be used to achieve efficient gas extraction to decrease the gas content and to eliminate the coal and gas outburst risk in the protected coal seam.

The permeability increasing in the protected coal seam could promote the gas desorption and the gas flow, which provide good conditions for the efficient gas extraction in the protected coal seam. The evolution of the borehole gas extraction flow in the protected coal seam with the advance of the working face in the protective coal seam is shown in Fig. 10.4 [28]. In this figure, curve 1 expresses the evolution of the borehole gas extraction flow in the lower #6 coal seam (protected coal seam) during the mining of the #5 coal seam (protective coal seam) at the No. 1 coal mine in the Nantong mine area. Curve 2 expresses the evolution of the borehole gas extraction flow in the lower #4 coal seam (protected coal seam) during the mining of

Table 10.1 The fracture development characteristics and the permeability evolution in the protected coal seam at different belts

Location of the coal seam	Zones	Fracture characteristics	Gas flow characteristics	Influence range	Expansion deformation	Permeability increasing magnitude
Overlying strata	Bending zone	Bedding fractures develop	Gas flow actively along the coal seam	Above the fracture zone, until to the ground surface	> 10‰	> 1000 times
	Fracture zone	Penetrating and bedding fractures both develop	Gas flow actively along the coal seam, and the gas could flow into the working face	The upper limit is 12–22 times the mining height above the gob		
	Caving zone	broken rock masses filled the gob loosely and many large fractures exist	Gas desorbs and flows into the working face	The upper limit is 4–6 times the mining height above the gob		
Underlying strata	Fracture zone	Penetrating and bedding fractures both develop	Gas flow actively along the coal seam, and the gas could flow into the working face	The lower limit is 15–20 m below the gob	> 3‰	> 300 times
	Deformation zone	Bedding fractures develop	Gas flow actively along the coal seam	The lower limit is 50–60 m below the gob		

Fig. 10.3 The permeability evolution in the protected coal seam with the advance of the working face in the protective coal seam

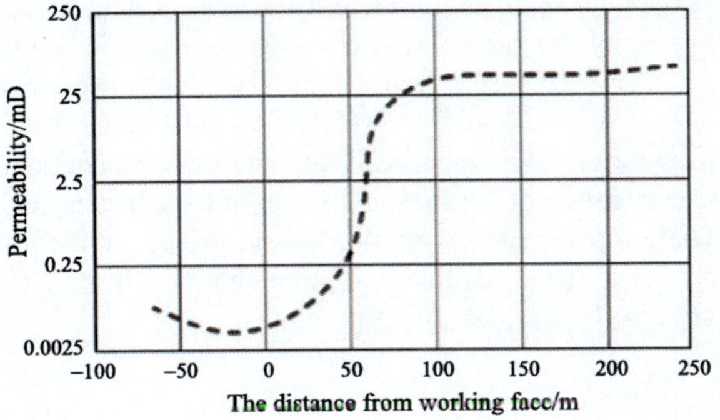

Fig. 10.4 The evolution of borehole gas extraction flows with the advance of the working face

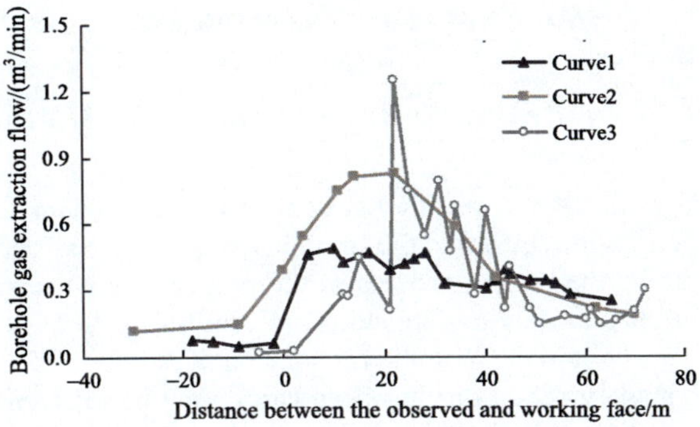

the #3A coal seam (protective coal seam) at the Yutianbao coal mine in the Nantong coal field. Curve 3 expresses the evolution of the borehole gas extraction flow in the lower protected coal seam during the mining of the #1 coal seam (protective coal seam) at the Hongshandian coal mine in the Lianshao mine area. As shown in Fig. 10.4, with the advance of working face in the protective coal seam, the gas extraction flow of the borehole increase significantly, which indicates that the stress unloading and permeability increasing effect could be achieved in the protected coal seam. However, the gas extraction flows of different boreholes are different because the gas geological conditions and the layer spacing between the protected and the protective coal seams are different in different coal mines. The maximum borehole gas extraction flow in the #6 coal seam is approximately 0.49 m^3/min at the No. 1 coal mine in the Nantong mine area, while that in the #4 coal seam could reach up to 0.83 m^3/min at the Yutianbao coal mine. At the Hongshandian coal mine, the maximum borehole gas extraction flow in the lower protected coal seam is approximately 1.25 m^3/min. Moreover, the efficient gas extraction period is longer, as a result enough time could be used to extract the stress relief gas.

After the mining of the protective coal seam, stress unloading and expansion deformation occur in the adjacent coal seam, which leads to the permeability increase. After the pressure gas extraction in the protected coal seam, the gas pressure and the gas content could be significantly decreased, while the coal mass would become hard. In general, the gas pressure could be decreased to below 0.5 MPa and the gas content could be decreased to below 6 m^3/t, while the Protodyakonov coefficient of the coal seam could increase by 48–100%. As a result, the coal and gas outburst risk could be successfully eliminated, and safe and efficient coal mining could be achieved, after the extraction of the stress relief gas in the adjacent coal seam (protected coal seam). The gas parameters evolution characteristics in the protected coal seam during the mining of the protective coal seam obtained in some mine areas of China are shown in Table 10.2 [1, 14, 28, 29].

10.2 Evolution Regulations of the Mining Induced Stress and Expansion Deformation

10.2.1 Evolution Regulations of the Mining Induced Stress

To obtain the stress relief gas extraction mechanism, the temporal evolution regulations of the stress and the strain in the adjacent coal seam should be well understood during the mining process of the protective coal seam. In this process, the elastic mechanic equations should be solved. Since the basic elastic mechanics equations were built in the past, the research on the solution of these equations under different boundary conditions has been carried out by different researchers. However, there are obvious limitations in the existing solving method. When the boundary conditions or the external stress condition is complex, the analytic solutions of the partial differential equations is difficult to be obtained, as a result the numerical simulation method is of great importance in solving these elastic mechanics problems.

Finite difference method is a numerical solution method, which has been used for a long time. In this method, the basic elastic mechanics equations and the boundary conditions, generally the differential equations, are expressed as the difference equations (algebraic equations) for simplification. With the progress of computer technology, this method has been widely used in solving the elastic mechanics problems.

In this chapter, the finite difference software $FLAC^{3D}$ is used to calculate the mining induced stress. In this method, the solution domain is divided into different differential mesh, and the continuous solution domain is replaced by finite difference point. Meanwhile, the parameters required to be solved (such as stress and strain) are stored in the grid point. Due to the fact that the differential term is replaced by the related difference quotient in the partial differential equation, the partial differential equation could be converted into the algebraic difference equation. By solving these difference equations, the numerical solutions of the variable in the grid could be obtained.

The $FLAC^{3D}$ software contains twelve kinds of material constitutive models. Among these models, the Mohr–Coulomb model is widely used in the researches

Table 10.2 Gas parameters evolution characteristics in the protected coal seam during the mining of the protective coal seam obtained in some coal fields of China

Coal mine	Protected coal seam	Mining height of the protective coal seam/m	Layer spacing/m	Expansion deformation/m	Permeability/mD			Gas pressure/MPa		Gas content/(m^2/t)		Pattern of the protective coal seam
					Before mining	After mining	Increasing magnitude	Original	Residual	Original	Residual	
Panyi coal mine in the Huinan coal field	C13	1.9	70	26.33	2.838×10^{-4}	8.172×10^{-1}	2880	4.4	0.5	13	5.0	Lower
Xieyi coal mine in the Huinan coal field	C13	1.1	19	4.0	–	–	–	4.5	0.5	16.5	4.3	Upper
Xieyi coal mine in the Huinan coal field	B11	2.0	70	–	–	–	–	4.5	0.85	14	5.5	Lower
Lier coal mine in the Huinan coal field	B8	1.7	11.4	–	4.075×10^{-3}	4.836×10^{-1}	118	0.8	0.27	5	2.7	Upper
Pansan coal mine in the Huinan coal field	C13	1.9	72	20.74	2.025×10^{-5}	5.250×10^{-2}	2592	4.1	0.51	10	3.8	Lower
Luling coal mine in the Huibei coal field	8	2.14	90	–	–	–	–	3.1	0.43	21	6.02	Lower
Haizi coal mine in the Huibei coal field	9	2.67	84	–	–	–	–	2.6	0.26	18.56	4.86	Lower

(continued)

Table 10.2 (continued)

Coal mine	Protected coal seam	Mining height of the protective coal seam/m	Layer spacing/m	Expansion deformation/m	Permeability/mD			Gas pressure/MPa		Gas content/(m^2/t)		Pattern of the protective coal seam
					Before mining	After mining	Increasing magnitude	Original	Residual	Original	Residual	
Hongling coal mine in the Shenyang coal field	12	1.4	16	7.2	3.500×10^{-4}	3.535×10^{-1}	1010	6–7	0.25	22.5	5.06	Upper
Cuimiao coal mine in the Zhengzhou coal field	= 1	0.5	18.5	1.5	1.175×10^{-3}	4.730×10^{-1}	403	0.75	0.15	13.09	4.66	Lower
No. 3 coal mind Yangquan coal field	3	6.14	136.5	7.84	–	–	–	1.3	–	18.17	5.67	Lower
Moxinpo coal mine in the Tianfu coal field	9	0.7	80	5.8	7.250×10^{-4}	4.750×10^{-1}	586	4.7	0.5	–	–	Upper
South coal mine in the Zhongliangshan coal field	1	0.7	6	6	–	–	–	0.78	0.15	–	–	Lower

related to the underground excavation. As a result, the Mohr–Coulomb model is used in the study of the stress evolution in the protected coal seam during the mining of the protective coal seam.

At the Panyi coal mine in the Huainan coalfield, the B11 coal seam is chosen as the lower protective coal seam to protect the C13 coal seam (the protected coal seam). Based on this engineering background, the stress evolution in the protected coal seam during the mining process of the protective coal seam is studied in this chapter. The strata conditions and the related mechanic parameters are shown in Table 10.3. The thickness of the B11 coal seam is 2 m, while that of the C13 coal seam is 6 m. The layer spacing between these two seams is approximately 67 m.

The boundary conditions in the simulation process are shown in Fig. 10.5. The buried depth of the C13 coal seam is approximately 500 m. According to the in-situ stress conditions in the Panyi coal mine, the upper side of the model is set as stress boundary with a normal stress of 9 MPa. Moreover, the bottom boundary is set as the fixed boundary, and the other sides of the model are given rolling boundaries with an imposed stress of 8 MPa.

Table 10.3 The strata conditions and the related mechanic parameters

Lithology	Density (kg/m^3)	Bulk modulus (GPa)	Shear modulus (GPa)	Internal friction angle (°)	Cohesion (GPa)	Tensile strength (GPa)	Thickness (m)
Medium grained sandstone	2400	5.28	3.80	20.90	4.27	2.00	20.00
Sandy mudstone	2500	5.56	3.18	26.00	2.00	3.50	20.00
C13 coal seam	1400	1.58	0.71	30.00	3.20	1.60	6.00
Siltstone	2500	13.3	7.60	32.00	2.40	1.80	6.00
Siltstone	2800	10.6	4.56	24.40	3.60	2.00	12.00
Medium grained sandstone	2400	5.28	3.80	20.90	4.27	2.00	12.00
Sandy mudstone	2500	13.5	4.40	26.10	2.99	2.40	15.00
Mudstone	2600	4.10	3.10	30.50	6.85	3.20	4.00
Fine sandstone	2800	3.98	4.04	25.60	4.58	2.00	12.00
Sandy mudstone	2800	7.52	4.08	30.50	3.76	2.40	6.00
B11coal seam	1400	1.58	0.71	30.00	3.20	1.60	2.00
Sandy mudstone	2500	5.56	3.18	26.00	3.50	3.50	14.00

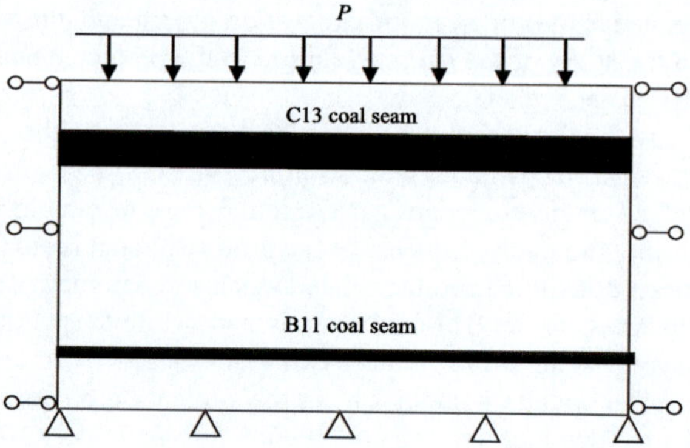

Fig. 10.5 The computational model and boundary conditions

According to the strata conditions and the related mechanic parameters in Table 10.3, the geometric model built in the FLAC^{3D} software is shown in Fig. 10.6. There are 359,856 zones and 377,685 grid points in the model. The length, width and height of the model are 540 m, 490 m and 139 m, respectively.

The length of the mined working face in the protective coal seam is 240 m, extending from 150 to 390 m in the *x*-direction. The width is 190 m, extending from 150 to 340 m in the *y*-direction. The mining height is 2 m. The working face is mined in the *x*-direction; the software iterates 250 steps every 2 m length mined. Moreover, the *x*-stress, *y*-stress and *z*-stress in the C13 coal seam were monitored carefully in the simulation process.

With the working face of the protective coal seam being mined at 80 m, 160 m and 240 m, the stress distributions in the *x*-direction, *y*-direction and *z*-direction of the protected coal seam are shown in Figs. 10.7, 10.8 and 10.9, respectively.

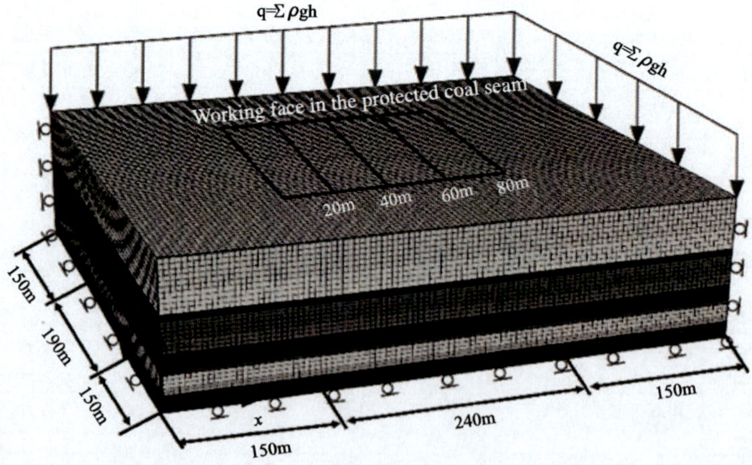

Fig. 10.6 Geometric model

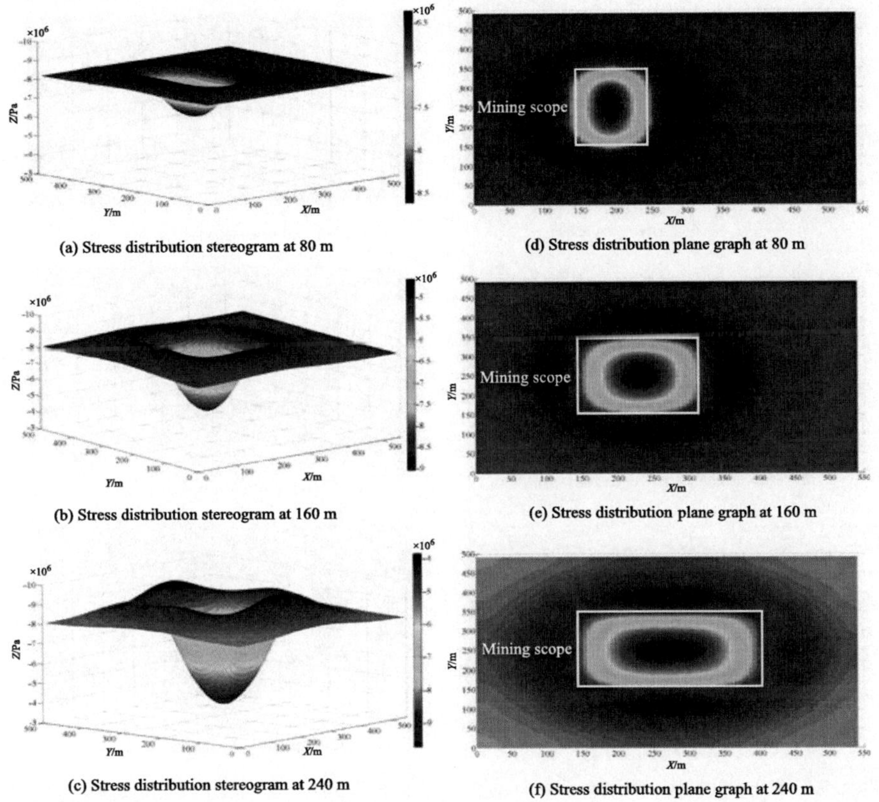

(a) Stress distribution stereogram at 80 m

(d) Stress distribution plane graph at 80 m

(b) Stress distribution stereogram at 160 m

(e) Stress distribution plane graph at 160 m

(c) Stress distribution stereogram at 240 m

(f) Stress distribution plane graph at 240 m

Fig. 10.7 Stress distribution of protected coal seam at the *x*-direction with the working face of the protective coal seam being mined 80, 160 and 240 m

The temporal evolution of three-dimensional stress around the protected coal seam is as follows:

(1) After the mining the workface in the protective coal seam, the stress concentration zone, stress unloading zone, and original stress zone emerged at the protected coal seam;
(2) With the mining of the protective coal seam, the stress unloading zone increased gradually;
(3) During mining of the protective coal seam, and in the direction of mining, the coal mass of the protected coal seam, originally in the stress concentration zone, became a stress unloading zone;
(4) First the coal mass of the protected coal seam would be loaded, and then it would be unloaded gradually, with the three-dimensional stress being reduced simultaneously.

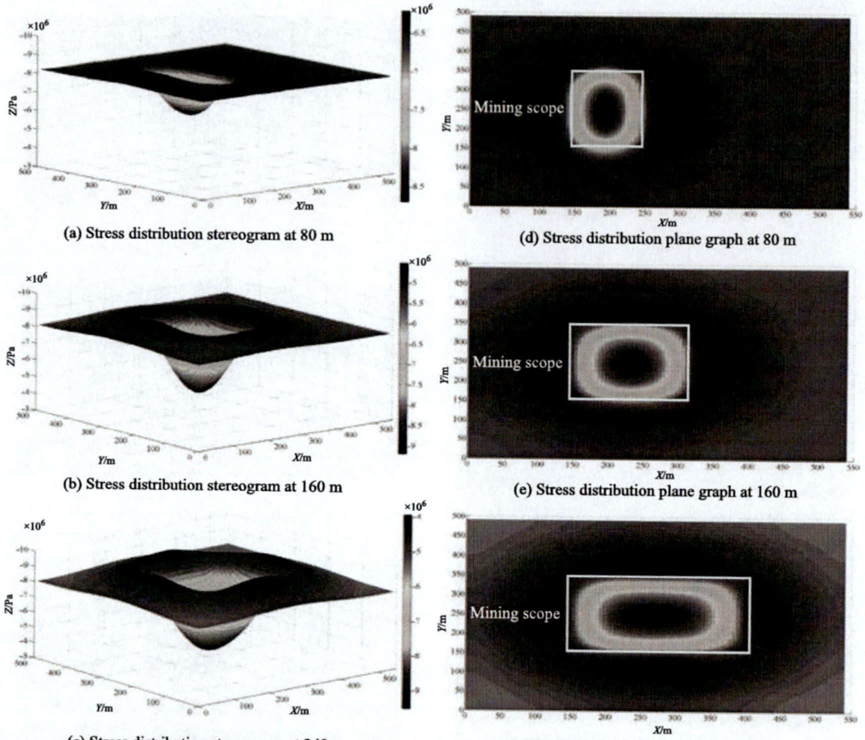

(a) Stress distribution stereogram at 80 m
(d) Stress distribution plane graph at 80 m
(b) Stress distribution stereogram at 160 m
(e) Stress distribution plane graph at 160 m
(c) Stress distribution stereogram at 240 m
(f) Stress distribution plane graph at 240 m

Fig. 10.8 Stress distribution of protected coal seam at the *y*-direction with the working face of the protective coal seam being mined 80, 160 and 240 m

Moreover, the above conclusions could also be seen from the three-dimensional stress monitoring results in the protected coal seam along the *x* direction during the mining of the protective coal seam. The monitoring results are shown in Fig. 10.10.

10.2.2 The Expansion Deformation

The expansion deformation in the protected coal seam could be used to describe the opening characteristics of the fractures. As a result, the expansion deformation is also used to describe stress unloading and the permeability increasing effort in the protected coal seam, which is defined as:

$$\varsigma = \frac{d_1 - d_2}{m} \times 1000 \tag{10.1}$$

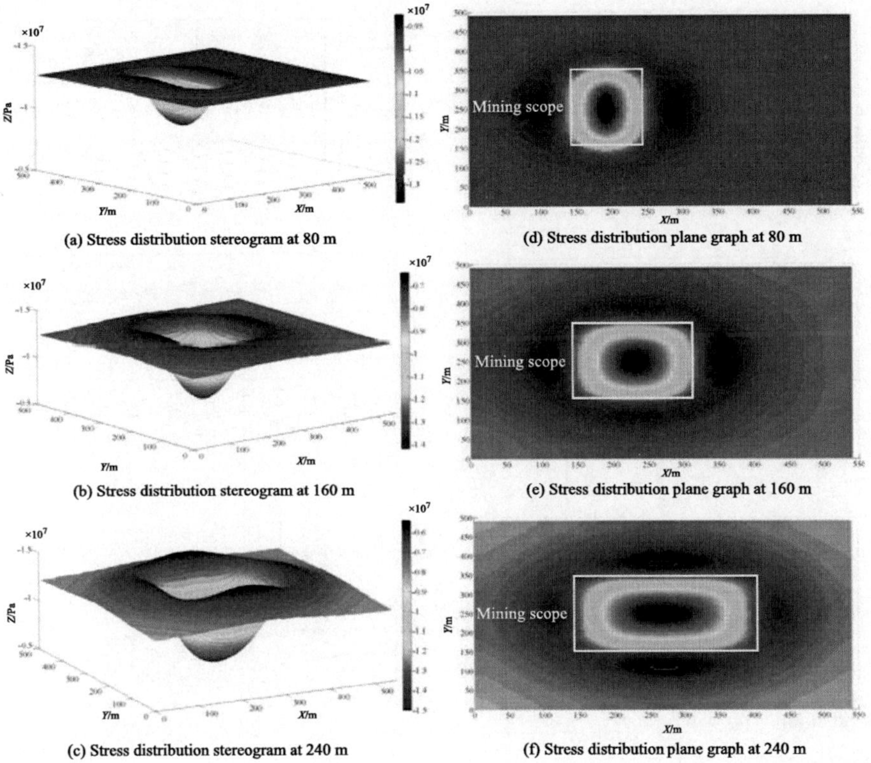

Fig. 10.9 Stress distribution of protected coal seam at the *z*-direction with the working face of the protective coal seam being mined 80, 160 and 240 m

where ς is the expansion deformation, ‰; d_1 is the vertical displacement of the floor; d_2 is the vertical displacement of the roof; and m is the thickness of the protected coal seam.

The greater the expansion deformation in the protected coal seam, the more significantly the fractures open, as a result the stress unloading and the permeability increasing effect is better. Generally speaking, when the expansion deformation is greater than 3 ‰, effective stress unloading and permeability increasing could be achieved in the protected coal seam.

During the mining process of the protective coal seam, the fracture evolution in the protected coal seam is related to the discrete characteristics of the coal and rock masses, as a result the discrete element software should be adopted in the simulation process. Discrete element method (DEM), firstly proposed by Cundall in the middle 70s of the last century, is specially used to solve the discontinuous medium problems. In this method, the discontinuous medium is represented as an assemblage of discrete blocks. The discontinuities are treated as boundary conditions between blocks; large displacements along discontinuities and rotations of blocks are allowed. Therefore,

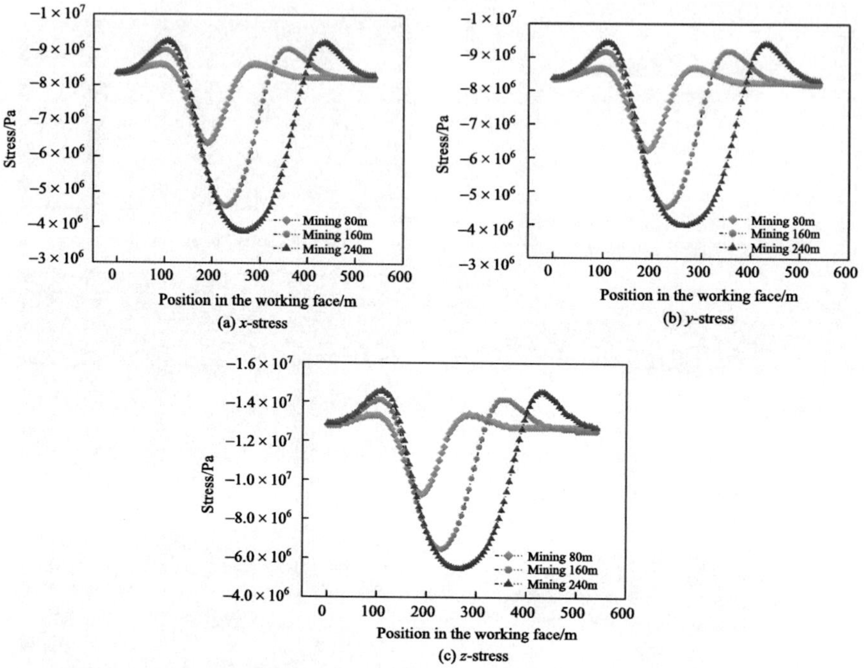

Fig. 10.10 The three-dimensional stress monitoring results in the protected coal seam along the *x* direction

the DEM could be used to well describe the mechanic property of the discontinuous medium. The solving process of the DEM is as follows: ① the solution space is divided into discrete element cell, and the adjacent two discrete element is connected by reasonable connecting element according to the practical problem; ② the relative displacement between different elements is the basic variable, and the normal and tangential forces could be obtained by the relationship between the force and the relative displacement; ③ the sum of forces and the sum of the force moments for each element could be calculated according to its force conditions, and thus the acceleration of each element could be calculated according to the Newton's Second Law of Motion; ④ integrate the acceleration of each element by time, and thus the velocity and displacement of each element could be obtained. Therefore, the physical quantities, such as the velocity, acceleration, angular velocity, line displacement, and so on, could be calculated for each element at any moment. Ever since this method was proposed, it has been widely used in the simulation of the deformation and failure of the discontinuous rocks. Therefore, the 3DEC discrete element software developed by the ITASCA Company is used to evaluate the fracture evolution and the expansion deformation in the protected coal seam.

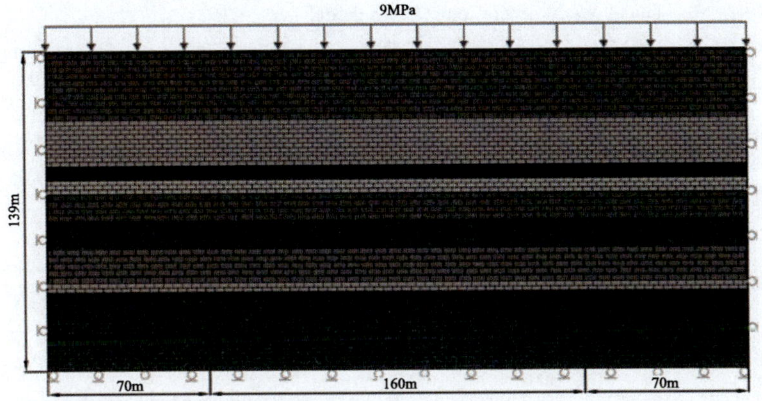

Fig. 10.11 The geometric model and boundary conditions

3DEC software has been widely used in the rock mechanic fields, such as mining engineering, geotechnical engineering, geological Engineering, architectural engineering, and so on. This software could be used to simulate the discontinuous properties of the coal and rock masses with many types of material constitutive models, for example the elastic, anisotropic, Mohr–Coulomb, Drucker-Prayer, bilinear plastic, strain softening and the creep models.

For simplification, a two-dimensional model is built based on the plane strain theory, according to the strata conditions at the Panyi coal mine in the Huainan mine area. The length of the model is set as 300 m, while the height of the model is set as 139 m, as shown in Fig. 10.11. The upper side of the model is set as stress boundary with a normal stress of 9 MPa. Moreover, the other boundaries are set as the roller boundary. The horizontal stress is set as 8 MPa.

The Mohr-Coulomb model for elements and the Coulomb slip model for joints are used are used in the simulation. The mechanical parameters of the main rock stratum and coal seams employed in the model are shown in Tables 10.3 and 10.4.

The length of the mined working face of the protective coal seam is 160 m, extending from 70 to 230 m in the *x*-direction. The mining height is 2 m. During the simulation process, the fracture evolution in the overlying rock stratum could be observed, and the vertical displacements of the roof and the floor of the protected coal seam are recorded. The overburden movement and the fracture evolution in the overlying rock masses are shown in Fig. 10.12.

With the working face of the protective coal seam being mined at 20 m, the rock masses in the overlying rock stratum don't collapse because the advance distance is shorter than the first roof weighting interval. However, plenty of fractures and sequential bed separation could be observed in the overlying rock masses, as shown in Fig. 10.12a. Moreover, with the advance of the working face, the rock masses gradually collapse and the mining induced fractures further develop, as shown in Fig. 10.12b and c. However, as the advance distance increases furthermore, the new generated fractures in the middle of the protected coal seam closes again with the

Table 10.4 Mechanical parameters of joints in the model

Lithology	Normal stiffness (GPa)	Tangential stiffness (GPa)	Cohesion (MPa)	Internal friction angle (°)	Tensile strength (GPa)
Medium grained sandstone	20.7	20.5	0.72	13.8	0.86
Sandy mudstone	32.7	35.0	0.40	10.5	0.52
C13 coal seam	8.42	8.56	0.26	12.2	0.31
Siltstone	78.1	80.2	0.56	15.4	0.62
Siltstone	27.8	30.6	0.80	12.0	0.92
Medium grained sandstone	14.8	15.6	0.72	13.8	0.86
Sandy mudstone	32.3	33.4	0.64	10.0	0.72
Mudstone	16.5	18.4	0.53	6.00	0.64
Fine sandstone	23.4	25.5	0.92	12.0	1.06
Sandy mudstone	43.2	45.8	0.59	14.2	0.68
B11 coal seam	8.42	8.56	0.26	12.2	0.31
Sandy mudstone	24.5	26.8	0.89	10.0	0.91

recovery of the in-situ stress, while the fractures at the both sides of the protected coal seam keeps opening, as shown in Fig. 10.12d.

During the mining process of the protective coal seam, the vertical displacements in the roof and floor of the protected coal seam are monitored, as shown in Fig. 10.13.

Substituting the vertical displacement monitoring results in Fig. 10.13 into Eq. 10.1, the expansion deformation of the protected coal seam could be obtained, as shown in Fig. 10.14.

As we can see from Fig. 10.14: ① with the advance of the working face in the protective coal seam from 0 to 80 m, the overlying rock masses of the protected coal seam don't collapse, as a result the expansion deformation in the protected coal seam increases gradually. With the working face of the protective coal seam being mined at 80 m, the expansion deformation could reach up to approximately 40‰; ② with the further increase of the advance distance, the expansion deformation decreases gradually with the compaction of the caved overlying rock masses of the protected coal seam. The expansion deformation curve present "M" shape. What's worse, the stress unloading zone gradually transformed into the stress concentration

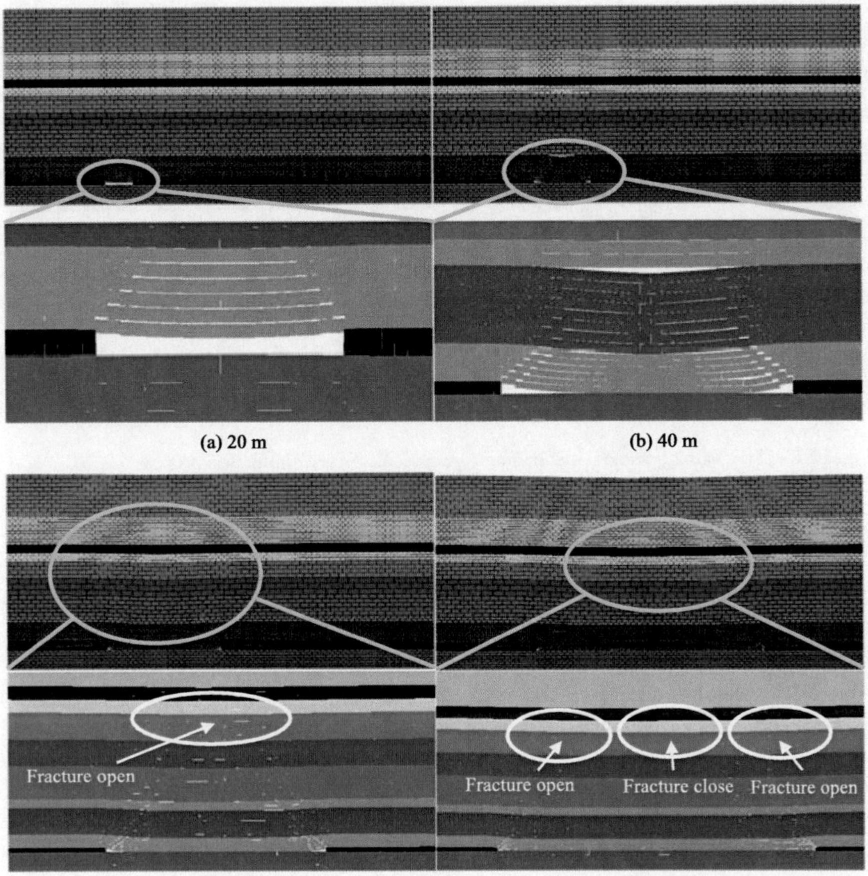

(a) 20 m (b) 40 m (c) 80 m (d) 160 m

Fig. 10.12 The overburden movement and the fracture evolution in the overlying rock stratum under different advance distance

zone, as a result the expansion deformation of the coal mass even decreases below 0; ③ after the first breakdown of the overlying rock masses of the protected coal seam, the rock masses breakdown periodically. Due to the fact that the periodic roof weighting interval is obvious smaller than the first roof weighting interval, the expansion deformation is much smaller in this stage, i.e., the maximum expansion deformation is just approximately 15‰.

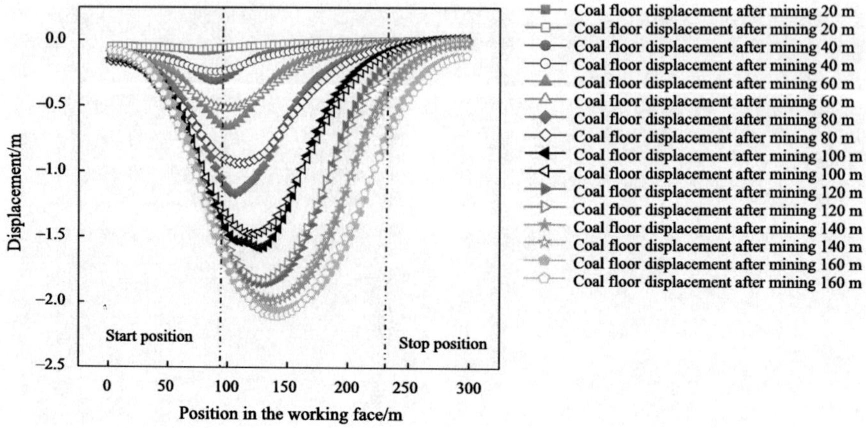

Fig. 10.13 The vertical displacements monitoring results in the simulation process

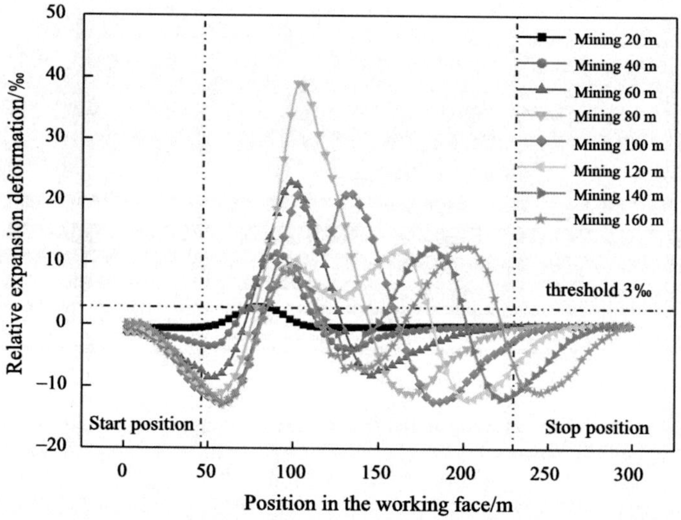

Fig. 10.14 The expansion deformation in the protected coal seam during the mining process of the protective coal seam

10.3 Temporal Evolution Regulation of the Permeability

10.3.1 The Basic Method for Forecasting the Permeability Evolution

The permeability of the coal seam is used as the most important parameter to evaluate gas flow ability and extraction production. As a result, a reasonable forecast of the permeability evolution in the protected coal seam after the mining of the protective coal seam is analyzed in detail in this chapter to evaluate the stress unloading effort and to guide the borehole design.

As we can see from the permeability evolution model considering the mining induced stress disturbance (Eq. 7.142), the permeability evolution after the mining of the coal seam is controlled by the change of the in-situ stress and the plastic failure of the coal mass. If we just consider the effect of stress change on permeability, the permeability in the pre-peak stage may equal to that in the post-peak stage because the stress value in the pre-peak stage may equal to that in the post-peak stage, which is certainly unsuitable. Therefore, the equivalent plastic deformation is used in the permeability model to determine the deformation state of the coal mass (the elastic stage or the post-peak stage). Moreover, the equivalent plastic deformation could also be used to describe the plastic failure degree of the coal mass. The greater the equivalent plastic deformation, the more severe the plastic failure. As a result, the physical essence of the equivalent plastic deformation could be considered as a damage factor to evaluate the plastic failure of the coal mass.

According to the above analysis, to obtain the permeability evolution in the protected coal seam after the mining of the protective coal seam, the evolution of the three-dimensional stress and the equivalent plastic deformation should be evaluated. In Sect. 10.2, the three-dimensional stress evolution has been obtained by the FLAC^{3D} software. Unfortunately, limited by the own characteristics of this software, the evolution of the equivalent plastic strain could not be obtained. However, using the COMSOL software, the evolution of the three-dimensional stress and the equivalent plastic strain could both be obtained [30]. For example, this software has been widely used in the permeability evolution in the coal mass before the heading face during the excavation of the coal roadway. Unfortunately, this software could not be in the numerical simulation of the large-scale multi-step excavation. As a result, this software could not be used in the numerical simulation of the mining of protective coal seam. In this chapter, a further simplification has been made on the permeability model proposed in Chap. 7. Thus, the permeability evolution in the protected coal seam could be obtained according to the three-dimensional stress simulation results obtained by the FLAC^{3D} software. In the subsequent revisions of this book, we will do more work to improve this content.

Based on the current technology condition and the three-dimensional stress evolution in Sect. 10.2, the temporal evolution of the permeability in the protected coal seam could be obtained based on the following two simplifications:

(1) Limited by the condition that the equivalent plastic strain could not be obtained in this book, the stress redistribution characteristics should be used to determine the deformation state of the coal mass. In this way, the error that the permeability in the post-peak stage may be equal to that in the pre-peak stage could be successfully avoided. As we can see from Sect. 10.2, the coal mass in the protected coal seam suffers loading and unloading during the mining process of the protective coal seam. In the unloading process, the volumetric stress decreases, as a result the deformation state of the coal mass could be determined by comparing the volumetric stress in the unloading process to its initial value. Therefore, the permeability equation in Chap. 7 could be determined (the permeability equation in Chap. 7 is a piecewise function).

When $\Theta > \Theta_0$, the coal mass is in the loading stage, as a result,

$$k = k_0 \exp[b_\sigma(\Delta\Theta)]. \tag{10.2}$$

When $\Theta < \Theta_0$, the coal mass is in the unloading stage. In this stage, the permeability equation should be further simplified.

(2) According to the above analysis, the equivalent plastic strain could not only be used to determine the deformation condition of the coal mass, but also be used to describe its damage degree, i.e., the equivalent plastic strain could be considered as the damage factor. Due to the fact that the equivalent plastic strain could not be obtained by the FLAC^{3D} software, the new damage factor should be defined. In the first simplification, the method to determine the deformation condition of the coal mass has been proposed. As a result, when Θ is smaller than Θ_0, the damage factor could be defined as follows:

$$k = \left(1 + \frac{\Theta_0 - \Theta}{\Theta}\xi\right) \cdot k_0 \tag{10.3}$$

Based on the above two simplifications, the temporal evolution regulation of the permeability in the protected coal seam could be reasonably forecasted based on the three-dimensional stress simulation results using the FLAC^{3D} software.

10.3.2 *Temporal Evolution Regulation of the Volumetric Stress and the Permeability*

With the working face in the protective coal seam being mined at 80 m, 160 m and 240 m, the evolution of the volumetric stress and the permeability in the protected coal seam are shown in Figs. 10.15, 10.16 and 10.17, respectively. In the calculation of the permeability, the logic function in the Matlab software is used to judge the deformation state of the coal mass. According to the volumetric stress value and the deformation state of the coal mass, the permeability evolution could be calculated according to Eqs. (10.2) and (10.3). In this process, the value of b_σ, ξ and k_0 are set as 0.05 MPa^{-1}, 2000 and 2.8375 mD, respectively.

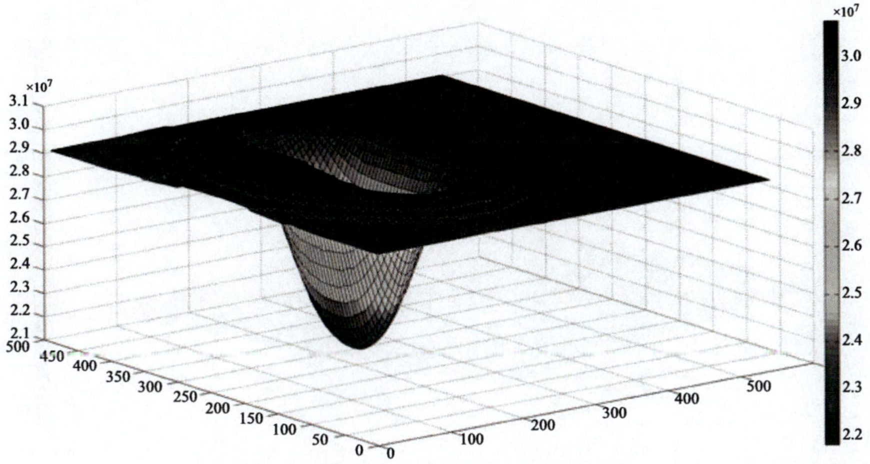

(a) The volumetric stress stereogram in the protected coal seam with protective coal seam mined at 80 m.

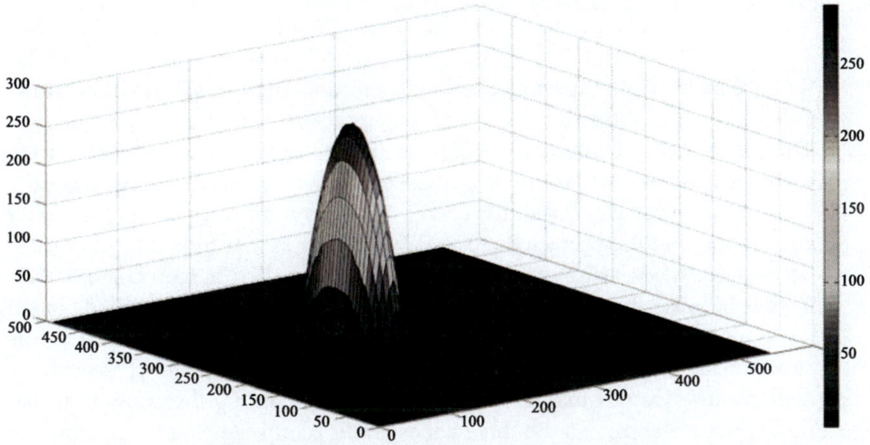

(b) The permeability ratio stereogram in the protected coal seam with the protective coal seam being mined at 80 m.

Fig. 10.15 The volumetric stress and permeability ratio stereograms in the protected coal seam with the protective coal seam being mined at 80 m

According to Figs. 10.15, 10.16 and 10.17, the temporal evolution regulations of the volumetric stress and the permeability could be concluded as follows:

(1) After the mining of the protective coal seam, the stress concentration zone, stress unloading zone, and the original stress zone emerges at the protected coal seam. Moreover, with the advance of the working face, the stress unloading zone increases gradually, and the stress concentration factor also increases. In

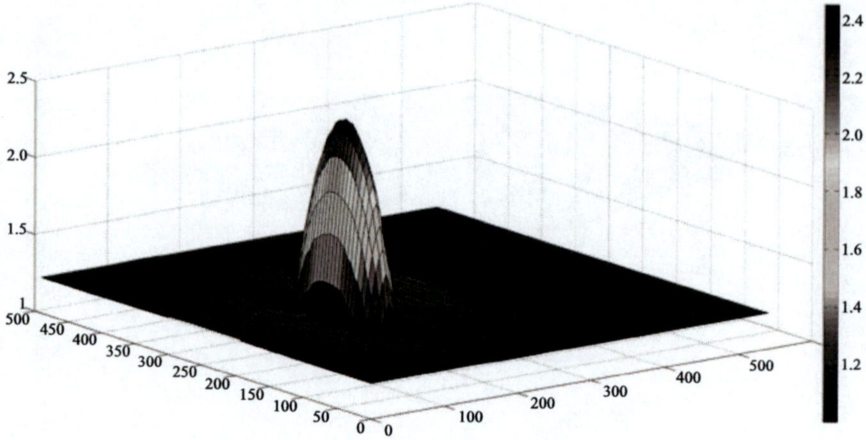

(c) The detailed permeability ratio stereogram in the protected coal seam with the protective coal seam being mined at 80 m.

Fig. 10.15 (continued)

the direction of mining, the stress concentration zone transforms into stress unloading zone during the mining of the protective coal seam.

(2) The permeability evolution presents an opposed regulation with that of the volumetric stress. With the mining of the protective coal seam, the permeability increasing zone, the permeability decreasing zone and the original permeability zone emerge at the protected coal seam. It is important to point out that the permeability decreasing zone is not obvious in Figs. 10.15b, 10.16b and 10.17b. This is because the permeability increases by 3–4 orders of magnitude in the stress unloading zone, while the permeability decreasing magnitude in the stress concentration zone is very small. To show the permeability decreasing zone clearly, we can magnify the details of permeability distribution. Under these circumstances, all the three zones could been seen clearly, as shown in Figs. 10.15c, 10.16c and 10.17c. With the advance of the working face in the protective coal seam, the permeability increases gradually in the permeability increasing zone. Furthermore, in the direction of mining, the permeability decreasing zone transforms into permeability increasing zone.

10.4 Stress Relief Gas Extraction and the Gas Flow Characteristics

At present, countries around the world all believe that the stress relief gas extraction is the most effective and economical method to reduce the coal mine gas disasters. According to the mine strata movement theory and the seepage theory, the strata in

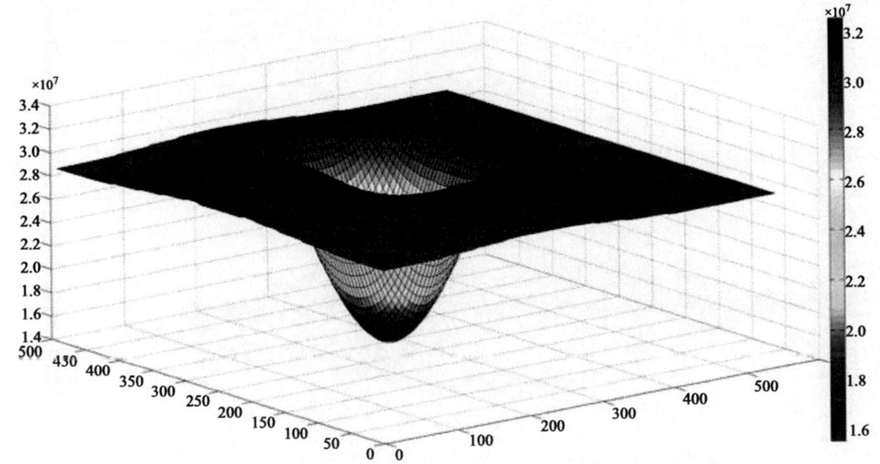

(a) The volumetric stress distribution stereogram in the protected coal seam with protective coal seam mined at 160 m

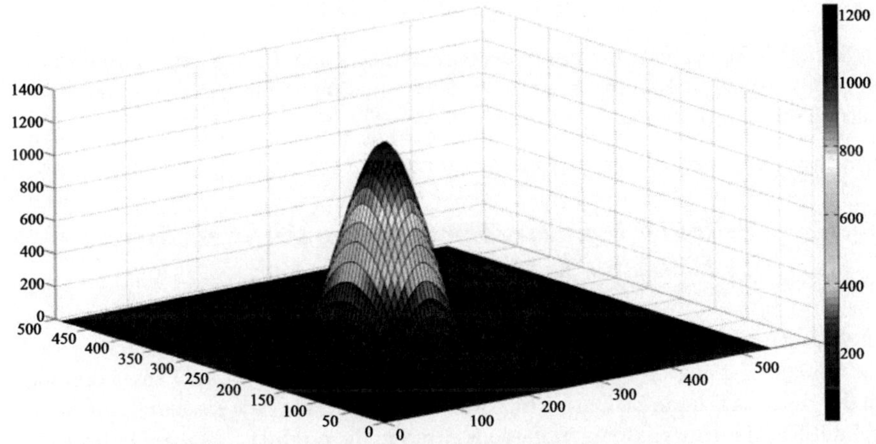

(b) The permeability ratio stereogram in the protected coal seam with the protective coal seam being mined at 160 m

Fig. 10.16 The volumetric stress and permeability ratio stereograms in the protected coal seam with the protective coal seam being mined at 160 m

the roof and floor deforms, collapses and moves during the mining of the protective coal seam, as a result the fractures open and new fractures generate, which could result in the permeability increasing in the protected coal seam. According to the previous research, after the mining of the protective coal seam, the permeability could increase by 3–4 orders of magnitude in the protected coal seam. The gas extraction simulation in the original low-permeability coal seam has been conducted

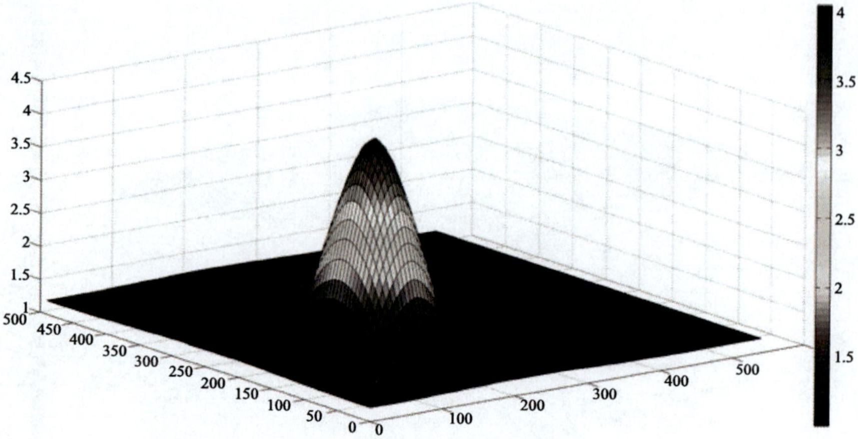

(c) The detailed permeability ratio stereogram in the protected coal seam with the protective coal seam being mined at 160 m

Fig. 10.16 (continued)

in Sect. 9.6. To evaluate the gas flow characteristics and the gas extraction effort in the stress relief coal seams, the numerical simulation of the gas extraction is also carried out in this chapter.

10.4.1 The Initial Boundary Conditions in the Stress Relief Gas Extraction Model

In this chapter, the pressure relief gas extraction simulation is conducted based on the background engineering conditions of the C13 coal seam at the Panyi coal mine in the Huainan mine area. During the simulation process, a new gas extraction model is built based on the gas extraction model used in the original coal seam in Sect. 9.6, and the geological model and the boundary conditions are the same as the previous, as shown in Fig. 9.21. Compared with the previous model, the changes in the new gas extraction model is the values of the initial gas pressure, the initial permeability and the initial porosity. During the mining process of the protective coal seam, the values of these three parameters change significantly, as a result the gas extraction would be rather difficult with that in the original coal seam. Furthermore, the related parameters used in the simulation process are list in Table 9.3.

According to the simulation results in Fig. 10.14 by the 3DEC software in Sect. 10.2, the average expansion deformation in the C13 coal seam could reach up to approximately 2.633%. Assuming that the coal matrix in the protected coal seam does not change during the mining of the protective coal seam, as a result the expansion deformation in the protected coal seam is completely caused by the

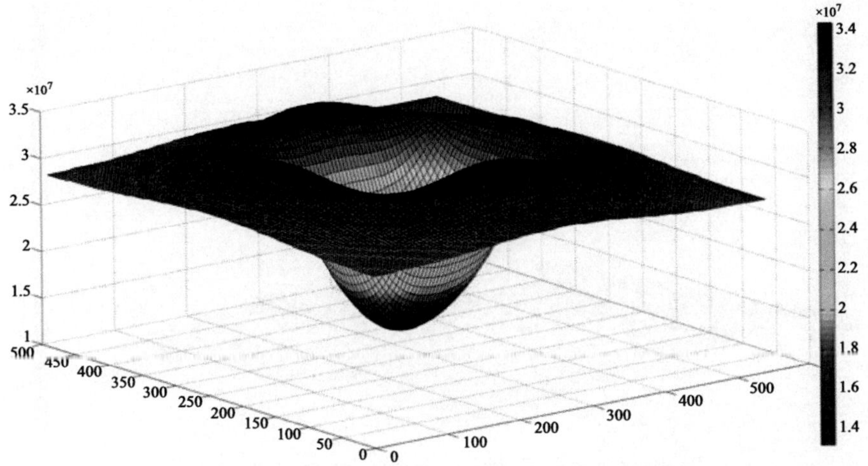

(a) The volumetric stress distribution stereogram in the protected coal seam with protective coal seam mined at 240 m

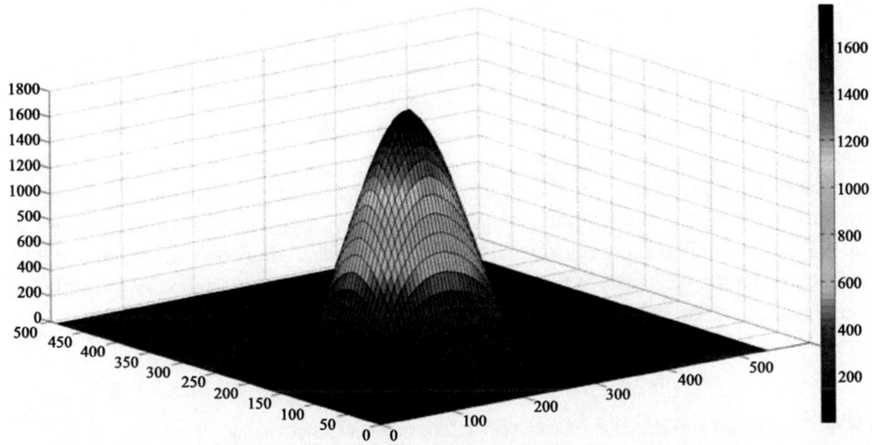

(b) The permeability ratio stereogram in the protected coal seam with the protective coal seam being mined at 240 m

Fig. 10.17 The volumetric stress and permeability ratio stereograms in the protected coal seam with the protective coal seam being mined at 240 m

opening of the fractures. Based on the above assumption, the porosity could reach up to 0.03833 in the protected coal seam after the mining of the protective coal seam. According to the field test results, the permeability in the C13 coal seam increases by approximately 2880 times after the mining of the protective coal seam, indicating that the stress unloading effort is rather obvious.

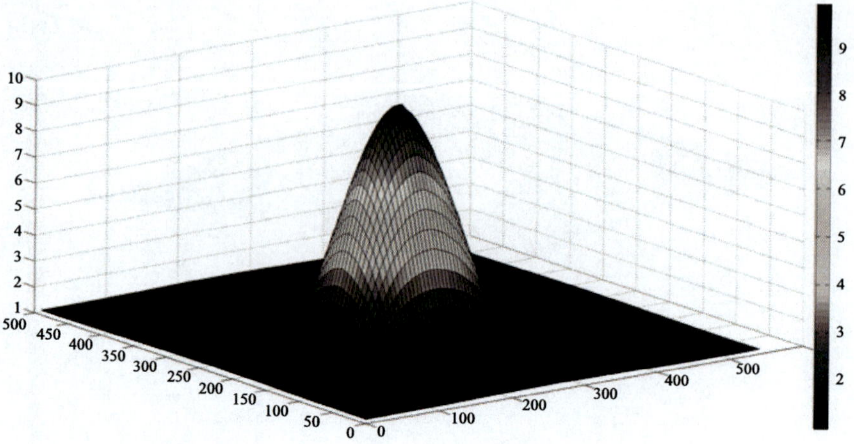

(c) The detailed permeability ratio stereogram in the protected coal seam with the protective coal seam being mined at 240 m

Fig. 10.17 (continued)

During the mining process of the B11 coal seam, the stress relief gas in the C13 coal seam could not be able to flow into the B11 coal seam under the sealing effect of the intermedia rock mass. As a result, the gas content in the C13 coal seam does not change. However, with the increase of the porosity in the C13 coal seam, the adsorbed gas desorbs and flows into the fractures, which could result in the decrease of the gas pressure. According to the relationship between the gas content and gas pressure in Eq. 4.13, the residual gas pressure in the C13 coal seam after the mining of the B11 coal seam is approximately 3.63 MPa when substitutes the new porosity of 0.03833, which decreases by 0.77 MPa when compared with the original gas pressure.

10.4.2 Stress Relief Gas Extraction Analysis

The gas pressure in fractures cloud charts under different gas extraction times (1, 10, 50, 100, 200 and 365 d) are shown in Fig. 10.18. Different color refers to different gas pressure. The minimum gas pressure is 25 kPa, while the maximum gas pressure is 3.63 MPa. As we can see from this figure, the gas pressure decreases significantly even after 1 day's gas extraction. The longer the gas extraction time, the larger the influence area. After 365 days' gas extraction, the gas pressure decreases significantly in the whole geometric model.

To quantify the extraction effect on the original and unloaded coal seams, a monitoring line *AB* was set to monitor gas pressure. The coordinates of the endpoints are as follows: *A* (50 m, 25.65 m) and *B* (100 m, 25.65 m). Under different gas extraction

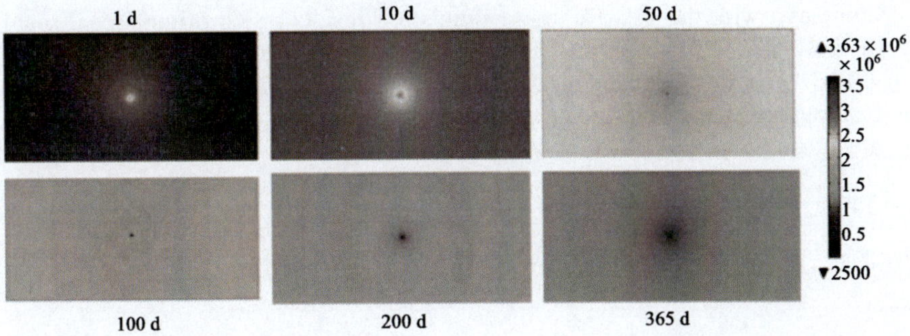

Fig. 10.18 The gas pressure cloud charts

times (1, 10, 50, 100, 200 and 365 d), the gas pressure in fractures monitoring results are shown in Fig. 10.19.

(1) With the increase of gas extraction time, the influence area increases gradually. Under the gas extraction time of 1 d, 10 d, 50 d, 100 d, 200 d and 365 d, the maximum gas pressures levels in the unloaded coal seam are 3.63 MPa, 3.15 MPa, 2.25 MPa, 1.8 MPa, 1.48 MPa and 1.2 MPa, respectively.

(2) With the increase of the gas extraction time, the decrease of gas pressure in the unloaded coal seam is increasingly dramatic. After 365 days' gas extraction, the gas pressure in unloaded coal seam decreases to approximately 1.2 MPa.

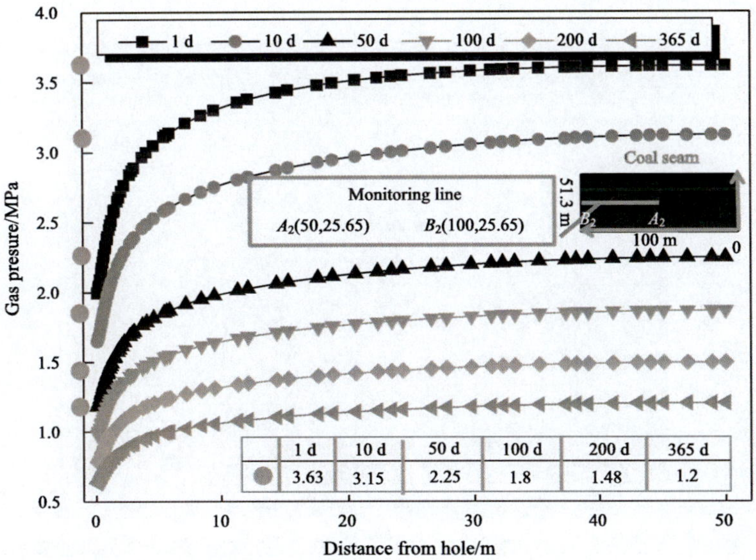

Fig. 10.19 The gas pressure monitoring results

Compared with the gas extraction simulation results in the original coal seam (Figs. 9.20 and 9.21), the gas pressure decreases much more dramatically in Figs. 10.18 and 10.19, indicating that the gas extraction efficiency of the boreholes in the low permeability coal seam could be significantly improved by unloading its in-situ stress.

10.4.3 Daily Gas Extraction Volume in Unloaded Coal Seam

After the stress unloading in the C13 coal seam in the Panyi coal mine of the Huainan mine area, the in-situ tested gas extraction volume in the first drilling field is shown in Fig. 10.20. The changing curve can be divided into three stages, namely, the growth, active and the attenuation stages. The first 40 days are in the growth stage. In this stage, the gas extraction volume increases with the gas extraction time. The 40–95th days are in the active stage. In this stage, the stress unloading effect is much better and the gas excavation volume reaches up to the maximum value. After 95 days, the gas extraction volume decreases with gas extraction time, indicating that this period is in the attenuation stage.

According to the gas extraction simulation in the unloaded coal seam, the gas pressure evolution is obtained, as a result the daily gas extraction volume could also be calculated, as shown in Fig. 10.21. We can see that daily maximum gas excavation volume in the unloaded coal seam reached 337.2 m^3/d which eventually stabilized to approximately 24 m^3/d. At the initial gas extraction stage, the gas extraction volume

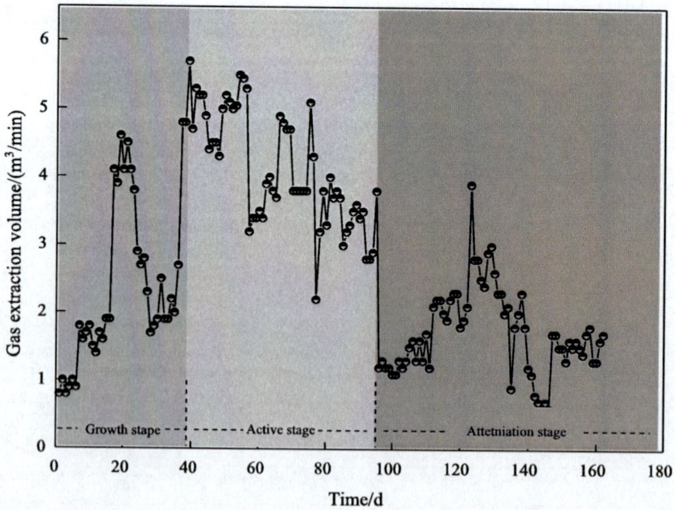

Fig. 10.20 The in-situ tested gas extraction volume in the first drilling field

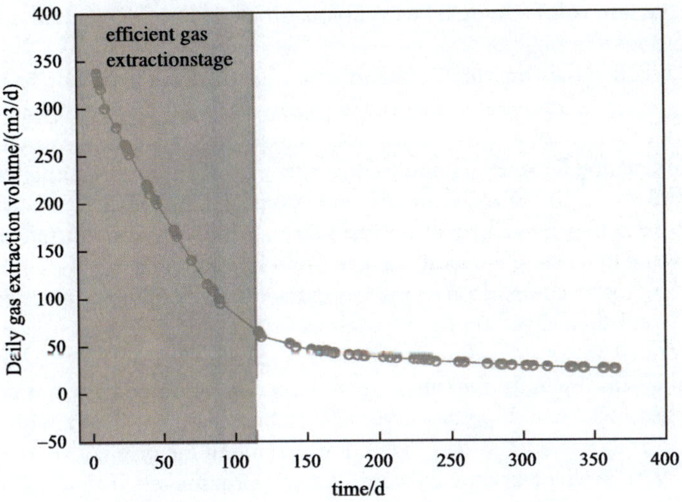

Fig. 10.21 The calculated gas extraction volume according to the simulation results

is rather huge although it decreases significantly with gas extraction time, making this stage the efficient gas extraction stage. This stage lasts approximately 120–125 d.

According to the gas extraction volume obtained in the simulation method as shown in Fig. 10.21, the gas extraction growth stage does not occur compared with the field gas extraction data (Fig. 10.20). Since most of the gas around the borehole is lost because the original equilibrium state of the coal seam around the borehole is disturbed in process of drilling the borehole. After drilling the borehole, the front end of the borehole must be sealed, and then the negative suction pressure is applied to the borehole. In the beginning phase of extraction, the gas extraction volume is very small as the most gas around the borehole has been lost during drilling, yet the extraction amount will increase as the gas of distal coal mass gradually flows into the coal mass around the borehole. Unlike the field test results, the numerical simulation results were obtained under an ideal state; the effect of borehole drilling on gas extraction is ignored. Therefore, the gas extraction volume is maximum at the beginning stage of extraction, and the early growth stage of the extraction curve is a default in numerical simulation method.

The field gas extraction data was based on a whole drilling field, which includes four boreholes that are evenly distributed. Moreover, the unit of the gas extraction volume is m^3/min, and the thickness of the coal seam is approximately 6 m. However, the simulated gas extraction data are based on a 1 m thick coal seam and one extraction borehole. The basic units of simulated and field data are different. Drawing comparisons between them requires determining the field gas extraction value of one borehole and converting the m^3/min unit into a m^3/d unit. The interaction between different boreholes should be disregarded during transformation, and the gas extraction volume should be evenly distributed between boreholes. The final results are shown in Fig. 10.22.

Figure 10.20 shows that the field extraction data are fluctuant due to relatively complex geological conditions and other factors such as the heterogeneity in the mining induced damage of the coal and rock masses, the changes in extraction pressure, the interaction between boreholes, the interferer of the mine ventilation system and artificial statistical errors. However, the curve of the simulated data is smooth, due to the numerical modeling of an ideal state. Although some differences were found between the data, the overall change trend is consistent.

According to the comparison of the gas extraction data in the original coal seam in Chap. 9 and that in the unloaded coal seam (Fig. 10.23), we can see that the mining induced stress unloading could have a great effect on the gas extraction. From Fig. 10.23, the daily maximum gas extraction volume in the unloaded coal seam reaches 337.2 m^3/d which eventually stabilizes to approximately 24 m^3/d. However, the daily maximum gas extraction volume in the original coal seam only reaches 0.82 m^3/d which eventually stabilizes to approximately 0.48 m^3/d. The daily gas extraction volume in the unloaded coal seam is far greater than that in the original coal seam. Moreover, the daily gas extraction volume declines rapidly with extraction time. According to the change rate of gas drainage amount per day, the high-efficiency extraction stage of the extraction curve is obtained. As can be seen from Fig. 10.23, the change rate is relatively steep in the efficient gas extraction stage, while it is relatively smooth outside the efficient gas extraction stage. Efficient gas extraction stages of the original and unloaded coal seam are approximately 75–80 d and 120–125 d respectively. Because it is difficult to define a specific value as the critical value to

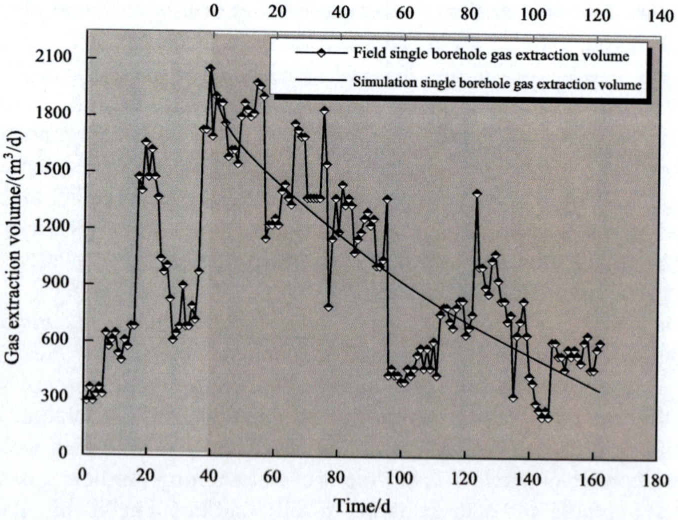

Fig. 10.22 The comparison of the single borehole gas extraction volume between the field test result and the numerical simulation result

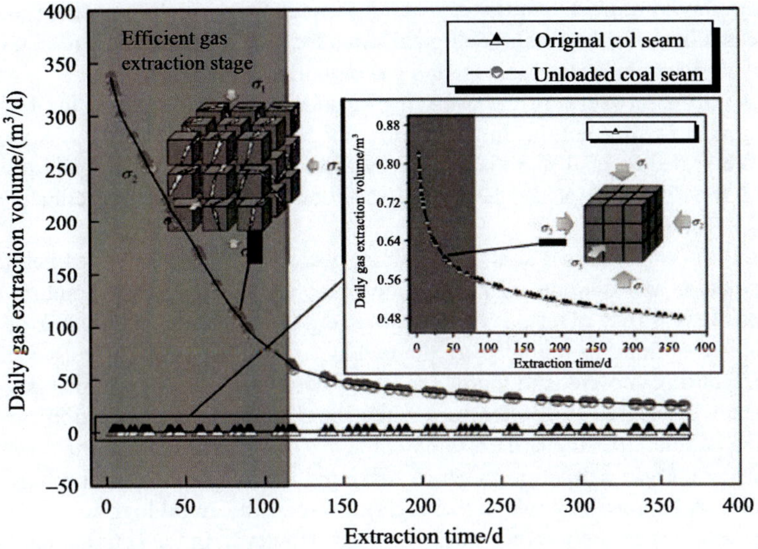

Fig. 10.23 The comparison of the gas extraction volume in the unloaded coal seam and the original coal seam

divide the efficient gas extraction stage, so the vague scopes of division are given respectively. The unloaded coal seam is more conducive to gas extraction, independent of maximum gas drainage volume per day and the high-efficiency extraction time.

10.5 Stress Relief Gas Extraction Method

As pointed out in Sect. 10.1, during the mining of the protective coal seam, the overlying and the underlying coal and rock masses move and deform, leading to the generation of the new fractures. With the increase of the distance to the protective coal seam, the development of the fractures is increasingly weak. After the mining of the protective coal seam, the overlying rock masses could be classified into a caving, a fracture and a bending zones, while the underlying rock masses could be classified into fracture and deformation zones. During the mining process of the lower protective coal seam, the mining condition of the protected coal seam should not be destroyed, as a result, the upper protected coal seam should be located in the fracture or the bending zone in the overlying rock masses. However, in the upper protective coal seam mining technology, the stress unloading and permeability increasing effect is rather weak beyond the fracture and the deformation zones in the underlying rock masses, as a result, the lower protected coal seam should also be located in these two zones. Furthermore, during the mining process of the protective coal seam, more attention should be paid to the layer spacing between the protective and the protected coal seams because the small geological structure may decrease the strength of the rock mass, which may lead to coal and gas outburst.

In the different zones of the overlying or underlying rock masses, the fractures development is different, leading to a different gas flow characteristic. As a result, when the protected coal seam is located in different zones, different gas extraction method should be adopted. These gas extraction methods could be concluded as shown in Table 10.4.

In the lower protective coal seam mining technology, new fractures paralleled, perpendicular and oblique to the bedding plane all develop in the fracture zone, as a result the stress relief gas could flow along the directions paralleled, perpendicular and oblique to the bedding plane under the effect of the negative suction pressure during the gas extraction process. Therefore, when the upper protected coal seam in located in the fracture zone, the following gas extraction methods could be adopted: the upward cross-borehole gas extraction method, the downward cross-borehole gas extraction method, strike high-level borehole, trend high-level borehole, strike long borehole, strike high-level roadway, trend high-level roadway, surface well gas extraction methods, and so on. However, in the bending belt of the overlying rock masses, the rock masses subside as a whole and only the fractures paralleled to the bedding plane develop, as a result the gas could flow easily along the direction paralleled to the bedding plane. Therefore, when the upper protected coal seam in located in the bending zone, the following gas extraction methods could be adopted: the upward net-like cross-borehole, the downward net-like cross-borehole and the surface well gas extraction methods.

In the upper protective coal seam mining technology, the lower protected coal seam should also be located in the fracture zone or the deformation zone in the underlying rock masses. In different zones, the development of new fractures is different. However, because the protected coal seam is located below the protective coal seam, the gas extraction method is seriously limited. As a result, the upward net-like gas extraction method is widely used. In the fracture zone, the fractures perpendicular or oblique to the bending plane widely develop and thus the stress relief gas in the protective coal seam could flow into the gob of the protected coal seam along with these fractures, leading to the gas overrun in the protected coal seam during the mining process of the proactive coal seam. As a result, if the protected coal seam is located in the fracture zone, the borehole spacing should not be too large. Generally, the borehole spacing in the protected coal seam should not be larger than the layer spacing between the protective and the protected coal seams. If the gas concentration is the upper protected coal seam is rather high, the high-level roadway gas extraction method could also be adopted to further enhance the gas extraction. Moreover, if the Y-shape ventilation system is adopted in the working face of the protective coal seam, cross-boreholes could also be constructed from the gob-side roadway to extract the stress relief gas. In addition, the gob burying pipe could also be applied to provide an auxiliary gas drainage.

In the field gas extraction process, different gas extraction methods should be adopted and the gas extraction parameters should be optimized according to the actual geological and mining conditions (Table 10.5).

10.5.1 Technology Synopsis of the Stress Relief Gas Extraction Method

1. Upward net-like cross-borehole gas extraction method

The upward net-like cross-borehole gas extraction method, the basic pressure relief gas extraction method, has been widely used in China for its notable and stable gas extraction effect. In this method, one or more rock roadway is arranged in the floor of the protected coal seam. In the rock roadway, a series of drilling fields are set, and a number of upward cross-boreholes are constructed from each drilling field to extract the pressure relief gas, as shown in Fig. 10.24.

The rock roadway in the floor is always arranged at 15–25 m below the protective coal seam along the strike of the working face. During the excavation of the rock roadway, the rock roadway may uncover the coal seam because of the ups and downs of the coal seam, leading to the occurrence of the coal and gas outburst accident or the gas overrun in the rock roadway. Therefore, the distance to the protective coal seam should be controlled precisely in this process. Moreover, in the lower protective coal seam mining technology, the rock roadway should not be destroyed during the mining of the protective coal seam. In the dip direction, the rock roadway should be arranged at the middle of the working face in the protected coal seam. In addition, the arrangement of the rock roadway should satisfy the condition that all the gas extraction boreholes are upward during the construction of the boreholes. If the width of the working face is long and the dip angle is great, more than one rock roadway could be arranged. Furthermore, the cross section of the rock roadway should be large enough to satisfy the requirement of rig moving and borehole construction.

In the rock roadway, a series of drilling fields, each with a length of 10 m are set at a certain distance. In the stress relief zone, the space of the drilling fields should be equal to the borehole spacing in the strike direction. Moreover, the cross section of the drilling fields should also satisfy the requirement of rig moving and borehole construction. In each drilling field, the boreholes are arranged in the fan pattern, and the borehole radius should be less than 90 mm. In the lower protective coal seam mining technology, the borehole spacing in the protected coal seam is usually set as 30–40 m. The boreholes should be drilled into the roof rock mass by at least 0.5 m. In the upper protective coal seam technology, the borehole spacing is related to the location of the protected coal seam. If the protected coal seam is located in the fracture zone, to prevent the gas abnormal emission in the protective coal seam, the borehole spacing should not be greater than the layer distance between the protective coal seam and the protected coal seam. If the protected coal seam is located in the

Table 10.5 Stress gas extraction method

Location of the protective coal seam	Location of the protected coal seam		Gas extraction method	Comments
Lower protective coal seam	Overlying rock masses	Bending zone	Upward net-like cross-borehole gas extraction method and surface well gas extraction method	One method or a combined methods could be selected according to the actual requirement
		Fracture zone	Strike high-level borehole gas extraction method, trend high-level borehole gas extraction method, gob-side roadway cross-borehole gas extraction method, gob burying pipe gas extraction method, strike long borehole gas extraction method, strike high-level roadway gas extraction method, trend high-level roadway gas extraction method and surface well gas extraction method	
Upper protective coal seam	Underlying rock masses		Upward net-like cross-borehole gas extraction method, strike high-level roadway gas extraction method and gob-side roadway cross-borehole gas extraction method	

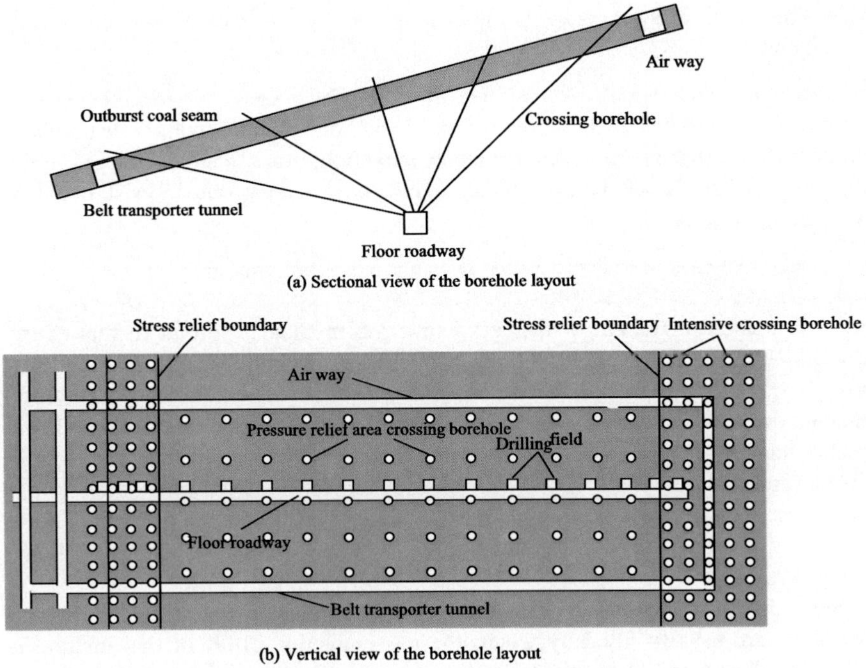

Fig. 10.24 The sketch map of the upward net-like cross-borehole gas extraction method

deformation zone of the protective coal seam, the borehole spacing is usually set as 20–30 m.

In the original zone or the insufficient stress relief zone near the open-off cut or the stopping line of the working face, the borehole spacing should be designed according to the initial permeability of the coal seam. According to the field gas extraction experience, the borehole spacing should be set to 5–10 m in these zones. The construction of the gas extraction boreholes should be completed before the mining of the protective coal seam, so that the stress relief gas could be extracted during the mining of the protective coal seam.

Compared with the surface well gas extraction method, the extra rock roadway should be constructed in the cross-borehole gas extraction method, as a result, the construction period is rather long and the construction cost is rather high. However, there are four obvious advantages in this method.

(1) The rock roadway and the gas extraction boreholes could be arranged and constructed flexibly according to the geological conditions, mining conditions and the gas occurrence conditions.
(2) The gas extraction effort is reliable, and the gas extraction design could be optimized according to the previous gas extraction effort.
(3) The rock roadway could also be used to determine the expansion deformation and the permeability evolution in the protected coal seam.

(4) The gas extraction period is rather long and the gas extraction effort is very good.

According to the field investigation results, the effective gas extraction stage could reach up to 3–4 months. In this stage, the gas extraction volume of one single borehole could reach up to 1 m^3/min. After the gas extraction, more than 65% of gas in the protected coal seam could be extracted, and the coal and gas outburst risk could be completely eliminated.

2. The downward netlike cross-borehole gas extraction method

The borehole arrangement of this gas extraction method is similar to that of the upward cross borehole gas extraction method. The only difference is that the location of the rock roadway is different. In the downward cross-borehole gas extraction method, the rock roadway is arranged in the roof rock masses of the protected coal seam while that is arranged in the floor rock masses of the protected coal seam in the upward cross-borehole gas extraction method. In the downward borehole gas extraction method, most of the boreholes are downward. However, part of the boreholes may also be upward in the steep or the extremely steep coal seams, as shown in Fig. 10.25.

Since that the construction of the downward borehole is rather difficult and these boreholes are usually filled by water, the gas extraction effort of this method is not good. Therefore, this method is seldom used in China, except for the case that there is an existing rock roadway in the roof rock masses of the protected coal seam. During the mining of the extremely steep upper protective coal seam, the gas extraction boreholes could be constructed from the roadway of the working face in the protective coal seam to extract the stress relief gas in the protected coal seam.

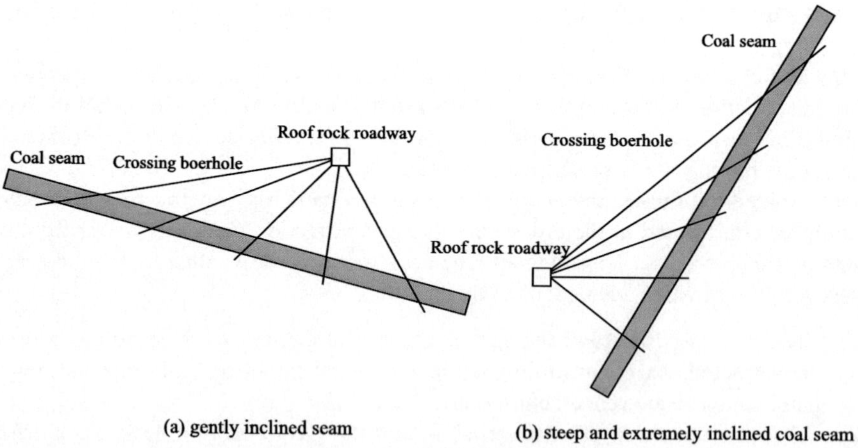

Fig. 10.25 Downward cross-borehole gas extraction method

10.5.2 Field Investigation of the Stress Relief Gas Extraction Effort

1. Engineering background

At the Haizi coal mine of the Huaibei mine area in China, the main minable coal seams are the #7, #8, #9 and #10 coal seams. Among these coal seams, the #10 coal seam is relatively stable, while the #7, #8 and #9 coal seams are instable or the extremely instable. Moreover, the #7, #8 and #9 coal seams are of coal and gas outburst risk, while the #10 coal seam is of no coal and gas outburst risk. Therefore, the #10 coal seam is chosen as the lower protective coal seam to be mined firstly to unload the stress and increase the permeability in the #7, #8 and #9 coal seams.

The experimental working face in the protective coal seam, the II1021 working face with a length of 640 m and a width of 140 m, is located in the II 102 mining district. Its elevation is – 577 to – 646 m. In this working face, the thickness of the #10 coal seam is 0.2–5 m, with an average value of 3.2 m. The #9 coal seam, with a thickness of 3.06 m, is located at 99.84 m above the #10 coal seam. The #8 coal seam, with a thickness of 0.51 m, is located at 103.67 m above the #10 coal seam. The #7 coal seam, with a thickness of 1.55 m, is located at 132.73 m above the #10 coal seam. Moreover, a thick igneous rock with a thickness of 120 m exists in 55 m above #7 coal seam. The generalized stratigraphy of coal strata in this working face is shown in Fig. 10.26.

The #7, #8 and #9 coal seams are located in the bending zone formed after the mining of the #10 coal seam. As a result, during the mining of the #10 coal seam, efficient stress unloading and permeability increasing could be achieved in the #7, #8 and #9 coal seams. Moreover, under the support of the thick igneous rock with a thickness of more than 100 m in the roof of the protected coal seam, the rock mass above the igneous rock would not deform and subside for a long time after the mining of the protective coal seam, and thus the fractures and the separations formed in the protected coal seam would keep opening for a long time, which provides good conditions for the extraction of the stress relief gas in the protected coal seam. Under such geological conditions, efficient gas extraction could be achieved by a few cross-boreholes.

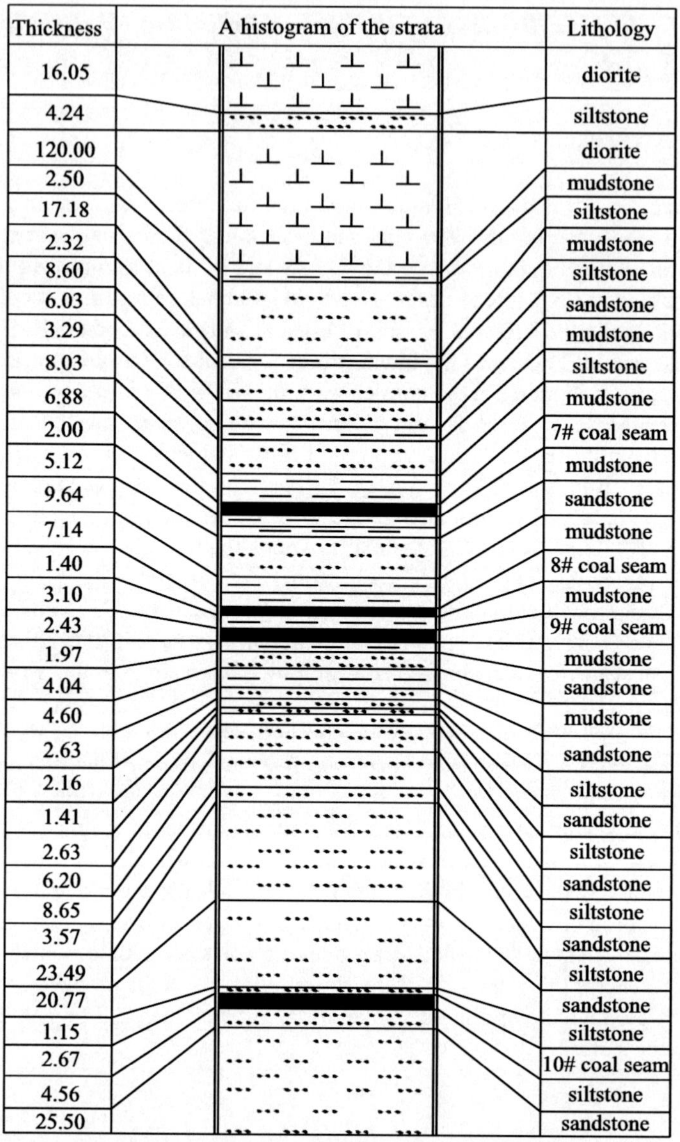

Fig. 10.26 The generalized stratigraphy of coal strata in the Haizi coal mine

2. The fracture development characteristics in the overlying rock mass of the protective coal seam

After the mining of the protective coal seam, the overlying rock masses break and deform. At different distances above the protective coal seam, the deformation and

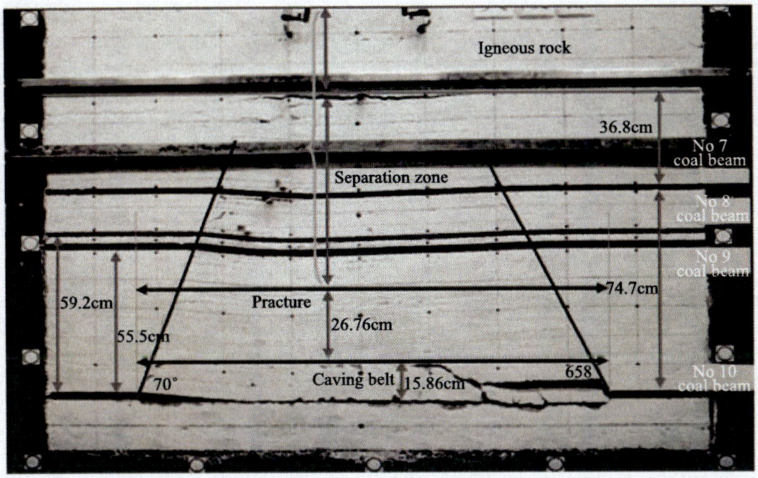

Fig. 10.27 The caving characteristics in the overlying rock mass of the #10 coal seam in the Haizi coal mine

the fracture development characteristics of the rock masses are different. Figure 10.27 shows the caving and deformation characteristics of the overlying rock masses after the mining of the #10 coal seam obtained by an similar simulation experiment in the laboratory. As shown in Fig. 10.27, three belts are formed in the overlying rock masses, i.e., a caving zone with a thickness of approximately 24 m, a fracture zone with a thickness of approximately 40 m and a bending zone. Below the igneous rock, a separation zone is formed in the bending zone. During the mining process of the #10 coal seam, the igneous rock mass does not bend, as a result, its overlying rock masses do not deform. In the fracture zone, the vertical fractures and the horizontal fractures both develop and connect with each other. However, in the bending zone, the rock masses move as a whole. The protected coal seam, located in the separation zone of the bending zone, has a great vertical expansion deformation. Under the support of the igneous rock mass, the overlying rock masses do not subside. However, with the continual mining of the protective coal seam, the igneous rock mass may break suddenly.

Figure 10.28 shows the separation development characteristics in the underlying rock masses of the igneous rock mass. The igneous rock plays as the main key stratum during the mining process of the protective coal seam, as a result, it controls the movement of its overlying rock masses. Under the support of the thick igneous rock mass, its overlying rock masses do not cave and subside, leading to the generation of the separation in the underlying rock masses. Therefore, during the mining process of the protective coal seam, plenty of separations could be formed in the protected coal seam located below the igneous rock mass. According to the numerical simulation results, the stress unloading and the permeability increasing effort could be much

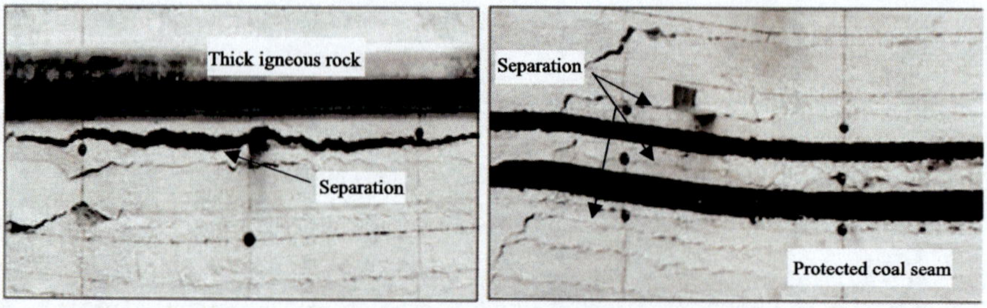

Fig. 10.28 The distribution characteristics of the separations under the thick igneous rock

more significant in the protected coal seams during the mining process of the protective coal seam under this particular geological condition. In the bending zone, the vertical fractures don't close, which connects the coal seams and the separations. Under the effect of the gas pressure difference, the stress relief gas flows into the separation zone though these vertical fractures, making this zone the gas concentration zone. The gas concentration in the separation zone provides good conditions for the gas extraction.

3. Stress relief gas extraction method

Under the geological conditions in the Haizi coal mine, the stress unloading effect is better and the effective stress unloading period is longer, as a result the efficient gas extraction could be achieved in the protected coal seam just by constructing a few cross-boreholes from the high-level drilling fields in the ventilation roadway or the high-level rock roadway of the working face in the protective coal seam, as shown in Fig. 10.29.

During the field experimental process, two long cross-boreholes are constructed from the No. 5 drilling field in the high-level rock roadway of the II1021 working

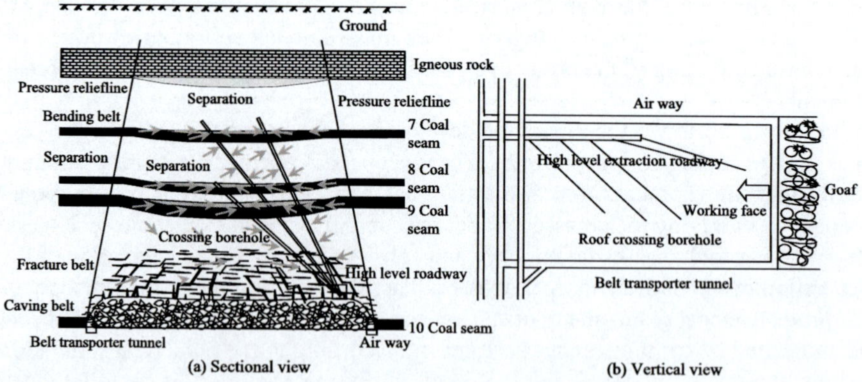

Fig. 10.29 The layout of the cross-borehole in the high-level rock roadway

face. The borehole radius is 133 mm. After the construction of the borehole, the casing pipes, each with a length of 2 m, are inserted into the borehole. The adjacent two pipes are connected by the screw thread. The radius of the casing pipe is 108 mm. The dip angle of the No. 1 borehole is 22.6°, and its length is 246.5 m. The dip angle of the No. 2 borehole is 21.8°, and its length is 225 m. The borehole spacing of these two boreholes is 28 m (Fig. 10.30).

4. Stress relief gas extraction effect

In this chapter, taking the gas extraction data of the No. 1 cross-borehole and the No. 2 cross-borehole in the No. 5 drilling field in the II1021 working face as an example, the gas extraction effect of the boreholes under this particular geological conditions are analyzed. The gas extraction parameters of the No. 1 borehole with the advance of the protective coal seam is shown in Fig. 10.31.

As shown in this figure, the gas extraction of the No. 1 borehole begins when the working face is 70 m behind the borehole. With the advance of the working face, the gas extraction concentration and the volume increase in the jumping style. When the working face is advanced to 4 m behind the borehole, the gas extraction data increases

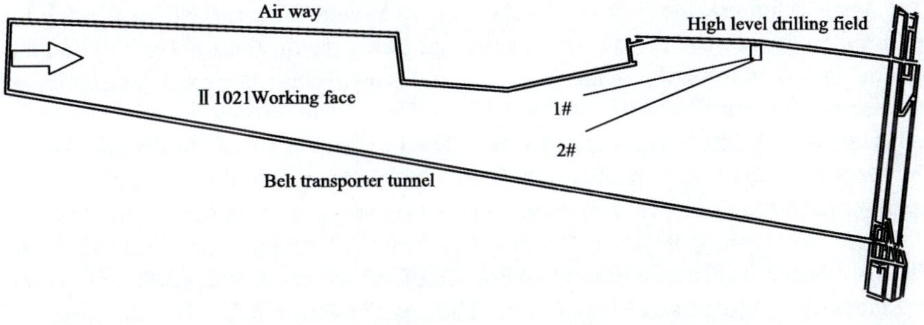

Fig. 10.30 The arrangement of the cross-boreholes in the II1021 working face

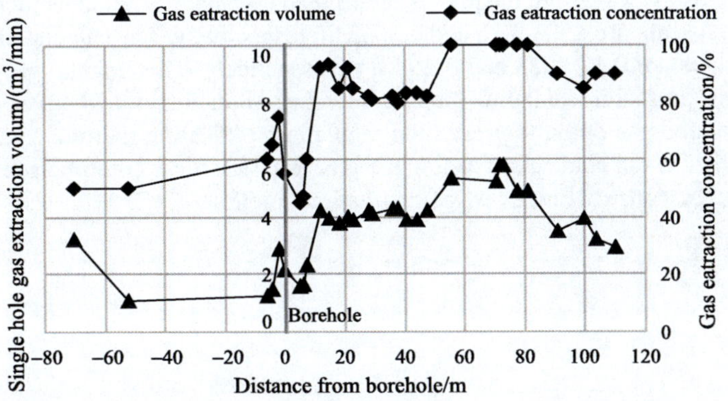

Fig. 10.31 The gas extraction volume of the No. 1 borehole

firstly, and the gas extraction volume reaches up to 2.85 m^3/min. When the working face is advanced ahead of the borehole, the gas extraction volume is approximately 2.11 m^3/min. When the working face is advanced to 12 m ahead of the borehole, the gas extraction concentration reaches up to 90% and the gas extraction volume reaches up to 4.2 m^3/min. This period is the gas extraction growth stage. During the advance of the working face from 12 to 100 m ahead of the borehole, the gas extraction data is relatively stable with a gas extraction concentration of 90% and a gas extraction volume of 4.0 m^3/min, as a result, this period is the active gas extraction stage. Moreover, during the advance of the working face from 56 to 81 m ahead of the borehole, the gas extraction concentration is nearly 100%, and the gas extraction volume is approximately 5–6 m^3/min, the gas extraction effort is rather notable. The efficient gas extraction stage is approximately 70 days. When the working face is advanced to 100 m ahead of the borehole, the gas extraction data is decreasing, indicating that this period belongs to the gas extraction attenuation stage. After the working face is advanced to 111 m ahead of the borehole, the gas extraction data is not countered. With the continual mining of the working face, the high-level drilling field caves, as a result, the gas extraction volume decreases to 0 gradually.

The gas extraction data of the No. 2 borehole shows the same regulation as that of the No. 1 borehole. After the caving of the No. 5 drilling field in the working face, some supplementary cross-boreholes constructed from the No. 1 parking lot are used to further extract the stress relief gas, as a result, the total gas extraction volume grows continually. In the gas extraction growth stage, the total gas extraction volume is 0.232 million, while that is 1.580 million m^3 in the active stage. However, the total gas extraction volume is 3.048 million m^3 in the attenuation stage. Under the support of the thick igneous rock mass, the fractures and the generations keep opening for a long time, leading to an attenuation stage of more than 8 months.

During the mining of the protective coal seam, the total gas extraction volume of the No. 2 borehole is approximately 4.86 million m^3, accounting for 60% of the total gas quantity in the pressure relief zone. The gas extraction effect is rather notable. According to the gas extraction volume, the gas extraction influence radius of a single borehole may be greater than 100 m.

After the gas extraction, the in-situ determined residual gas contents in the #7, #8 and #9 coal seam are 4.48, 3.26 and 4.86 m^3/t, respectively. The calculated residual gas pressures are 0.27, 0.28 and 0.26 MPa, respectively. The residual gas contents and the gas pressures are below the critical values (8 m^3/t and 0.74 MPa) specified in the coal and gas outburst prevention regulation, indicating that the coal and gas outburst risk in the coal seams below the igneous rock mass could be successfully eliminated by constructing a few gas extraction boreholes.

References

1. Yu, Q. (1992). *Coal mine gas control*. Xuzhou: China University of Mining and Technology Press. (in Chinese).
2. State administration of work safety. AQ 1050–2008 Technical criterion of protective coal seam exploitation. 2008 (in Chinese).
3. Khodot, B. B. (1966). *Coal and gas outburst*. Beijing: China Industry Press. (in Chinese).
4. Yu, B. (1986). *Understanding and practice of protective coal seam mining technology*. Beijing: China coal industry publishing house. (in Chinese).
5. Yu, B. (2005). *Technical manual of gas disaster prevention and utilization in coal mine*. Beijing: China coal industry publishing house. (in Chinese).
6. Gas group in China University of Mining and Technology. Prevention and control of coal and gas outburst disaster. Beijing: China coal industry publishing house, 1979 (in Chinese).
7. Zhou, S., & Lin, B. (1999). *The theory of gas flow and storage in coal seams*. Beijing: China coal industry publishing house. (in Chinese).
8. Liu, Z., Yuan, L., Dai, G., et al. (2004). Study on the characteristics of the gob roof cracks in stopes and its application in gas drainage. *Journal of Anhui University of Science & Technology, 24*(4), 10–15. (in Chinese).
9. Cheng, Y., Zhou, D., Yu, Q., et al. (2006). Research on extraction and emission laws of gas for pressure-relief in protecting coal seams. *Journal of Mining & Safety Engineering, 23*(1), 12–18. (in Chinese).
10. Tu, M., Huang, N., & Liu, B. (2007). Research on pressure-relief effect of overlying coal rock body using far distance lower protective seam exploitation method. *Journal of Mining & Safety Engineering, 24*(4), 418–421. (in Chinese).
11. Yu, Q., Cheng, Y., & Jiang, C. (2004). Principles and applications of exploitation of coal and pressure relief gas in thick and high-gas seams. *Journal of China University of Mining & Technology, 33*(2), 127–131. (in Chinese).
12. Cheng, Y., Yu, Q., & Yuan, L. (2003). Application of safe and high-efficient exploitation system of coal and gas in coal seams. *Journal of China University of Mining & Technology, 32*, 471–475. (in Chinese).
13. Cheng, Y., & Yu, Q. (2003). Application of Safe and high-efficient exploitation system of coal and gas in coal seams. *Journal of China University of Mining & Technology, 32*, 471–475. (in Chinese).
14. Yuan, L. (2004). *Gas extraction theory and technology in the soft and low-permeability coal seam group*. Beijing: China coal industry publishing house. (in Chinese).
15. Wang, H. (2008). *Pressure relief functional principle of stope underlying coal-rock mass and application in gas extraction of protected coal seam, safety science and engineering*. Xuzhou: China University of Mining and Technology. (in Chinese).
16. Wang, L. (2009). *Fracture evolution and seepage characteristics of distant pressure relief coal-rock masses and application in gas extraction under an extremely thick igneous rock, safety science and engineering*. Xuzhou: China University of Mining and Technology. (in Chinese).
17. Liu, H. (2010). *Fully coupled model and engineering application for deformation and pressure-relief gas flow of remote coal and rock mass due to mining, safety science and engineering*. Xuzhou: China University of Mining and Technology. (in Chinese).
18. Liu L. Rational protection range of lower protective layer and application in pressure-relief gas extraction, safety science and engineering. Xuzhou: China University of Mining and Technology (in Chinese).
19. Wang, L., Cheng, Y., Li, F., et al. (2008). Fracture evolution and pressure relief gas drainage from distant protected coal seams under an extremely thick key stratum. *International Journal of Mining Science and Technology, 18*, 182–186.
20. State administration of work safety. (2016). *Safety regulations for coal mine*. Beijing: China Coal Industry Publishing House. (in Chinese).
21. State administration of work safety. (2009). *Provisions for the prevention and control of coal and gas outbursts*. Beijing: China Coal Industry Publishing House. (in Chinese).

22. Tu, X. (1965). On the effectiveness of working the upper protective stratum. *Journal of China Coal Society, 1*(3), 17–32. (in Chinese).
23. Wang, H., Cheng, Y., Wu, D., et al. (2010). Gas emission and parameter optimization of gas extraction in mining face of short distance protective seam. *Journal of China Coal Society, 35*(4), 590–594. (in Chinese).
24. Liu, H., Cheng, Y., Song, J., et al. (2009). Pressure relief, gas drainage and deformation effects on an overlying coal seam induced by drilling an extra-thin protective coal seam. *International Journal of Mining Science and Technology, 19*(6), 724–729.
25. Cheng, Y., Fu, J., & Yu, Q. (2009). Development of gas extraction technology in coal mines of China. *Journal of Mining & Safety Engineering, 26*(2), 127–139. (in Chinese).
26. Li, S., Wang, L., Lin, H., et al. (2013). Similar simulation experiment and analysis on evolution characteristics of "three zones"in overburden strata of stope. *Mining Safety & Environmental Protection, 3*(40), 17–20. (in Chinese).
27. Wu, R. (2011). *Study on the scope of the "three zones" of gas pressure relief and extraction in coal seam group mining*. Xuzhou: China University of Mining and Technology Press. (in Chinese).
28. Yu, B. (1986). *Technical manual of gas disaster prevention and utilization in coal mine*. Beijing: China coal industry publishing house. (in Chinese).
29. You, H., Li, B., & Zhang, F. (2008). *Comprehensive gas control technology for the fully mechanized top coal caving working face in Yangquan coalfield*. Beijing: China coal industry publishing house. (in Chinese).
30. An, F., Cheng, Y., Wang, L., et al. (2013). A numerical model for outburst including the effect of adsorbed gas on coal deformation and mechanical properties. *Computers & Geotechnics, 54*(10), 222–231.

Chapter 11
Application of Coal Mechanics in Coal and Gas Outbursts

Learning Outcomes

- After working through the material in this chapter, you should know the following:
- General laws of coal and gas outbursts;
- Representative hypothesis of coal and gas outbursts;
- Physical process of coal and gas outbursts;
- Influence of gas, in situ stress, and coal properties on outbursts;
- Role of rapid gas desorption in the development stage of outbursts.

11.1 Introduction to Coal/Gas Outbursts

11.1.1 General Laws of Coal/Gas Outbursts

Coal and gas outburst is one of the major disasters that threaten safe production in coal mines. Outburst is a complex dynamic failure process that emits a large amount of coal (rock) and gas into the production space in a short time [1]. Since the first reported outburst incident occurred in the Issac coal mine, Loire coal field, France in 1834, > 30,000 coal and gas outbursts have occurred in the world, causing great loss. Among them, China is the country with the highest risk of outbursts because of the complex geological conditions and the great depth and high gas content of coal-bearing strata. Many people die in outbursts each year and a lot of money is spent on saving lives and properties. According to the mass size of rock and coal being injected, outbursts can be divided into small outbursts (< 100 t), medium outbursts (100–500 t), large outbursts (500–1000 t), and extraordinary accidents (> 1000 t).

Although there are many factors affecting outbursts, some general laws can be observed based on the records of outburst accidents.

Y. Cheng et al., *Coal Mechanics*,
https://doi.org/10.1007/978-981-16-3895-4_11

1. Outbursts have regional distribution characteristics

Previous records have indicated that the area with a risk of outbursts accounts for 10% of the total area of a coal seam, characterized by zonal distribution. This may be attributed to the zonal distribution of geological structures. Tectonic movement may change the stress conditions of a coal seam, form structural coal, and exert an effect on the storage and transportation of gas; thus, tectonic movement has an important influence on the occurrence and distribution of outbursts.

Most outbursts were clearly identified with relation to tectonic types, and most of them occurred in the tectonic zone such as fault, fold and heavily structured zones. Geological structures can break intact coal, reduce coal strength, and increase the specific surface area and pore volume of coal, thus resulting in a higher gas desorption ability. Meanwhile, the ring-like distribution characteristic of geological structures could exist in areas with faults, oblique shafts, metamorphic coal, and non-metamorphic coal caused by magma intrusion as well as other geological structure intersections. In addition, the stress concentration and superposition area formed by excavation works (e.g., at the mining face near the coal pillar, start–end line, and stress concentration zone before the intersection of two tunnels and mining faces) all possess a ring-like distribution characteristic. These areas have a high risk of outbursts, and usually have a high frequency of powerful outbursts. However, with increasing depth, the geostress and gas pressure increase significantly, and the above characteristics become less obvious.

2. Outburst risk of a coal seam increases with depth

For the same coal seam in the same mining area, as the mining depth increases, the geostress and gas pressure as well as the outburst risk will also increase. This characteristic can be observed by obtaining data on a high number of outbursts (including outburst coal seams), a greater intensity of outbursts, and a larger scale of outbursts. The locations where outbursts occur increase from one point in shallow areas to several points in the middle, and even to a large plane in the deep.

The relationship between outbursts and the mining depth is not linear because the geostress does not have a linear relationship with depth. In shallow areas, the horizontal stress is larger than the vertical stress, and the ratio of these two stresses is not constant but decreases with depth. In addition, there are too many geological structures in an outburst coalmine, which change the regional stress distribution.

3. Outburst risk increases with the thickness of a coal seam

The number and intensity of outbursts increase as the thickness of a coal seam increases. Thick coal seams or coal seams containing thin parting layers have a greater outburst risk in comparison to single coal seams, and usually occur more often with a large intensity and a shallower outburst depth. For the same coal seam, the outburst risk increases with increasing thickness.

In addition, according to previous statistics, the location where an outburst occurs usually has a soft coal seam with a bedding disorder. When the thickness of the soft seam increases suddenly, an outburst usually occurs. The outburst risk generally disappears when the thickness of the soft coal seam is < 0.3 m.

4. Relationship between roadway types and outbursts

Outbursts mainly occur in heading faces or mining faces, and rarely occur in goafs, finished tunnels, and stress unloading areas. Based on Chinese outburst records, according to the possibility of outbursts (low to high), the locations can be sorted as plane roadway, mining face, up-working face, cut, down-working face, and large drilling holes among others locations. According to the intensity of an outburst (strong to weak), the locations can be sorted as cut, down-working face, up-working face, plane roadway, and mining face. An outburst in the mining face is usually a sudden squeeze of coal and belongs to the sudden extrusion type.

With respect to roadways, although the number of outbursts occurring in cuts is not large, they are very intense. Most extraordinary outburst accidents occur in these locations. For example, on October 20, 2004, there was an extraordinary outburst accident at the Daping Coalmine (Zhengzhou Coal Industry Group Co. Ltd.) in Henan Province. The outburst coal was 1849 t and the gas volume was 250,000 m^3. The outburst was located on the heading face of the No. 21 down railway tunnel (612 m from the ground surface), which is a strongly squeezed reverse fault with a 307° strike (NW 53°), 217° dip (SW 37°), 49° dip angle, and 10 m drop, as shown in Fig. 11.1.

5. Most outbursts are induced by mining activity

Mining activity may induce outbursts. In particular, falling coal and vibration activity can cause a change in stress and make the dynamic stress work on the new exposed coal body from fragmented coal particles. More than half of the outbursts in China occur during blasting activity. Their intensities are strong and can involve an injection mass of approximately 100 t. The number of outbursts during wind-induced falling coal activity is small; however, the intensity is normally relatively strong. In recent years, with the development of mechanized coal mining, the outbursts caused by mechanized cutting activity have become frequent, and now exceed those caused by wind-induced falling coal. In addition, support, drilling, and other operational activities may also cause outbursts. In fact, few outbursts occur in the case of no operations.

6. There are usually some omens before an outburst

Although outbursts are short, they may be preceded by some clear omens. Sound omens include strange sounds such as stuffy thunder, gun, machine guns, firecrackers, and buzzing in the coal body. Structural omens are evident if the coal seam bedding becomes disordered, coal becomes soft, dark, dull, and dry, and there is an increase in coal dust. Ground stress omens include an increased support pressure and expansion of the coal wall, forming cracks and small pieces of coal. In addition, gunpowder cannot be put into the blast hole because of its deformation. Gas omens are those when the gas concentration increases, gas emission quantity becomes unstable, drilling becomes difficult, and the temperature of the coal wall varies.

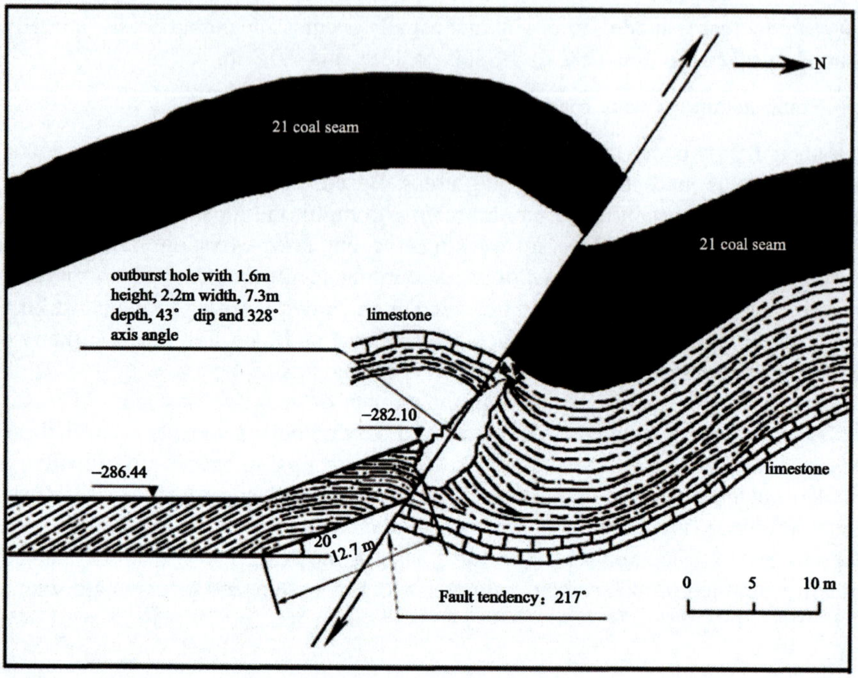

Fig. 11.1 Profile of the outburst at Daping Coalmine

7. The shape of the outburst hole differs between accidents

The locations and shapes of outbursts are different. Most holes are located in the upper roadway and upper corners of a workface. Typical outbursts may have a small entrance and a large internal volume (except coal extrusions or coal slips), and are usually of a pear shape or oval shape (Fig. 11.2). Sometimes there is a gap between the roof and the deposited coal, and a layer of pulverized coal is covered on the surface. Along the direction of the roadway, the hole angle (i.e., the angle between the center line and the horizontal plane) is approximately 40°–45°, and few holes exist in the horizontal direction. Occasionally, holes are not clear, and only some movement of the coal body can be seen. The actual volume of a hole is always less than the volume of the discharged coal. For more than 30 outburst holes in Chongqing, the actual volume of the hole was half the size of the thrown coal, and in some cases the ratio was 1/10. French statistics have shown that the volume of the outburst hole is 1/4–1/3 the size of the injected coal volume. The largest outburst in the world occurred in Gagarin Mine in the Soviet Union, when the hole volume was 4/5 the size of the injected coal volume.

8. Outburst coal exhibits a sorting phenomenon

Fig. 11.2 Outburst hole shape following the Daping outburst accident

One important characteristic of outbursts is that the coal will be sorted very well along the roadways, whereby the size of coal pieces will decrease with increasing distance from the outburst hole. Sometimes, wind will blow pieces of coal between the roof and outburst hole. There is generally a lot of powder on the floor or coal wall in areas far from the outburst hole. Figure 11.3 shows the coal distribution along the roadway following the Daping outburst, and reveals that large coal lumps began to appear as the outburst hole was approached.

9. Dynamic effects in outbursts

The normal dynamic power may overturn the tub, destroy the stent, and transport boulders, thus causing air waves and noise. In extraordinary outbursts, huge rocks may be carried by dynamic power for a long distance. Figure 11.4 shows a large coal rock, which was carried by the Xingxing outburst flow. Another case occurred at the Sanhuiyi Coalmine, which is operated by Tianfu Company. When opening the cutting of the No. 6 coal seam, a 1 t rock was thrown > 120 m and then underwent two 90° turns, while a 3 t rock was also thrown > 50 m. The huge power generated in outbursts can easily destroy underground facilities. Figure 11.5 is a photograph of a deformed mining car after the Xinxing outburst.

During outbursts, the outburst coal mass can range from several tons to tens of thousands of tons, and the released gas can be millions of cubic meters. In extraordinary outbursts, the gas emission volume over short periods can be hundreds of thousands or even millions of cubic meters, and the gas emission per ton of coal can be 100–800 m^3, which is approximately 5–30 times the gas content of a coal seam. The released gas will generally move in the wind direction under normal conditions. However, for large or extraordinary outbursts, coal powders and gas may

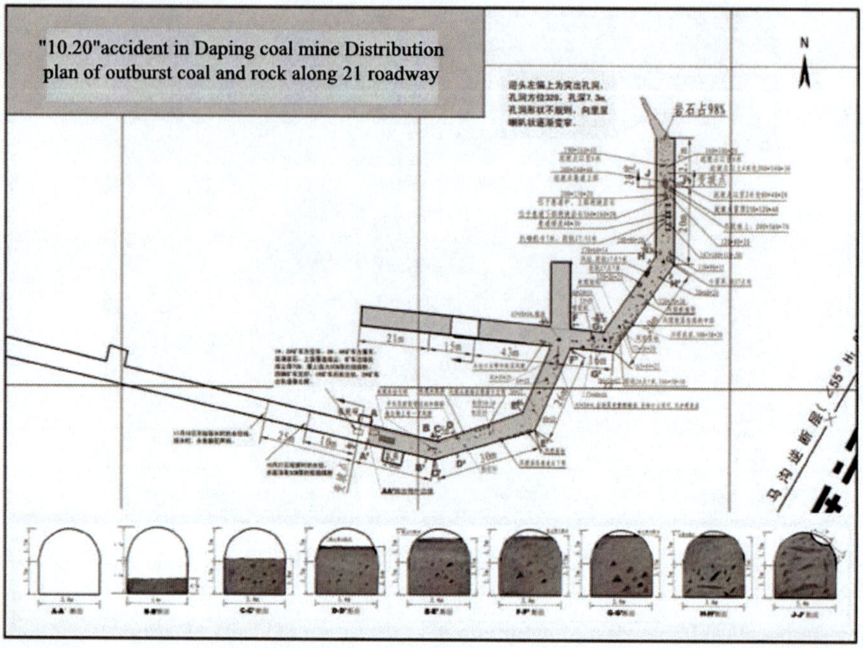

Fig. 11.3 Coal distribution along the roadway following the Daping outburst

Fig. 11.4 Outburst rock from the Xinxing outburst

move opposite to the wind direction, filling tunnels of several thousands of meters in length.

10. Characteristics of extraordinary outbursts

Extraordinary outbursts usually have the following characteristics. ① They largely occur when dangerous coal seams encounter mining. ② Most extraordinary outbursts are caused by blasting activity, and there may between several minutes to several hours between the start of an outburst and the end of blasting activity. ③ The volume

Fig. 11.5 Deformed mining car after the Xinxing outburst

of released gas is extremely large in extraordinary outbursts ④ Many extraordinary outbursts occur in areas with a changeable coal seam thickness, which is caused by geological structures. ⑤ The soft sub-coal seam is usually seen in extraordinary outbursts. Alternatively, the original coal seam is soft and easy to break. These coal seams have a dispersed structure, are mostly granular or powder-like, and are very broken. In areas with geological structures, they often produce highly crumpled coal with destroyed bedding and joints.

11.1.2 Research Methods

Coal and gas outbursts are complicated phenomena whose internal laws are still unclear. Previous scholars tended to obtain the excitation conditions, process, and mechanism of outbursts based on statistical methods, theoretical analysis, and experimental stimulation.

Outbursts have been studied for hundreds of years. In engineering practice, people have summarized the statistical laws from the viewpoint of stress, gas, and the physical mechanism of coal. Such information is very helpful for preventing outbursts. Chinese and foreign scholars have tried to obtain a critical index to illustrate the risk of outburst. The outburst index in the Bulli Coalmine in Australia has a methane content of 9.4 m^3/t and a carbon dioxide content of 6.4 m^3/t. If small outbursts of < 20 tons are permitted, the critical index is a methane content of 12 m^3/t and a carbon dioxide content of 10 m^3/t [2]. If there are no geological structures, the critical content of carbon dioxide is 10 m^3/t [3]. A similar desorption index (DRI) of 900 based on the above index successfully predicted the outburst risk in the southern marine coal area, which verifies the correctness of the above index [4]. The Former Federal Republic of Germany used the gas desorption volume as the outburst index, whose value ranged from 7 to 9 m^3/t [5]. In China, scholars such as Wang Youan and

Yang Qiluan considered that the outbursts occurring at the shallowest depth generally have a gas content of > 10 m^3/t and a gas pressure of > 0.7 MPa [6]. The Chinese provision of coal and gas outburst prevention and control states that the gas content and gas pressure of the coal mining face (before excavation) should be reduced to the corresponding value at the depth required to produce outbursts. If there is no such index, a gas content of 8 m^3/t and a gas pressure of 0.74 MPa (relative pressure) are recommended.

In the statistics of outburst locations, most sites have a direct relationship with the geological structures, which will change the original stress conditions and produce tectonic coal, thereby influencing the storage and transport of gas. Thus, geological structures have a key impact on the distribution of outbursts. Regarding the statistics of outburst coal, most outbursts involve deformed coal, which has a relatively small mechanical strength, low permeability, and rapid desorption rate. These characteristics increase the risk of outbursts. Studies on the structure of tectonic coal have indicated that the small force between the tectonic coal molecules determines its low strength and high adsorption capacity [7]. Extrusion zones and thick coal seams are the major areas of tectonic coal, and the areas with the highest outburst risk [8]. Furthermore, studies on the stress environment surrounding the outburst area have also demonstrated the control of tectonic stress on outbursts. Ancient and modern tectonic stress fields as well as their evolution have important effects on gas generation, migration, and occurrence [9, 10]. The area of high outburst risk is usually that experiencing high stress.

Regarding the theoretical analysis of outbursts, scholars have put forward many hypotheses, which are the simplification of outburst processes and conditions. These hypotheses help people understand the occurrence, development, and termination mechanism of outbursts. Previous hypotheses were mainly qualitative analyses, while quantitative analyses began much later. Because of the large differences in the stages of an outburst and the main related physical processes, the theories and models used in outburst studies also differ. Before an outburst starts, the primary difficulty is describing the interaction of the stress distribution. In the development stage, the main difficulties are determining how to describe the continuous stripping damage to coal and the two-phase flow of coal/gas.

Compared to other questions on rock mechanics, it is more complicated to study outbursts because they are related to the interaction of coal mechanics and gas. It is a multi-disciplinary problem of rock mechanics, adsorption theory, porous media mass transfer, and seepage mechanics. When building numerical models of outbursts, not only do we need to consider the coal material and structural instability, but also the gas adsorption, diffusion, and seepage influence by coal mechanics. The trigger mechanism of outbursts gained early research attention. In the 1980s, related numerical simulation research was reported. For instance, Paterson [11] established a model involving gas diffusion, seepage flow, and coal elastic medium models, and treated the gas pressure as the volume force to simulate the impact of gas on the force of coal. Paterson also analyzed the distribution of two-dimensional gas flow and stress. He believed that there was a large gas pressure gradient in the early stage of outbursts, which leads to tension failure and the formation of outbursts. This theory explains the role of gas in outbursts.

In recent years, with the development of studies on the interaction between coal and gas, coal mechanics, and gas migration, related numerical model research of outbursts has also developed. On the one hand, the damage and plastic softening characteristics of coal materials are taken into account. On the other hand, research on the coupling of gas and coal mechanical responses is also developing. For example, Choi and Wold used coal as a plastic pore medium and simulated the early stage of outbursts using the SIMED and FLOMEC software [12]. Then, the authors used a two-dimensional geometric model to proceed with an outburst stimulation with consideration of the coal damage and fracture propagation. Their results showed that the coal desorption time had little influence on the outburst risk. In another study, by linking and executing FLAC3D and COMET3 sequentially, Xue incorporated coal–gas coupling by transferring the calculation result between the two codes and computing the pore pressure, main stress, and distribution of yield zones in the early stage of outbursts during roadway excavation [13]. The author also analyzed the influence of gas pressure, mining depth, and coal body mechanics on outbursts [14]. Yu studied the one-dimensional movement of an outburst face, and discussed the seepage process during constant-speed evacuation [15]. He finally obtained the following outburst criteria: ① a possible condition of geostress, ② a breeding condition of the excavation or gas, and ③ the trigger conditions of the exudation speed [16]. Zheng developed a theory to explain the steady-state pushing of the failure front observed in a one-dimensional simulation experiment of coal/gas outbursts [17]. Zhao et al. studied the unified mechanism of the impact pressure and outburst features, and then put forward an outburst mathematical model. The authors obtained the instability criterion and the control equation, and used the finite element method to simulate the probabilities of the stress yield area and elastic area to analyze the possibility of outbursts [18, 19]. In addition, Tang used RFRA software to simulate outburst trigger factors. He regarded the elastic modulus and permeability of coal as functions of damage parameters, considering the heterogeneity of the mechanical properties of coal and the loss of damage during loading [20]. The simulation results showed the formation of outbursts in the coal/rock medium (from crack initiation, expansion, and interaction with thrown coal) under the combined action of the gas pressure, ground stress, and mechanical properties of coal and rock [21–23].

Owing to the difficulty in reproducing disasters and monitoring the related parameters, coal/gas outburst simulation experiments can systematically investigate the outstanding conditions, influencing factors, processes, and characteristics to verify the correctness of the theoretical analysis. The physical simulation of outbursts is an important way to study the outburst mechanism. The experimental device develops gradually from the initial one-dimensional experiment to the current three-dimensional experimental device. Many scholars have done a considerable amount of work in this area, including three-dimensional simulations [24], delayed outbursts [25], and cross-cut uncovering [26, 27]. Among them, National Engineering Research Center for Coal and Gas Control at China University of Mining and Technology used their self-developed real triaxial coal/gas outburst simulation test system to conduct a study on the outburst spalling development mechanism, the outburst instability process, and the influence of local gas enrichment on outbursts, as illustrated in Figs. 11.6 and 11.7.

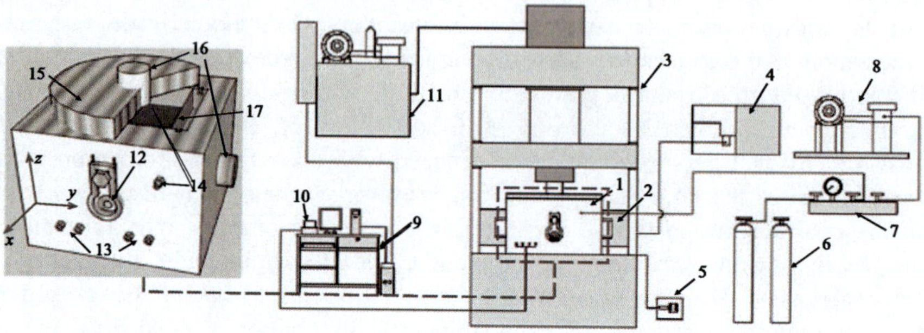

1: Triaxial coal/gas outburst simulation test system; 2: horizontal load jack; 3: axis pressure tester; 4: horizontal loading pump station; 5: temperature control system; 6: high-pressure cylinder; 7: three-way valve; 8: vacuum pump; 9: console: 10: data acquisition instrument; 11: pump station; 12: outlet; 13: measuring hole; 14: pore; 15: upper cover; 16: piston; 17: pressure-bearing plate

Fig. 11.6 Illustration of the real triaxial coal/gas outburst simulation test system [86]

Fig. 11.7 Photographs of the real triaxial coal/gas outburst simulation test system [86]

11.2 Mechanical Mechanism of Coal/Gas Outbursts

11.2.1 Hypothesis of Coal/Gas Outbursts

In 1834, coal and gas outbursts were recorded for the first time in the Ake Coalmine, France. Since then, a lot of work has been done by scholars from different countries to obtain the conditions and laws of outbursts. Many hypotheses have been proposed to explain the formation mechanism of coal and gas outbursts. In general, these hypotheses can be divided into three categories: the gas-dominant theory, stress-dominant theory, and multiple-factor theory [1].

Among these, gas-dominant theory considers that the high-pressure gas stored in the coal seam is the leading factor for an outburst. A representative hypothesis called the "gas pocket" hypothesis postulates that there is an abnormal "gas pocket" area with a high gas pressure and high gas content in the coal seam. In addition, the coal in this area is soft and contains many fractures and pores with a greater capacity for gas storage. Once the "gas pocket" is exposed, the soft coal is broken and thrown out by the high-pressure gas [28]. Another theory of gas-dominant considers that methane exists in coal as an unstable compound, whereby its rapid decomposition causes an outburst. Odintsev proposed that dissolved gas rapidly desorbs from coal before expanding and causing the failure of coal [29].

Stress-dominant theory holds that an outburst is the result of high stress in coal, which includes deformation potential theory, stress concentration theory, stress superposition theory, and the uneven displacement of roof theory. Kitdybinski proposed that there was a protective belt/degassing belt, a high-pressure gas belt/active belt, and a support pressure belt in front of the work face, and postulated that an outburst is the extrusion failure of coal in the active belt [30].

However, according to the multiple-factor theory, an outburst is the combined effect of many factors, such as stress, gas, and the mechanical properties of coal. Because of its comprehensive consideration of the power and resistance of outbursts, this theory has been widely acknowledged by scholars. This type of theory was first proposed by Nie Lipinski of the former Soviet Union in the 1950s, which held that an outburst was the combined effect of stress and gas. Then, in the mid-1950s, Skochinski of the former Soviet Union put forward that outbursts were caused by the comprehensive effect of stress, gas, and the physical properties of coal. In the late 1970s, based on the study of the adsorption, permeability, and mechanical properties of coal, Hoddot carried out many outburst simulation experiments, and proposed the "energy hypothesis". He also calculated the elastic energy, gas potential energy, gas expansion energy, and crushing work of coal, and gave the energy condition for outbursts [31]. Subsequently, the "energy hypothesis" has played an important role in the study of the outburst mechanism. Hanes et al. believed that both stress and gas play important roles in outbursts, but that one factor sometimes dominates [32].

In addition, Sato and Fujii analyzed the large seismic waves of the Sunagawa Coalmine, and believed that earthquakes could induce outbursts and play a major role in the outburst process [33]. Chen considered that there was a choking effect for gas flow when the width of the fracture area in front of the working face was less than the critical scale, which caused the mutation of gas pressure and led to a domino effect prior to the occurrence of an outburst [34].

For many years, a lot of research has been conducted on the outburst mechanism by Chinese scholars. Through a study on the mechanical properties of coal containing gas under the three-dimensional stress state, it was postulated that there is a rheological characteristic for coal containing gas; thus, the rheological hypothesis of outbursts was proposed [35]. Jiang and Yu undertook many outburst experiments, and explained the dynamic process and characteristic phenomenon of an outburst from the viewpoint of "spherical shell destabilization" [36]. This was termed the "spherical shell destabilization" hypothesis, and was subsequently used for outburst prevention [37]. According to the energy dissipation in the outburst process, the elastic energy dissipates first to break the coal, which creates the condition for the release of gas energy; hence, the initial gas expansion energy plays a decisive role in outbursts [38].

Zhang proposed the unified instability theory, and considered that both gas outbursts and rock outbursts involve a dynamic instability process under an unstable state. They presented an energy criterion as an instability criterion for gas outbursts and rock outbursts [39]. Li believed that coal and gas two-phase fluid forms during the outburst process, whereby the two-phase fluid accumulates outburst energy under compression conditions, and an outburst occurs when the outburst energy is greater than the outburst energy dissipation [40]. However, Guan et al. thought that outbursts are similar to an eruption of volcanic magma, whereby a large amount of gas rushes out to break the coal and release gas expansion energy [41]. In addition, Lyu proposed an outburst mechanism for key layer-stress walls [42], while the outburst mechanism for compression disseminated values of coal and (out of control) gas was proposed by Ma and Yu [43].

11.2.2 Division of Outburst Process

As shown in Fig. 11.8, an outburst generally involves the following four stages: preparation stage, trigger stage, development stage, and termination stage. There is a significant difference between the physical processes of different stages.

(1) Preparation stage: All kinds of outburst factors (e.g., outburst energy, stress, and coal gas) continuously accumulate with the quasi-static failure of coal, and the critical condition or critical threshold is satisfied gradually.
(2) Trigger stage: Disturbed by external activities, coal in a critical state enters rapid failure, a small amount of coal is stripped, and an initial outburst hole forms. This stage provides a space condition for the subsequent development of the outburst.
(3) Development stage: This stage is a complex transfer process involving the gradual deepening of the damage surface of an outburst into the internal side. It

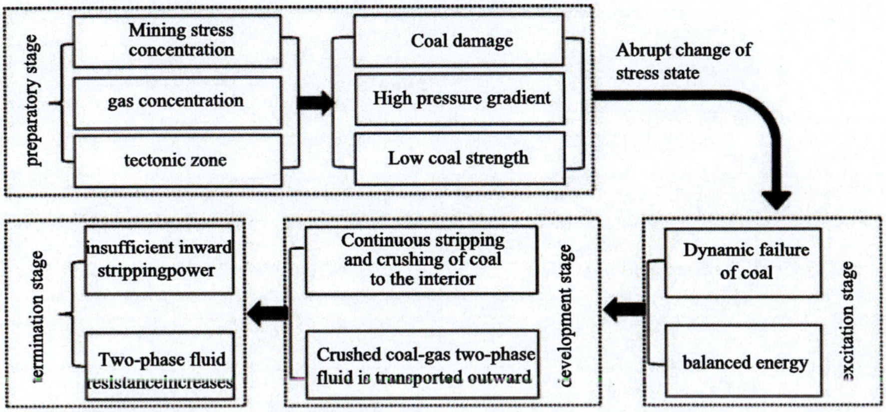

Fig. 11.8 Coal and gas outburst process

can be subdivided into three periods (a growth period, stable period, and decay period), and the coal may experience two kinds of failure: shear failure and tensile failure [44]. Evidently, the cycle process of coal failure can be described as follows: ① When coal is suddenly exposed, stress transfers quickly, resulting in concentrated stress forms in front of the exposed surface. ② Under concentrated stress, coal experiences shear failure, and many coal fractures develop, causing a significant weakening of coal strength. ③ A high gas pressure gradient forms because of the quick migration of coal gas. It has a tensile effect on the coal body. The macroscopic crack extends in the direction perpendicular to the gas pressure gradient, and a "spherical shell" spallation appears. ④ Due to the supply of desorption gas, a high gas pressure can be maintained in the macroscopic crack, and the coal spallation is formed and separated from the exposed surface. ⑤ The broken coal is consistently transported gas flow to the working space. ⑥ The damage surface of outburst goes deeper into the internal side.

(4) Termination stage: The outburst energy reduction (or outburst resistance) increases, and as a result, the outburst energy, stress, and coal gas cannot maintain the transportation of outburst coal [45, 46]. The outburst ceases, while the coal returns to a static balance.

11.3 Coal and Gas Outburst Preparation and Trigger

11.3.1 Theoretical Model

In the mining process, there is a preparation process before an outburst occurs, and the trigger of an outburst is the sudden instability of coal (rock) under the effect of external activities and gas. Therefore, the study of the physical process before an

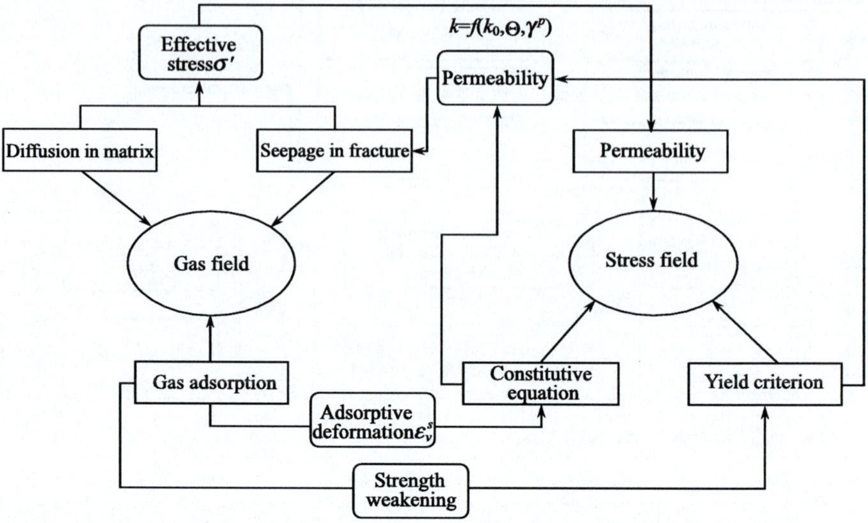

Fig. 11.9 Theoretical model for the physical process of outburst preparation

outburst occurs is the foundation of analyzing outburst preparation and triggers. The physical process involves the evolution of the stress and gas fields, and the interaction between them, as illustrated in Fig. 11.9. In addition, there are two types of effects of gas on coal: free gas affecting the effective stress of coal, and adsorbed gas causing the swelling deformation of coal. Meanwhile, the stress and deformation of coal influence coal permeability and gas migration.

To analyze the physical process, a theoretical model involving gas migration, coal stress–strain, and yield failure can be built. The governing equations for gas migration and permeability evolution were introduced in Chaps. 7 and 9. Therefore, an evolution model for the yield failure of coal should be established.

To solve the problem of singular points prone to appear in numerical calculations using the Mohr-Coulomb criterion. Here, we selected the Mohr–Coulomb matching DP yield criterion for the failure criterion of the coal mass [47]:

$$F = \alpha_{DP} I_1 + k_{DP} - \sqrt{J_2} \tag{11.1}$$

where I_1 is the first stress invariant; and J_2 is the second deviator stress invariant. In addition, a_{DP} and k_{DP} match the Mohr-Coulomb criterion in the case of the plane strain correlation flow rule, which can be identified by the cohesion c and friction angle φ:

$$\alpha_{\mathrm{DP}} = \frac{\sin\varphi}{\sqrt{3}\sqrt{3+\sin^2\varphi}} \tag{11.2}$$

$$k_{DP} = \frac{3c\cos\varphi}{\sqrt{3}\sqrt{3+\sin^2\varphi}} \tag{11.3}$$

It is well known that gas adsorption reduces the strength of coal, and multiple mechanisms may be involved in multiple sorption forms of gas in coal. The adsorbed gas lowers the surface tension of coal and reduces the crack initiation threshold stress according to Griffith's failure criterion, thus decreasing the coal strength. However, the absorbed gas dissolved in coal acts as a plasticizer and alters the coal structure. Although the evolution of mechanical properties after gas adsorption has not yet been quantified, it can be inferred that the strength reduction is proportional to the adsorption quantity. The mechanical parameter of coal can be expressed as follows:

$$\omega = \alpha(p_m)\omega_n \tag{11.4}$$

where ω is the mechanical parameter of coal, ω_n is the mechanical parameter before adsorbing gas, and α is the weakening coefficient.

Equations (7.130) and (11.1) constitute the constitutive equations and yield criterion for coal (rock). Once coal is in plastic failure, several mechanical parameters are irreversibly changed, and it experiences an apparent strain softening. The softening property of coal after the stress peak greatly reduces the load-carrying capacity of the coal structure and accelerates the occurrence of instability.

The coal mass is an elastic–plastic material with a strain-softening property; hence, a strain-softening model is appropriate [48, 49]. The mechanical parameters are defined to evolve with the softening parameter [50]:

$$\omega = \begin{Bmatrix} \omega_0 - (\omega_0 - \omega_r)\gamma^p/\gamma^{p^*}, & 0 < \gamma^p < \gamma^{p^*} \\ \omega_r, & \gamma^p \geq \gamma^{p^*} \end{Bmatrix} \tag{11.5}$$

where ω_0 represents the original parameters before the stress peak, ω_r represents the residual parameters at the residual zone, γ^p represents the softening parameter, and γ^{p^*} represents the transition value of the softening parameter from which the residual behavior begins. A common softening parameter is the equivalent plastic shear strain, defined as [51]:

$$\gamma^p = \sqrt{\frac{2}{3}\left(\varepsilon_1^p\varepsilon_1^p + \varepsilon_2^p\varepsilon_2^p + \varepsilon_3^p\varepsilon_3^p\right)} \tag{11.6}$$

where ε_1^p, ε_2^p, and ε_3^p are principal plastic strains.

Research has shown that cohesion decreases with the strain softening of coal, while the friction angle is less affected [52, 53]. Therefore, it can be considered that the decrease in cohesion after the stress peak results in the softening of the coal. Combined with Eq. (11.5), the decrease in cohesion with the change in the softening parameter can be expressed using Eq. (11.7):

Table 11.1 Model parameters

Parameter	Value
Elastic modulus (*E*/GPa)	3.5
Poisson's ratio (ν)	0.3
Friction angle (φ)	35°
Initial value of cohesion (*c*/MPa)	11.3
Residual value of cohesion (*c*∗/MPa)	3.8
Transition value of softening parameter (γ^{p^*})	0.04

$$c = \left\{ \begin{matrix} c_0 - (c_0 - c_r)\gamma^p/\gamma^{p^*}, & 0 < \gamma^p < \gamma^{p^*} \\ c_r, & \gamma^p \ge \gamma^{p^*} \end{matrix} \right\} \tag{11.7}$$

The solid module of COMSOL Multiphysics finite element code was used to analyze the mechanical behavior of coal. To verify the model, the corresponding numerical simulation was carried out based on the mechanical test results of coal samples from the Daning Coalmine, Shanxi Province, China. The confining pressure of the mechanical test was 2 MPa. The numerical simulation parameters were obtained from the mechanical test curve, and are listed in Table 11.1. Based on the Mohr–Coulomb criterion, the relation between cohesion, friction angle, peak stress, and confining pressure is as follows:

$$\sigma_1 = \frac{2c\cos\varphi}{1 - \sin\varphi} + \frac{1 + \sin\varphi}{1 - \sin\varphi}\sigma_3 \tag{11.8}$$

where σ_1 is the maximum principal stress, which is the peak stress of a coal sample in a triaxial compression test, and σ_3 is the minimum principal stress, which is the confining pressure in a triaxial compression test.

According to the peak stress and confining pressure, the initial cohesion can be obtained. In addition, the cohesion at the plastic flow stage can be calculated if the plastic flow stage is considered as a plastic yield strain state. The curve obtained by numerical simulation is compared with the mechanical test curve in Fig. 11.10. It can be observed that the strain-softening model exhibited the softening characteristic of coal after the stress peak. The curve obtained by numerical simulation was in good agreement with the mechanical test results. Therefore, the strain-softening model can well describe the mechanical behavior of coal.

11.3.2 Instability Criterion for the Coal and Gas Outburst Trigger

The dynamic failure of a coal body is the visual display of an outburst. Thus, the dynamic instability caused by the accelerating expansion of coal fractures can be considered as direct evidence of the outburst trigger. However, it is still difficult to

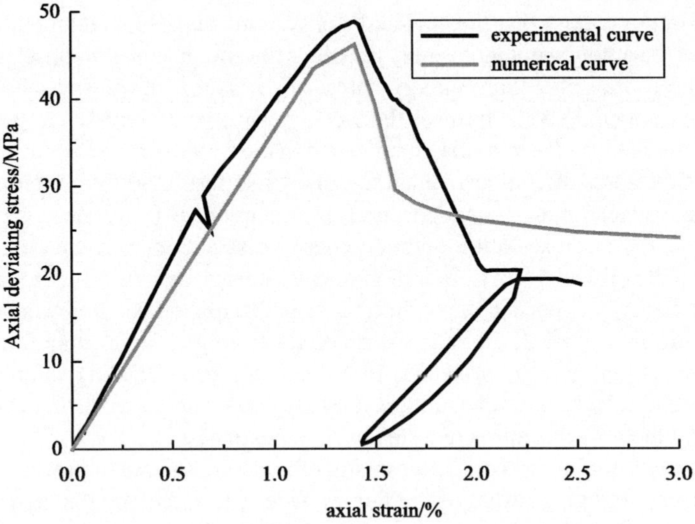

Fig. 11.10 Experimental and model fitting results of a triaxial compression test

obtain the criteria for coal failure, coal fracture expansion, and dynamic instability. Therefore, in this work, from the viewpoint of energy, we analyze the energy principle of coal and gas systems, and obtain the instability criterion.

As mentioned above, the outburst trigger is the result of the evolution of outburst energy, and the precondition of outburst is the instability failure of coal. The analysis of the energy outburst is aimed at the outburst potential energy of coal and gas systems in front of the mining face. Figure 11.11 reveals that the outburst energy consumption

Fig. 11.11 Cumulative energy state of the coal–gas system

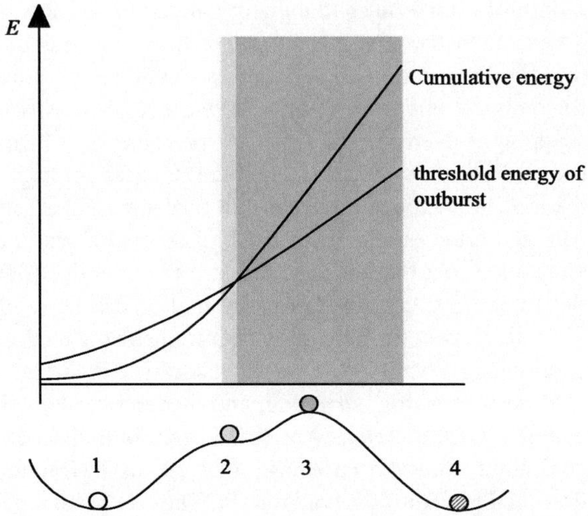

is the maximum energy that the coal and gas system can hold, and an outburst will occur when the outburst potential energy exceeds the energy consumption. Therefore, there are three kinds of outburst energy states: steady state, critical state, and unstable state. When the coal gas in front of the working face is effectively emitted, or the coal intensity is high, the outburst energy state in the failure area is in state 1 (steady state). In this state, the outburst potential energy is less than the outburst energy dissipation, and an outburst cannot occur. If the energy state is in state 2, the outburst potential energy is close to the outburst energy dissipation, and the coal and gas system is in the critical state; the addition of extra energy by mining activities causes the outburst energy to exceed the outburst energy dissipation, leading to an outburst. If the system is in state 3, the outburst potential energy is larger than the outburst energy dissipation, and the system is in an unstable state, leading to an outburst. After the rapid release of outburst energy, less residual energy exists in the coal–gas system, and the system returns to a steady state (state 4).

In the actual mining process, after taking effective measures to prevent outburst, the gas content of coal in front of the mining face is low, and the outburst potential energy is less than the outburst energy dissipation. While there are no effective measures to prevent outbursts, a high gas pressure exists in front of the mining face, and the energy state is state 2 or state 3; hence, an outburst may occur after mining activities. However, some tectonic areas are often encountered in the mining process. In these areas, the weakening of coal strength decreases the outburst energy dissipation, and the difficulty of gas extraction increases as the permeability of coal decreases, causing the accumulation of outburst potential energy; the system is in state 2 or state 3, and an outburst occurs more easily.

Based on the analysis of outburst energy conditions, we propose an energy method to analyze the outburst tendency. This method can be used to judge the outburst risk of coal seams under different occurrence conditions.

With the exception of the strength failure being a precondition for the dynamic failure of a coal mass, the energy accumulated for an outburst initiation should be greater than the energy consumed. Based on this consideration, it can be deemed that the preparation of an outburst involves the accumulation of potential energy for outburst initiation relative to the energy dissipated in coal mass crushing and stripping. Given that that the gas pressure is 2–3 orders of magnitude lower than the coal stress, it is easy to find that the gas promotes the failure of coal; however, the leading factor of failure is still coal stress. Gas pressure aggravates the failure of coal and crushes and strips the broken coal mass in outbursts. Therefore, it can be concluded that the outburst starts in the coal destruction zone and that the ratio of the potential energy to energy dissipation can be used to evaluate the outburst risk.

With respect to the energy requirement for coal and gas outbursts, the potential energy for outburst initiation (E_i) should be greater than the energy dissipation in the coal mass crushing, stripping, and movement (E_w) (i.e., $E_i > E_w$) [31]. As shown in Fig. 11.12, the energy released in an outburst includes the elastic energy of the coal mass, the internal energy of free gas in fractures, and the internal energy of desorbed gas from the coal matrix. The energy that participates in outburst initiation includes the elastic energy stored in the coal mass (E_e) and the internal energy of

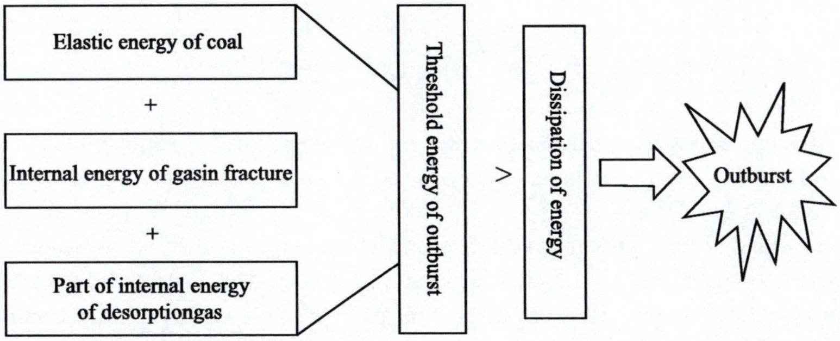

Fig. 11.12 Diagram of outburst energy

partial gas (E_g) due to the short time of outburst initiation. The involved gas occupies a small fraction of the coal gas. The energy dissipation in movement and in broken coal during the follow-up process is provided by the sequentially desorbed gas from the coal matrix. Therefore, the potential energy for outburst initiation is defined as $E_i = E_g + E_e$. the elastic energy stored in coal (E_e) and the energy of partially adsorbed gas (E_b), and is dissipated during the crushing and stripping of a coal mass; hence, $E_w = E_b$ can be assumed.

The unit volume strain energy of coal under the condition of linear elasticity can be expressed using Eq. (11.9):

$$E_e = \int_0^{\varepsilon_{ij}} \sigma_{ij} d\varepsilon_{ij} = \frac{1}{2}\sigma_{ij}\varepsilon_{ij} \tag{11.9}$$

After yielding, the coal undergoes plastic strain, but the unloading stress–strain curve is similar to that in the elastic stage. Thus, the elastic energy of the coal can be expressed by Hooke's law. Under a one-dimensional stress condition, the elastic energy per unit volume of coal mass is given as follows:

$$E_e = \frac{\sigma_i^2}{2E} \tag{11.10}$$

For the condition of three-dimensional stress with the generalized Hooke's law, the elastic energy per unit volume of coal mass is expressed by Eq. (11.11):

$$E_e = \frac{1}{2E}\left[\sigma_1^2 + \sigma_2^2 + \sigma_3^2 - 2\nu(\sigma_1\sigma_2 + \sigma_2\sigma_3 + \sigma_3\sigma_1)\right] \tag{11.11}$$

The potential energy for outburst initiation by coal gas through expansion work can be calculated using Eq. (11.12):

$$E_g = \frac{p_0 V_0}{\eta - 1}\left[\left(\frac{p}{p_0}\right)^{\frac{\eta-1}{\eta}} - 1\right] \tag{11.12}$$

where V_0 represents the volume of coal gas involved, and η represents the adiabatic coefficient. The gas expansion in an outburst is a polytropic process, and η is approximately 1.3 [28].

The amount of coal gas involved in outburst initiation is difficult to measure. In addition to free gas, research by Valliappan and Zhang demonstrated that desorbed gas from the coal matrix plays an important role [54]. As dual-porosity media, the gas in fractures is deemed to participate entirely in the outburst initiation, whereas the gas in the coal matrix participates partially. The potential energy for outburst initiation by coal gas is calculated as follows:

$$E_g = \frac{p_{f0} V_{f0}}{n - 1}\left[\left(\frac{p_f}{p_0}\right)^{\frac{\eta-1}{\eta}} - 1\right]\rho_c + \frac{p_{m0} V_{m0}}{n - 1}\left[\left(\frac{p_m}{p_0}\right)^{\frac{\eta-1}{\eta}} - 1\right]\rho_c \tag{11.13}$$

where V_{f0} represents the coal gas content in the fracture, and V_{m0} represents the participant coal gas content from the coal matrix.

The participant coal gas from the coal matrix varies and is related to the gas content, coal particle size, and diffusion coefficient [55]. The desorbed gas from the coal matrix during outburst initiation can be estimated by the desorption ratio (λ), and the participant coal gas from the coal matrix ($V_{m\infty}$) can be expressed as follows:

$$V_{m0} = \lambda V_{m\infty} \tag{11.14}$$

where $V_{m\infty}$ represents the total desorbed gas of the coal matrix.

If the ambient pressure of gas desorption is estimated according to atmospheric pressure, the total desorbed gas can be calculated based on the difference in the gas content of the coal matrix before outburst initiation and the gas content with the gas pressure as the standard pressure. The participant's internal energy changes from Eqs. (11.13) to (11.15):

$$E_g = \frac{p_{f0}}{\eta - 1}\frac{p_f \phi_f}{p_0}\left[\left(\frac{p_f}{p_0}\right)^{\frac{\eta-1}{\eta}} - 1\right] + \frac{p_{m0}}{\eta - 1}\rho_c \lambda\left(\frac{p_m \phi_m}{p_0 \rho_s} + \frac{V_L p_m}{p_L + p_m} - \frac{p_0 \phi_m}{p_0 \rho_s} - \frac{V_L p_0}{p_L + p_0}\right)\left[\left(\frac{p_m}{p_0}\right)^{\frac{\eta-1}{\eta}} - 1\right] \tag{11.15}$$

The desorption ratio is estimated using the following method. The outburst initiation is a transient process, and the desorption amount of coal gas over a short period is linearly related to the square root of the desorption time, and is expressed as follows [56]:

$$\frac{V_t}{V_\infty} = \frac{6}{\sqrt{\pi}}\sqrt{\frac{Dt}{r_p^2}} \tag{11.16}$$

where V_∞ represents the total desorbed gas, D represents the diffusion coefficient, and r_p represents the diffusion radius. The adsorption time τ is related to the diffusion coefficient and matrix size. The cubic model of the coal matrix can be estimated using Eq. (11.16) [57]:

$$\tau = \frac{L^2}{3\pi^2 D} \tag{11.17}$$

where L represents the space between fractures, cm. If r_p is $L/2$, $\sqrt{\frac{Dt}{r_p^2}}$ can be obtained from Eq. (11.15). According to the observation records of the coal and gas outburst test site in Zhongliangshan Mine, China in 1977, the first thud in sonic record occurred without a change in other parameters at 1.5 s after the inducing blast, and then three thuds occurred at 2.5, 3.5, and 4 s [58]. The continuing time of outburst initiation is the average of the recorded occurrence time of the later three thuds. For the adsorption time in the model, the desorbed gas for outburst initiation was 0.677% of the total desorbed gas, and the desorption ratio was 0.677%.

The coal crushing work is equivalent to the surface energy of the newly increased surface area of crushed coal, and can be expressed by Eq. (11.18):

$$E_b = s \cdot w \cdot \rho \tag{11.18}$$

where s represents the increased surface area, and w represents the specific energy.

Studies on the specific energy of coal have shown that soft coal is proportional to the firmness coefficient and has an outburst danger with less specific energy than hard coal. The specific energy of coal was estimated to range from 1.07×10^{-3} to 2.88×10^{-3} J/cm^2 [59, 60]. The additional specific surface area of a coal mass that was crushed by the hammer drop method ranged from 113 to 525 cm^2/g, and the specific surface area of a fragment from the rock burst test was 200 cm^2/g [61]. In this book, a specific energy of 1×10^{-3} J/cm^2 and an additional specific surface area of 150 cm^2/g are used, and the surface energy consumed in outburst initiation is taken as 0.15 MJ/t.

Outburst initiation requires the accumulation of more potential energy than is consumed, and can occur in the failure zone. Hence, the potential energy accumulated in the coal failure zone is related to the energy necessary for coal mass crushing, and can be analyzed to indicate the outburst tendency by using the following ratio:

$$R_0 = \frac{\int_{V_p} e_i \mathrm{d}V}{\int_{V_p} e_b \mathrm{d}V} = \frac{\int_{V_p} (e_e + e_g) \mathrm{d}V}{\int_{V_p} e_b \mathrm{d}V} \tag{11.19}$$

where V_p is the plastic zone. When R_0 is > 1, the potential energy for outburst initiation is sufficient and indicates the danger of an outburst.

11.3.3 Numerical Model Calculation

Equations (9.19), (9.24), (7.128), (7.130), and (11.1) are the governing equations for the coal gas field and stress–strain field. The evolution of mechanical parameters, coal porosity, and permeability follow Eqs. (11.7), (7.143), and (7.142), respectively. The relationship between coal gas and stress–strain is presented in Fig. 11.13.

Because the governing equations are complex mathematical models involving the coupling of a multi-physics field and multivariable differential equations, it is difficult to obtain an analytical solution; however, a numerical solution can be found with the help of the COMSOL Multiphysics finite element code. In the process of model solving, the solid module in COMSOL Multiphysics can be adopted for the calculation of coal stress–strain, while gas migration can be determined using the PDE module. The interaction between the solid module and PDE module is realized through the transfer of parameters.

The failure of gas containing coal is a multi-physical field process, and there is a coupling relationship between the stress field, seepage field, and temperature field. Owing to the small changes in temperature near the mining face, the temperature field is simplified into a constant temperature field. Therefore, the multivariable differential equations include the stress field, seepage field, and their coupling effects.

The solving methods of the multi-physics field can be divided into two methods: direct coupling calculation and sequential coupling calculation. The direct coupling calculation is a simultaneous calculation of variables in a multi-physics field, which is closest to the real physical process. However, because of the simultaneous calculation of variables, multiple equilibrium conditions need to be satisfied; thus, nonlinearity is strong and the calculation of these variables is difficult, which leads to a convergence issue. In general, sequential coupling calculations are often used when the dependence between the physical fields is not strong. The sequential coupling calculation is

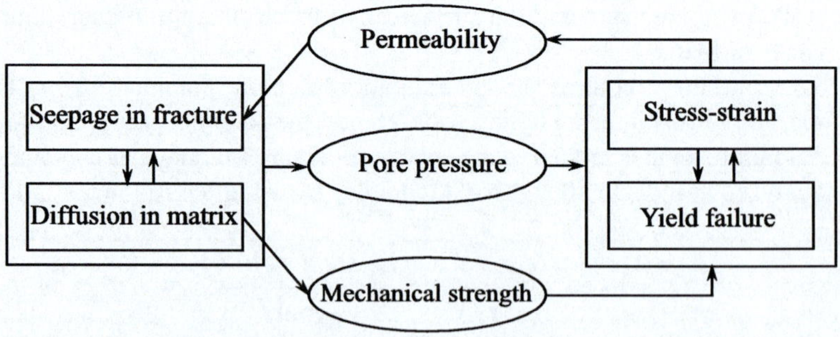

Fig. 11.13 Interaction of coal and coal gas

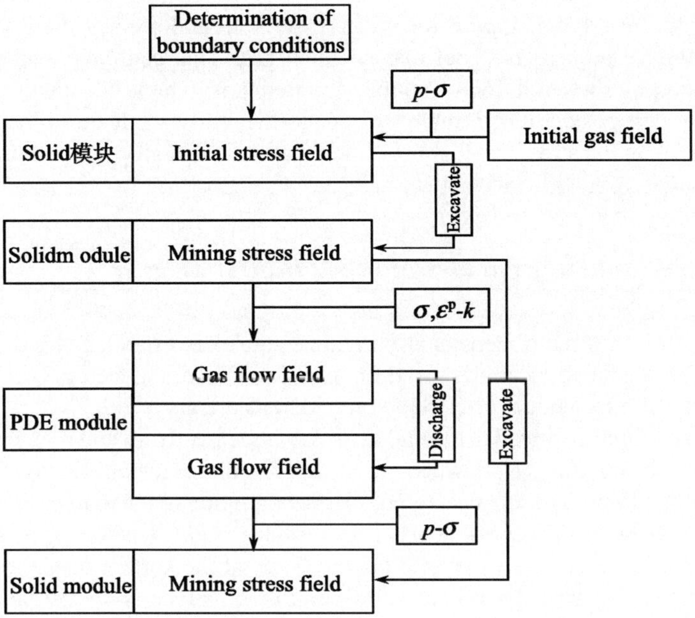

Fig. 11.14 Flowchart of the solution procedure

used to separate the physical fields in a specific order before calculating the results of the preceding physical field as the calculation conditions for the later physical fields. Compared with the direct coupling method, this type of analysis method is efficient and easy; therefore, it is widely applied in the analysis of multi-physics fields.

As shown in Fig. 11.14, there is a coupling relationship between the gas field and stress field, and it is a sequential process for the two fields over time. The existence of gas will affect the force of the coal body; that is, the initial gas field will affect the initial stress field. Because the evolution of the gas field is a relatively lengthy process, the changes in the stress field are shorter, which can quickly return to the stress balance. Therefore, the evolution of the gas field has little influence on the stress field, and the stress field is not recalculated in the gas migration process after excavation. However, the change in the stress field causes the failure of coal, which affects the condition for gas migration. In the next step of excavation, the stress field is influenced by gas migration.

The coupling relationship between the stress field and seepage field leads to a complex failure process for coal that contains gas. This coupling relationship is characterized by the evolution of coupling parameters. The influence of the stress field on the seepage condition is mainly that of permeability, while the influence of the seepage field on the stress field is that of pore pressure and adsorption deformation.

11.3.4 Influencing Factors of the Outburst Trigger

In this section, a three-dimensional numerical model is established based on the theoretical model and instability criterion, and the influencing factors (gas, tectonic stress, and coal mechanical properties) on outbursts are analyzed.

A coal seam was included in the middle of the model, and roof and floor rocks were included in the geometry, as shown in Fig. 11.15. The finite element mesh consisted of 74,692 tetrahedrons. The x–z face ($y = 0$) at the beginning of the roadway excavation was treated as the symmetry boundary. The x–y face ($z = 21.5$) at the top of the model was the load boundary. The x–y face ($z = -21.5$) at the bottom of the model was fixed. The two y–z faces ($x = 0$; $x = 100$) and the x–z face ($y = 150$) at the sides were the slip boundaries or the stress load boundaries according to the ground stress conditions. The excavation direction was in the y direction, and the roadway was 10 m long in the model.

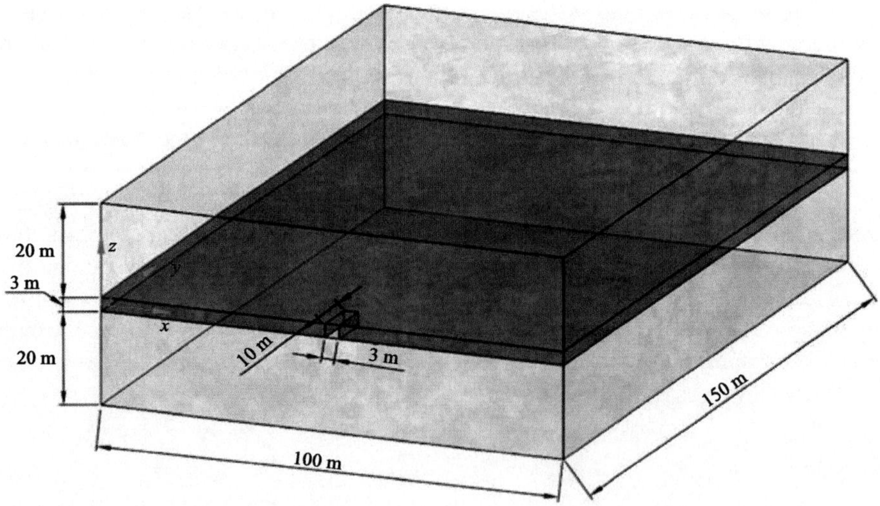

Fig. 11.15 Geometric model

Table 11.2 Model parameters

Parameter	Value	Parameter	Value
Elastic modulus of coal (E_c/GPa)	2.3	Bulk density of rock (ρ_r/(kg/m^3))	2500
Poisson ratio of coal (ν_c)	0.3	Elastic modulus of rock (E_r/gGPa)	20
Bulk modulus of the coal matrix (K_m/GPa)	7.5	Cohesion of rock (c_r/MPa)	20
Bulk modulus of the coal skeleton (K_s/GPa)	8.9	Friction angle of rock (φ_r/(°))	40
Maximum adsorption-induced volume strain for CH_4 (ε^s_{Lv})	0.012	Poisson ratio of rock (ν_r)	0.3
Langmuir pressure of adsorption-induced volume strain for CH_4 (p^s_L/MPa)	1	Temperature of coal seam (T/K)	293
Maximum weakening coefficient (α)	0.3	Initial permeability of coal seam (k_0/mD)	0.0025
Initial cohesion of coal without gas (c_c/MPa)	0.92	Initial fracture porosity (ϕ_f/(%))	0.1
Residual cohesion of coal without gas (c^*_c/MPa)	0.6	Matrix porosity (ϕ_m/(%))	6
Friction angle of coal (φ_c/(°))	30	CH_4 Langmuir volume (V_L/(m^3/t))	20
Transition value of coal softening parameter (γ^{p*}/%)	2	CH_4 Langmuir pressure (P_L/(MPa))	1
Bulk density of coal (ρ_c/(kg/m^3))	1300	Adsorption time (τ/d)	11.7

The mechanical properties of the roof and floor rocks were simplified as hard elastoplastic materials. The coal seam was assumed to be a strain-softening elastoplastic material. The mechanical and gas migration parameters were obtained from previous studies and are provided in Table 11.2.

In the tectonic area, the horizontal stress of the gravity stress field increased with the effect of tectonics and formed stress field types as: $\sigma_H > \sigma_v > \sigma_h$ and $\sigma_H > \sigma_h > \sigma_v$. With an overlying stratum load at a depth of 400 m and an average density of 2500 kg/m^3 (10 MPa) as the reference stress, the ground stress of the stress field types above was loaded in the model (Table 11.3). The ground stress included the gravity stress field (case 1), the stress field influenced by tectonics (cases 2 and 3), and the stress field strongly influenced by tectonics (cases 4 and 5). The slip boundary was set as the gravity stress field in case 1.

The difference between the coal seam roadway and the rock roadway is that the coal gas affects the stress state of the coal mass surrounding the roadway. With various stress and coal seam conditions, the gas pressure required for an outburst will differ. A relative gas pressure of 0.74 MPa is recommended in China. Coal seams without coal gas and with a gas pressure of 0.8 and 1.5 MPa (9.3 m^3/t and 12.7 m^3/t, respectively) are considered here. Under different stress conditions, the coal gas distributions

Table 11.3 Conditions of ground stress and coal gas in the model

Case	Ground stress in each direction			Initial gas pressure in coal seam (MPa)		
	X direction	*Y* direction	*Z* direction	I	II	III
1	Slip boundary		1	0	0.8	1.5
2	0.5	1.5	1			
3	1.5	0.5	1			
4	1.5	2	1			
5	2	1.5	1			

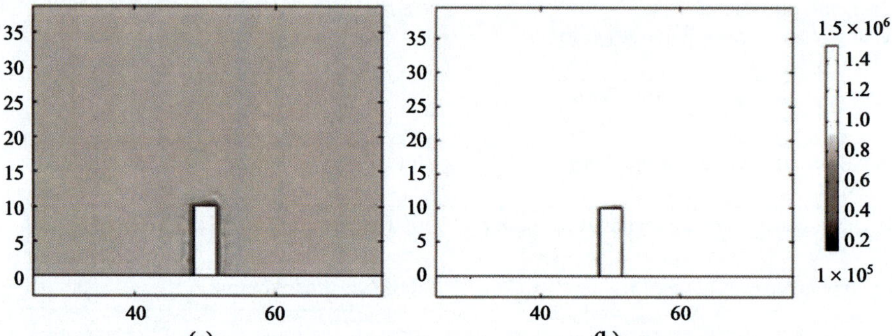

Fig. 11.16 Distribution of coal gas pressure in fractures surrounding the roadway

surrounding the roadway were found to be the same (Fig. 11.16). Figure 11.16a shows the fracture pressure distribution surrounding the roadway under an initial gas pressure of 0.8 MPa, while Fig. 11.16b presents that under an initial gas pressure of 1.5 MPa. The numbers beside the frame are coordinate values. The values in the legends represent the gas pressure (in Pa) in the fractures. The gas distribution was obtained after 0.5 d of gas migration to the coal wall (permeability of 0.0025 mD).

The effect of horizontal stress on roadway stability is focused on the deformation and failure of coal walls and roofs [62–64]. Coal and gas outbursts occur in the coal mass surrounding the roadway, and most outbursts occur in front of the heading face during excavation. In this section, the deformation and plastic zone surrounding a roadway with various ground stress conditions are analyzed, and the outburst risk in the front of the roadway is studied.

Figure 11.17 reveals that the deformation of the coal wall was affected by the coal gas and stress conditions. The tectonic stress increased the average and maximum deformation of the coal wall in front of the roadway, and the increase was affected by the distribution of tectonic stress. The coal mass deformation in front of the roadway was greater when the maximum horizontal stress was parallel to the direction of the roadway; however, as the tectonic stress increased, the effect of the stress distribution decreased. In addition, coal gas had an important effect on the coal wall. For the

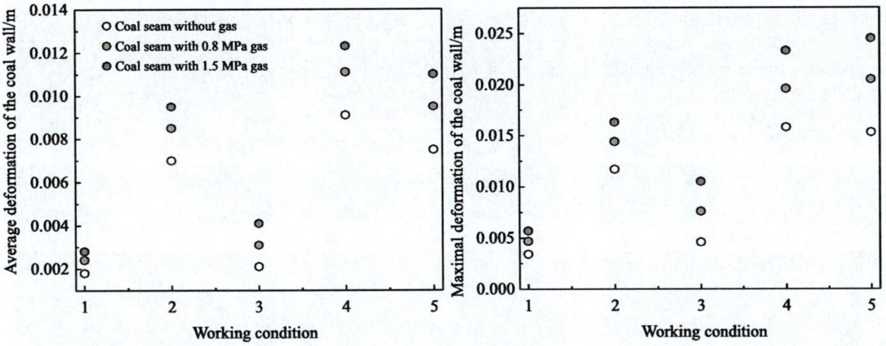

Fig. 11.17 Deformation of the coal wall in front of the roadway

various stress conditions, the deformation of the coal mass increased as the coal gas increased.

The size of the coal mass plastic zone under different stress conditions and various gas conditions is detailed in Table 11.4. The coal gas surrounding the roadway had an obvious effect on the plastic zone with respect to pore pressure and strength reduction, causing a larger area to have the same ground stress. The change in the ground stress condition affected the magnitude of the coal mass plastic zone and altered the distribution of the plastic zone. The plastic zone distribution for the middle cross-section of the coal seam with various ground stress fields is shown in Fig. 11.18. The red zone is the plastic zone, and the blue zone is the elastic zone. With the alteration of coal mass failure, the outburst risk would be affected.

Figure 11.19 shows that the energy accumulation for outburst initiation was affected by the coal gas and stress conditions. Coal gas significantly increased the potential energy for outburst initiation, and, without coal gas, the accumulated energy had difficulty reaching the energy requirement for an outburst. Coal gas is the main factor influencing outburst initiation. Additionally, tectonic stress has an important effect on the outburst tendency and is enhanced by tectonics (Fig. 11.19). For example, when the initial gas pressure was set to 0.8 MPa in the model, tectonics could decide the occurrence of an outburst. This result verifies the previous ground stress measurement results, whereby the horizontal stress was greater in the outburst area.

Table 11.4 Plastic zone of the coal mass in front of the roadway

Coal seam	Case 1/m^3	Case 2/m^3	Case 3/m^3	Case 4/m^3	Case 5/m^3
Coal seam without gas	3.40	4.44	3.70	3.41	4.57
Coal seam with initial gas pressure 0.8 MPa	5.44	5.55	5.13	4.93	5.67
Coal seam with initial gas pressure 1.5 MPa	6.32	6.85	6.39	5.86	6.62

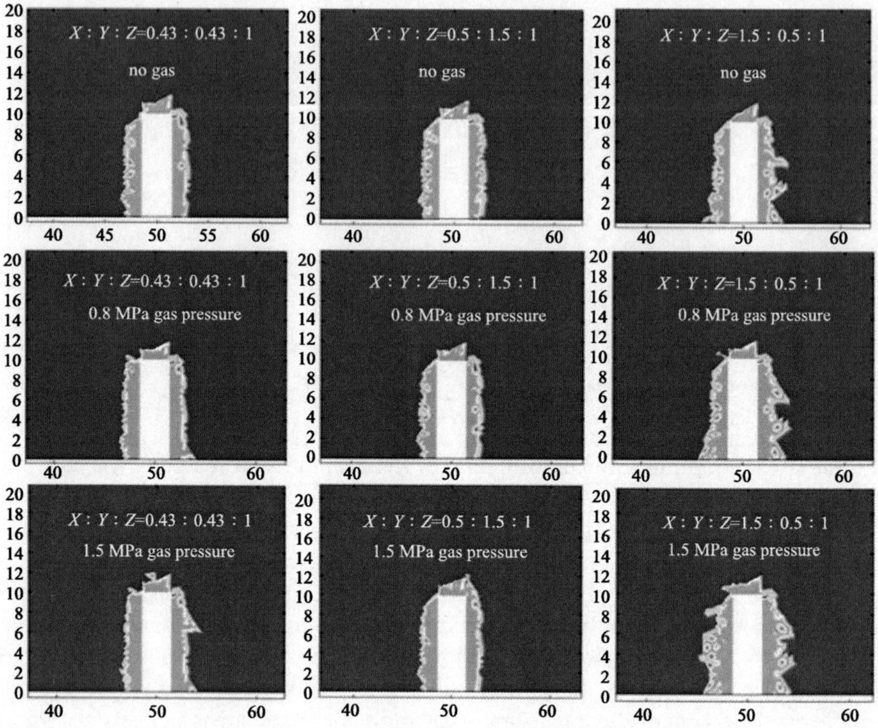

Fig. 11.18 Plastic zone of the coal mass surrounding the roadway

Fig. 11.19 Effect of coal gas and tectonic stress on the energy required for outburst initiation in front of the roadway

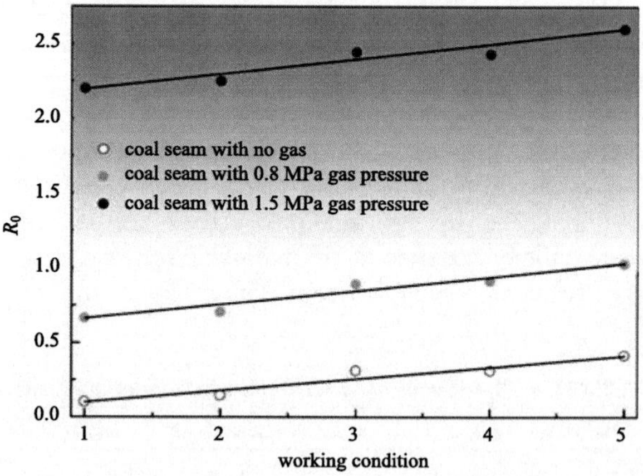

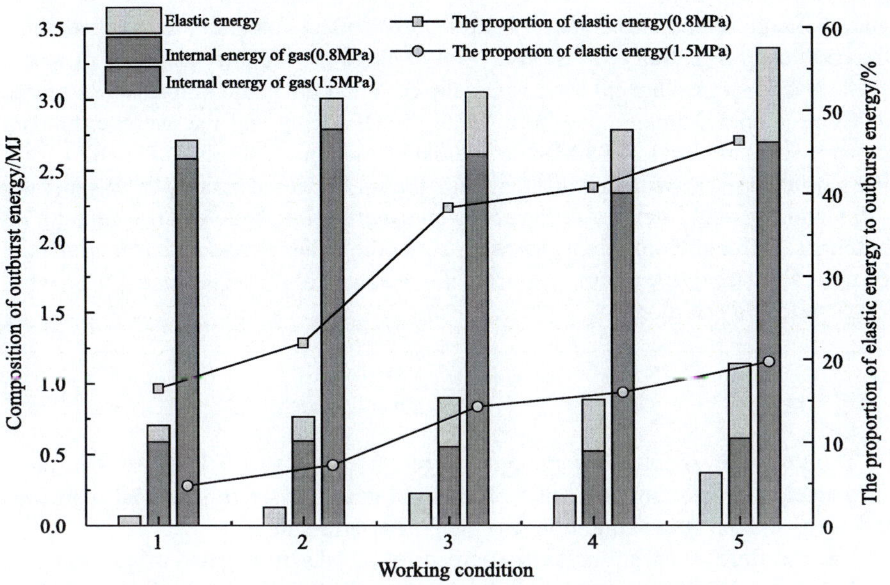

Fig. 11.20 Elastic energy for outburst initiation in front of the roadway

Furthermore, the stress distribution affected the outburst. More potential energy was accumulated when the maximum horizontal stress was perpendicular to the roadway.

To further analyze the effect of tectonic stress, the energy composition for various tectonic stresses was studied, as shown in Fig. 11.20. The result revealed that the proportion of elastic energy increased as the tectonic stress increased, with the proportion being greater when the maximum horizontal stress was perpendicular to the roadway. The elastic energy for outburst initiation, which is increased by tectonic stress, enhanced the outburst tendency. Moreover, the distribution of the horizontal stress could alter the elastic energy. Additional elastic energy was held during excavation when the maximum horizontal stress was perpendicular to the roadway, indicating a greater tendency for an outburst.

11.3.5 Influence of Mechanical Properties on Outburst Tendency

The mechanical properties of coal are a major factor in coal failure, and are also the main factor for the elastic energy and energy dissipation in the outburst process. There is an obvious regularity of soft stratification for most coal seams with outburst tendency, and outbursts often occur in tectonic areas. Therefore, the coal strength is the key factor influencing the mechanical properties of an outburst. The coal strength can be expressed by the uniaxial compressive strength. However, the soft nature of coal seams means that it is difficult to collect an intact coal sample; thus, the coefficient of solidity "f" is often used to judge the outburst tendency of a coal seam. Based on mechanical test results, the common uniaxial compressive strength of Chinese coal seams ranges from 4.9 to 49 MPa [65], and the average uniaxial compressive strength is 4.85 MPa, while the coefficient of solidity "f" is 0.51 [66]. For most coal seams with outburst tendency, the coefficient of solidity "f" is generally < 0.5 and the coal intensity is lower. To analyze the influence of coal strength on outbursts, different coal intensities were set in the solving model, and the uniaxial compressive strength was obtained from the mechanical properties using the Mohr–Coulomb yield criterion:

$$\sigma_c = \frac{2c\cos\varphi}{1-\sin\varphi} \tag{11.20}$$

The uniaxial compressive strength ranged from 1.73 to 6.93 MPa. The other parameters are listed in Table 11.5. The initial gas pressure for the coal seam was 0.8 MPa, and the stress condition was the gravity stress field.

Under different coal strengths, the acreage of the plastic area and the deformation of the coal wall exhibited obvious changes in front of the working face. Figure 11.21 shows that the acreage of the plastic area sharply decreased as the coal strength increased; that is, increasing the coal strength could significantly reduce the failure acreage of coal. In addition, as the coal intensity increased, the deformation of the coal wall greatly decreased, and the average deformation and maximum deformation of the coal wall in front of the working face decreased obviously (Fig. 11.22).

Considering that an outburst occurs in the failure area, the coal strength has a great influence on the outburst potential energy. As the coal strength increases, the outburst potential energy greatly decreases (Fig. 11.23). Therefore, the existence of soft coal causes the accumulation of outburst energy and leads to a decrease in energy dissipation for coal crushing. Thus, the possibility of the dynamic instability of coal further increases.

Owing to the softening characteristics of coal (rock), the load capacity of coal drops significantly after yielding, and the mechanical properties tend to be stable again at the residual flow stage. Therefore, not only will the initial and residual coal

Table 11.5 Mechanical parameters of various coal strengths

Case	Coal seam condition			
	Elastic modulus of coal (*E*/GPa)	Initial cohesion of coal (*c*/MPa)	Residual cohesion of coal (*c*/MPa)	Uniaxial compressive strength (σ_c/MPa)
1	1.5	0.5	0.35	1.73
2	2	0.8	0.6	2.77
3	2.5	1.5	1	5.20
4	3	2	1	6.93

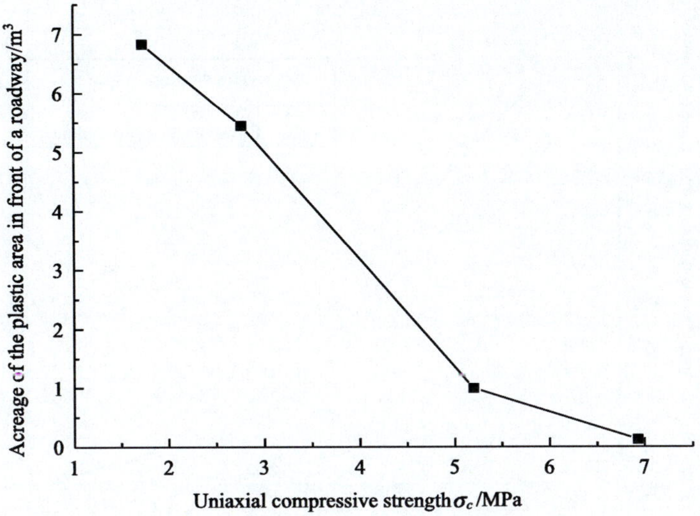

Fig. 11.21 Acreage of the plastic area in front of the roadway under different coal intensities

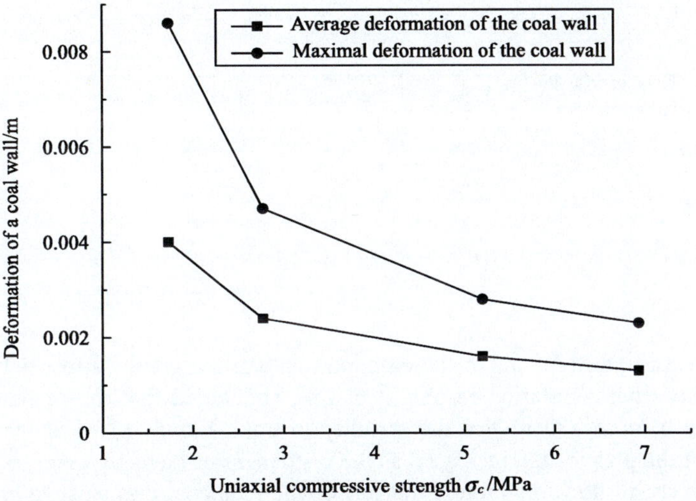

Fig. 11.22 Deformation of a coal wall under different coal intensities

intensities affect the stability of coal, but the brittleness of coal will also affect the failure of coal, which has an impact on an outburst. To analyze this problem, different brittleness degrees of coal were chosen. The pre-peak and residual mechanical properties were identical, but the plastic strain from the stress peak to the residual flow stage differed, that is, the transition value of the softening parameter from which the residual behavior began was different (Table 11.6). The initial gas pressure for the

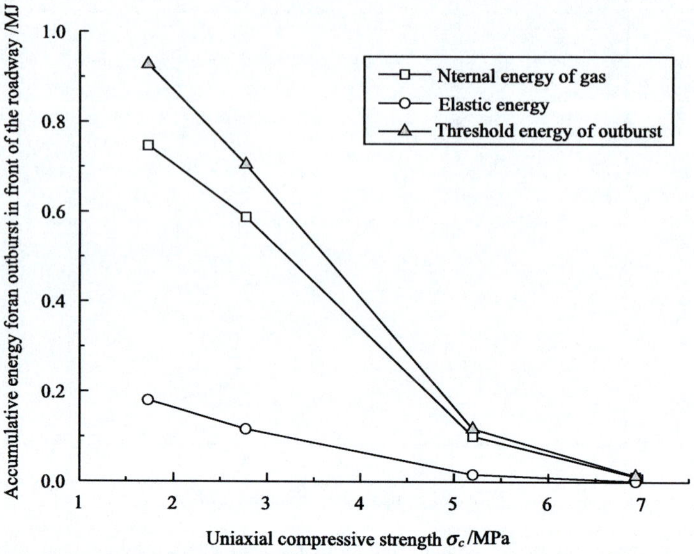

Fig. 11.23 Accumulative energy for an outburst in front of the roadway under different coal intensities

Table 11.6 Transition value of the softening parameter

Case	Transition value of the softening parameter (γ^{p^*} /%)
1	0.002
2	0.01
3	0.02
4	0.1
5	0.2

coal seam was 0.8 MPa, and the stress condition was the gravity stress field.

Based on the calculation results, it can be concluded that the increase of the brittleness of coal will reduce the stability of coal. Figure 11.24 shows that the acreage of the plastic area increased as the brittleness of the coal increased, and the acreage of the plastic area rapidly increased when the coal was close to brittleness. Similarly, the deformation of the coal wall increased as the brittleness of the coal increased (see Fig. 11.25). However, compared with the coal intensity, the brittleness of coal had less influence on the failure acreage and deformation of the coal wall.

In Fig. 11.26, although the brittleness of the coal affected the accumulation of outburst potential energy, the influence was very small. When the coal was close to brittleness, brittleness had some influence on the elastic energy and gas energy, mainly the elastic energy in the failure area. However, for coal with a certain plasticity, the increase in coal plasticity will have little influence on the accumulation of outburst potential energy.

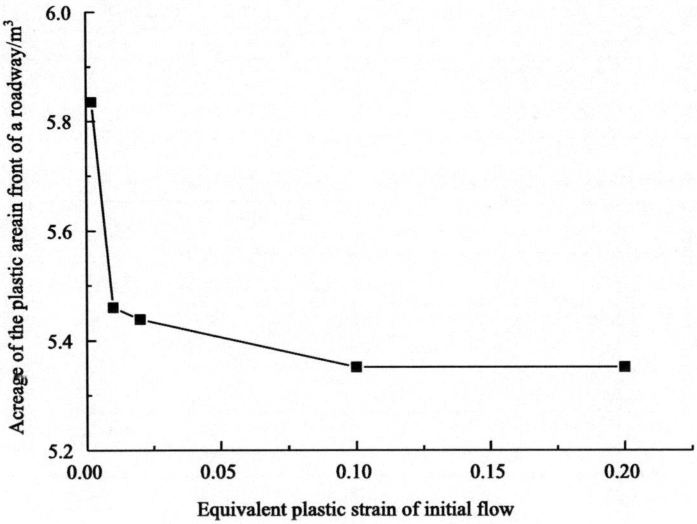

Fig. 11.24 Acreage of the plastic area in front of a roadway with the variation of the brittleness property of the coal mass

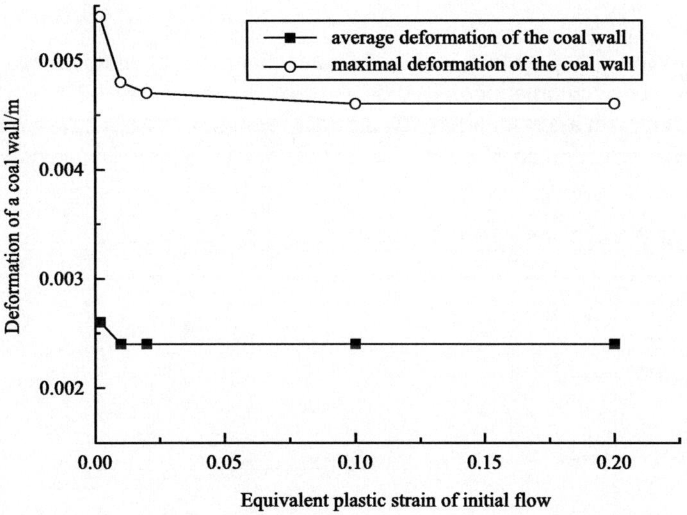

Fig. 11.25 Deformation of a coal wall with the variation of the brittleness property of the coal mass

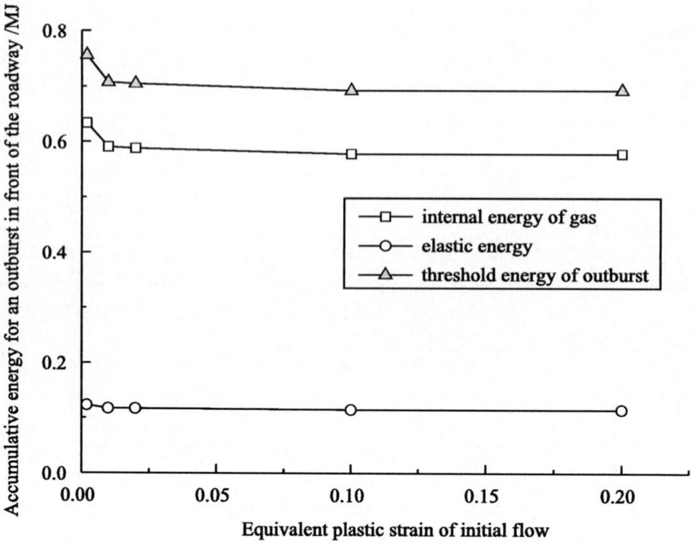

Fig. 11.26 Accumulative energy for an outburst in front of the roadway as the brittleness property of the coal mass increases

The analysis of the outburst energy indicates that the brittleness of coal has little influence on the accumulation of outburst potential energy (Fig. 11.27). Therefore, for a coal seam with a certain plasticity, the coal strength will have a greater influence on the outburst trigger than the brittleness of coal. It can be inferred that the reduction

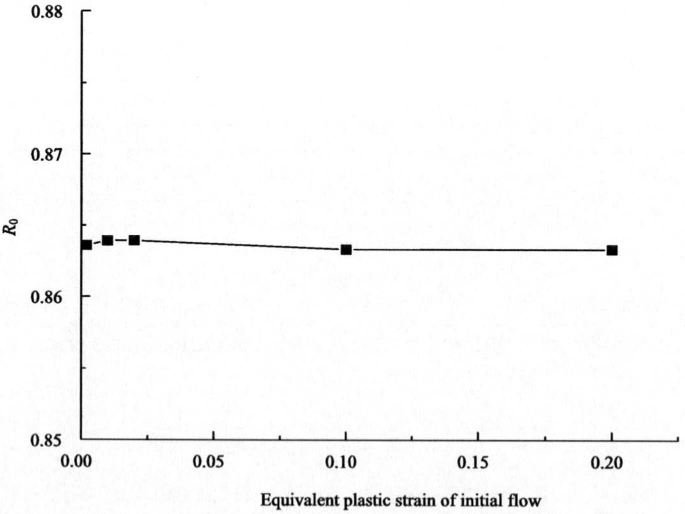

Fig. 11.27 Effect of the brittleness property of a coal mass on the outburst energy

in coal strength is the main reason for the influence of the mechanical properties on the outburst tendency.

In addition, the mechanical properties of coal not only affect the outburst potential energy, but also have an impact on energy consumption. The energy dissipation in the coal crashing process includes surface energy, heat energy, and sound energy. Among these, the surface energy for the newly added surface of coal is the main form of energy dissipation. Therefore, the crushing energy of coal can be expressed by the energy dissipation for the newly added surface. There are two main factors for coal crushing. On the one hand, the crushing of coal is related to the coal load; on the other hand, it is related to the coal strength and degree of fragmentation. The crushing test of coal shows that there is a proportional relationship between the crushing energy and newly added surface. There is also an important relationship between the crushing energy ratio and the coefficient of solidity "f", whereby the former increases as the latter increases [75, 76]. When the crushing degree is high, coal is soft and the crushing energy needs to be reduced to a certain degree. However, if the need for crushing energy is higher, an increase in the outburst energy condition will greatly reduce the outburst tendency.

Based on the analysis of the factors influencing gas, tectonic stress, and mechanical properties, we found that the gas energy is the main source of outburst energy, while tectonic stress has an important influence on the accumulation of outburst energy, which increases the outburst tendency. In addition, mechanical properties have a great influence on outbursts as well, and are mainly reflected by the influence of coal intensity and crushing energy on an outburst. Under the same conditions, soft coal can accumulate more outburst energy, and the weakening of the coal intensity will greatly reduce the energy demand for coal crushing, which increases the outburst tendency.

11.4 Coal and Gas Outburst Development

11.4.1 Analysis of the Coal and Gas Outburst Development Process

After the outburst trigger, an initial outburst hole forms, which exposes more of the coal body and provides the spatial conditions for the development of outbursts. Under the effect of stress and a high gas pressure gradient, the coal near the exposed surface is unstable and thrown out. After the coal that is exposed earlier experiences unstable failure, gas is emitted. New coal is exposed, and a new gas pressure gradient appears near the newly exposed surface. The unstable failure of coal is then transferred progressively to the internal coal.

Figure 11.28 shows that stress and redistribution are experienced with the alternating process of the exposed surface. The radial stress of the coal near the exposed face is unloaded to zero, and the stress state of the coal approximates the unloading

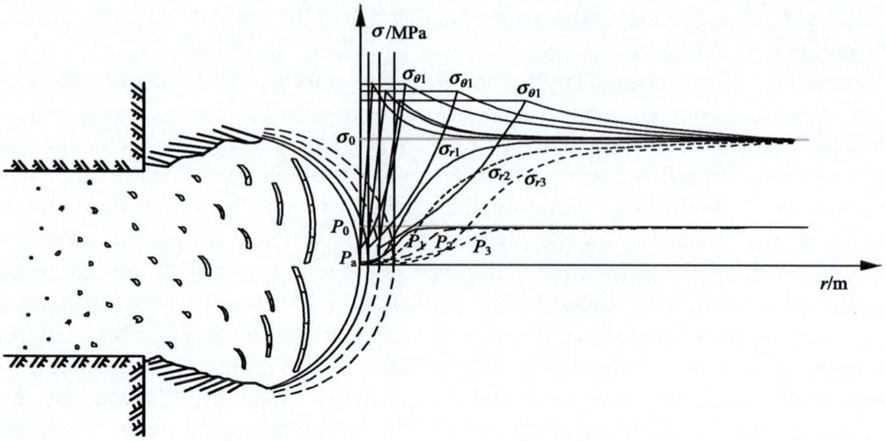

Fig. 11.28 Stress and redistribution with the alternating process of the exposed surface [37, 67, 68]

stress state. With increasing distance from the exposed surface, the radial stress gradually recovers to the original value, and the tangential stress returns to the initial value after experiencing a stress concentration; hence, the stress state changes to the original stress state. Coal is seriously damaged in the area near the exposed surface, and many coal fractures form, which cause the rapid migration of gas. At high stress concentrations, the coal pores and fractures are closed under high stress, poor coal permeability causes a rapid increase in gas pressure, and a high gas pressure gradient forms. With the exposed surface transfer to the internal coal, the stress concentration area is also gradually transferred to depth. However, owing to the large amount of energy dissipation in the outburst process, the stress reduction and gas emission cause a decrease in the stress concentration degree and the gas pressure gradient. When the remaining outburst energy cannot afford the development of an outburst, the outburst ends and the coal returns to a stable state [68].

The outburst development process is essentially the instability failure process of coal because coal fractures develop and extend under the effects of stress and gas. There are two failure types for gas containing coal: outburst and spallation. The spallation phenomenon occurs prior to an outburst [69, 70].

Before spallation, coal experiences plastic shear failure in the stress concentration area [68] and many "X"-shaped fractures appear in the coal. However, due to the unloading of radial stress, the tangential stress dominates the extension of the "X"-shaped fractures. According to the Griffith theory, fractures will extend along the direction of tangential stress to a certain degree, as illustrated in Fig. 11.29. Therefore, these fractures can be simplified to "I"-shaped fractures.

After the exposure of coal, the radial stress is rapidly unloaded to zero, the stress state of coal approximates to an unloading stress state, and a high gas pressure forms near the exposed surface (Fig. 11.30). The high-pressure gas is sealed in a fracture, and a tensile stress appears near the tip of the fracture, causing the fracture to extend

Fig. 11.29 Sketch of the extension of "X"-shaped fractures

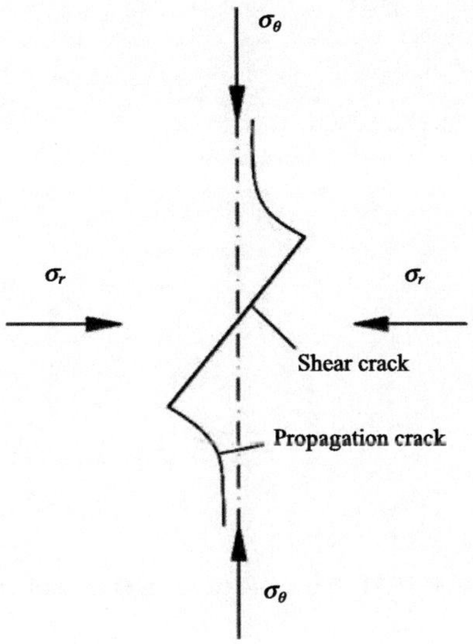

Fig. 11.30 Stress analysis of a coal cell body near the exposed surface

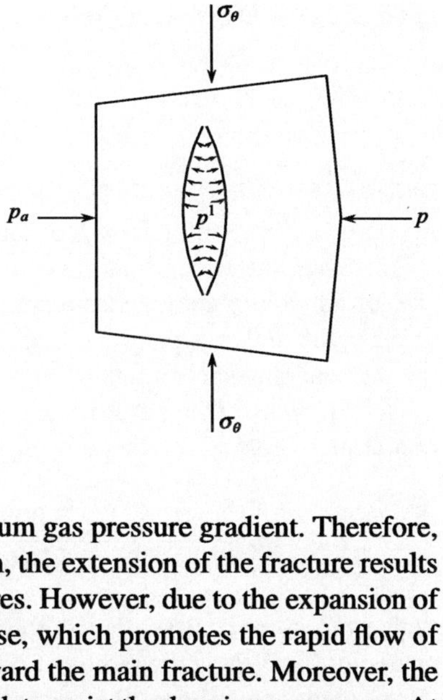

along the direction perpendicular to the maximum gas pressure gradient. Therefore, coal experiences tensile failure [70]. In addition, the extension of the fracture results in a connection with the pores and small fractures. However, due to the expansion of the fracture space, the gas pressure will decrease, which promotes the rapid flow of free gas from the pores and small fractures toward the main fracture. Moreover, the adsorbed gas near the fracture will quickly desorb to resist the drop in gas pressure. At this stage, with the extension of the fracture, a high gas pressure can be maintained in the fracture; thus, the fracture continues to extend along the tangential. When there are several such fractures within the same sphere surface (Fig. 11.31), their extension causes them to gradually close and communicate with each other. Then, an arc fracture forms, which divides the coal into a spherical coal plate with a certain thickness [71]; finally, the spallation phenomena appear in the coal.

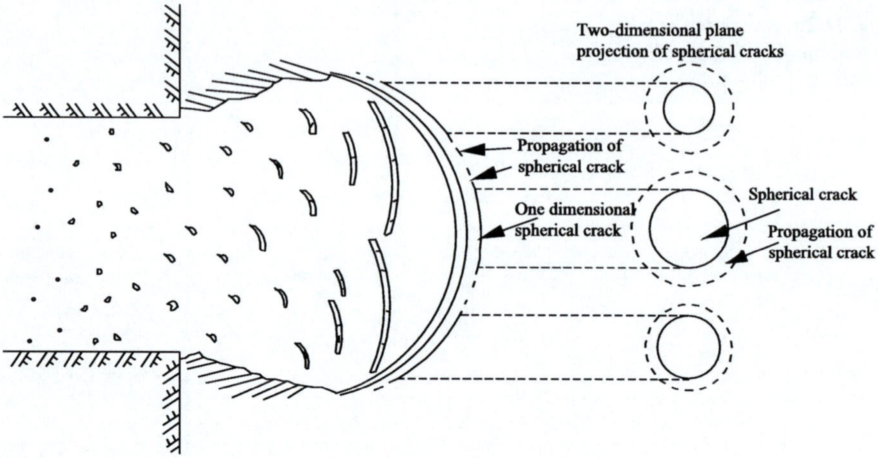

Fig. 11.31 Sketch of the expansion and transfixion of fractures

According to fracture mechanics [45], the extension condition for a coal fracture under the tensile action of high gas pressure is given by Eq. (11.20):

$$p_{if} - p_a \geq K_{\mathrm{Ic}}\sqrt{\pi}/(2\sqrt{c_l}) \tag{11.21}$$

where p_{if} is the gas pressure in the coal fracture, MPa; p_a is the atmospheric pressure in front of the exposed surface, MPa; K_{Ic} is the fracture toughness of coal, MPa · $\mathrm{m}^{0.5}$ and c_l is the initial length of the coal fracture, m.

However, the strength of the coal plate decreases as the size increases. When the size of the coal plate reaches a certain value, the coal plate strength is insufficient to resist the high gas pressure, and short cracks are produced in the coal, which can cut the coal plates into smaller fragments. Then, the high-pressure gas quickly flows outside. Driven by the gas flow, the coal will be stripped and discarded. The unstable condition of coal in this stage is expressed as follows [27]:

$$p_{if} - p_a \geq \frac{3Ed_h^2[1 - 0.00875(\theta - 20°)]}{10R_a^2}(1 - 0.000175\frac{R_a}{d_h}) \tag{11.22}$$

where d_h is the coal spallation thickness, m; θ is half of the field angle for coal spallation, and R_a is the radius of coal spallation, m.

As mentioned above, the formation of coal spallation is a coupling process of stress, gas, and the mechanical properties of coal. It involves coal failure under mechanical action, gas migration, and their coupling effect. The unstable failure of coal, along with the fact that coal is thrown out by the gas flow, provides a space condition for the dynamic failure of coal in the next cycle. At the same time, the rapid release of gas causes the gas pressure in front of the newly exposed surface to decrease rapidly. Thus, a high gas pressure gradient forms near the newly exposed surface, which provides the power for the tensile failure of coal. Finally, the unstable failure of coal is transferred progressively to the internal coal. However, in the transfer process of coal failure, the gas is in a dynamic state of change. On the one hand, the extension of coal fractures causes the rapid release of free gas; on the other hand, the variation of the adsorption equilibrium state leads to the quick desorption of adsorbed gas to supplement the free gas. In addition, the dynamic changes in gas will lead to the enhancement of coal strength [72]. With the release of gas, the gas pressure gradient near the exposed surface slows down, and the main power for coal failure reduces gradually. Therefore, the failure degree gradually decreases during the outburst development process from crushing to becoming intact.

After an outburst, an outburst hole is left in the affected coal. Considering that coal fails with spherical coal spallation, the edges of the outburst will be spherical. During the outburst, the coal that has failed is driven by the gas and moves forward. The coal is loosened, leading to the shrinkage of the outburst hole. Moreover, in the late stage of an outburst, due to the decreased outburst energy, it is not enough to throw out all the stripped coal, and a large number of smaller fragments remain in the outburst hole. Therefore, the actual volume of the outburst hole is always less than the calculated volume. In some cases, the outburst holes have been completely blocked.

In the outburst process, the propagation velocity of the outburst face and the coal spallation thickness are related to the gas pressure gradient and coal strength near the exposed face [69, 70]. However, in the propagation process, the gas pressure gradient continuously decreases and the coal strength is enhanced with the desorption of adsorbed gas. These two aspects lead to a decrease in the propagation velocity and increase in the coal spallation thickness.

11.4.2 Verification of the Outburst Development Process

To verify the theoretical analysis of coal and gas outburst development processes, several outburst simulation experiments were conducted under different experimental conditions. Coal and gas outbursts were simulated in a triaxial coal and gas outburst simulation system [73, 74].

The triaxial coal/gas outburst simulation system was composed of an outburst simulation experimental chamber, a loading system, a temperature control system, a gas injection/vacuum pumping system, and a data collection system. The simulation system is illustrated in Fig. 11.6. The axial direction of the outburst port is defined as the x axis, the lateral direction as the y axis, and the vertical direction as the z axis (Fig. 11.6). Using the loading system, the specimen can be loaded as σ_x, σ_y, and σ_z independently. By being transmitted through a rigid bearing plate, the concentrated load is transformed into a uniformly distributed load before loading into the specimen. Diagrams of the loading system and the outburst simulation experimental chamber are shown in Fig. 11.7. The main performance parameters for the system were: ① σ_z

≤ 80 MPa, σ_x and σ_y ≤ 27 MPa; ② gas pressure ≤ 10 MPa; and ③ temperature ≤ 60 °C.

The coal was collected from the No. 10 coal seam at the Wolonghu Coalmine in Anhui Province, China. The coal was broken and screened for pulverized coal with a particle size greater than 60 mesh. The outburst experiment specimens used molded coal with physical properties similar to tectonic coal [75]. The specimens were molded in an outburst simulation experiment chamber. For these specimens, 6% water was added to the materials before molding. The molding pressure was 64 MPa and was applied for 40 min. The final specimen size was 250 mm × 250 mm × 240 mm.

The evolution of stress and gas pressure during the outburst process was collected by the data collection system. In the specimen preparation process, two groups of stress and gas pressure sensors were buried in the center of the specimen. Among these, sensor group #1 was set at a distance of 50 mm from the front wall of the experimental chamber, and sensor group #2 was 70 mm behind sensor group #1 (see Fig. 11.32).

Before testing the specimens, each was outgassed into a vacuum for 48 h. Through the air intake pipe (Fig. 11.6), CO_2 was injected to reach the desired gas pressure and this pressure was maintained for a minimum of 48 h to ensure that the specimen was

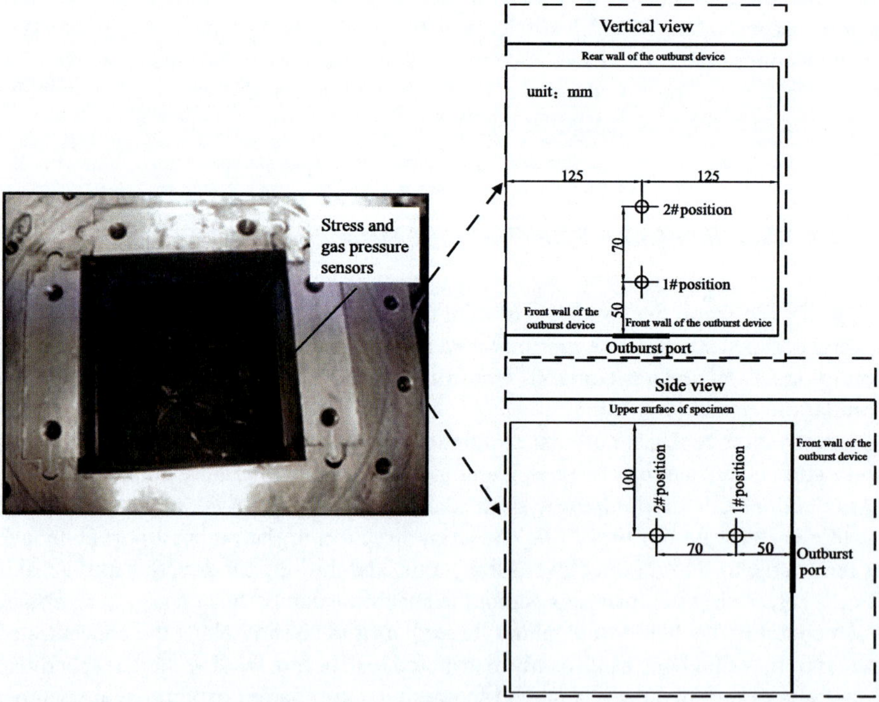

Fig. 11.32 Arrangement of sensors for stress and gas pressure

saturated with gas. Then, the specimen was loaded as σ_x, σ_y, and σ_z by the loading system. After equilibration, the outburst port was opened, the specimens were tested, and the results were recorded. After testing, polyurethane molds were made from outburst holes.

Six outburst simulation experiments were conducted under different experimental conditions. Field monitoring data of coal pillar loading suggest that the average confining pressure in a typical underground coal pillar is approximately 4 MPa and that the confining pressure typically ranges from 0 to 10 MPa. Therefore, the triaxial loading pressures for σ_x, σ_y, and σ_z were 5.5 MPa, 5.5 MPa, and 4.8 MPa, respectively. The gas pressures were in the range of 0.25–0.5 MPa. For safety, CO_2 was used rather than CH_4 in all the experiments. The outburst experiments were conducted at atmospheric pressure, and the experimental scheme is listed in Table 11.7.

The simulation experiments were conducted according to the experimental scheme. The results are presented in Table 11.8. Based on the dynamic phenomenon of outbursts, the results could be divided into three classes: outburst, stripping, and no outburst. There is a gas pressure threshold for an outburst, which was 0.35–0.40 MPa for the experiments. When the gas pressure exceeds the threshold, the outburst occurs and the outburst intensity (note: here, we assume that the absolute intensity of an

Table 11.7 Experimental scheme of a coal/gas outburst simulation

Experiment number	Gas pressure (MPa)	Ambient temperature (°C)	Triaxial stress		
			σ_z(MPa)	σ_x (MPa)	σ_y.(MPa)
#1	0.25	25	4.8	5.5	5.5
#2	0.30	25	4.8	5.5	5.5
#3	0.35	25	4.8	5.5	5.5
#4	0.40	25	4.8	5.5	5.5
#5	0.45	25	4.8	5.5	5.5
#6	0.50	25	4.8	5.5	5.5

Table 11.8 Results of the coal/gas outburst simulation experiments

Experiment number	Results of the outburst	Mass of filled coal (kg)	Absolute intensity of the outburst (kg)	Relative intensity of the outburst (%)	Distance of the outburst (m)
#1	No outburst	20.934	0	0	0
#2	Stripping	20.967	0.009	0.043	0.10
#3	Stripping	21.078	0.026	0.12	0.21
#4	Outburst	21.198	4.126	19.46	14.47
#5	Outburst	21.039	5.295	25.17	16.98
#6	Outburst	20.920	6.366	30.43	18.14

outburst is equivalent to the mass of the outburst [76]) increases with increasing gas pressure.

The evolution of stress and gas pressure during the outburst process were monitored by the two groups of sensors, and the results of experiment #5 are displayed in Fig. 11.33. It can be observed that the outburst process was a quick propagation process. The stress and gas pressure monitored by sensor group #1 changed prior to those monitored by sensor group #2. The results showed that the stress monitored by sensor group #1 decreased from 4.8 to 0 MPa in 90 ms. The gas pressure also dropped rapidly to 0 MPa in a short time, whereby time taken for this pressure drop was shorter for sensor group #1 than for sensor group #2. The rapid unloading of stress led to a sudden change in the coal stress state, and the high gas pressure gradient near the exposed surface was the result of rapid gas migration. The time difference between sensors #1 and #2 proves that the outburst process was a quick propagation process.

Outburst holes are shown in Fig. 11.34. When an outburst did not occur (Fig. 11.34a), only the coal near the exposed surface of the specimen was stripped and no outburst hole was evident. When an outburst occurred (Fig. 11.34b–d), there was an obvious outburst hole with spherical edges. It was observed that some smaller fragments that were not thrown out by the escaping gas remained inside the hole. The holes from outbursts produced under different experimental conditions had different shapes. Molds of the outburst holes were made using polyurethane. V_0 represents the actual volume of the hole as measured by determining the volume of the polyurethane mold (by immersing the mold in water), and V_1 represents the calculated volume of the hole determined by the density of the molded coal and the outburst intensity. The volume of a hole clearly increased with increasing gas pressure, and V_0 was always less than V_1, as described in Sect. 11.4.1 (Fig. 11.35).

After an outburst, the upper surface of the specimen was cleaned so that the spall from the outbursts could be examined (Fig. 11.36). When an outburst did not occur, the coal remained almost intact (Fig. 11.36a). However, when an outburst occurred (Fig. 11.36b–c), there were many arcuate cracks in the coal behind the outburst hole

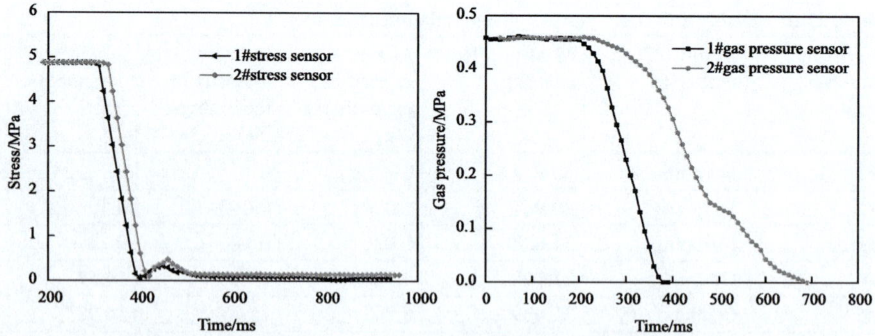

(a) Evolution of gas pressure (b) Evolution of stress

Fig. 11.33 Evolution of stress and gas pressure during the outburst process

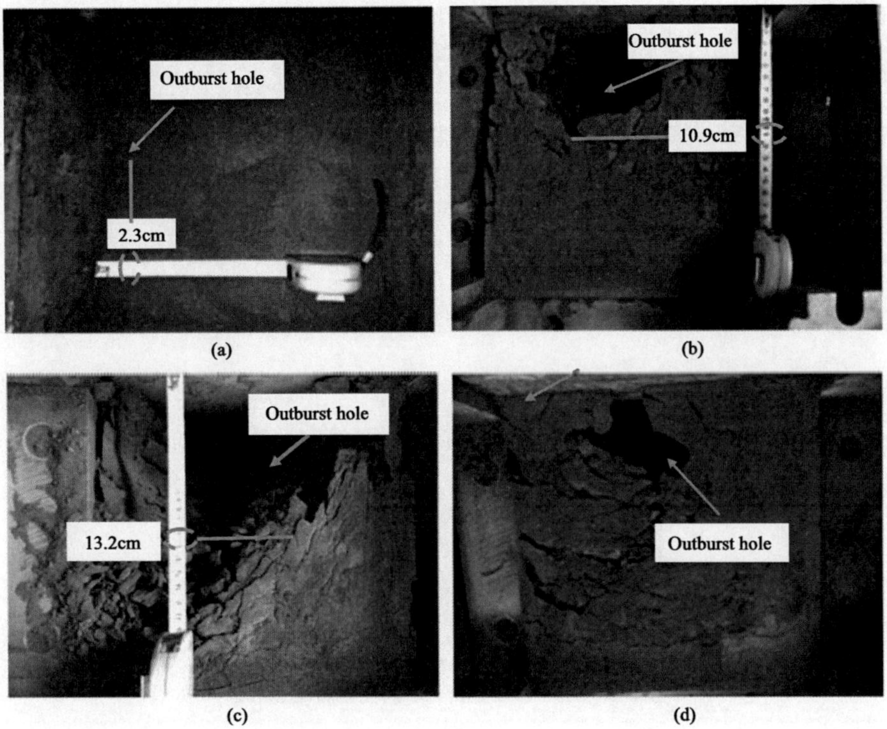

Fig. 11.34 Outburst hole characteristics. Experiment **a** #3; **b** #4; **c** #5; **d** #6

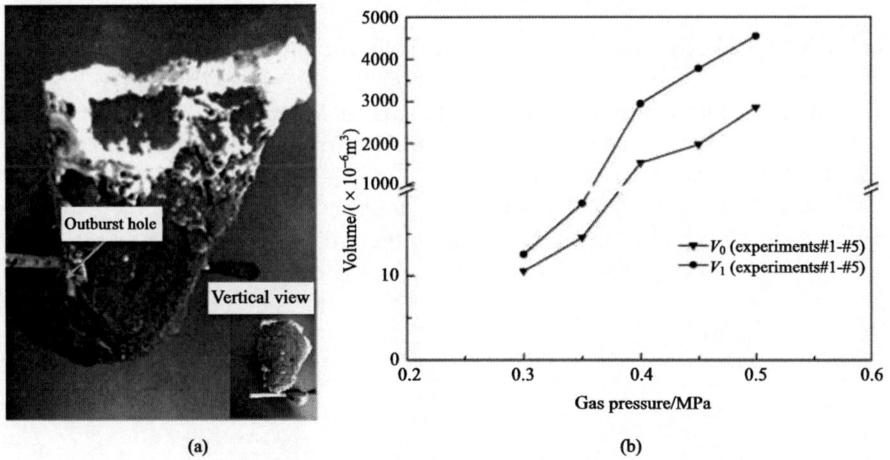

Fig. 11.35 **a** Outburst hole model of experiment #5, and **b** the relationship between the volume of an outburst hole and gas pressure

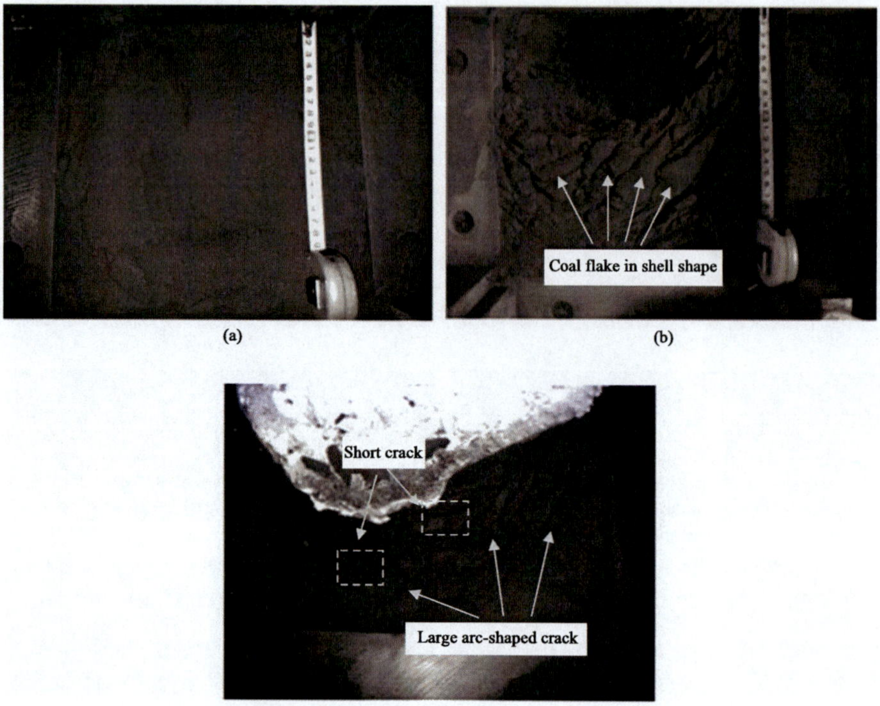

(a) (b) (c)

Fig. 11.36 Spall characteristics of coal after an outburst. Experiment **a** #3; **b** #4; **c** #5

and arcuate cracks parallel to the edges of the hole. The coal was broken into spherical shell plates with a specific thickness. Additionally, short cracks were produced in the coal that could cut the coal plates into smaller fragments. These short cracks were widely distributed near the outburst hole. The fragmentation degree of the coal decreased away from the outburst hole.

Table 11.9 lists the outburst propagation parameters, among which L_0 represents the maximum length of the outburst hole in the x direction; t represents the total

Table 11.9 Parameters related to outburst propagation

Experiment number	L_0/m	t/s	v/(m/s)	L_1/m	N	d/m
#1	0	–	–	–	–	–
#2	0.012	–	–	–	–	–
#3	0.023	–	–	–	–	–
#4	0.109	1.477	0.074	0.141	11	0.0128
#5	0.132	1.65	0.080	0.118	10	0.0118
#6	0.167	1.69	0.099	0.083	9	0.0092

time of an outburst and is obtained by recording the beginning/ending time of an outburst; L_1 represents the width of the spall area; N represents the number of arc cracks; L_0/t is defined as the propagation velocity of the outburst (v); and L_1/N is defined as the average thickness of the spall (d). Based on the experiment results, we verified that the propagation velocity of the outburst increased and the average thickness of the spallation reduced with increasing gas pressure.

11.5 Effect of Rapid Desorbed Gas on Transporting Coal

11.5.1 Energy Demand for Transporting Coal in the Development Stage of Outbursts

In investigations of outbursts, specialists usually consider the final gas emission amount as the gas exhausted during the period from when the concentration starts to rise to the time when the gas has decreased to the normal level. This period typically lasts for several days to dozens of days, whereas the effective outburst dynamic process only lasts for a few tens of seconds. Therefore, gas emission statistics primarily pertain to low-concentration gases released from nearby cracks or outburst coal itself after an outburst ends. It is important to calculate the volume of gas-transporting coal mass as this greatly contributes to further clarifying the outburst mechanism.

1. Outburst energy

Because of the complexity and variability of outbursts, it is more feasible to describe this phenomenon from the perspective of energy analysis. Significant research conducted by Hodot, Gray (1980), Valliappan and Zhang, Jiang and Yu, and Wen et al. provided a detailed analysis of outburst energy [37, 54, 76–78]. It is generally concluded that the gas expansion energy plus the elastic energy of coal is transferred to the coal crushing energy, transport energy, and remaining kinetic energy of gas [Eq. (11.23)].

$$E_e + E_g = W_3 + W_4 + W_5 \tag{11.23}$$

where W_3 is the coal crushing energy, MJ; W_4 is the coal–gas transport energy, MJ; and W_5 is the remaining kinetic energy of the gas, MJ.

Based on the previous analysis, the crushing and conveying processes can be assumed to be two successive independent activities:

$$E_g = W_4 + W_5 \tag{11.24}$$

(1) Gas expansion energy

The gas expansion occurring over a short time period can be simplified as an adiabatic process [54], which satisfies the following equation:

$$p_1 V_1^{\eta} = p_0 V_0^{\eta} \tag{11.25}$$

where P_1 and P_0 represent the gas pressure and atmospheric pressure, respectively, MPa; V_1 and V_0 are the gas volumes under P_1 and P_0, respectively, m^3; and n is the adiabatic coefficient. For gas mixtures composed mainly of methane, n is usually taken as 1.3.

The gas expansion energy can be expressed as follows:

$$E_g = \frac{p_0 V_0}{\eta - 1}\left[\left(\frac{p_1}{p_0}\right)^{\frac{\eta-1}{\eta}} - 1\right] \tag{11.26}$$

Considering that gas adopts a free state and an adsorbed state, we separated Eq. (11.26) into the components shown in Eq. (11.27):

$$W_1 = \frac{p_0}{\eta - 1}\left(V_0^a + V_0^f\right)\left[\left(\frac{p_1}{p_0}\right)^{\frac{\eta-1}{\eta}} - 1\right] = W_a + W_f \tag{11.27}$$

where $V_0{}^a$ and $V_0{}^f$ represent the volume of adsorbed gas and volume of free gas under P_0, respectively, m^3; w_a and w_f represent the energy contributions of adsorbed gas and free gas, respectively, MJ.

It is feasible to obtain data regarding the initial gas pressure P_1 and the initial free gas volume $V_1{}^f$ in an outburst investigation, whereas the contributive gas volume V_0 is difficult to determine. Therefore, Eq. (11.25) can be modified as follows to be more practical:

$$\begin{aligned} W_1 &= \frac{p_0 V_0}{\eta - 1}\left[\left(\frac{p_1}{p_0}\right)^{\frac{\eta-1}{\eta}} - 1\right] = \frac{p_0}{n - 1}\left(V_1^a + V_1^b\right)\left(\frac{p_1}{p_0}\right)^{\frac{1}{\eta}}\left[\left(\frac{p_1}{p_0}\right)^{\frac{\eta-1}{\eta}} - 1\right] \\ &= W_a + W_f \end{aligned} \tag{11.28}$$

(2) Transport energy of coal

The transport energy calculation varies with the final shape of the coal dump [31]. When only horizontal displacements exist, the transport energy can be expressed using Eq. (11.29):

$$W_4 = Sm[g(f\cos\alpha \mp \sin\alpha)] \times 10^{-6} \tag{11.29}$$

where S represents the distance over which coal is ejected, m; m is the coal mass, kg; f is the friction coefficient; g is the gravity acceleration, kg/m2; and α is the coal seam angle.

When $\alpha = 0$, Eq. (11.29) can be simplified as Eq. (11.30):

$$W_4 = fmgS \times 10^{-6} \tag{11.30}$$

Coal gas transport is a complex process, and it is difficult to find an accurate formula that incorporates all factors, such as the coal collision loss and heat loss. Thus, the result of Eq. (11.29) will be smaller than the actual value. However, if these smaller results still require adsorbed gas to provide energy, a participative role of the adsorbed gas will become a necessity.

(3) Remaining kinetic energy of gas

Ideally, the coal-gas transport process on a straight roadway can be treated as a homogeneous powder flow occurring in a pipe. Previous studies have shown a strong relationship between the gasesolid flow rate and the corresponding flow state [79]. Generally, following a decrease in the flow rate, the states of suspension flow, stratified flow, sedimentation flow, plug flow and, finally, blockage will occur sequentially. Normally, there exists a critical blockage velocity that characterizes the flow state from the blockage to the plug flow. When the speed decreases below this critical speed, the gas will not have sufficient power to support the transport of coal, and coal will begin to form a deposit. The gas desorbed after this moment will not be involved in an outburst. The remaining kinetic energy of the gas, which is termed the "ineffective gas", can therefore be calculated using the critical blocking flow speed.

Regarding the critical blocking flow speed of coal powders in pipes, Xiong et al. proposed a formula that is appropriate for a small scale and provided a good fit to laboratory data [80]. However, Cong considered that this formula would generate a large error at the industrial scale, and thus established a more reasonable transport equation in his doctoral dissertation [81]:

$$v_r = \frac{G}{A\rho_b} \tag{11.31}$$

where v_r is the blockage critical velocity, m/s; G is the mass of coal per second, kg/s; A is the cross-sectional area, m^2; and ρ_b is the natural bulk density, kg/s.

Hence, the remaining kinetic energy of gas (W_5) can be calculated using Eq. (11.32):

$$W_5 = \frac{1}{2}m_g v_r^2 \times 10^{-6} = \frac{1}{2}m\left(\frac{G}{A\rho_b}\right)^2 \times 10^{-6} = \frac{1}{2}V_0\rho_{0g}\left(\frac{G}{A\rho_b}\right)^2 \times 10^{-6} \tag{11.32}$$

where mg is the outburst coal mass, kg; and ρ_0 is the gas density under P_0 (usually 0.1 MPa), kg/m^3.

By introducing all the aforementioned factors into Eq. (11.24), we can obtain the final energy conversion during the transport, as shown in Eq. (11.33):

$$\frac{p_0 V_0}{\eta - 1}\left[\left(\frac{p_1}{p_0}\right)^{\frac{\eta-1}{\eta}} - 1\right] = \left[f\,Smg + \frac{1}{2}V_0\rho_0\left(\frac{G}{A\rho_b}\right)^2\right] \times 10^{-6} \tag{11.33}$$

2. Desorption speed demand for conveying outburst coal for a real case

(1) Overview of the Zhongliangshan outburst

The Zhongliangshan outburst was a test conducted on November 4, 1977, and is the sole field simulation with relatively detailed outburst data existing in China. After this test and upon consideration of the high risk to coalmine workers, the Chinese government introduced rigorous controls to prevent outburst accidents, and underground field tests were inevitably prohibited. In the Zhongliangshan outburst, the outburst coal mass was 817 t, and the released gas was 38,540 m^3 (the volume of gas after the outburst, which did not contribute to the outburst, was also included). The outburst lasted for 39 s. Two pressure inspection holes (#1 and #2), one flow inspection hole (#3), and one temperature inspection hole (#4) were established. Figure 11.37 displays the variations of pressure, gas flow, and temperature [28].

From Fig. 11.37, it can be seen that 1.52 s after the outburst commenced, the pressure in hole #2 (2 m to the outburst front) began to decrease, and when the time advanced to 6 s, a similar decrease was observed in hole #1 (14 m to the outburst front). The pressure hysteresis indicates that the outburst gas release first occurred at

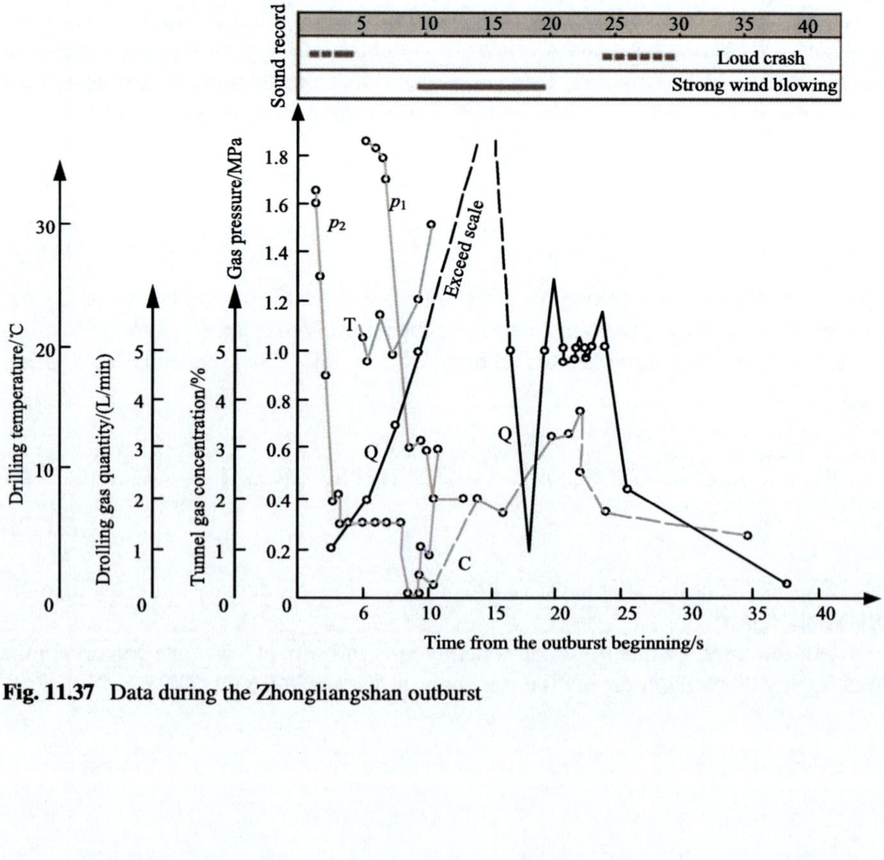

Fig. 11.37 Data during the Zhongliangshan outburst

Table 11.10 Calculation parameters

Outburst coal mass (t)	817	Outburst coal rank	Coking coal
Gas released (m^3)	38,540	Porosity (%)	6
Outburst period (s)	39	Roadway cross-section area (m^3)	7
Outburst distance (m)	167	Initial pressure (MPa)	1.75
Coal bulk density (kg/m^3)	700	Methane density (30 °C, 0.1 MPa)/(kg/m^3)	0.6375
Coal apparent density (kg/m^3)	1300	Friction coefficient	0.5
Adiabatic coefficient	1.3		

Note The initial outburst pressure is the average value of holes #1 and #2

the outburst front and then proceeded to a further point only a few seconds after the coal was crushed. Finally, the flow became a gas–solid transport powered by a static head of 0.3–0.6 MPa. This phenomenon supports the theory that coal is crushed by great stress and then discharged by an enormous amount of gas.

(2) Gas volume demand for the Zhongliangshan outburst

Several parameters [77, 82, 83] based on observational data are summarized in Table 11.10.

First, Eqs. (11.28) and (11.32) can be modified into simpler forms, as shown in Eqs. (11.34) and (11.35).

$$W_1 = \frac{P_0 V_0}{n-1}\left[\left(\frac{P_1}{P_0}\right)^{\frac{n-1}{n}} - 1\right] = \lambda_1 V_0 \tag{11.34}$$

$$W_5 = \frac{1}{2} V_0 \rho_0 \left(\frac{G}{A\rho_b}\right)^2 = \lambda_2 V_0 \tag{11.35}$$

where λ_1 and λ_2 can be regarded as the gas expansion energy factor and the remaining kinetic energy factor, respectively.

Using the data above, the expansion energy of free gas, the transport energy of coal, and the remaining kinetic energy of the gas were calculated (Table 11.11).

Clearly, λ_1 was much greater than λ_2, which means that the remaining kinetic energy of the gas could be ignored during the transport process, and that all the gas expansion energy was converted into the transport energy of coal. In addition, approximately 5.61×10^8 J of desorbed gas energy, or 1799 m^3 (at 30 °C and 0.1 MPa) of adsorbed gas, would be required for this transport, which is nearly 6.3 times greater than the energy of free gas (Fig. 11.38). However, because of the age and irreproducibility of the data, an exact description of the Zhongliangshan

Table 11.11 Results of the different energies involved in the Zhongliangshan outburst

Coal transport energy (J)	Free gas expansion energy			Gas remained kinetic energy			Supply desorption gas
	Free gas volume (m^3)	Gas expansion energy (J)	Gas expansion energy factor (λ_1)	Blockage critical velocity (m/s)	Remained kinetic energy (J)	Remained kinetic energy factor (λ_2)	Energy (J)
6.67×10^8	341	1.06×10^8	3.12×10^5	4.28	1986	5.83	5.61×10^8

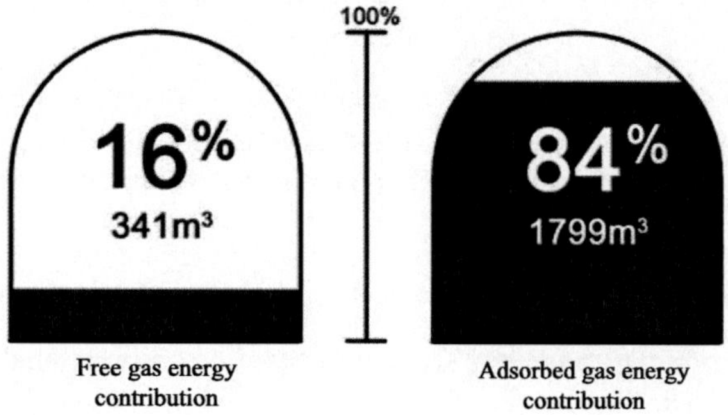

Fig. 11.38 Energy contribution to the outburst

outburst is difficult to obtain. Certain parameters and conditions need to be established empirically to obtain a reliable answer. Although it is not accurate, the results provide strong evidence that a certain percentage of adsorbed gas is involved in outbursts and usually plays a crucial role.

11.5.2 Source of Rapid Desorbed Gas

From the analysis above, we can conclude that fast desorption of adsorbed gas necessary during an outburst. Studies worldwide mostly attribute this fast desorption to the pre-crush of outburst coal. Hu collected coal samples from some outburst accidents for further diameter distribution analysis. The results provided reliable evidence that many coal powders exist with diameters of < 100 μm, and in some situations, this proportion can reach 53.8% [77]. Concerning the relationship between the desorption speed and coal particle diameter, researchers have preferred to analyze the change via laboratory tests, and have hypothetically proposed the general role that rapid desorption will play in outbursts. For this reason, it is difficult to estimate the required gas desorption speed and the proper diameter range for a real outburst. Efforts to clarify the true speed and diameter are essential and valuable. For a better illustration, we define the coal diameter with the lowest gas desorption speed that transports the outburst coal a certain distance as the transport critical diameter (TCD).

1. Desorption tests of different coal diameters

As it is impossible to obtain an original coal sample from the Zhongliangshan outburst (it occurred too long ago and is not repeatable), a coal sample from the Machang outburst accident was taken as a replacement for further desorption tests. Both are coking coal and originated from the Longtan Formation in Southwest China. Firstly, the coal sample was crushed into seven diameter ranges: 0.01–0.08 mm, 0.08–0.2 mm, 0.2–0.25 mm, 0.25–0.5 mm, 0.5–1 mm, 1–3 mm, and 3–4 mm. Then, via a process of exhausting-equilibrating-desorbing, the volume of the released gas under conditions of 30 °C and 1.75 MPa was measured, and a graph describing the relationship between gas desorption and time was plotted (Fig. 11.39). The rate of gas desorption obviously increased with decreasing diameter, as did the extent of the increase. An extremity diameter divided the entire desorption stage: the fast desorption zone, and the normal or slow desorption zone. This result is consistent with the observations of other related studies [84, 85]. Zhou believed that the extremity diameter D_e ranged from approximately 0.5–10 mm (i.e., essentially at the millimeter level). The existence of the extremity diameter is determined by the pore-fracture system of coal. Because the flow difficulty in the fracture system is much smaller than that in the pore system, the damage to the fracture zone has little influence on the transport of gas. When the diameter decreases to a certain value, below the matrix size, a large number of pores will become exposed to air, and the gas flow begins to increase dramatically. Based on the relationship between the average desorption

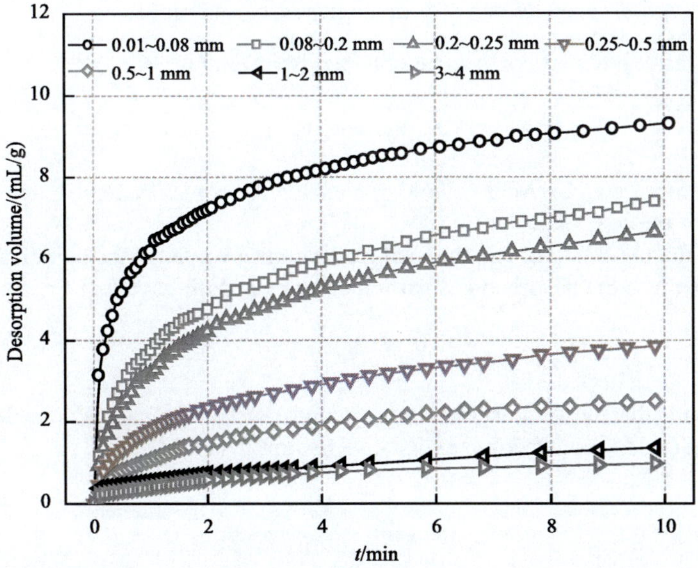

Fig. 11.39 Desorption volume variation over time

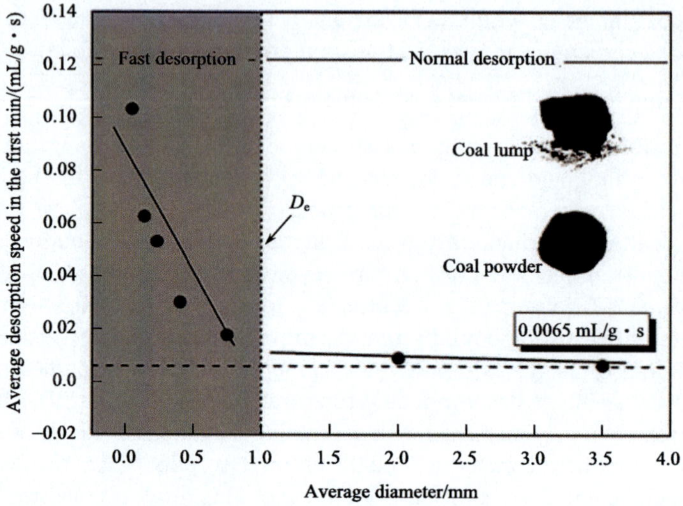

Fig. 11.40 Relationship between the average desorption speed in the first minute and the average diameter

speed in the first minute and the average diameter (Fig. 11.40), it can be deduced that the extremity diameter of the test sample was approximately 1 mm. When the diameter was < 1 mm, the rate of desorption increased sharply.

2. TCD estimation in outbursts

(1) Desorption speed demand for the transport of outburst coal

The free gas of outburst coal can be obtained from Eq. (11.36):

$$V_f = \phi_p V_c \tag{11.36}$$

where V_c represents the total volume of coal and the pores inside, m^3; and ϕ_p is the porosity of the coal, %.

Using Eq. (11.33) to obtain the total gas volume involved in the outburst (V_a), we can then deduce the adsorbed gas participating in an outburst (V_m):

$$V_m = V_a - V_f \tag{11.37}$$

Assuming that the outburst lasts for t_o seconds, we can easily obtain the desorption speed required for conveying outburst coal using Eq. (11.38):

$$v_a = \frac{V_a}{mt_1} \tag{11.38}$$

Based on Eq. (11.38), the supply desorption gas volume is 1799 m^3 (at 0.1 MPa, 30 °C), and the average rate of desorption in 39 s is 0.05646 mL/g·s.

(2) Desorption speed of normal-diameter coal

Many researchers preferred to use a diameter of 3 mm as the dividing line of the fracture system and pore system, and often selected 1–3 mm as their diameter choice. Various studies on the desorption volume and pressure have been conducted to establish an outburst risk evaluation index, from which we can obtain a simple relationship between the first minute desorption volume and the initial pressure [67]:

$$Q_1 = Ap^B \tag{11.39}$$

swhere A and B are two fitting parameters.

Table 11.12 provides the desorption data obtained using a normal diameter from eight sections collected over the last few years. These desorption tests were all conducted using the same method as that mentioned above, except for the initial equilibrium pressure. However, the relationship between Q_1 and P is always available. Based on this relationship, we can easily obtain the desorption speed of various types of coal under 1.75 MPa. Table 11.12 shows that the desorption speed of normal particles ranged from 10^{-3} to 10^{-2} mL/g s, which are much lower values than the desorption speed required for conveying outburst coal. With respect to the coal rank, there is a "U"-shaped variation. The desorption speed decreases gradually to a certain point, and then increases with the coal rank (Fig. 11.41). The medium-rank coal has the smallest desorption speed compared to the others.

(3) Estimation of the TCD in the conveying process

It is believed that diffusion dominates the gas flow in coal particles with diameters of < 3 mm. Therefore, the fracture-pore system can be simplified into a single-pore

Table 11.12 First minute desorption speed for 1–3 mm coal particles from different sections

Section	Chinese coal classification	$R_{0,max}$	Relationship between the first minute desorption and the initial pressure	First minute desorption under 1.75 MPa/ (mL/g s)
Dalong	Long flame	0.57	$Q_1 = 0.1719P^{0.8403}$	0.004584
Renlou	Gas-fat	0.92	$Q_1 = 0.1360P^{0.6900}$	0.003333
Shuangliu	Coking	1.20	$Q_1 = 0.16180P^{0.3734}$	0.003309
Tunlan	Coking	1.87	$Q_1 = 0.14739P^{0.6910}$	0.003616
Jinhuangzhuang	1/3 coking	–	$Q_1 = 0.2157P^{0.5436}$	0.004873
Bailongshan	Anthracite	2.60	$Q_1 = 1.0935P^{0.5151}$	0.02431
Daning	Anthracite	2.77	$Q_1 = 1.0491P^{0.8294}$	0.02781
Wolonghu	Anthracite	2.75	$Q_1 = 0.6734P^{0.5173}$	0.01499

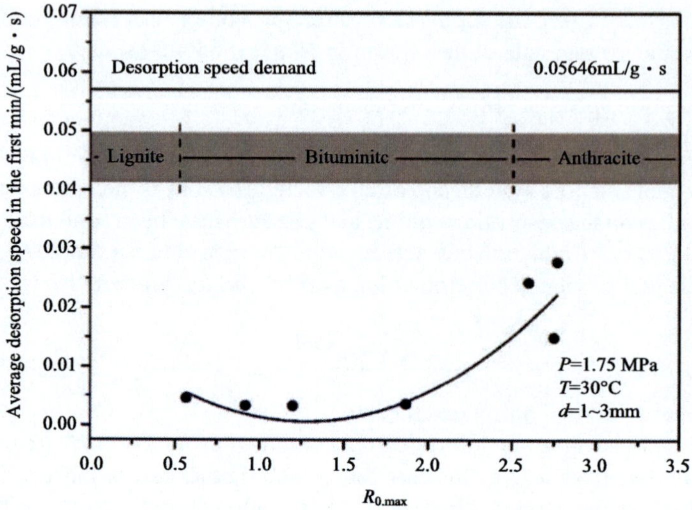

Fig. 11.41 Variations in gas desorption speed of normal-sized particles with coal rank under a pressure of 1.75 MPa

system, and the diffusion model can be implied directly. The single-pore diffusion model can be expressed by Eq. (11.40):

$$\frac{Q_t}{Q_\infty} = 1 - \frac{6}{\pi^2}\sum_{n=1}^{\infty}\frac{1}{n^2}e^{-Dn^2\pi^2t'/r^2} \tag{11.40}$$

where D is the diffusion coefficient, t is the diffusion time, d represents the coal diameter (m), and Q_t and Q_∞ are the desorption volume at time t and the final desorption volume, respectively (m^3).

For short times, Eq. (11.19) can be simplified as Eq. (11.41):

$$\frac{Q_t}{Q_\infty} = \frac{6}{\sqrt{\pi}}\left(\frac{D}{r^2}t'\right)^{\frac{1}{2}} \tag{11.41}$$

Assuming that the D value is an irrelevant parameter with t'; we take a derivation for t', and a formula describing the relationship between the diameter and the desorption speed will be obtained, as shown in Eq. (11.42):

$$v_t = \frac{3Q_\infty}{r}\sqrt{\frac{D}{\pi t'}} \tag{11.42}$$

The average desorption speed can be deduced from Eq. (11.43):

$$\overline{v_t} = \frac{6Q_\infty}{r}\sqrt{\frac{D}{\pi t'}} \tag{11.43}$$

Hence, for short outbursts, the average desorption speed at any time can be expressed by Eq. (11.44):

$$\overline{v_t} = \frac{1}{r} f[D(r), Q_\infty(r)] \tag{11.44}$$

Regarding the adsorption capacity (Q_∞), a common consensus is that a decrease in diameter can hardly damage the micropore system in coal, which dominates the adsorption capacity or Langmuir volume V_L. That is, Q_∞ will hardly change with diameter. Guo et al. tested the isothermal adsorption of coal with different diameters and obtained several coincident Langmuir curves [86]. The same result can also be found in studies by other scientists [87–89]. It is generally acknowledged that crushing occurs easily on the weak side of coal, which consists of many macropores and mesopores. This damage not only increases the external area exposed to the air but also contributes little to the internal pore area. In addition, the smallest crush powder diameter in an outburst is assumed to be approximately 10 μm, whereas the upper limit of the micropore diameter is 10 nm. This diameter difference also leads to the impossibility of an adsorption capacity change. Moreover, based on the assumption that coal is an isotropic material, the diffusion coefficient D will not vary with the diameter under the same concentration difference. For an ideal situation, the parameters Q_∞ and D can be treated as stable values that are not influenced by the diameter. Therefore, the average desorption speed and diameter conform to the following proportional relationship:

$$\frac{v_1}{v_2} = \frac{r_2}{r_1} \tag{11.45}$$

The ratio of the diameters equals the ratio of the volumes and surface area [Eq. (11.45)]. Because the total coal volume does not change during crushing, the decrease in diameter only leads to an increase in the surface area. The breaking up of coal shortens the diffusion distance and increases the exposure area, such that the desorption speed increases:

$$\frac{V}{S} = \frac{\frac{4}{3}\pi r^3}{4\pi r^2} = \frac{1}{3}r \tag{11.46}$$

In fact, the inhomogeneity of coal affects the accuracy of Eq. (11.45). Watanabe and Yang believed that there should be a diameter factor *n* in this proportional relationship [85, 90], as shown in Eq. (11.26). For the test sample of Machang, *n* was 0.65.

$$\frac{v_1}{v_2} = \left(\frac{r_2}{r_1}\right)^n \tag{11.47}$$

Based on this equation, the critical coal diameter in the conveying process can be estimated from the measured desorption speed for a certain diameter.

For the Zhongliangshan outburst, the supply of desorption gas was 1799 m^3 (at 0.1 MPa and 30 °C), and the average desorption speed in 39 s was 0.05646 mL/g s. Considering that the 0.51 mm rate of desorption of Machang coal was 0.02263 mL/g s, we can expect the TCD in the Zhongliangshan outburst to have been approximately 100 μm. Because the results are average values, a certain fraction of coal with a particle size less than the calculated TCPS exists, which would benefit the energy accumulation and contribute to a higher possibility of outbursts. Limited to the objective conditions, we make several assumptions and simplifications to obtain an easy answer. The crash energy loss, heat loss, coal crushing time, and final coal dump shape are generally neglected or strictly simplified. However, the general range of the TCD has some main points.

3. Verification of coal powder existence

Mounting accident investigations in China have proven the widespread existence of coal powders in outbursts. At the edge of the coal dump, with a slope angle smaller than the natural rest-angle, exists a large quantity of coal powders without a concave–convex sense. These powders can be easily blown up by a light wind. When there is water nearby, a thin coal layer floats on the water surface. Figure 11.42 shows coal

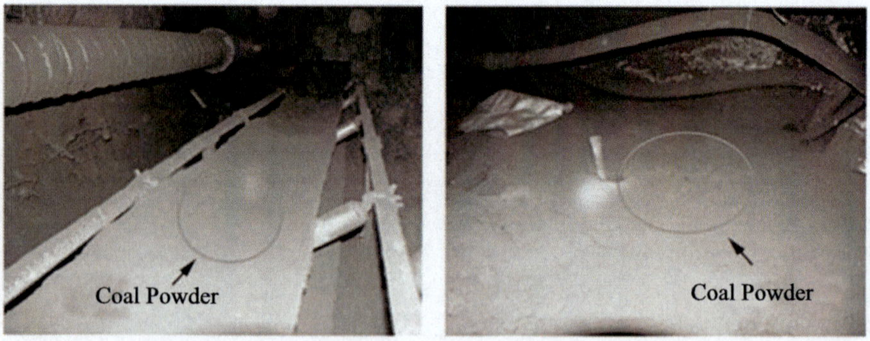

(a) Coal powder on the belt conveyor (b) Coal powder on the tunnel

Fig. 11.42 Coal dump at the Machang outburst

powders appearing after the Machang outburst.

Hu conducted an exploration of the coal particle composition after four outbursts at the Zhongliangshan Coalmine in the 1970s and 1980s (Table 11.13 and Fig. 11.43). The weight distributions of the different diameter ranges of four coal samples from these outbursts are shown in Fig. 11.43. The coal samples had a large proportion of small-sized coal powders ($d < 1$ mm), comprising 54.1, 34.2, 33.9, and 34.1% of the total size fractions for each outburst, respectively, while the < 100 μm size fraction comprised 25.4%, 4.3%, 3.5%, and 6.6%, respectively. Coal lumps ($d > 10$ mm) also contributed 20.0, 27.2, 16.2, and 30.8 of the total size fractions for each accident. The entire diameter distribution was a non-Gaussian distribution with a variation of > 10 orders of magnitude. Hu used the Rosin–Rammler model to fit the data and found that the uniformity coefficient was considerably less than 1, which indicates

Table 11.13 Coal size distribution of the four outbursts that occurred in the 1970s and 1980s at Zhongliangshan coalmine

No	Size range/mm					突出地点
	< 0.1 (%)	0.1–1.0 (%)	1–5 (%)	5–10 (%)	> 10 (%)	
1	25.4	26.0	27.6	1.0	20.0	K_2 coal seam at the open-off cut of the 5th cross-cut, + 280 m elevation
2	4.3	29.9	24.6	14.1	27.2	K_1 coal seam at the Xisi half rising cross-cut, + 280 m elevation
3	3.5	30.4	30.5	19.4	16.2	K_{10} coal seam at the 5th cross-cut, + 280 m elevation
4	6.6	27.5	16.9	18.2	30.8	K_{10} coal seam at the Xisi half rising cross-cut, + 280 m elevation

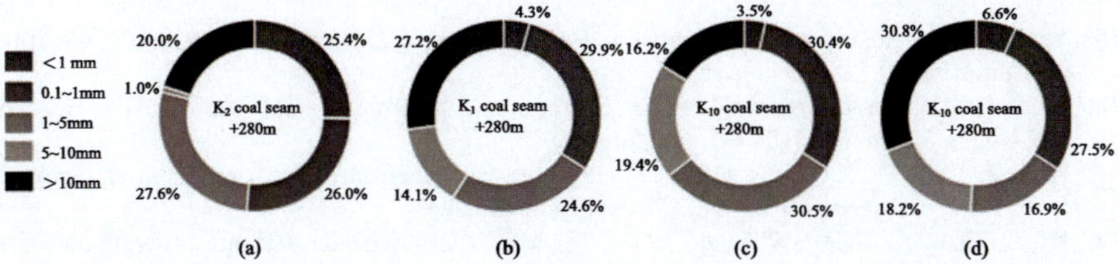

Fig. 11.43 Coal size distribution of the four outbursts that occurred in the 1970s and 1980s at Zhongliangshan Coalmine. **a** K_2 coal seam at the open-off cut of the 5th cross-cut, + 280 m elevation; **b** K_1 coal seam at the Xisi half rising cross-cut, + 280 m elevation; **c** K_{10} coal seam at the 5th cross-cut, + 280 m elevation; **d** K_{10} coal seam at the Xisi half rising cross-cut, + 280 m elevation

the extreme uniformity of the outburst coal size distribution. The order of magnitude was similar to the former calculation. It is unavoidable that there are some errors, but we must admit the existence of coal powders with millimeter and micrometer sizes.

References

1. Yu, B. (1985). *Series of coal and gas outburst mechanism.* China Coal Industry Publishing House. (in Chinese).
2. Lama, R D. (1995). Safe gas content threshold value for safety against outbursts in the mining of the Bulli seam. In R. D. Lama (Ed.), *International Symposium-cum-Workshop on Management and Control of High Gas Emissions and Outbursts in Underground Coal Mines* (pp. 175–189). Wollongong.
3. Lama, R. D. (1996). Assessment of threshold values for safety against outbursts of gas and coal in the Bulli/seam at Appin Colliery. Kembla Coal and Coke Pty Limited.
4. Black, D., Aziz, N., Jurak, M., et al. (2009). Outburst threshold limits-are they appropriate. In *Coal Operators' Conference.* University of Wollongong.
5. Wu, Y. (1988). Status of coal and gas outburst prevention for coal mine in West Germany. *Coal Engineer, 2*, 58–65. (in Chinese).
6. Wang, Y., & Yang, Q. (1988). Prediction of coal and gas outburst danger. *Safety in Coal Mines, 04*, 35. (in Chinese).
7. Zhang, Y., Zhang, Z., & Cao, Y. (2007). Deformed-coal structure and control to coal-gas outburst. *Journal of China Coal Society, 03*, 281–284. (in Chinese).
8. Shao, Q., Wang, E., Wang, W., et al. (2010). Control to coal and gas outburst of tectonic coal distribution. *Journal of China Coal Society, 02*, 250–254. (in Chinese).
9. Han, J., Zhang, H., Zhu, Z., et al. (2007). Controlling of tectonic stress field evolution for coal and gas outburst in Fuxin basin. *Journal of China Coal Society, 09*, 934–938. (in Chinese).
10. Zhu, X., & Xu, F. (1994). The controlling effect of tectonic stress field and its evolutions on coal and gas outburst. *Journal of China Coal Society, 03*, 304–314.
11. Paterson, L. (2008). A model for outbursts in coal. *International Journal of Rock Mechanics and Mining Sciences and Geomechanics Abstracts, 23*(4), 327–332.
12. Choi, S., & Wold, M. (2001). A mechanistic study of coal and gas outbursts. In *Dc Rocks E U. s. symposium on rock mechanics.*
13. Xue, S., Wang, Y., Xie, J., et al. (2011). A coupled approach to simulate initiation of outbursts of coal and gas-model development. *International Journal of Coal Geology, 86*(2–3), 222–230.
14. Xue, S., Yuan, L., Wang, Y., et al. (2014). Numerical analyses of the major parameters affecting the initiation of outbursts of coal and gas. *Rock Mechanics and Rock Engineering, 47*(4), 1505–1510.
15. Yu, S. (1988). Steady advance of coal and gas outburst. *Acta Mechanica Sinica, 02*, 97–106. (in Chinese).
16. Yu, S. (1992). One-dimensional flow model for coal-gas ourbursts and initiation criterion. *Acta Mechanica Sinica, 04*, 418–431. (in Chinese).
17. Zhen, Z., Chen, L., & Ding, Y. (1993). Steady advance of one-dimensional gas outburst fractured array surface. *Science China, 04*, 377–384. (in Chinese).
18. He, J., Zhao, Y., Zhang, W., et al. (1993). The soften analysis and instability study on outburst of coal and methane. *Engineering Mechanics, 02*, 79–87. (in Chinese).
19. Zhao, Y. (1993). The numerical simulation on effect of methane pressure to outburst. *Chinese Journal of Rock Mechanics and Engineering, 04*, 328–337. (in Chinese).
20. Xu, T., Tang, C., Yang, T. H., et al. (2006). Numerical investigation of coal and gas outbursts in underground collieries. *International Journal of Rock Mechanics and Mining Sciences, 6*(43), 905–919.

21. Xu, T., Tang, C., Song, L., et al. (2005). Numerical simulation of coupled gas flow in failure process of gassy coal-rock. *Chinese Journal of Rock Mechanics and Engineering, 10*, 1667–1673. (in Chinese).
22. Xu, T., Tao, T., Tang, C., et al. (2005). Numerical simulation of outburst process of gas-coal and rock. *China Safety Science Journal, 01*, 111–113. (in Chinese).
23. Tang, C., & Liu, H. (2002). Numerical approach on outburst in crosscutting in coal seam containing gas. *Chinese Journal of Rock Mechanics and Engineering, 21*(10), 1467–1472. (in Chinese).
24. Cai, C. (2004). Experimental study on 3D simulation of coal and gas outbursts. *Journal of China Coal Society, 01*, 66–69. (in Chinese).
25. Xu, J., Tao, Y., Yin, G., et al. (2008). Development and application of coal and gas outburst simulation test device. *Chinese Journal of Rock Mechanics and Engineering, 11*, 2354–2362. (in Chinese).
26. Xu, J., Zhao, H., Xu, J., et al. (2009). Experimental study of simulation of coal and gas outbursts. *Chinese Journal of Rock Mechanics and Engineering, 08*, 1674–1680. (in Chinese).
27. Jiang, C. (1994). *Study on spherical shell instability mechanism of dynamic phenomena in crosscut gas containing coal*. China University of Mining and Technology. (in Chinese).
28. Yu, Q., & Cheng, Y. (2012). *Coal mine gas control*. China University of Mining and Technology Press. (in Chinese).
29. Odintsev, V. N. (1997). Sudden outburst of coal and gas-failure of natural coal as a solution of methane in a solid substance. *Journal of Mining Science, 33*(6), 508–516.
30. Kidybinski, A. (1980). Significance of in situ strength measurements for prediction of outburst hazard in coal mines of Lower Silesia. In *Proceedings The occurrence, prediction and control of outbursts in coal mines* (pp. 193–201). The Australasian Institute of Mining and Metallurgy.
31. Hoddt, B. B. (1966). *Coal and gas outburst*. China Industry Press. (in Chinese).
32. Hanes, J., Lama, R. D., & Shepherd, J. (1983). Research into the phenomenon of outbursts of coal and gas in some Australian collieries. In *5th ISRM congress. International society for rock mechanics*.
33. Sato, K., & Fujii, Y. (1989). Source mechanism of a large scale gas outburst at Sunagawa Coal Mine in Japan. *Pure and Applied Geophysics, 129*(3), 325–343.
34. Chen, K. P. (2011). A new mechanistic model for prediction of instantaneous coal outbursts-dedicated to the memory of Prof. Daniel D. Joseph. *International Journal of Coal Geology, 87*(2), 72–79.
35. Zhou, S., & He, X. (1990). Rheological hypothesis of coal and methane outburst mechanism. *Journal of China University of Mining and Technology, 02*, 4–11. (in Chinese).
36. Jiang, C., & Yu, Q. (1995). Spherical shell instability hypothesis of coal and gas outburst mechanism. *Safety in Coal Mines, 02*, 17–25. (in Chinese).
37. Jiang, C., & Yu, Q. (1998). *Spherical shell instability mechanism for coal and gas outburst and control technique*. China University of Mining and Technology Press. (in Chinese).
38. Jiang, C., & Yu, Q. (1996). Rules of energy dissipation in coal and gas outburst. *Journal of China Coal Society, 02*, 173–178. (in Chinese).
39. Zhang, M., Xu, Z., Pan, Y., et al. (1991). A united theory on coal (rock) burst and outburst. *Journal of China Coal Society, 04*, 48–53. (in Chinese).
40. Li, F. (1989). Discussion on the mechanism of coal and gas outburst mechanism—Two-phase fluid hypothesis. *Safety in Coal Mines, 11*, 29–35. (in Chinese).
41. Guan, P., Wang, H., & Zhang, Y. (2009). Mechanism of instantaneous coal outbursts. *Geology, 37*(10), 915–918.
42. Lyu, S., & He, J. (1999). Key layer and stress dike mechanism of coal and gas outbursts. *Journal of Chongqing University (Natural Science Edition), 22*(6), 80–85. (in Chinese).
43. Ma, Z., & Yu, Q. (2006). The pilot study on outburst mechanism for compression disseminated values of coal and gas out of control. *Journal of China Coal Society, 03*, 329–333. (in Chinese).
44. Cai, M., Kong, G., & Jia, L. (1997). Criterion of energy catastrophe for rock project system failure in underground engineering. *Journal of University of Science and technology Beijing, 04*, 325–328. (in Chinese).

45. Xie, H., Ju, Y., & Li, L. (2005). Criteria for strength and structural failure of rocks based on energy dissipation and energy release principles. *Chinese Journal of Rock Mechanics and Engineering, 17*, 3003–3010. (in Chinese).
46. Choi, X., & Wold, M. (2004). Study of the mechanisms of coal and gas outbursts using a new numerical modeling approach. In *Coal Operators' Conference*. University of Wollongong.
47. Drucker, D. C., Prager, W., & Greenberg, H. J. (1952). Extended limit design theorems for continuous media. *Quarterly of Applied Mathematics, 9*(4), 381–389.
48. Jaiswal, A., & Shrivastva, B. K. (2009). Numerical simulation of coal pillar strength. *International Journal of Rock Mechanics and Mining Sciences, 46*(4), 779–788.
49. Fama, M. E. D., Trueman, R., & Craig, M. S. (1995). Two-and three-dimensional elasto-plastic analysis for coal pillar design and its application to highwall mining. *International Journal of Rock Mechanics and Mining Sciences and Geomechanics Abstracts. Pergamon, 32*(3), 215–225.
50. Alonso, E., Alejano, L. R., Varas, F., et al. (2003). Ground response curves for rock masses exhibiting strainsoftening behaviour. *International Journal for Numerical and Analytical Methods in Geomechanics, 27*(13), 1153–1185.
51. Hajiabdolmajid, V., & Kaiser, P. (2003). Brittleness of rock and stability assessment in hard rock tunneling. *Tunnelling and Underground Space Technology, 18*(1), 35–48.
52. Zhang, F., Sheng, Q., Zhu, Z., et al. (2008). Study on post-peak mechanical behavior and strain-softening model of three gorges granite. *Chinese Journal of Rock Mechanics and Engineering, S1*, 2651–2655. (in Chinese).
53. Lu, Y., Wang, L., Yang, F., et al. (2010). Post-peak strain softening mechanical properties of weak rock. *Chinese Journal of Rock Mechanics and Engineering, 03*, 640–648. (in Chinese).
54. Valliappan, S., & Wohua, Z. (2015). Role of gas energy during coal outbursts. *International Journal for Numerical Methods in Engineering, 44*(7), 875–895.
55. Gray, I. (2006). Coal mine outburst mechanism, thresholds and prediction techniques. ACARP Report C.
56. Smith, D. M., & Williams, F. L. (1984). Diffusional effects in the recovery of methane from coalbeds. *Society of Petroleum Engineers Journal, 24*(05), 529–535.
57. Mora, C. A., & Wattenbarger, R. A. (2009). Analysis and verification of dual porosity and CBM shape factors. *Journal of Canadian Petroleum Technology, 2*(48), 17–21.
58. Measurement of coal and gas outburs. *Safety in Coal Mines* (01), 45–50. (1979). (in Chinese).
59. Cai, C., & Xiong, Y. (2005). Theoretical and experimental study on crushing energy of outburst-proneness coal. *Journal of China Coal Society, 01*, 63–66. (in Chinese).
60. Cai, C., & Wang, Y. (1988). Experimental study on crushing work of coal under impact and non-dangerous coal impact crushing. *Safety in Coal Mines, 07*, 13–18. (in Chinese).
61. Li, D., Jia, X., Miao, J., et al. (2010). Analysis of fractal characteristics of fragment from rockburst test of granite. *Chinese Journal of Rock Mechanics and Engineering, S1*, 3280–3289. (in Chinese).
62. Gadde, M. M., & Peng, S. S.(2003). *Effect of in-situ stresses on the stability of coal mine development workings*. West Virginia University Libraries.
63. Singh, R. N., Porter, I., & Hematian, J. (2001). Finite element analysis of three-way roadway junctions in longwall mining. *International Journal of Coal Geology, 45*(2–3), 115–125.
64. Gale, W. J. (1991). Strata control utilising rock reinforcement techniques and stress control methods, in Australian coal mines. *Mining Engineer, 150*, 247–253.
65. Qian, M., & Shi, P. (2003). *Mining pressure and strata control*. China University of Mining and Technology Press. (in Chinese).
66. Su, C., Tang, X., & Ni, X. (2012). Study on correlation among point load strength, compression and tensile strength of coal samples. *Journal of Mining and Safety Engineering, 04*, 511–515. (in Chinese).
67. Guo, P. (2014). *Research on laminar spallation mechanism of coal and gas outburst propagation*. China University of Mining and Technology. (in Chinese).
68. Liu, J. (2014). *Experiment research on evolution process and mechanism of coal and gas extrusion*. China University of Mining and Technology. (in Chinese).

69. Fang, J., Yu, S., & Tan, Q. (1995). A lamination separation and fragmentation model of coal and gas outburst. *Journal of China Coal Society, 02*, 149–153. (in Chinese).
70. Ding, X., Yu, S., Ding, Y., et al. (1989). Mechanism of continuous failure of coal under gas seepage. *Science China, 06*, 600–607. (in Chinese).
71. Wu, A., Jiang, C., & Wang, F. (2014). The rule of crack-like coal body's structure evolution and destruction during the coal and gas outburst. *China Mining Magazine, 09*, 107–111. (in Chinese).
72. Ates, Y., & Barron, K. (1988). The effect of gas sorption on the strength of coal. *Mining Science and Technology, 6*(3), 291–300.
73. Tu, Q., Cheng, Y., Guo, P., et al. (2016). Experimental study of coal and gas outbursts related to gas-enriched areas. *Rock Mechanics and Rock Engineering, 49*, 3769–3781.
74. Tu, Q., Cheng, Y., Wang, L., et al. (2015). Study on dynamic process of coal and gas outburst. *Coal Science and Technology, 06*, 71–75. (in Chinese).
75. Skoczylas, N., Dutka, B., & Sobczyk, J. (2014). Mechanical and gaseous properties of coal briquettes in terms of outburst risk. *Fuel, 134*, 45–52.
76. Xu, J., Liu, D., Peng, S., et al. (2010). Experimental research on influence of particle diameter on coal and gas outburst. *Chinese Journal of Rock Mechanics and Engineering, 06*, 1231–1237. (in Chinese).
77. Wen, G. (2003). Study on energy of coal and gas outburst. *Mining Safety and Environmental Protection, 06*, 1–3. (in Chinese).
78. Zhen, Z. (2004). *Analysis of the mechanism of coal and gas outburst from the analysis of magnitude and dimension.* (in Chinese).
79. Overview, O. M. (1996). pneumatic transport of solids. *Powder Technology, 3*(88), 309–321.
80. Xiong, Y., Guo, X., Gong, X., et al. (2009). Blockage critical state of pulverized coal dense-phase pneumatic conveying in horizontal pipe. *Ciesc Journal, 60*(6), 1421–1426.
81. Cong, X. (2013). *Study on relationship between flow patterns and pipeline pressure signals in dense-phase pneumatic conveying of pulverized coal.* East China University of Science and Technology. (in Chinese).
82. Sun, W., & Hu, D. (2012). Analysis of factors influencing coking coal's bulk density. *Baosteel Technology, 02*, 10–14. (in Chinese).
83. Su, X. (1990). The coal facies analysis of the main coal seams in the Longtan formation of low permain from Zhongliangshan and Nantong mining region, Sichuan. *Journal of Jiaozuo Mining Institute, 03*, 49–57. (in Chinese).
84. Cao, G. (2007). Study on gas desorption law of clastic coal core. *China Mining Magazine, 16*(12), 119–123. (in Chinese).
85. Yang, Q. (1987). Experimental study on coal gas diffusion law. *Safety in Coal Mines, 02*, 9–16. (in Chinese).
86. Guo, J., Kang, T., Kang, J., et al. (2014). Effect of the lump size on methane desorption from anthracite. *Journal of Natural Gas Science and Engineering, 20*, 337–346.
87. Liu, Y. (2011). *Study on gas emission rules, mechanics and dynamic model from coal particle.* Henan Polytechnic University. (in Chinese).
88. Zhang, T., Xu, H., Li, S., et al. (2009). The effect of particle size on adsorption of methane on coal. *Journal of Hunan University of Science and Technology(Natural Science Edition)* (01), 9–12. (in Chinese).
89. Zhang, X., Sang, S., Qin, Y., et al. (2005). Isotherm adsorption of coal samples with different grain size. *Journal of China University of Mining and Technology, 04*, 427–432. (in Chinese).
90. Ewen, W., & Wen, X. (1985). Gas desorption rate at the initial stage of coal seam gas outburst index-An investigation of the Kt value method. *Safety in Coal Mines, 05*, 56–63. (in Chinese).